GROUNDWATER QUANTITY AND QUALITY MANAGEMENT

SPONSORED BY
Groundwater Management Technical Committee
of the Groundwater Council of EWRI

Environmental and Water Resources Institute (EWRI)
of the American Society of Civil Engineers

EDITED BY
Mustafa M. Aral
Stewart W. Taylor

ASCE

EWRI
ENVIRONMENTAL &
WATER RESOURCES
INSTITUTE

Published by the American Society of Civil Engineers

Cataloging-in-Publication Data on file with the Library of Congress.

American Society of Civil Engineers
1801 Alexander Bell Drive
Reston, Virginia, 20191-4400

www.pubs.asce.org

ISBN 978-0-7844-1176-6
Manufactured in the United States of America.

Table of Contents

Acknowledgments

The book on Groundwater Quantity and Quality Management is the end product of an effort which required some time to complete and the organizational contribution of two individuals. The second editor is the one who initiated the idea long time ago and the first editor is the one who gave it a final push for successful completion of the project. Both editors express their deep appreciation to all authors who enthusiastically embraced the idea of the preparation of this book and who spend much time, effort, and resources to write their contributions. However, this book not only reflects the collective efforts and dedication of all these individual contributors, but it also reflects the contributions of thousands of people who have long been engaged in the development and application of groundwater management technologies both in the field, on paper and also on computers. Without the efforts of these broader contributors we would not have the information base we have summarized in this book. We all owe a debt of gratitude to all these people for their unselfish work and devotion to the advancement of the science and technology of groundwater quantity and quality management for the benefit of the society. On behalf of all the authors of this book, the editors dedicate this book to these scientists and technology specialists who laid the groundwork which we rely on and routinely use today.

Finally, the editors' families are acknowledged for their unwavering support of this project.

Mustafa M. Aral
Stewart W. Taylor
June 2010

Contributors

David Ahlfeld	*University of Massachusetts, USA.*
Mohammad N. Almasri	*An-Najah National University, Palestine.*
Mustafa M. Aral	*Georgia Institute of Technology, USA.*
Natalie L. Cápiro	*Tufts University, USA.*
T. Prabhakar Clement	*Auburn University, USA.*
Bithin Datta	*Indian Institute of Technology, India.*
Anirban Dhar	*Indian Institute of Technology, India.*
Randall W. Gentry	*University of Tennessee, USA.*
Rao S. Govindaraju	*Purdue University, USA.*
Mohamed M. Hantush	*Environmental Protection Agency, USA.*
Latif Kalin	*Auburn University, USA.*
Wonyong Jang	*Georgia Institute of Technology, USA.*
Jagath J. Kaluarachchi	*Utah State University, USA.*
Phillip J. King	*New Mexico State University, USA.*
Jiang Li	*Morgan State University, USA.*
William Miller	*William J. Miller Engineers, Inc., USA.*
Kurt D. Pennell	*Tufts University, USA.*
George F. Pinder	*University of Vermont, USA.*
Zhuping Sheng	*Texas A&M University, USA.*
Venkataraman Srinivasan	*University of Illinois, USA.*
Stewart W. Taylor	*Bechtel Corporation, USA.*
Frank T.-C. Tsai	*Louisiana State University, USA.*
William W.-G. Yeh	*University of California Los Angeles, USA.*

CHAPTER 1

INTRODUCTION

Stewart W. Taylor
Geotechnical & Hydraulic Engineering Services
Bechtel Corporation, Frederick, Maryland

1.1 Groundwater Management

Groundwater is an essential resource both in the U.S. and the rest of the world. As a water-supply source, groundwater has several advantages when compared to surface water: it is generally of higher quality, better protected from chemical and microbial pollutants, less subject to seasonal and perennial fluctuations, and more uniformly spread over large regions than surface water (Zektster 2006). Water-use data compiled by Kenny et al. (2009) for the U.S. indicate that groundwater supplies 33%, 98% and 42% of the water for public supply, domestic use and irrigation use, respectively. About 67% of the fresh groundwater withdrawals in 2005 were for irrigation, while about 18% were for public supply. Outside of the U.S., groundwater is the only water supply for some countries (e.g., Denmark, Malta, Saudi Arabia), while groundwater use exceeds 70% of the total water consumption and is the main source of municipal domestic and drinking water supply in most European countries (Zektser 2006). While its "in-stream" value, as is the case with rivers, has not been widely acknowledged, the critical role groundwater plays in maintaining important surface water systems, riparian and other types of vegetation as well as vital ecosystems is also increasingly recognized (Villholth and Giordano 2007). Estimates from a water balance model indicate that about 36% of river runoff is contributed by groundwater on a global basis (Döll et al. 2002), whereas in the U.S. the groundwater contribution to streamflow, estimated to be 52%, is even greater (Winter et al. 1998). Clearly, groundwater is one our most important natural resources.

Given the demands and desirability of groundwater as a water resource, it comes as no surprise that the resource has been overexploited both domestically and abroad. The consequences of overuse vary widely but can include water level decline and depletion of aquifers as a result of groundwater pumping, depletion of surface water resources due to groundwater-surface interactions, land subsidence as a result of groundwater development, and saltwater intrusion into coastal aquifers. In addition waste management practices have led to chemical and microbial contamination of shallow aquifers, degrading water quality and rendering the water supply unsuitable for public and domestic use. Similarly, applications of fertilizers, pesticides, and herbicides for agricultural purposes have adversely impacted the water quality of shallow aquifers on regional scales. Effective management of the resource, from both quantity and quality perspectives, is vital to the health of our economies, environment, and quality of life.

The scope of groundwater management, as described in the literature, has matured, evolved and increased over time. Bear (1979) describes groundwater management in the context of systems analysis wherein decision variables (e.g., location, rate and time of pumping) are assigned values with the aim of modifying state variables (e.g., water levels, depth of subsidence, solute concentrations), the goal being to alter the system state in a way that achieves certain objectives. These objectives must be quantifiable for the purpose of comparing the outputs of the system (e.g., costs, pumping) and selecting the best policy. In this systems analysis context, solving groundwater management problems entails coupling predictive simulation models for groundwater flow and/or solute transport with optimization algorithms that select decision variables that minimize or maximize an objective function subject to constraints on the state variables. ASCE (1987) approaches the topic of groundwater management primarily from the perspective of groundwater resource development, providing overviews of groundwater hydrology, the details of planning for groundwater management, and a process for selection and implementation of management plans. While ASCE (1987) focuses primarily on the management of groundwater quantity, consideration is given to groundwater quality aspects as well, recognizing that groundwater development can in of itself degrade groundwater quality over time (e.g., saltwater intrusion) and that consideration must be given to point and non-point sources of pollutants. Fetter (2001) addresses this topic in the context of groundwater development as well as protection of water quality in aquifers with a focus on policy and laws regulating both the quantity and quality of groundwater. Fetter (2001) notes that in developing a groundwater resource we are capturing part of the natural discharge of the system, implying that interactions between groundwater and surface water can be an important element of the groundwater management problem. More recently, the FAO (2003) documents a groundwater crisis (with falling water tables and polluted aquifers), particularly in developing countries that are dependent on groundwater for irrigation and where access and use are perceived as private irrespective of the legal status of the groundwater. In many developing countries the use of mechanized pumping has induced widespread drawdown, including the depletion of important shallow aquifers, and the disposal of human and industrial waste and the percolation of pesticides and herbicides have degraded many aquifers beyond economic remediation. In such areas, the FAO (2003) notes that the approach to groundwater management must consider the relationship of the economic and social dynamic to the specific aquifer hydrodynamics.

The small sample of the groundwater management literature reviewed above suggests that this topic is quite broad and includes elements of forecasting coupled with decision analysis driven by policy. Forecasting the response of a groundwater system to pumping in terms of yield, water levels, and water quality requires a theoretical understanding of groundwater flow and solute transport. In turn, flow and transport models that make use of the theory are required to predict aquifer response, from both quantity and quality perspectives. Such models may be analytical or numerical. Furthermore, such models require parameters that must be determined from field or laboratory measurements and calibrated to be useful as predictive tools. Depending on site-specific considerations, models may need to account for interaction between groundwater and surface water, or density-dependent groundwater flow when dealing with saltwater intrusion in coastal aquifers. Given a management policy, models may be used to predict the response of the system to any proposed operation policy, and to obtain from it the new state of the system, given its initial one. The policy may take the form of enforceable legal constraints in developed countries, or programs that modify human behavior via economic or social incentives in developing countries. In either case, decision analysis can be used to help identify the optimal locations, rates, and time of pumping or recharge that fulfills the management objectives and policy.

Given that the water quality in many aquifers has been degraded as a consequence of waste disposal practices, the restoration of groundwater has become an important component of groundwater management. Depending on the nature of the contamination, remediation technologies have been developed that can reduce contaminant concentrations to levels that would allow the aquifer to be used as a water supply source. These remediation technologies may be physically, chemically, or biologically based, but their successful implementation requires that they be compatible with site contaminants and subsurface characteristics. Application is facilitated through the use of mechanistic groundwater flow and solute transport models that represent the key processes governing the attenuation of contaminant concentrations.

1.2 Purpose, Scope, and Organization of Book

The main objective of this book is to provide the groundwater community an overview of groundwater quantity and quality management with a focus on the interrelationship between quantity and quality. Based on the discussion above, the following areas are targeted: (1) modeling groundwater flow and solute transport for the purpose of forecasting, including the physical and chemical characterization needed to establish model parameters and support model calibration; (2) developing groundwater resources with consideration given to groundwater-surface water interaction and saltwater intrusion; and (3) remediating groundwater resources by physical, chemical, and biological means.

Accordingly, the present chapter establishes the framework for groundwater management and introduces the reader to the key components required to manage groundwater quantity and quality.

Chapter 2 introduces the hydrologic cycle, provides definitions for the vadose zone and unsaturated zone, and discusses the physics of flow through porous media. The scope of the latter discussion includes porous media properties, balance laws, constitutive relationships, single- and multi-phase flow, fracture flow, and the auxiliary conditions required to solve groundwater flow problems. Regional flow and surface water-groundwater interactions are discussed as well. The aim of this chapter is to provide the reader with the theoretical background necessary to make quantitative forecasts related to groundwater quantity.

Chapter 3 provides an overview of the main concepts of groundwater contamination, including sources and types of groundwater contaminants, fate and transport processes in groundwater, and mathematical modeling. The equations governing solute transport are developed, and the reaction processes of adsorption and biological, chemical, and radiological decay are discussed. Analytical solutions for various boundary and initial conditions are presented. Numerical solutions are described. The objective of Chapter 3 is to provide the reader with the theoretical underpinnings needed to forecast changes in groundwater quality as result of introducing contaminants.

Chapter 4 describes analytical solutions to contaminant transport equations with a focus on solutions to reactive transport problems. Solutions to both single- and multi-species reactive transport equations in one- and multi-dimensional systems are reviewed from a historical perspective. The general solution to the multi-species reactive transport equation is discussed in detail, including special cases solutions involving a zero initial condition, identical retardation factors, zero advection velocity, steady state transport, and zero dispersion coefficient. The solutions presented in this chapter offer efficient means of simulating the fate and transport of reactive contaminants.

Chapter 5 outlines the field and laboratory methods that are commonly used to physically and chemically characterize groundwater systems. Surface and downhole geophysical methods are discussed. Test drilling and piezometer and well installation are described, and piezometer tests and aquifer pumping tests are reviewed. An overview of groundwater sampling and analysis for the purpose of characterizing water quality and contamination levels is provided. Commonly used industry standards for conducting subsurface investigations are identified. The aim of this chapter is to identify the investigative methods that are commonly used to characterize the various parameters that appear in the groundwater flow equation (Chapter 2) and solute transport equations (Chapter 3).

Chapter 6 discusses the use of models for groundwater management. The steps commonly used to select, construct, and calibrate a site-specific groundwater model are described. The simulation-optimization approach, which involves the coupling of a predictive simulation model with an optimization algorithm, to solving groundwater management problems is discussed. The intent of this chapter is to provide the reader an overview of how groundwater models can be used to facilitate groundwater management.

Chapter 7 summarizes recent progress in model calibration and model uncertainty assessment in groundwater modeling. The main focus is on identifying the parameter structure of a heterogeneous hydraulic conductivity field via solution to the inverse problem. Zonation and interpolation methods are discussed. A generalized parameterization method of parameter structure identification is introduced, which integrates the traditional zoning and interpolation methods, but avoids the shortcomings of each individual method. The parameter structure identification problem is discussed along with the problems of interpolation point selection and model. To resolve the non-uniqueness problem, Baysian model averaging is proposed to consider multiple methods for parameter structure identification.

Chapter 8 explores the concepts of well yield, perennial yield, and mining yield as they relate to the development of groundwater resources. Effects of groundwater development are described, including water level decline, depletion of surface water, brackish and saltwater intrusion, and land subsidence. Regional-scale development of groundwater, conjunctive use of surface water and groundwater, artificial recharge, and coastal and brackish groundwater development are discussed as well. Several case studies are described that illustrate the effective development and management of groundwater in various hydrogeological settings.

Chapter 9 presents the basic concepts and principles underlying the phenomena of groundwater and surface water interactions. Fundamental equations, analytical and numerical solutions describing stream-aquifer interactions are presented in hillslope and riparian aquifer environments. Analytical and numerical techniques for solving flow routing problems in channels with contiguous alluvial aquifers are discussed. Classical and recent developments in modeling stream depletion due to groundwater pumping are described. A case study on watershed-scale modeling of groundwater and surface water interaction is presented.

Chapter 10 discusses saltwater intrusion into coastal aquifers and various approaches to managing saltwater intrusion. The governing equations for density-dependent flow are presented and numerical solutions methods are described. The descriptive-predictive management model, which combines a simulation model with and optimization model, is discussed. Three multi-objective models for managing saltwater intrusion are presented. These models are solved for a representative coastal aquifer by constructing a physical simulation model and then using this model to train an artificial neural network model. This approximate simulation model is then linked with a genetic algorithm optimization model to develop solutions for each of the multi-objective models. The role of monitoring data for improving management strategies is also discussed.

Chapter 11 deals with the application of *in situ* air sparging and thermal venting for groundwater remediation. The physical, chemical, and biological processes associated with air sparging are described, and the contaminants and site conditions conducive to air sparging are identified. Multi-phase flow and contaminant transport models for the air sparging process are developed, results for a representative contaminant plume are presented, and field applications are discussed. Injection of hot air or steam, electrical resistance heating, and radio frequency heating for the thermally-enhanced venting of low- to semi-volatile organic compounds from the unsaturated zone are described. A multi-phase heat transport model for thermally-enhanced venting is presented and applied to simulate the removal of non-aqueous phase liquid (NAPL) source in the unsaturated zone. Results of field applications of thermally-enhance venting are discussed.

Chapter 12 focuses on *in situ* remediation of source zones comprised of petroleum hydrocarbons and chlorinated solvents, the most widespread contaminant classes. NAPL source zones are described, and NAPL migration, constituent properties, and dissolution are discussed. Hydraulic controls and pump-and-treat systems used to limit and/or control the migration of dissolved-phase contamination from NAPL source zones are described and their limitations are highlighted. *In situ* technologies for the remediation of dissolved-phase contamination associated with NAPL source zones are discussed, including permeable reactive barriers and *in situ* chemical oxidation. An overview of *in situ* methods for NAPL source zone remediation is presented that addresses surfactant enhanced aquifer remediation, cosolvent flushing, and thermal treatment. The effectiveness of various source zone treatment technologies are discussed along with the benefits of combining *in situ* remediation strategies.

Chapter 13 presents methods used to bioremediate groundwater systems contaminated by petroleum hydrocarbons and chlorinated solvents. The biodegradation mechanisms are discussed for both classes of contaminants. Active bioremediation system designs for groundwater systems are described along with the approach of monitored natural attenuation. Analytical and numerical methods for modeling the biodegradation of petroleum hydrocarbons and chlorinated solvents are addressed.

Finally, Chapter 14 discusses the current state of the groundwater quantity and quality management research field. The evolution of environmental management paradigms over the last several decades is described. The role of health effects in shaping future groundwater and environmental management philosophy is discussed, along with the need to better integrate the engineering, social, basic and health sciences in addressing complicated issues related to incorporating environmental health issues into our groundwater management philosophy. Lastly, potential applications of resilience analysis and complex systems theory to groundwater management problems are discussed.

1.3 Future Trends

Recent studies suggest that we are on the verge of a freshwater crisis wherein demand relative to supply is projected to lead to water scarcity for significant percentages of the global population in the relatively near future. It has been estimated that each person on the earth needs a minimum of about 1,000 m^3 of water per year for drinking, hygiene, and growing food for sustenance; whether people get enough water to meet their needs depends primarily on where they live as the distribution of global water resources varies widely (Rogers 2008). Currently, about 1.6 billion people, or about 20% of the global population, live in river basins characterized by absolute physical water scarcity (de Fraiture et al. 2007). Projections to 2025 suggest that the freshwater resources of more than half the countries across the globe will undergo stress or experience shortages (Rogers 2008). Models analyzing the vulnerability of water resources to future climate change, population growth and migration, and industrial development indicate that trends in water scarcity can be attributed to climate change and population growth. Climate change alone is predicted to produce a mixture of response, both positive and negative, that is highly region dependent, whereas the increase in population to about 8 billion people in combination with climate change is expected to lead to a pandemic increase in water scarcity (Vorosmarty et al. 2000).

Given that groundwater is an integral component of the water supply, future increases in demand will present significant groundwater management challenges. In fact, the pressure on the groundwater resource is expected to be even greater relative to surface water as groundwater supplants surface water for all types of use, that is domestic, agricultural (crop and livestock) and industrial. This change is being driven by groundwater's inherent beneficial properties in terms of both quality and quantity combined with easy access through better and cheaper drilling and pumping techniques (Villhoth and Giordano 2007). Temporal trends indicate that groundwater withdrawals are increasing most rapidly in developing countries such as India, Bangladesh, and China, whereas groundwater withdrawals for the U.S. and Western Europe are projected to be relatively stable (Shah 2005). World-wide trends further suggest that it is crucial to focus on agricultural use of groundwater for the simple reason that the volumes used in this sector significantly exceed industrial and domestic uses at the global scale (Burke 2002). Regarding agricultural use of groundwater, the FAO (2003) projects that: (1) access to groundwater will continue to allow intensification of agricultural production in response to changing patterns of demand; (2) the scope for managing agricultural demand for groundwater is limited, particularly where rural communities are trying to escape from poverty; and (3) the overexploitation of aquifers by agriculture will force users into economic and social transitions (moving off the land or transferring resources or users rights to competing users).

While it is known that management based on strict control of groundwater development and use (demand management) is generally difficult to implement and enforce in developing countries, it should be recognized that intensive groundwater exploitation in agriculture today is a common global phenomenon across quite different socio-economic settings, resulting in the same type of physical or environmental impacts. Villholth and Giordano (1997) note that the two fundamental human drivers for groundwater development for irrigation, survival and profit, may be considered as extremes of a continuum governed primarily by the stage of regional development. One end of the extreme may represented by the subsistence farmer on a small tract in a developing country, while the other end may be exemplified by the large-scale commercial farmer in the U.S. farming thousands of acres. The large-scale users exert the greatest abstraction pressure on the resource per person; however, when many small-scale users are conglomerated within larger areas, like in parts of China and India, the aggregate effect may be similar. In both cases, overexploitation of the groundwater could occur with associated adverse impacts. A key difference in the examples provided above is that an infrastructure exists in developed countries to manage and enforce groundwater development and use, whereas the "open access" nature of groundwater, sense of entitlement, and lack of enforceable policies in developing countries represents a core dilemma in groundwater management.

The future trends in groundwater use highlighted above suggest that groundwater scientists and engineers will play a very important role in developing solutions to the complex groundwater management challenges that we are dealing with now and for the foreseeable future. The management solutions will need to consider both the socio-economic and sociopolitical characteristics of any groundwater-using society as well as the physical, chemical, and biological characteristics of the groundwater resource. The goal in writing this book is to contribute to the development of those solutions.

1.4 References

American Society of Civil Engineers (ASCE) (1987). *Ground water management,* 3rd edition, ASCE Manuals and Reports on Engineering Practice No. 40, ASCE, New York.

Bear, J. (1979). *Hydraulics of groundwater*, McGraw-Hill, New York.

Burke, J. J. (2002). "Groundwater for irrigation: Productivity gains and the need to manage hydro-environmental risk." *Intensive use of groundwater: Challenges and opportunities*, R. Llamas and E. Custodio, eds., Balkema, Lisse, the Netherlands, 59-92.

de Fraiture, C., Wichelns, D., Rockstrom, J., Kemp-Benedict, E., Eriyagama, N., Gordon, L. J., Hanjra, M. A., Hoogeveen, J., Huber-Lee, A., and Karlberg, L. (2007). "Looking ahead to 2050: Scenarios of alternative investment approaches." *Water for food, water for life: A Comprehensive Assessment of Water Management in Agriculture*, D. Molden, ed., Earthscan, London, UK, IWMI, Colombo, Sri Lanka, 91-145.

Döll, P., Lehner, B., and Kaspar, F. (2002). "Global modeling of groundwater recharge." *Proc., Third Int. Conf. on Water Resources and Environment Research, Vol. I*, G. H. Schmitz, ed., Technical University, Dresden, Germany, 322–326.

Fetter, C. W. (2001). *Applied hydrogeology,* 4th edition, Prentice Hall, Upper Saddle River, N.J.

Food and Agriculture Organization (FAO) of the United Nations (2003). *Groundwater management: The search for practical approaches*, FAO Water Report 25, Rome.

Kenny, J. F., Barber, N. L., Hutson, S. S., Linsey, K. S., Lovelace, J. K., and Maupin, M. A. (2009). *Estimated use of water in the United States in 2005*, Circular 1344, U.S. Geological Survey, Reston, Va.

Rogers, P. (2008). "Facing the freshwater crisis." *Scientific American*, August.

Shah, T., (2005). "Groundwater and human development: Challenges and opportunities in livelihoods and environment." *Water Science & Technology*, 51(8), 27-37.

Villholth, K. G., and Giordano, M. (2007). "Groundwater use in a global perspective – Can it be managed?" *The agricultural groundwater revolution: Opportunities and threats to development*, M. Giordano and K. G. Villholth, eds., CAB International, Wallingford, U.K., 393-402.

Vorosmarty, C. J., Green, P., Salibury, J., and Lammers, R. B., (2000). "Global water resources: Vulnerability from climate change and population growth." *Science* 289, 284-288.

Winter, T. C., Harvey, J. W., Franke, O. L., and Alley, W. A. (1998). *Ground water and surface water: A single resource*, Circular 1139, U.S. Geological Survey, Reston, Va.

Zektser, I. S. (2006). "Introduction." *Ground water resources of the world and their use*, I. S. Zekster and L. G. Everett, eds., NGWA Press, Westerville, Ohio, 13-15.

CHAPTER 2

GROUNDWATER HYDROLOGY

George F. Pinder
University of Vermont

2.1 Hydrologic Cycle

Groundwater is the subsurface component of the hydrologic cycle illustrated in Figure 2.1. The amount of water contained in each of these components is provided in terms of km^3. The transport of water between them is provided as a volumetric flux in units of km^3/yr. For example, the amount of water contained in the atmosphere is reported as 12,900 km^3 and the volumetric flux, or direct interaction, between water in the atmosphere and that in the soil is 116,000 km^3/yr. Of the 116,000 km^3/yr that reaches the soil, 46,000 km^3/yr (39.6%) enters the groundwater system while the remainder returns to the atmosphere via evapotranspiration. Of the 46,000 km^3/yr that reaches the groundwater system, 43,800/km^3/yr, or 95.2 %, is discharged to rivers, lakes, and marshes while 2,200 km^3/yr is discharged to the ocean. The contribution of water to rivers, lakes, and marshes due to precipitation is negligible when compared to groundwater.

Oceans are the primary recipient of the water from rivers, lakes and marshes (98%), and the remainder is lost to the atmosphere via evaporation. The remaining water is stored in glaciers. Glaciers receive water from the atmosphere in terms of precipitation and discharge it to the oceans in the form of ice and meltwater.

Groundwater represents only 0.39% of the world's water, but 49% of the world's fresh water. As such it is, and will continue into the future to be essential in providing potable water for private, municipal, agricultural and industrial uses.

2.2 The Near Surface Environment

In this book we are primarily interested in that portion of the earth's subsurface from which man derives potable water. Figure 2.2 is a diagrammatic representation of the near surface environment of primary interest. The curve to the right of the soil column describes the water saturation as a function of depth below land surface. The saturation is formally defined as the ratio of the volume of water in a sample to the volume of void space in the sample, that is $S_w = V_w / V_v$.

According to the saturation versus depth curve, we find that below a depth defined by a hypothetical surface called the water table, an environment where all the pores are filled with water, that is $S_w = 100\%$. Above the water table the saturation begins to decrease as the elevation increases. However, the saturation remains high in this zone called the capillary fringe, sometimes reaching 100%. The key factor that separates water in this zone from that in the saturated zone is that water in this zone is under tension. It is held in place by capillarity and exhibits a fluid pressure below atmospheric. On the other hand, water in the saturated zone always has pressure in excess of atmospheric.

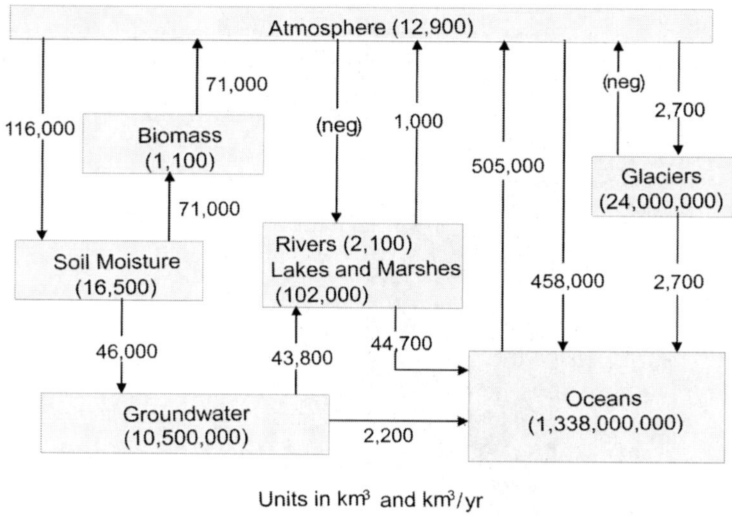

Figure 2.1: The hydrologic cycle. Reprinted with permission from *Physical Hydrology*, 2nd ed., by S. Lawrence Dingman, published by Waveland Press. Copyright c 2002, 1994 by S.L. Dingman.

Above the capillary fringe, in the vadose zone, the water saturation initially decreases very rapidly with increasing elevation and then approaches an apparent asymptote. Water is also under tension in the vadose zone.

2.3 Physics of Flow through Porous Media

The role of the groundwater professional in the investigation of, and utilization of groundwater supplies is to provide, in addition to elements of engineering design, quantitative as well as qualitative analysis. To provide this input, the groundwater professional requires a clear understanding of the physical processes that govern groundwater occurrence and flow behavior. In this section we discuss these principals.

2.3.1 Porous Media Properties

Porous media are considered to be materials that are composed of at least one solid and one fluid phase. Thus soil is a porous medium since it contains soil particles and either air, water or both. Dense rock, on the other hand, is not a porous medium, although it may be composed of different minerals, because it has no fluid phase.

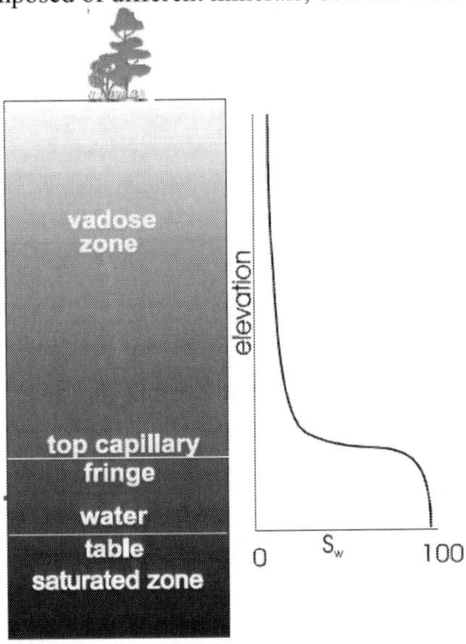

Figure 2.2: Diagrammatic representation of water in the shallow subsurface. It is assumed that the system is in dynamic equilibrium, that is that it represents, in some sense, a steady state.

For our purposes a porous medium of interest must also have void-space continuity. A porous medium that is composed of isolated openings surrounded by impermeable material is of little or no interest. Thus, volcanic rock may have many void spaces due to trapped gas present during cooling of the lava, but since they are not necessarily connected void spaces, they may not be useful for the transmission of fluids. A porous medium of practical interest from a groundwater perspective must have continuous void space such that fluids can flow.

The void volume, or pore volume, we will describe using the concept of porosity. The porosity, ε is defined as the volume of voids, V_v in a soil sample (porous medium) over the total soil-sample volume, V that is,

$$\varepsilon \equiv \frac{V_v}{V} \tag{2.1}$$

A commonly used related measure is the void ratio, defined as

$$e \equiv \frac{V_v}{V_s} \tag{2.2}$$

where V_s is the volume of solids in the sample. One can easily show that these two measures are related as $e = \varepsilon/(1-\varepsilon)$.

While the geometry of the pores affects the flow of fluids in a porous medium just as the geometry of a pipe affects the fluid dynamics of pipe flow, pore geometry is not readily measurable. As a surrogate for pore geometry, the size of the grains in a granular soil is measured.

Although perhaps counter-intuitive, the size of the grains in a uniform sand made up of spherical particles does not affect the porosity. Only the packing arrangement of the grains is of importance in such uniform (homogeneous) porous media.

However, if there are multiple grain sizes, the porosity does change. The cubic packed spheres shown in Figure 2.3 would have a porosity of 48% in the absence of the smaller spheres. However, the porosity is substantially reduced when the smaller spheres are added because the smaller spheres occupy the otherwise available void space.

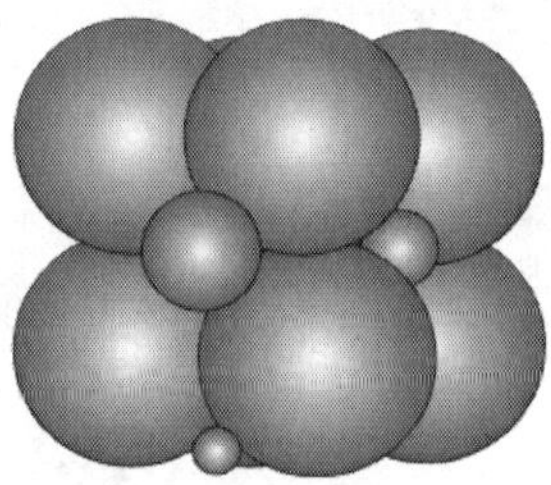

Figure 2.3: In this packing arrangement involving two grain sizes, the small grains tend to occupy spaces left between larger grains to yield a smaller percentage of void space in the sample.

Since soil porosity, and to some degree soil permeability, are dependent upon the grain size distribution, soils tend to be classified based upon grain size. The soil sample is analyzed using a two-phase process, sieving and hydrometer separation.

In sieving, the sample is placed in the top of a series of sieves that have a pan located at the base to collect material smaller than the hole diameter of the sieve with the smallest openings. The sieve with the largest openings is located on the top of the stack. The hole sizes of the sieves decreases uniformly from top to bottom. The sieve stack is shaken to allow the sample grains to separate such that a given sieve collects particles that are larger than the hole size of the sieve and smaller than the hole size of the overlying sieve. The amount of material contained in each sieve is weighed and the data recorded.

The residual soil, that is that soil fraction collected in the pan, is normally that with a grain diameter smaller than 0.002 mm. This material is further subdivided using the hydrometer method. In this approach the soil sample is made into an emulsion with water and the rate of fall of the particles determined using a measure of the density of the suspension and Stoke's law. A determination of the weight of various particle sizes less than 0.002 mm is determined.

From the sieving and hydrometer analysis data a plot such as found in Figure 2.4 is generated. The percent by weight passing each screen, and therefore smaller than the corresponding screen opening, is recorded. Thus the curve represents the cumulative grain size distribution by weight. For example, in Figure 2.4, approximately 27% of the grains in the 'clayey sandy silt' are smaller than 0.005 mm. Along the horizontal axis the soil type designation is defined in terms of the grain size. Silt, for example, is defined as soil with grain size boundaries of 0.002 mm and 0.02 mm in this figure.

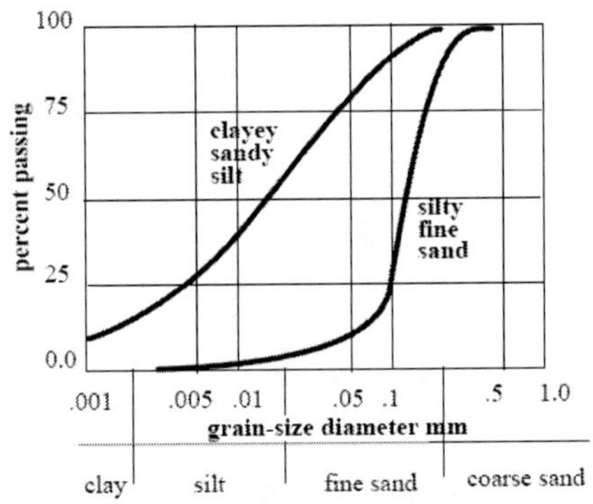

Figure 2.4: The grain-size distribution indicates the soil classification of a sample and its degree of gradation.

Several soil classification strategies have been proposed. The classification strategy used by the U.S. Department of Agriculture is presented in Figure 2.5. Based upon the percentage of silt, sand and clay present in a sample, the sample is located on the triangular plot. The plot region is subdivided into areas defining different soils. The sample plotted in Figure 2.5 is classified as a clay with 50% clay, 30% silt and 20% sand.

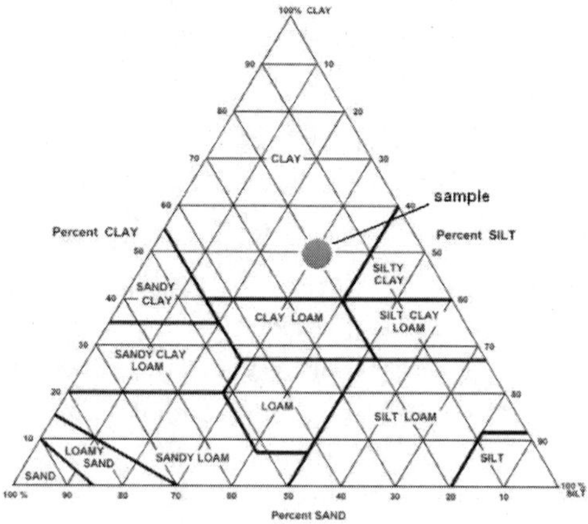

Figure 2.5: U.S. Department of Agriculture soil classification chart.

2.3.2 Balance Laws

The balance laws for groundwater flow are obtained formally through integration from the microscopic level (the level of observation of a fluid particle) to the macroscopic level (the level of observation of the porous medium). When the equations are to apply to problems of very large scale, on the order of tens of meters, then a further averaging to achieve a megascopic level equation may be required.

The point of departure is the point mass balance equation, which can be written as

$$\frac{\partial \rho}{\partial t} + \nabla \cdot (\rho \mathbf{v}) = 0. \tag{2.3}$$

where ρ is the fluid density and $\mathbf{v}$ the fluid velocity. This equation holds for a fluid or solid irrespective of whether or not the fluid or solid are associated with a porous medium. The porous medium, or soil, can be composed of several fluid phases as well as a solid phase. Air, water and oil are all fluid phases. To identify the phase to which Equation (2.3) is applicable we multiply it by a pointer function $\gamma_\alpha(\mathrm{x},t)$ defined for the phase of interest. The subscript α denotes the phase, that is $(\alpha = l, s)$ is the case of a porous medium with only liquid water l and solid s phases. The pointer function has the properties

$$\gamma_\alpha(\mathrm{x},t) == \begin{cases} 1 & \mathrm{x} \in dV_l \\ 0 & \mathrm{x} \in dV_s \end{cases} \tag{2.4}$$

where dV_l is the volume of a sample volume of soil, dV, that is liquid (water) and dV_s is the volume of dV that is solid (grains).

The choice of the sample size must be such that it is representative of the porous medium at the scale of observation of interest. It must be large enough that a statistically representative volume of each phase is present, yet small enough that heterogeneities such as sand lenses are not masked in the averaging process. When Equation (2.3) is multiplied by γ_l it is thereafter identified with the liquid phase because γ_l has a value of unity in the α phase and zero in all other phases.

To make the relationship applicable over the entire volume of interest, V, we must sum up the amount of liquid present in V. Integration over this volume achieves this goal, that is

$$\int_V \frac{1}{dV}\int_{dV}\left[\frac{\partial\rho}{\partial t}+\nabla\cdot(\rho\mathbf{v})\right]\gamma_l dv_\xi dV = 0. \tag{2.5}$$

Equation (2.5) states that we first integrate over the sample volume dV and account for the volume of liquid in that sample. We then proceed to integrate over the whole volume of interest V. The result is an equation that is applicable throughout V. Inherent in the mathematical steps involved in the above formulation is the requirement that dV be a moving average. In other words, we have overlapping volumes of dV. The result of the moving average concept is a new average value of state variables such as ρ that are smoothly varying over V.

Given the volumes dV and V are chosen arbitrarily, one can move the integrand from the integral to provide the following equation

$$\frac{\partial\varepsilon\rho^l}{\partial t}+\nabla\cdot\left(\varepsilon\rho^l\mathbf{v}^l\right)=0 \tag{2.6}$$

where the superscripts on the variables indicate that they are averages over the volume dV. In other words, we have arrived, using the averaging concepts, to a new equation that looks very much like the point Equation (2.3), but for the use of averages and the addition of the porosity ε. However this equation is applicable only to the liquid phase in a porous medium. The variables ρ^l and $\mathbf{v}^l$ represent the average fluid density and fluid velocity. A similar equation could be written for the solid grain phase.

If there is mass exchange between phases, such as may occur in an oil-water system, there is an additional term added to this equation.

2.3.3 Constitutive Relationships

2.3.3.1 *Darcy's Law*

Equation (2.6) is one equation with two unknowns (assuming porosity ε is known). To augment the system of equations we introduce constitutive (experimental) relationships. Perhaps the best known constitutive relationship in groundwater hydrology is Darcy's law. Many variants on Darcy's law exist and we will use three in this and later sections of this presentation. The formulation presented in Darcy's classic paper published in 1856 (Darcy 1856) is

$$q = -K\frac{dh}{dl} \tag{2.7}$$

where q is the specific discharge, that is $\varepsilon \mathrm{v}^l$, h is the hydraulic head, and l is the coordinate direction along the length of the column experiment (the vertical direction in Darcy's experiment). The parameter K is the proportionality constant that relates the hydraulic head gradient to the specific discharge and is called the hydraulic conductivity. The relationship can be extended to multiple dimensions in which case it becomes

$$\mathbf{q} = -\mathbf{K}\cdot\nabla h \tag{2.8}$$

where

$$\nabla(\cdot) = \frac{\partial(\cdot)}{\partial x}\mathbf{i} + \frac{\partial(\cdot)}{\partial y}\mathbf{j} + \frac{\partial(\cdot)}{\partial z}\mathbf{k} \tag{2.9}$$

and $(\mathbf{i},\mathbf{j},\mathbf{k})$ are the unit vectors in the three spatial-coordinate directions. The specific discharge now is a vector and the hydraulic conductivity a tensor. When the principal directions of the hydraulic conductivity tensor are aligned with the coordinate directions, the hydraulic conductivity matrix is diagonal and the three components of the flux are given by

$$\begin{Bmatrix} q_x \\ q_y \\ q_z \end{Bmatrix} = \begin{bmatrix} K_{xx} & 0 & 0 \\ 0 & K_{yy} & 0 \\ 0 & 0 & K_{zz} \end{bmatrix} \begin{Bmatrix} \partial h/\partial x \\ \partial h/\partial y \\ \partial h/\partial z \end{Bmatrix} \tag{2.10}$$

where K_{xx} is, for example, the hydraulic conductivity in the x coordinate direction. When the diagonal components of **K** are equal the medium is isotropic, when unequal the medium is anisotropic (has directional properties). If $\mathbf{K} = \mathbf{K}(\mathrm{x})$, that is it varies from point to point, the medium is called heterogeneous. If it does not vary spatially, it is called homogeneous.

2.3.3.2 *Heterogeneity*

Heterogeneous hydraulic conductivity reflects the natural variation in subsurface material properties. Heterogeneity can occur on a variety of scales. At the smallest scale it is attributable to the variability in depositional environments. The size of water-transported particles is directly proportional to the energy of the transporting water. Thus one would expect to find coarse materials, such as gravels, in the headwaters of streams originating in mountainous terrain. Similarly clay deposits would be associated with quiet off-shore environments such as identified with lakes, seas or the oceans. However, especially in high energy environments, a variety of grain sizes can be found in deposits. For example stream-bed deposits may contain deposits which have a variety of grain sizes. Such deposits are called poorly sorted or well graded, depending upon whether you are reading geological or engineering literature.

When suspended sediments in a high-energy environment suddenly encounter a low energy environment they form deposits that grade from fine to coarse. A common example is a deltaic deposit encountered where a river empties into a lake or other quiescent water body. Coarse material carried by the river cannot be suspended in the quite water environment and settles to form a stratified deposit of materials of various grain sizes.

The variability within a deposit, such as a river bed, is on the scale of centimeters to meters. However, variability due to changes in the depositional environment, for example that due to changes between river deposits and lake deposits, are on the scale of meters to kilometers. Each represents heterogeneity, albeit at a different scale.

2.3.3.3 *Compressibility*

The second constitutive relationship needed to close the system of equations that describe groundwater flow relates to the compressibility of the soil and the flowing fluid. When the groundwater fluid pressure is reduced due, for example, to pumping, both the fluid and the soil structure responds. The fluid expands as a result of its elasticity so that the volume occupied by a given mass of fluid increases. On the other hand, the soil responds by contracting. The reduction in pressure results in a decrease in support provided to the soil by the fluid. The grains rearrange to accommodate to this new environment. The result is consolidation, a topic addressed in detail in soil mechanics literature. In theory, the grains also expand, but this effect is small relative to the deformation of soil due to grain rearrangement.

When water expands or soil consolidates, the net effect is an apparent increase in water volume. While this effect is relatively unimportant in problems of small scale, that is on the scale of meters, it is important in large scale problems on the scale of kilometers. Compressibility is also more important in compressible soils, such as clay, relative to more rigid soil formations and rock.

The term in Equation (2.6) that involves compressibility is the time derivative, which can be chained out to yield

$$\frac{\partial \varepsilon \rho^l}{\partial t} = \varepsilon \frac{\partial \rho^l}{\partial t} + \rho^l \frac{\partial \varepsilon}{\partial t} \tag{2.11}$$

The first term on the right hand side of Equation (2.11) represents the compressibility of the fluid, in our case water, whereas the second term represents the change in porosity due to soil deformation (consolidation).

Constitutive relationships that express terms A and B in terms of the state variable of interest to us, in this case hydraulic head, are needed. Consolidation is described by

$$\frac{\partial \varepsilon}{\partial t} = \frac{\partial \varepsilon}{\partial p}\frac{\partial p}{\partial t} = C_v \frac{\partial p}{\partial t} \tag{2.12}$$

where we have used chaining and the definition $C_v \equiv d\varepsilon/dp$. A similar relationship written for the fluid yields

$$\frac{\partial \rho^l}{\partial t} = \frac{d\rho^l}{dp}\frac{\partial p}{\partial t} = \rho^l \beta \frac{\partial p}{\partial t} \tag{2.13}$$

where β is the fluid compressibility.

2.3.4 Single-phase Flow Equation

We are now at the point where we can write an equation for saturated (single-phase) groundwater flow. Because Equations (2.12) and (2.13) are in terms of fluid pressure, and we wish to use hydraulic head as our state variable, we introduce the relationship

$$\frac{\partial p}{\partial t} = g\rho^l(p)\frac{\partial h}{\partial t}. \tag{2.14}$$

Substitution of Equations (2.8), (2.12), (2.13) and (2.14) into (2.6) yields

$$g\rho^l(p)\rho^l(\varepsilon\beta + C_v)\frac{\partial h}{\partial t} - \nabla\cdot\left[\rho(\mathbf{K}\cdot\nabla h)\right] = 0 \tag{2.15}$$

or

$$S_s\frac{\partial h}{\partial t} - \nabla\cdot(\mathbf{K}\cdot\nabla h) = 0 \tag{2.16}$$

where $S_s \equiv g\rho^l(p)(\varepsilon\beta + C_v)$ is the specific storage and the spatial gradient in ρ^l and the grain velocity have been neglected. Equation (2.15) or (2.16) is the classical groundwater flow equation.

2.3.4.1 *Density-concentration Relationships*

In problems wherein the fluid density is dependent upon either concentration or temperature, Equation (2.16) takes on a different, albeit equivalent, form. In this case Darcy's law is written

$$q = -\frac{\mathbf{k}}{\mu}\left(\nabla\left[p - \rho^l gz\right]\right) \tag{2.17}$$

where $\mathbf{k}$ is the intrinsic permeability, μ is the fluid viscosity, z is elevation, and g is gravitational acceleration. Substitution of Equations (2.17), (2.12) and (2.13) into Equation (2.6) yields

$$\rho^l\left(C_v + \varepsilon\beta\right)\frac{\partial p}{\partial t} - \nabla\cdot\frac{\mathbf{k}}{\mu}\cdot\left(\nabla\left[p - \rho^l gz\right]\right) = 0 \tag{2.18}$$

When density is a function only of pressure (or hydraulic head), Equations (2.16) and (2.18) are the same (note that $\mathbf{K} = \left(\mathbf{k}\rho g/\mu\right)$). However, Equation (2.18) can be used when $\rho = \rho(\mathrm{T})$ or $\rho = \rho(c)$ where T is temperature and c is concentration. Notice that since the state variable in Equation (2.18) is pressure, boundary and initial conditions will need to be expressed in terms of pressure.

Since the fluid density ρ^l depends upon either temperature, concentration, or both, a constitutive equation relating ρ^l to T and c will be needed. Also required is an equation describing the movement of the solute or energy (a mass or energy transport equation). These equations are discussed in a later chapter.

2.3.4.2 *Aquifer Parameters*

Equations (2.16) and (2.18) have several parameters that must be specified. The hydraulic conductivity can be determined using a permeameter in the laboratory for bench-scale samples. However, such determinations are not usually representative of the hydraulic conductivity measured in the field because of the small sample size and the natural variability found in nature.

To obtain estimates at the field scale pump tests or slug tests are often employed. In a pump test a well is pumped at a known rate and the water levels in the pumped well and neighboring observation wells are recorded over time. Using a simplified mathematical model of the aquifer, a value of the hydraulic conductivity that reproduces the observed water level response over time is determined. A value can be obtained for each observation well and in some instances the pumping well. When geological formations of lower hydraulic conductivity bound the aquifer from either above or below, the hydraulic conductivity of these bounding formations can sometimes be determined.

In a slug test, a known volume of water is either added to or removed from a well and the water level response of the well recorded over time. The introduction of the slug raises the water level which subsequently decays. The rate of decay is used in conjunction with a mathematical model of the aquifer to obtain the hydraulic conductivity. In an alternative strategy, a cylinder of known volume is placed in the well below the water level. The water level in the well initially rises and then decays as in the case with the introduction of a known quantity of water. The response to the physical slug is then recorded and the results analyzed. The procedure can also be used in reverse, that is a known quantity of water or a physical slug can be removed resulting in an initial decline in water level followed by a recovery period.

Both of these methods are discussed in more detail in a later chapter.

The storage coefficient can be determined using a similar procedure, provided adequate 'transient' data are available. Of special importance are the very early time data.

The parameters appearing in Equation (2.18) can be determined from the hydraulic conductivity and the storage coefficient if one assumes small changes in fluid density.

2.3.5 Multiphase Flow Equations

As illustrated in Figure 2.2, between the land surface and the water table there exist the partially saturated zone. Water and air coexist in this zone, that is there are two fluid phases present. The equations that define the flow of such fluids are called the multiphase flow equations. Multiphase equations stem from Equation (2.6). The multiphase form is

$$\frac{\partial n_\alpha \rho^\alpha}{\partial t} + \nabla\cdot\left(n_\alpha \rho^\alpha \mathbf{v}^\alpha\right) = 0 \tag{2.19}$$

where α now refers to the air, water and solid phases, that is $(\alpha = a, w, s)$ and $n_\alpha = V_\alpha / V = S_\alpha \varepsilon$. Chaining this equation yields

$$S_\alpha \varepsilon \frac{\partial \rho^\alpha}{\partial t} + \rho^\alpha \frac{\partial \varepsilon S_\alpha}{\partial t} + \nabla\cdot\left(\varepsilon \rho^\alpha \mathbf{v}^\alpha\right) = 0 \tag{2.20}$$

Chaining further and introducing a modified form of Darcy's law one obtains

$$\begin{aligned} &S_\alpha(p^\alpha)\varepsilon \frac{\partial \rho^\alpha}{\partial t} + \rho^\alpha S_\alpha(p^\alpha)\frac{\partial \varepsilon}{\partial t} + \rho^\alpha \varepsilon \frac{\partial S_\alpha(p^\alpha)}{\partial t} \\ &-\nabla\cdot\left(\rho \frac{k k_{r\alpha}(p^\alpha)}{\mu_\alpha}\left(\nabla[p^\alpha - \rho^\alpha g z]\right)\right) = 0. \end{aligned} \tag{2.21}$$

Examination of Equation (2.21) reveals that the fundamental form of the equation is similar to that describing saturated groundwater flow, Equation (2.15), but for the addition of a time derivative of the saturation and the coefficient $\mathrm{k}_{r\alpha}\left(p^{\alpha}\right)$. This coefficient is called the relative permeability and is responsible for modifying the effective permeability to account for the effect of phases other than the α phase that occupy the pores. In essence, the phase coexisting in the pores with the α phase tends to block the pores and to thereby inhibit the movement of the α phase through the pores. The relative permeability has a range of $0 \le \mathrm{k}_{r\alpha} \le 1$. Because the relative permeability depends upon p^{α} and is a coefficient multiplying a derivative of p^{α}, the equation contains non-linear terms in the state variable p^{α}.

The saturation S_{α} is a non-linear function of p^{α}. Thus there is a non-linear time derivative term as well.

The illustrative relationship between S_{α}, $\mathrm{k}_{r\alpha}$, and p^{α} is determined experimentally, usually at the bench scale. A typical $S_{\alpha} - p^{\alpha}$ curve is found in Figure 2.6. If we consider two fluids, water and air, then we can define the capillary pressure $P_c \equiv p^{a} - p^{w}$. Since in a partially-saturated medium the water pressure is always less than atmospheric, the value of P_c is always positive. In fact the lower the absolute pressure of the water phase, the higher the value of P_c.

Let us assume the system is fully saturated with water, thus we are at a value of P_c between zero and **A** on the vertical axis of Figure 2.6. As the capillary pressure is increased from zero, the porous medium initially remains saturated and continues to remain so until the pressure **A** is reached. At this point the saturation begins to decrease as the capillary pressure increases. This continues until the curve approaches an asymptote. This limiting value of saturation is called the residual water saturation. Below this saturation the water exists only as discrete unconnected pockets.

If one introduces water into the system such that the saturation increases the resulting behavior is described by the imbibition curve in Figure 2.6. As the capillary pressure approaches zero, the saturation does not return to unity. Rather it reaches a maximum at the point *I*. The value of the saturation at *I* is the irreducible air saturation. Air below this saturation can be removed only by dissolution into the water phase. If the capillary pressure is again increased, the resulting curve would not, in general, follow the imbibition curve, but rather create a new third curve. The creation of a set of curves, called scanning curves, means that the system has memory. Its behavior depends upon its prior and existing states. Because one can have a multiplicity of saturation values for a given pressure value, and the value selected depends upon system history, the system exhibits hysteresis.

The role played by the saturation-capillary pressure curve becomes evident if we chain out the third term in Equation (2.21), that is

$$\rho^{\alpha}\varepsilon\frac{\partial S_{\alpha}\left(p^{\alpha}\right)}{\partial t}=\rho^{\alpha}\varepsilon\frac{dS_{\alpha}\left(p^{\alpha}\right)}{dP_{c}}\frac{\partial P_{c}}{\partial t} \tag{2.22}$$

The right-hand side of Equation (2.22) contains the slope of the pressure-saturation curve, that is $\left(dS_{\alpha}\left(p^{\alpha}\right)/dP_{c}\right)$. Thus in solving for unsaturated flow one uses the curves in Figure 2.6 interactively, solving for values of P_c and then determining the non-linear coefficients from these curves.

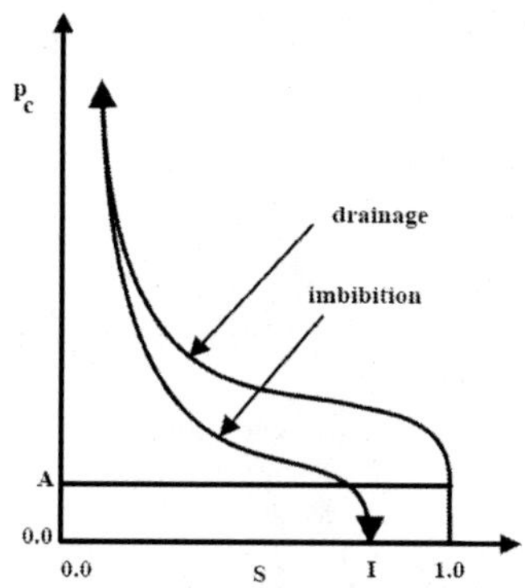

Figure 2.6: The capillary pressure-saturation relationship looks different on imbibitions than on drainage. This is called hysteresis.

The non-linear, relative permeability-capillary pressure relationship must also be determined experimentally. An illustrative example of such a curve in provided in Figure 2.7. The upper panel shows that at residual saturation of water, the relative permeability is zero. The water is no longer connected and cannot flow. As the saturation increases, so also does the relative permeability until a maximum value of one is reached. At this point the intrinsic permeability is the dictating parameter.

The lower panel shows that the saturation is related to the capillary pressure. Thus, in combination, we see from these panels that the relative permeability is in fact a function of the capillary pressure with the saturation as a linking concept. The solution of Equation (2.21) is therefore achieved by determining the saturation, saturation-capillary pressure slope, and relative permeability using the capillary pressure and then iterating to a solution.

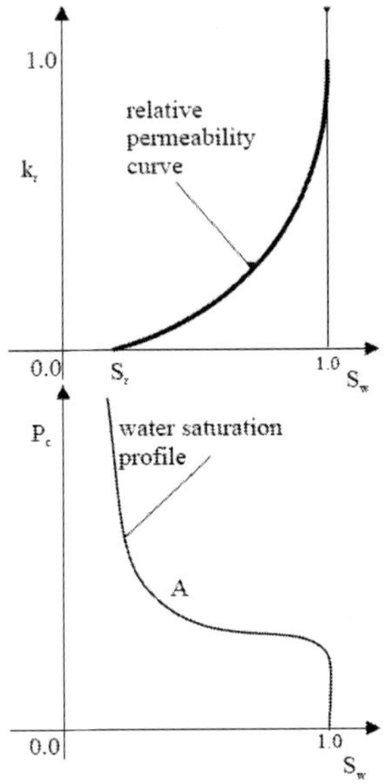

Figure 2.7: The top panel shows the relative permeability curve for water in an air-water system and the lower curve provides a corresponding saturation-capillary pressure relationship.

2.3.5.1 *Unsaturated Flow*

Water flow in the partially saturated zone can be achieved using the above formulation where $\alpha = w, a$. However, in practice one often assumes the air to be immobile such that the air-pressure equation disappears. In this instance one solves only for the water pressure using the water flow equation, that is $\alpha = w$.

2.3.5.2 *Non-aqueous Phase Flow*

Contaminants that form a separate phase have become a problem of increasing significance. Petroleum products form a Non-Aqueous Phase Liquid, or NAPL, that is less dense than water and tends to float above the water table. Such contaminants are also often distributed through the capillary fringe. Such contaminants are called Lighter-than-water Non-Aqueous Phase Liquids, or LNAPL.

The corresponding heavier than water contaminants, such as many of the chlorinated hydrocarbons, are called Denser-than-water Non-Aqueous Phase Liquids or DNAPLs.

NAPL is described using the equations developed above in this section. In this case $\alpha = n, w$ where n is the NAPL phase. In this case, two equations must be solved and the state variables would be the water and NAPL pressures (one can alternatively solve for one pressure and one saturation). In this case NAPL plays the role of the air. However, unlike the case of water and air, there is an important transfer of mass from the NAPL phase to the water phase. NAPL dissolves in water, albeit slowly, and this transfer of mass may contaminate the water phase. Thus, for the case of NAPL flow, an additional mass transfer term in needed in the governing equation to account for the dissolution phenomenon.

If the NAPL resides in the partially-saturated zone, then three phases may be present, air, water and NAPL. In this case three equations are required for the determination of three state variables, p_a, p_w, and p_n.

2.3.6 Fracture Flow

Groundwater flow in fractured media is encountered primarily in rock formations such as granite or limestone, but may also occur in relatively impermeable sediments such as clay. The distinguishing feature of fracture flow is that the water is transported primarily via fractures, while the intervening blocks provide long-term supply. Thus the system operates on two time scales, rapid movement in the fractures and slow movement in the porous blocks. When the pressure drops in the fractures water moves from the blocks to the fractures and when the pressure in the fractures rises, water moves from the fractures into the blocks. The resulting transfer of water is illustrated in Figures 2.8 and 2.9.

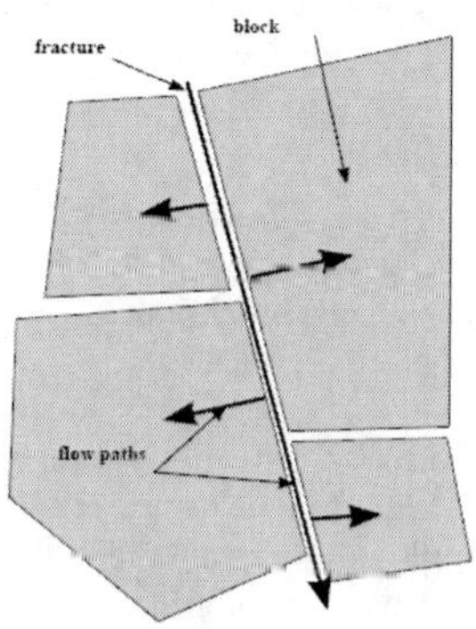

Figure 2.8: Diagrammatic representation of water movement from fractures to blocks under conditions where the pressure in the fracture exceeds that in the block.

One approach to understanding such a system is to think of the fractures and the blocks as separate and distinct. Flow in the fractures is dictated by the geometry of the fractures and flow in the blocks by the geometry of the blocks. In such a model, the equations that govern the flow in the fractures are similar to those identified with pipe flow and those in the blocks are the porous medium equations presented above.

The two sets of equations are coupled along the block faces such that there is a conservation principle governing the movement of water between the fractures and the blocks.

Because knowledge of the geometry and roughness characteristics of fractures is not generally known, the use of the dual discrete model described above has rather limited application. A more common approach to fracture flow is to think of the fractures and the blocks as two pseudo phases, that is the fractures as one pseudophase and the blocks as another. Governing equations for the combined system can then be obtained using the same averaging concepts used in arriving at the groundwater flow equations in Section 2.3.2. Following this logic, one must define a suitable sample volume of the fractured medium, that is a representative elementary volume.

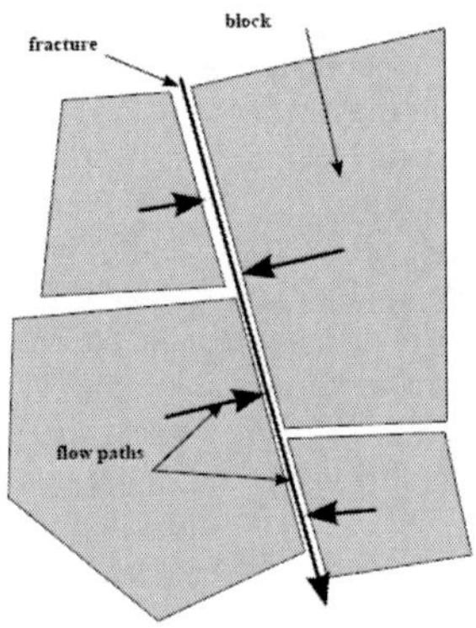

Figure 2.9: Diagrammatic representation of a fractured porous medium where water movement is from blocks into fractures due to a lower pressure in the fractures than in the blocks.

While the concept of a suitable sample volume seemed plausible in our earlier discussion, when the material exhibits a relatively limited range of pore sizes, it is more difficult to imagine a sample volume that is suitable for both the blocks and the fractures. In essence, there must be one sample size that is meaningful for both the fractures and the blocks. One cannot use one sample size for the fractures and another for the blocks.

If such a sample size can be identified, and this determination must be made on a case-by-case basis by the analyst, then the theory provided above for multiphase flow is applicable to fractured media. The phases, in this case are the pseudophases fractures, f, and blocks, b. Therefore one proceeds by writing a flow equation for the blocks and one for the fractures, that is $\alpha = f, b$. Each equation will have its own set of constitutive relationships that will lead to parameters such as hydraulic conductivity and storativity. Thus there will be two parameter sets and two sets of state variables. Movement between the blocks and the fractures will be accommodated using a mass exchange term, such as introduced above, to accommodate dissolution of NAPL into water. Because there are two sets of parameters, more specifically two sets of porosities (one for the blocks and one for the fractures) this is often called a dual porosity formulation.

The concept can be extended to consider multiphase flow in fractured media. Now each pseudophase must accommodate multiple real phases. While conceptually straight forward, the resulting system of governing equations is quite cumbersome.

2.3.7 Boundary and Initial Conditions

The equations presented above are generally applicable, irrespective of the geometrical characteristics of the system under consideration. To uniquely define the system requires the specification of boundary and initial conditions. In the case of groundwater flow, the use of hydraulic head as the state variable in the boundary and initial conditions is the most straight forward. The reason that it is an attractive choice is that it is the state variable most often measured in the field. In the case of Equations (2.18) and (2.21), the boundary and initial conditions must be defined in terms of fluid pressure.

Groundwater flow boundary conditions can have three forms: Dirichlet, also known as constant head or type one; Neumann, also known as constant flux or type two; and Robbins, also known as induced flux or type three. A conceptual model of a problem employing three different boundary condition segments is presented in Figure 2.10. Mathematically, the boundary conditions are stated as:

$$h(\mathrm{x}) = h_0(\mathrm{x}) \qquad \mathrm{x} \in \partial\Omega_1 \quad \textit{Dirichlet} \tag{2.23}$$

where h_0 is the specified head along the boundary segment $\partial\Omega_1$ of the modeled domain Ω,

$$\frac{\partial h(\mathrm{x})}{\partial \mathrm{n}} = \left.\frac{\partial h(\mathrm{x})}{\partial \mathrm{n}}\right|_0 \qquad \mathrm{x} \in \partial\Omega_2 \quad \textit{Neumann} \tag{2.24}$$

where $\partial h(x)/\partial \mathrm{n}|_{x=0}$ is the specified outward normal gradient to, and along the boundary of, the segment $\partial\Omega_2$, and

$$\alpha h(\mathrm{x}) + \gamma \frac{\partial h(x)}{\partial \mathrm{n}} = C_0(\mathrm{x}) \qquad x \in \partial\Omega_3 \quad \textit{Robbins} \tag{2.25}$$

where C_0 is a specified function value along the boundary segment $\partial\Omega_3$. The parameters α and γ are known functions. Since the entire boundary must be defined by a boundary condition the sum of the above segments must equate to the whole, that is

$$\partial\Omega = \partial\Omega_1 + \partial\Omega_2 + \partial\Omega_3 \tag{2.26}$$

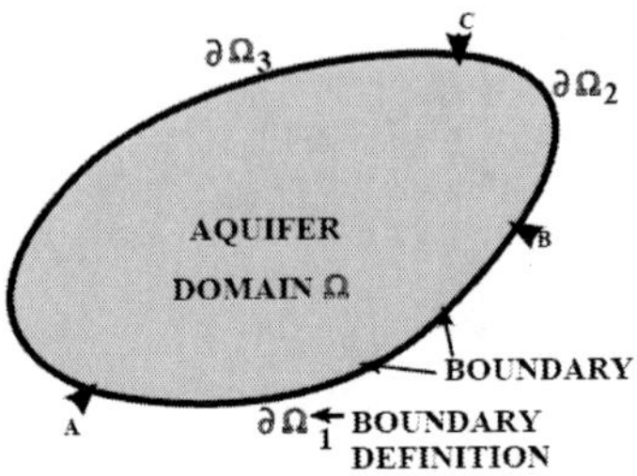

Figure 2.10: Diagrammatic sketch showing boundary conditions along $\partial\Omega$ enclosing a region Ω.

Equation (2.23) is used to specify the head along the boundary. It can be a function of time if the time-dependent behavior is known. Tidal behavior of the water surface of an estuary could, for example, be described by this kind of boundary specification.

Equation (2.24), when combined with the hydraulic conductivity or, in the case of a two-dimensional areal model, transmissivity, provides a statement of the flow across the boundary. In other words, in terms of flow per unit length of boundary we have:

$$\mathbf{q}_{\mathrm{n}}(\mathrm{x}) = -\mathbf{K} \cdot \frac{\partial h(\mathrm{x})}{\partial \mathbf{n}} \tag{2.27}$$

where $\mathbf{q}_{\mathrm{n}}$ is the specified flux across the boundary $\partial\Omega_2$.

The Robbins, or third-type boundary condition specified in Equation (2.25) is used to describe what is called a leakage condition when used in a three-dimensional model. It can also be used to represent a physical boundary at a long distance from a model boundary when it is used in a two or three-dimensional model setting. The form used for both of these conditions is

$$\mathbf{K} \cdot \frac{\partial h(\mathrm{x})}{\partial \mathbf{n}} = \boldsymbol{\kappa}\left(h_0(\mathrm{x}) - h(\mathrm{x})\right)\mathrm{n} \tag{2.28}$$

where h_0 is the head external to the model. For example, h_0 might be the elevation in a lake on the top of the model in the case of a leakage situation. In this case the coefficient $\boldsymbol{\kappa}$ has the meaning of the resistivity to flow across the boundary. Alternatively, h_0 could be considered as the head in the aquifer at some distance from the boundary and the coefficient $\boldsymbol{\kappa}$ in this case would be used as a surrogate for the distance to the location of h_0 relative to $\partial\Omega_3$, i.e. $\boldsymbol{\kappa} \equiv \mathbf{K}/L$ where L is the distance from the model boundary to the real boundary where h_0 is observed.

2.3.8 Initial Conditions

Equation (2.16) contains a first-order time-derivative term. Thus it is necessary, from a mathematical perspective, to define the state of the system at the beginning of a model simulation. Since the dependent variable is the head, one is required to specify the head everywhere in Ω at the initial time $t_0(\mathrm{x})$.

In general, the role of the initial conditions in a groundwater flow problem is relatively minor. The reason is that for problems defined at a scale of practical importance, the head quickly adjusts to the boundary conditions. Because the system quickly readjusts from the initial state to that consistent with the boundary conditions, exact specification of the initial conditions is normally unnecessary. Often a uniform value of zero can be provided and the problem allowed to equilibrate with respect to the boundary conditions and predefined stresses before new hydrodynamic stresses are applied.

2.3.9 Aquifer Properties and Dynamics

2.3.9.1 *Regional Flow Systems*

In Section 2.2 we introduced the concept of saturated and unsaturated flow in the near-surface environment. At this scale, water flow is essentially vertically downward. Now we will investigate groundwater movement on a regional scale. At this scale water movement is not necessarily vertical but rather responds to large-scale hydrodynamic conditions.

In his classic paper, 'The Theory of Ground-Water Motion,' Hubbert (1940) presented a qualitative representation of a regional groundwater flow field which is reproduced here as Figure 2.11. In this cross-sectional view one observes a water table with maximum elevation at point *A*, bounded on each side by streams located at points *B* and *C*. The dashed line identified as the water table approximately mimics the shape of the land surface. This observation is consistent with observations of water tables found in a humid environment. The solid lines with arrows affixed are flow lines and the dashed lines are equipotential lines. In an isotropic and homogeneous medium, the flow lines are orthogonal to the equipotential lines.

Precipitation enters the lands surface and moves vertically down to the water table. At the highest point, *A*, the water continues to move vertically downward until it encounters the impermeable boundary at the base of the cross-section. At that point it moves horizontally to the left and right until encountering location *B* and *C*. At this point, assuming that the geometry of the water table shown between points *B* and C is replicated to the left and right of points *B* and C, the water moves vertically upward to discharge into the stream beds. The flow lines originating at the water table at points other than $A - E$ are not exactly vertical. Because of the bounding flow line described above, the other flow lines between the water table and those discussed above are deflected towards the stream discharge location.

Of particular interest are the equipotential lines that have been hypothesized for this system. Notice that Hubbert has drawn a horizontal dashed line at the point where the equipotential lines intersect the water table. The horizontal line indicates the magnitude of the fluid potential for the corresponding equipotential line. At the higher elevations one observes that the fluid potential decreases with depth (consider the vertical flow line at A for example). However, at low elevations, the opposite is true. If one moves from the stream bed down the line located at B the fluid potential increases. Thus one observes that in a recharge area the fluid potential decreases with depth and in a discharge area the fluid potential increases with depth.

Consider drilling a well at the location indicated with a w. Assume that the well screen is open only at the bottom of the well. The fluid potential at that location is higher than the elevation of the top of the well. The consequence of this is a flowing well. Devoid of pumping water will discharge from the well.

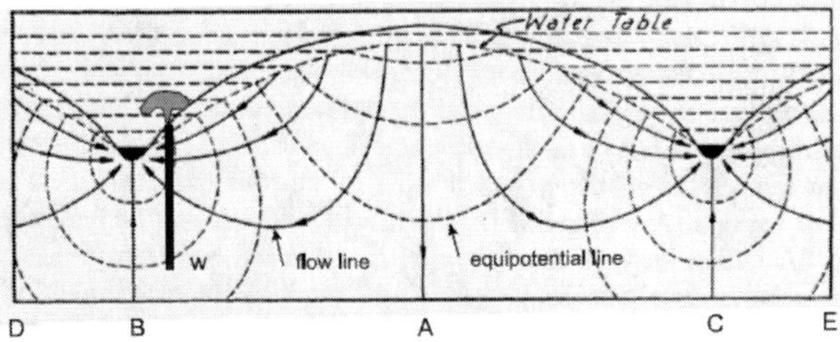

Figure 2.11: Regional groundwater flow as represented in Hubbert (1940).

Toth (1963) examined the regional flow concept using a mathematical model based upon field observations he had made in Alberta, Canada. Whereas the concept presented by Hubbert was based upon an infinite series of symmetric basins, Toth superimposed an undulating topography on a regionally inclined plane. As is evident from an examination of his computed flow field (see Figure 2.12), his analysis was limited in having to assume a rectangular geometry for the aquifer cross section. Nevertheless, he was able to demonstrate that under the assumed water-table topography, flow systems of different scale could be realized. Shallow flow systems with discharge points at the locations indicated by the arrows in Figure 2.12 are analogous to those hypothesized by Hubbert. However at greater depth there exists a regional flow system that extends from the right hand side of the figure to the left (note the streamline extending downward along the right side to the base, left along the base to the left side of the modeled aquifer and then up the left side). Moreover, there are a series of streamlines that are intermediate between the regional and local flow systems. These streamlines do not discharge locally, but their trajectories reflect the influence of the local flow systems.

2.3.9.2 *The Influence of Nonhomogeneity*

In the preceding examples, the aquifer has been assumed to be homogeneous, that is characterized by hydraulic conductivity that exhibits no spatial variability (see section 2.3.3.2). Aquifers, however, are seldom homogeneous. In general the hydraulic conductivity will vary with depth. To examine the influence of geological layering, or lithology, on regional flow, Freeze and Witherspoon (1967) modeled a layered aquifer in cross-section using numerical methods.

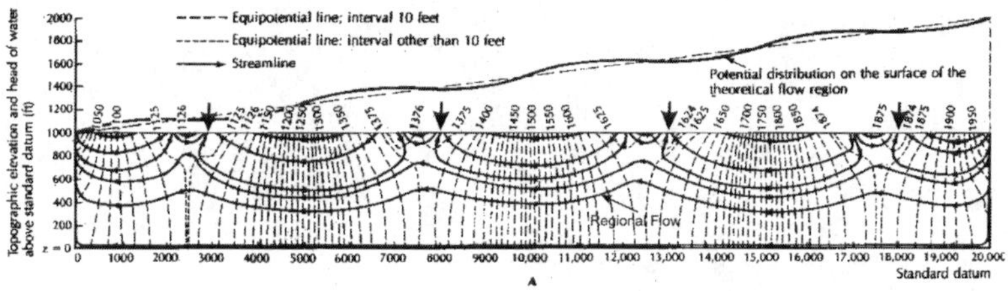

Figure 2.12: Regional flow systems as calculated by Toth (1963).

In Figure 2.13 flow paths and equipotential lines calculated for a cross section with geological layering are presented. The upper layer has a hydraulic conductivity of unity, that is $K = 1$. The lower layer has a hydraulic conductivity of ten. The water table slopes downward from right to left generating a regional flow field. Although in the recharge area flow tends to be primarily downward, and in the discharge area primarily upward, along nearly the entire length of the flow field there is a downward component of flow from the less permeable into the more permeable aquifer. In the more permeable aquifer, flow tends to be nearly horizontal except at the recharge and discharge points.

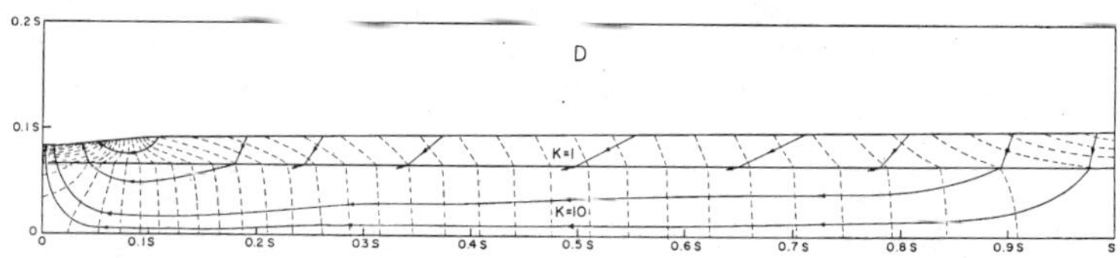

Figure 2.13: Influence of geological layering (nonhomogeneity) on regional flow (Freeze and Witherspoon 1967).

Lower permeablity units over the more permeable units, are called aquitards, since they tend to inhibit the flow of groundwater to permeable aquifers. Aquitards that are of very low permeability are called aquicludes. The observation that water tends to flow vertically in aquitards and horizontally in aquifers plays a significant role in the theory of well hydraulics, a subject to be discussed in a later section. An aquitard that provides practically useful amounts of water to an aquifer is called leaky.

Aquifers bounded by confining beds are called confined aquifers. As is evident from Figure 2.13, the water table elevation does not, in general, represent the fluid potential in a confined aquifer. Aquifers that have no overlying confining beds are called unconfined aquifers. The fluid potential of unconfined aquifers is approximately represented by the elevation of the water table.

An aquifer that is confined from above and below by aquitards normally remains saturated. The fluid potential surface generally remains above the top of the aquifer.

If there exist two aquifers separated by an aquitard it is possible, on rare occasions, to encounter perched water. This phenomenon is observed when the fluid potential in the lower aquifer drops quickly to an elevation below the top of the lower aquifer. Thus a partially saturated zone develops in the lower aquifer. But, due to the low permeability of the aquitard, water from the upper aquifer may not effectively drain, such that it remains saturated. Water existing in an aquifer which has beneath it a partially saturated zone is called perched water.

2.3.9.3 *Influence of Anisotropy*

In section 2.3.3.1 we introduced the concept of anisotropy, that is directionally dependent hydraulic conductivity. In general aquifers are not isotropic, but tend to be more permeable in the horizontal direction than in the vertical. Higher values of hydraulic conductivity are a natural consequence of sediment deposition and consolidation. Freeze and Witherspoon examined the impact of anisotropy on regional groundwater flow (Freeze and Witherspoon, 1967). The results of their investigation are found in Figure 2.14. The upper panel (panel A) represents regional flow in an isotropic aquifer given a specified water table configuration. The lower panel (panel B) shows the flow paths and equipotential lines for the case of horizontal hydraulic conductivity, K_h, of 10 and vertical conductivity, K_v, of unity. One observes that the equipotential lines are less vertical in panel B and that they are no longer orthogonal to the flow lines. The latter observation is the direct consequence of Equation (2.10), wherein the diagonal elements of the K tensor have different values.

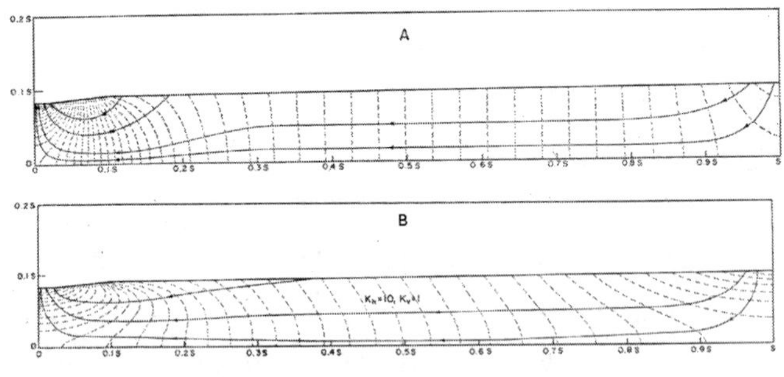

Figure 2.14: Impact of anisotropy on regional groundwater flow (Freeze and Witherspoon 1967).

2.3.10 Surface Water-Groundwater Interactions

The regional flow patterns presented in section 2.3.9 are determined in large part by the discharge points provided by the surface water bodies. In essence surface water bodies act as sources and sinks in regional groundwater flow. The portion of the surface water hydrograph identified as groundwater discharge varies depending in the long term upon the regional hydrology and in the short term on recent rainfall patterns. When, after a long dry period, groundwater discharge constitutes nearly all of stream flow, this discharge is called base flow.

Groundwater can both enter and leave a surface water body. For example a flood peak will tend to cause water to move from the surface water body into the groundwater. Similarly, after the flood peak passes, water retained in the river banks (bank storage) will return to the stream.

In Figure 2.15 is presented an analysis of bank storage via a numerical model. A flood wave has passed along this stretch of river and the water table elevation in the adjacent aquifer reflects this event. At a distance of approximately 500 ft from the river bank there is an elongated ridge in the water table. The ridge extends along and parallel to the river. It diminishes with distance downstream. The source of this peak is the flood wave that has moved down the river in the recent past. Initially water moved from the river into the adjacent aquifer causing the water table to rise. After passing, the water stored in the river bank began to return to the river. The head near the river dropped most quickly in response to the drop in river stage. The response of the water table further from the river lagged that nearer the bank and therefore is higher in this figure. The existence of a peak in the water table (rather than a plateau) is due to the fact that the impact of the flood wave has not yet propagated the full width of this flood plain aquifer.

In Figure 2.15 we capture the bank storage effect as a snapshot in time. Another way to view this phenomenon is to observe the impact of bank storage on the stream flow over a period of time at a specific location. Figure 2.16 shows the decomposed stream flow hydrograph and the effect bank storage has had upon it. On the vertical axis is plotted the change in stream flow from the steady state flow existent at the beginning of the study. The arrival of the flood peak is reflected in the increase in discharge which peaks at about 10,000 s. The top curve, identified as 'flood hydrograph, no leakage', shows how the stream flow responds when there is no bank storage effect (no leakage into the aquifer). The middle curve identified as 'flood hydrograph, with leakage' is the discharge experienced by the stream when bank storage effects are included. The bottommost curve is the net negative flow lost from the stream due to water moving into the river bank. The top curve plus the bottom curve yields the middle curve. While this is but one numerical example, it illustrates the general elements of the bank-storage effect.

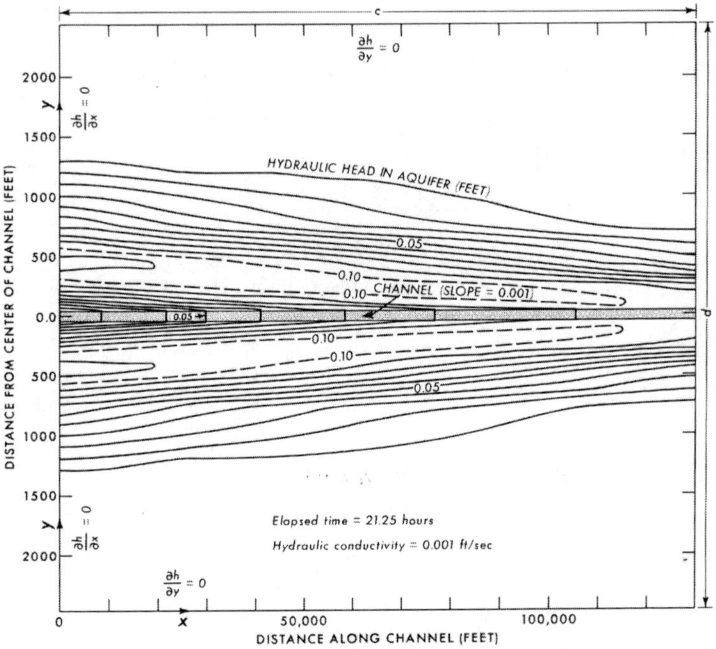

Figure 2.15: Bank storage effect calculated using a numerical model.

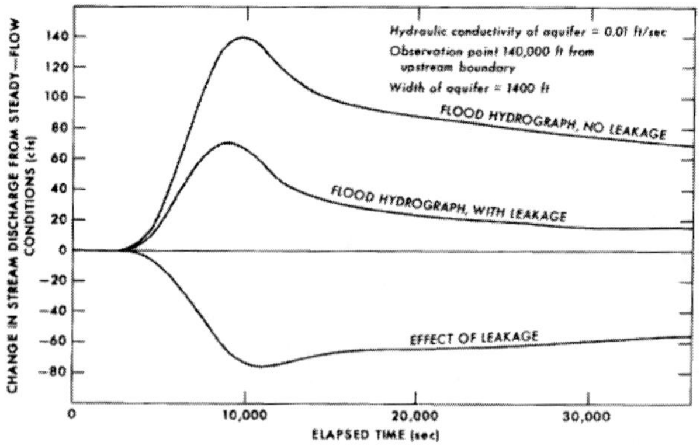

Figure 2.16: Flood hydrograph separation as calculated using a numerical model.

2.4 References

Darcy, H. (1856). *Les fontaines publiques de la ville de Dijon*, Victor Dalmont, Paris.

Dingman, S. L. (1994). *Physical hydrology*, Macmillan, New York.

Dingman, S. Lawrence (2002). *Physical hydrology*, 2nd ed., Waveland Press.

Freeze, R. A., and Witherspoon, P. A. (1967). "Theoretical analysis of regional groundwater flow: 2. Effect on water-table configuration and subsurface permeability variation," *Water Resources Research*, 3(2), 623-634.

Hubbert, M. K. (1940). "The theory of ground-water motion." *Journal of Geology*, 48(8), 785-944.

Toth, J. A. (1963). "A theoretical analysis of ground-water flow in small drainage basins," *Journal of Geophysical Research*, 68(16), 4795-4811.

CHAPTER 3

GROUNDWATER QUALITY: FATE AND TRANSPORT OF CONTAMINANTS

Mohammad N. Almasri[1] and Jagath J. Kaluarachchi[2]
[1]An-Najah National University, Nablus, Palestine
[2]Utah State University, Logan, Utah, USA

3.1 Introduction

With the increasing water demand due to the population growth and rapid urbanization, management of water resources is becoming a challenging task. One of the most important aspects of water resources management is prevention and control of groundwater contamination (Katsifarakis 2000). Groundwater contamination is a critical issue because of the importance of groundwater supply and quality in sustainable development (Adriano et al. 1994). Groundwater is an important natural resource in the U. S.; for example, 37% of public-supply withdrawal or 14,874 million gallons per day in 1995 is from groundwater (Solley et al. 1998).

In general, contaminants found in groundwater are associated with adverse health, social, environmental, and economic impacts (OTA 1984). Health concerns are present because some contaminants are associated with cancer, liver and kidney damage, and damage to the central nervous system. In addition, the health impacts associated with exposure to many of the contaminants found in groundwater encompass significant uncertainty. Exposure to contaminated groundwater may occur accidentally since contaminated water might be colorless, odorless, and tasteless. Environmental impacts include degradation of water quality due to the interactions between soil, surface water, and air with contaminated groundwater. For instance, contaminants from groundwater may enter surface water bodies through baseflow. This interaction can produce harmful effects to vegetation, fish habitat, and wildlife. The economic ramifications of groundwater contamination include the significant costs of detecting, preventing, and mitigating contamination. Furthermore, economic damages may also include the costs of developing alternative water supplies, lowered property values, the costs incurred from possible mortality or morbidity, and the decrease in agricultural and industrial yield as a result of using the impaired water.

One of the major problems in groundwater contamination is that it occurs in the subsurface surreptitiously. Contaminants are not easily observed because their detection is usually based on obtaining groundwater samples from monitoring wells, which sample a relatively small volume of the aquifer. Detection may be difficult because contaminated groundwater is commonly odorless and colorless. Furthermore, the concentrations at which contaminants pose a health risk is very low for many contaminants, making their detection and quantification difficult analytically. The noticeable impacts of groundwater contamination are present long after the initial incident that caused the contamination (Pye and Kelley 1984).

Severity of groundwater contamination may depend on one or a combination of the following situations (Pye and Kelley 1984): (i) the use of contaminated groundwater as a potable water source; (ii) the contaminant concentration relative to the maximum contaminant level; (iii) the number of persons affected by the contamination; (iv) the area of the aquifer contaminated; and (v) the degree by which the contaminants pose hazards, which is a function of toxicity of the contaminant, persistence, concentration, and contaminant movement in the aquifer.

It is essential to understand the fate and transport processes of groundwater contaminants for the reliable evaluation of the extent of groundwater contamination, the assessment of the corresponding hazards and risks, and the design of effective remediation systems and protective alternatives. Such processes include, but are not limited to, advection, mechanical dispersion, molecular diffusion, and chemical and biological reactions (National Research Council 1984). Typically, mathematical models need to be developed and employed in characterizing and managing groundwater contamination. Some principal areas of model applications are (Barcelona et al. 1990; Rifai and Bedient 1994): (i) assessment of the extent of groundwater contamination; (ii) assessment of the consequences of current practices on groundwater quality; (iii) prediction of future conditions under current practices; and (iv) evaluation of the efficiency of probable remedial actions, protection alternatives, and management options aimed at minimizing groundwater contamination.

This chapter provides a general overview of the main concepts of groundwater contamination, including sources and types of groundwater contaminants, fate and transport processes in groundwater, and mathematical modeling.

3.1.1 Sources of Groundwater Contamination

According to Domenico and Schwartz (1990), there are three key features that differentiate groundwater contamination sources: (i) degree of localization; (ii) loading history; and (iii) types of contaminants.

Sources of groundwater contamination can be classified into point and non-point sources (Domenico and Schwartz 1990; Fitts 2002). Point sources include leaking underground pipelines or tanks, wastewater lagoons, septic systems, or leaking barrels of chemical wastes. These sources are identifiable, located at single locations, and have local impacts on groundwater quality. Conversely, non-point sources are diffuse, originating from many smaller sources, largely distributed, lack a well-defined single point of origin, and have a regional influence on groundwater quality. Examples of non-point sources involve agricultural activities such as fertilizer and manure applications, polluted atmospheric deposition including dry fallouts, precipitation, and runoff from roads and parking lots.

The loading history defines how the concentration of a contaminant or its rate of application changed with time at the source location (Domenico and Schwartz 1990). Figure 3.1 depicts four example scenarios of loading history functions used to identify contaminant loadings. Figure 3.1(a) depicts an example of a pulse loading of a contaminant at a constant concentration over a short time. A long-term continuous loading at constant concentration is also depicted in Figure 3.1(b). Since the majority of contamination sources vary temporally on a seasonal or monthly basis, such as in agricultural practices, the loading history can be defined as continuous with variable concentration as shown in Figure 3.1(c). In situations where contaminant is subject to decay, the loading history may have the declining exponential form depicted in Figure 3.1(d).

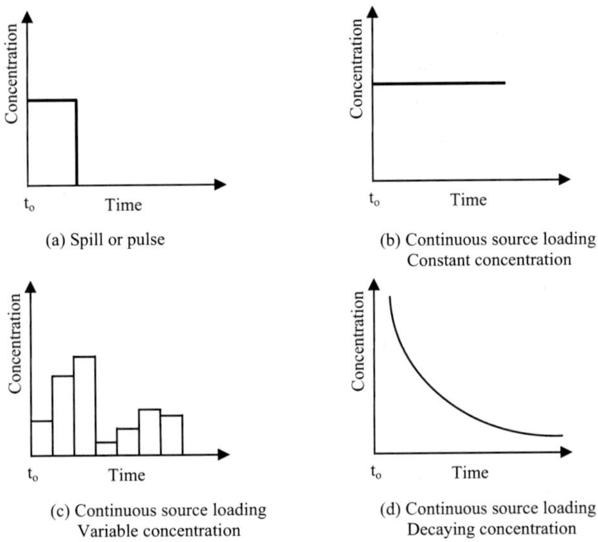

Figure 3.1: Examples of functions used to characterize contaminant loading from various sources.

A common practice in groundwater engineering is the reconstruction of the loading history of groundwater contamination, spatially and temporally, from the current plume spatial distribution. This is essential for the designation of the parties responsible for the contamination and who may be in turn responsible for remediation (Skaggs and Kabala 1994).

The possible pollutants of groundwater are almost unlimited with a great diversity of sources. Most contaminants originate from unmanaged disposal of wastes on or into the ground. Disposal methods include wastes in permeation ponds, spreading on the ground surface, landfills, seepage drains, and disposal and injection wells (Todd 1980). Inadvertent releases of products from aboveground and underground storage may also be contaminant sources. There are four main mechanisms by which groundwater contamination occurs (Barcelona et al. 1990): infiltration, direct migration, inter-aquifer exchange, and recharge from surface water. Todd (1980) classified the sources of groundwater contamination into four categories: municipal, industrial, agricultural, and miscellaneous sources.

The OTA (1984) identified 33 sources of groundwater contamination and grouped them into 6 categories, which are summarized in Table 3.1.

Category I includes sources that are designed to discharge polluting substances, such as septic tanks, cesspools, injection wells, and land application of wastewater via spray irrigation or sludges. The basic premise in designing these systems is to exploit the capacity of the soil to assimilate the wastewater. Septic tanks have been reported to discharge domestic wastewater rich with ammonium to groundwater (Cox and Kahle 1999). Wastewater injection wells are another major potential source of contamination. In essence, injection wells are designed to discharge liquid wastes into subsurface zones below any drinking water aquifers. Injection wells can cause groundwater contamination in cases of defective well construction and design, possible migration of contaminants to usable aquifers, and probable faults and fractures of confining beds.

Category II includes sources that are designed to store, treat, or dispose of substances, but are not designed to release contaminants to the subsurface. These sources include landfills, open dumps, residential disposal, surface impoundments, waste tailings, waste piles, material stockpiles, graveyards, animal burial, aboveground and underground storage tanks, containers, open burning and detonation sites, and radioactive waste disposal sites.

Category III includes sources designed to retain substances during transport or transmission. These sources principally consist of pipelines and material transport and transfer operations. Pipelines may be used for transmitting wastewater as well as gas, petroleum, etc. When leaking, sanitary sewer pipelines may contaminate the groundwater with bacteria, viruses, nitrate, and chloride.

Category IV includes sources that discharge substances as a consequence of other planned activities. This category is comprised mainly of agriculture-based activities, such as irrigation return flow, pesticide and fertilizer applications, and animal feeding operations. This category also encompasses other sources, such as urban runoff, percolation of atmospheric pollution, and surface and underground mine-related drainage.

Category V includes sources that result from the inadvertent creation of conduits in the subsurface and alteration of groundwater flow patterns. These sources may be created by the installation of oil and gas wells, geothermal and heat recovery wells, water supply wells, monitoring and exploration wells, and construction dewatering wells. In general, such wells when improperly abandoned, unsuitably completed, or constructed of inappropriate materials, can provide vertical conduits for contaminants to reach groundwater from the surface or from underlying contaminated strata. Also, construction excavations that strip the soil overburden from bedrock can also expose conductive fractures, allowing the entry of surface contaminants.

Lastly, ***Category VI*** includes naturally occurring sources that can cause groundwater contamination as a result of human activity. Examples of sources in this category are groundwater-surface water interactions, natural leaching, salt-water intrusion, and brackish water upconing. In groundwater-surface water interaction, aquifers are recharged from the surface water when the stream stage is higher than the water table. If the surface water is contaminated, then contaminants may reach the aquifer with recharge. Additionally, cones of depression of wells located close to contaminated surface water bodies may provoke surface water to enter the aquifer. In coastal areas, excessive pumpage beyond the aquifer's sustainable yield induces salt water to move into freshwater aquifers. Irrigation can result in the leaching of natural constituents from the soil and bring about groundwater contamination.

3.1.2 Types of Groundwater Contaminants

Domenico and Schwartz (1990) classified contaminants according to the reaction type and means of occurrence. Accordingly, the major group of contaminants include: (i) radionuclides; (ii) trace elements; (iii) nutrients; (iv) other inorganic species; (v) organic contaminants; and (vi) microbial contaminants. All these contaminants have the potential to pose significant risk to human health as long as these contaminants are present at concentrations in excess of the maximum contaminant level (MCL). In the following sections, additional discussion of these contaminant types is provided.

3.1.2.1 *Radioactive contaminants*

Most groundwater has naturally-occurring radioactive substances. The radionuclides present and their associated activities vary from region to region depending on the geologic origins of the parent bedrock. Human activities also contribute to the radionuclide inventory in groundwater. Fallout from aboveground nuclear weapons testing has resulted in the presence of tritium and fission products in groundwater. Releases may also occur during the various steps of the nuclear fuel cycle. The mining of uranium ore can result in groundwater contamination from tailings and mine drainage (OTA 1984). The processing and reprocessing of fuel generates solid and liquid radioactive waste that may be inadvertently released. Releases from nuclear power plants, which use nuclear fuel to produce electricity, has contributed to groundwater contamination. Finally, the storage of spent fuel and disposal of other radioactive waste can result in groundwater contamination.

The health risks linked to the exposure of radioactive contaminants and radiation are very well recognized. An exposed person can be affected by cancer or genetic defects that can affect the offspring. Other health problems include cataracts, nonmalignant skin damage, depletion of bone marrow, and infertility (Domenico and Schwartz 1990).

Table 3.1: Sources of groundwater contamination (OTA 1984).

Sources
Category I—Sources designed to discharge substances
Subsurface percolation (e.g., septic tanks and cesspools)
Injection wells
Hazardous waste
Non-hazardous waste (e.g., brine disposal and drainage)
Non-waste (e.g., enhanced recovery, artificial recharge, solution mining, and in-situ mining)
Land application
Wastewater (e.g., spray irrigation)
Wastewater byproducts (e.g., sludge)
Hazardous waste
Non-hazardous waste
Category II—Sources designed to store, treat, and/or dispose of substances; discharge through unplanned release
Landfills
Industrial hazardous waste
Industrial non-hazardous waste
Municipal sanitary
Open dumps, including illegal dumping (waste)
Residential (or local) disposal (waste)
Surface impoundments
Hazardous waste
Non-hazardous waste
Waste tailings
Waste piles
Hazardous waste
Non-hazardous waste
Material stockpiles (non-waste)
Graveyards
Animal burial
Aboveground storage tanks
Hazardous waste
Non-hazardous waste
Non-waste
Underground storage tanks
Hazardous waste
Non-hazardous waste
Non-waste
Containers
Hazardous waste
Non-hazardous waste
Non-waste
Open burning and detonation sites
Radioactive disposal sites
Category III–Sources designed to retain substances during transport or transmission
Pipelines
Hazardous waste
Non-hazardous waste
Non-waste
Material transport and transfer operations
Hazardous waste
Non-hazardous waste
Non-waste
Category IV—Sources discharging substances as consequence of other planned activities
Irrigation practices (e.g., return flow)
Pesticide applications
Fertilizer applications
Animal feeding operations
De-icing salt applications
Urban runoff
Percolation of atmospheric pollutants
Mining and mine drainage
Surface mine-related
Underground mine-related
Category V—Sources providing conduit or inducing discharge through altered flow patterns
Production wells
Oil (and gas) wells
Geothermal and heat recovery wells
Water supply wells
Other wells (non-waste)
Monitoring wells
Exploration wells
Construction excavation
Category VI—Naturally occurring sources whose discharge is created and/or exacerbated by human activity
Groundwater-surface water interactions
Natural leaching
Salt-water intrusion/brackish water
Upconing (or intrusion of other poor-quality natural water)

A detailed illustration of contamination sources can be found in Todd (1980), OTA (1984), National Research Council (1984), Roy (1994), Domenico and Schwartz (1990), Fetter (1999) and Fitts (2002).

3.1.2.2 *Trace metals*

Trace metals are a major group of contaminants that include aluminum, antimony, arsenic, barium, beryllium, boron, cadmium, chromium, cobalt, copper, gold, iron, lead, lithium, manganese, mercury, molybdenum, nickel, selenium, silver, strontium, thallium, tin, titanium, uranium, vanadium, and zinc (Domenico and Schwartz 1990).

Trace metals can be toxic and harmful to humans even at low concentrations because these substances tend to accumulate in the body. The main sources of trace metals in groundwater are: (i) mining effluents; (ii) wastewater from industrial and urban areas; (iii) runoff and solid wastes; (iv) agricultural activities, such as fertilizer and manure applications; and (v) relic fuels.

3.1.2.3 *Nutrients*

The common nutrients are ions or organic compounds containing nitrogen or phosphorus. Undoubtedly, the common nitrogen species in groundwater include nitrate (NO_3) and, to a minor extent, ammonium (NH_4). Nitrate contamination of groundwater is considered one of the most serious contamination problems worldwide because of the diffuse nature of the sources and the difficulties in its control and management (Latinopoulos 2000).

Although the issue of nitrate pollution of groundwater has been studied extensively for many decades and many protection alternatives have been developed (Mercado 1976), recent studies show the persistent existence of nitrogen in both soils and groundwater (Nolan et al. 2002). Elevated nitrate concentration in drinking water can cause *methemoglobinemia* in infants and stomach cancer in adults (Lee et al. 1991; Wolfe and Patz 2002). Nitrate may indicate the presence of bacteria, viruses, and protozoa in groundwater if the source of nitrate is animal waste or effluent from septic systems. Agricultural activities are probably the most significant anthropogenic source of nitrate contamination in groundwater (Follett and Delgado 2002).

Non-point sources of nitrogen from agricultural practices include fertilizers, dairy farms, manure application, and leguminous crops (Almasri 2003; Almasri and Kaluarachchi, 2004a, b; 2005a, b). For instance, the extensive use of fertilizers on row crops is considered to be the main non-point source of nitrate leaching to groundwater, particularly on sandy soils. In urban areas, infiltration from fertilized lawns may also contribute nitrate to groundwater, mainly when overirrigated. Movement of nitrate to groundwater from animal waste application can be a more serious concern than with row crops, especially in areas with a high density of dairy farms. Elevated nitrate concentration in groundwater is common around dairy operations, barnyards, and feedlots (Paul and Zebarth 1997; Harter et al. 2002). Poultry manure also has been shown to cause an increase in nitrate concentration in groundwater (Hii et al. 1999). In addition to agricultural practices, non-point sources of nitrogen include dissolved nitrogen in precipitation, irrigation flows, and dry deposition. Point sources of nitrogen are shown to contribute to nitrate pollution of groundwater. The major point sources include septic tanks and dairy lagoons. Many studies have shown high concentrations of nitrate in areas with septic tanks and dairy lagoons (Keeney 1986; Cantor and Knox 1984; Arnade 1999; MacQuarrie et al. 2001).

The issue of phosphorus contamination of groundwater is considered of less importance than that of nitrate. This is due to the low mobility and solubility of phosphorus compounds in groundwater since phosphorus tends to adsorb onto solids.

3.1.2.4 *Other inorganic species*

This group of contaminants includes metals present in non-trace quantities, such as calcium, magnesium, sodium, and nonmetals, the latter including ions containing carbon and sulfur or other species such as chloride and fluoride (Domenico and Schwartz 1990). Many of these ions are main contributors to the salinity of groundwater. High concentrations of these ions make the water unfit for human consumption as well as for many industrial usages. For ions such as sodium, high concentrations may cause acute disruption in blood and cell chemistry. The potential sources of major ion salinity include: (i) saline brine produced with oil; (ii) leachate from sanitary landfills; and (iii) industrial wastewater.

3.1.2.5 *Organic contaminants*

Organic compounds have carbon as the key element and usually hydrogen and oxygen as elements in their structure. Dissolved organic matter is always present in natural groundwater but in lower concentrations compared to inorganic compounds (Freeze and Cherry 1979). In general, organic compounds in groundwater might be dissolved in water or exist as a separate, immiscible liquid.

Organic compounds, as a result of man-made activities, are of great concern. Large quantities of organic compounds especially unrefined petroleum products are used extensively worldwide. An important group of organic compounds is the soluble aromatic hydrocarbons that enter groundwater following a spill of petroleum fuels or lubricants. This group includes benzene, ethylbenzene, toluene, and para-xylene, meta-xylene, and ortho-xylene. This group of organics is called the BTX (benzene, toluene, and xylene), or BTEX (benzene, toluene, ethylbenzene, and xylene) (Domenico and Schwartz 1990).

Chlorinated solvents represent another important group of man-made organic chemicals that are common groundwater contaminants. These are typically halogenated aliphatic compounds that are moderately soluble in water. The most ubiquitous examples include trichloroethene (TCE) and tetrachlroethene, also known as perchloroethylene (PCE). Because these solvents enjoyed widespread industrial use, historical disposal practices and inadvertent releases have resulted in many cases of groundwater contamination. Most chlorinated solvents are denser than water and relatively immiscible. When released in the groundwater, they sink as a separate liquid phase through the groundwater due to gravity, leaving trace quantities in the pore space as they descend, and pool on clay lenses or bedrock surfaces they cannot penetrate. This separate phase is often referred to as dense, non-aqueous phase liquid or DNAPL. Pooled or residual DNAPL, once in the subsurface, can serve as a long-term source of groundwater contamination due to its slow rate of dissolution.

Furthermore, most of these solvents are biologically recalcitrant in shallow, aerobic groundwater systems, and therefore tend to persist in the subsurface. When the factors cited above are combined, it is easy to understand why chlorinated solvent contamination plumes are so prevalent.

In general, exposure to water contaminated with organic compounds may cause cancer, liver damage, impairment of cardiovascular function, depression of the nervous system, brain disorders, and various lesions.

3.1.2.6 *Biological contaminants*

The main biological contaminants that can be found in groundwater include bacteria, viruses, and parasites (OTA 1984). These contaminants can cause serious health problems, such as typhoid, bacillary dysentery, cholera, gastroenteritis, tuberculosis, and hepatitis. The primary sources of biological contaminants in groundwater are the introduction of human or animal fecal material via septic tank leakage and cesspools, land application of dairy and poultry manure, and animal feedlots.

3.1.3 Drinking Water Standards

The Safe Drinking Water Act passed in 1974 and amended in 1986 and 1996, gives the United States Environmental Protection Agency (US EPA) the authority to set drinking water standards. Drinking water standards are regulations that the US EPA sets to control the level of contaminants in the drinking water and apply to all public water systems.

There are two types of standards: primary and secondary. Primary standards are known as the Maximum Contaminant Levels (MCLs) and are health-based enforceable standards. The MCLs are feasible and based upon treatment technologies and cost affordability. Secondary standards include those parameters concerning the aesthetic quality of the water and are non-enforceable guidelines. The US EPA sets the MCL close to the Maximum Contaminant Level Goal (MCLG). The MCLG is the maximum level of a contaminant in drinking water at which no identified adverse effects on human health would occur.

For carcinogenic contaminants, the MCLG is set at zero signifying that no amount of contaminants is acceptable. Nevertheless, since zero cannot be measured, the MCL is based on the lowest measurable level. This MCL is known as the Practical Quantification Limit (PQL). Thus for known or probable carcinogens, the MCL is not necessarily a safe level but may instead be the lowest level that can be feasibly measured.

3.2 Transport Processes

The mass conservation statement for a dissolved species in a groundwater system is (Freeze and Cherry 1979):

$$\begin{pmatrix} \text{Net rate of} \\ \text{change of solute} \\ \text{mass within} \\ \text{the system} \end{pmatrix} = \begin{pmatrix} \text{Flux of} \\ \text{solute out of} \\ \text{the system} \end{pmatrix} - \begin{pmatrix} \text{Flux of} \\ \text{solute into} \\ \text{the system} \end{pmatrix} + \begin{pmatrix} \text{Loss or gain} \\ \text{of solute mass} \\ \text{due to reactions} \end{pmatrix}$$

The physical processes that control the contaminant flux into and out of the groundwater system are advection, mechanical dispersion, and molecular diffusion. Loss and gain of contaminant mass in the system are attributed to the chemical, biological and radiological reactions. In the following subsections, the basic concepts related to advection, mechanical dispersion, and molecular diffusion are illustrated.

3.2.1 Advection

Transport by *advection* or *convection* entails the movement of contaminants by the flowing groundwater (Fetter 1999). In the advective transport, contaminants travel at a rate equal to the average linear velocity of the groundwater. When advection is the only means of contaminant transport, mass will remain within groundwater flow paths and contaminant spreading is characterized by these paths (Domenico and Schwartz 1990). Mathematically, the advective flux is expressed by the following equation under the assumption that contaminant mass does not alter the water density sufficiently to impact the groundwater flow pattern (Fetter 1999)

$$J = vn_eC \tag{3.1}$$

where: J is the advective mass flux [ML^2T^{-1}]; v is the average linear velocity of groundwater [LT^{-1}]; n_e is the effective porosity through which flow occurs; and C is the contaminant concentration [ML^{-3}]. The average linear velocity can be determined from the following equation (Fetter 1994)

$$v = -\frac{K}{n_e}\frac{dh}{dl} \tag{3.2}$$

where K is the hydraulic conductivity [LT^{-1}] and dh/dl is the hydraulic gradient.

The product of the average linear velocity and time elapsed since contaminants were introduced into the groundwater system defines the position of the advective front. This is a sharp front, where the contaminant concentration behind the front equals that of the moving groundwater while on the other side of the front the concentration equals the background value. This is known as plug flow and is depicted in Figure 3.2. The vertical line in Figure 3.2 represents the advective front without considering the effects of dispersion and diffusion (Fetter 1999).

The one-dimensional advective transport equation is given as (Fetter 1999)

$$\frac{\partial C}{\partial t} = v \frac{\partial C}{\partial x} \tag{3.3}$$

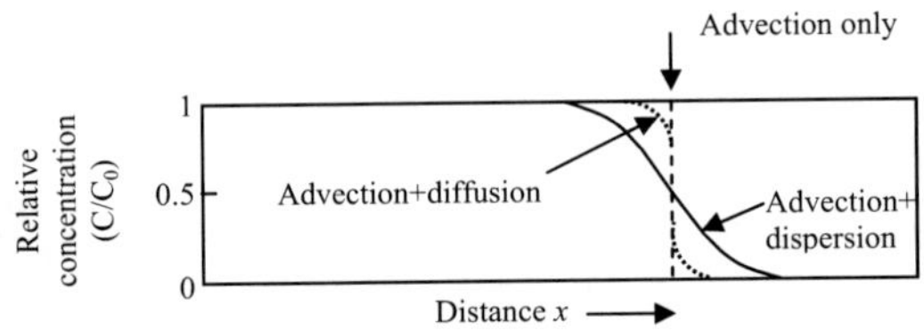

Figure 3.2: Advective transport and the influence of longitudinal dispersion and diffusion on the transport of a solute in one-dimensional flow.

3.2.2 Diffusion

The movement of contaminants from an area of high concentration to an area of lower concentration is called *molecular diffusion* or *diffusion*. Molecular diffusion is driven by the random motion of molecules in solution. This diffusive flux is directly proportional to the diffusion coefficient and the concentration gradient (Ingebritsen and Sanford 1998). This is known as *Fick's first law* and can be expressed in a one-dimensional system as in the following:

$$F = -D_d \frac{dC}{dl} \tag{3.4}$$

where: F is the diffusive mass flux [ML^2T^{-1}]; D_d is the diffusion coefficient for open water [L^2T^{-1}]; C is the contaminant concentration [ML^{-3}]; dC/dl is the concentration gradient [ML^{-4}]. The negative sign signifies that the diffusive flux is in the direction of decreasing contaminant concentration.

In situations where the concentrations change with time, *Fick's second law* can be applied. In a one-dimensional system in which contaminant transport is by molecular diffusion only, Fick's second law is (Fetter 1999)

$$\frac{\partial C}{\partial t} = -D_d \frac{\partial^2 C}{\partial x^2} \tag{3.5}$$

where $\partial C/\partial t$ is the rate of change of concentration [$ML^{-3}T^{-1}$].

The above equation considers molecular diffusion at a microscopic scale. In the subsurface, diffusion occurs within the porous medium and molecular diffusion is constrained relative to that in open water because the molecules have to follow long pathways around the grains in the subsurface. As such, the effective diffusion coefficient is introduced to represent molecular diffusion in a porous medium. This is termed D^* and determined from the following relationship (Ingebritsen and Sanford 1998)

$$D^* = \frac{n_e}{\tau} D_d \tag{3.6}$$

where τ is the *tortuosity* which is defined as the ratio of the actual length of the flow path of the diffused contaminants over the straight line distance (Appelo and Postma 1996). In essence, D^* is the effective diffusion coefficient at the macroscopic scale and is always less than D_d.

As can be concluded from Equation (3.6), molecular diffusion increases with increasing porosity and decreases with increasing tortuosity or the length of the path. Figure 3.2 depicts the influence of diffusion on the relative concentration of an injected contaminant in a one-dimensional system.

Typical values of the diffusion coefficient are in the order of 10^{-9} m^2/sec. As noted by Ingebritsen and Sanford (1998), diffusion coefficients generally decrease with increasing charge and decreasing ionic radius. The analytic solution to Equation (3.5) is (Fetter 1999)

$$C(x,t) = C_o erfc\left(\frac{x}{2(D^* t)^{0.5}}\right) \tag{3.7}$$

where C_0 is the initial concentration and *erfc* is the complementary error function.

3.2.3 Mechanical dispersion

Dispersion describes the gradual spreading of contaminants beyond the region they are expected to occupy when considering advection alone. The phenomenon of dispersion is attributed mainly to the discrepancy between contaminant velocities and the average linear velocity caused by subsurface heterogeneity (Bear 1979; Domenico and Schwartz 1990; Zheng and Bennett 1995). According to Fetter (1994), this discrepancy is attributed to three key factors: (i) the velocity of groundwater flow is higher in the center of pores than at the edges; (ii) the contaminants can pass through different flow paths to reach the same location; and (iii) compared to small pores, groundwater moves faster in large pores. Therefore, contaminants will dilute due to the mixing with non-contaminated water. Mixing, called *mechanical dispersion*, may occur along the flow path and is called *longitudinal dispersion* and is attributed to the abovementioned key factors. Since the groundwater flow path may diverge at the pore scale, the contaminant will spread in directions normal to the flow path. This mixing is called *lateral* or *transverse dispersion* (Fetter 1994). The difference between the straight dashed line and the solid line in Figure 3.2 depicts the influence of longitudinal dispersion on the transport of a contaminant.

The coefficient of mechanical dispersion equals the product of the average linear velocity and a parameter related to subsurface properties called the dispersivity. If *i* is the principle direction of groundwater flow, then the following definitions apply (Fetter 1999):

$$\text{Coefficient of longitudinal mechanical dispersion} = a_i v_i \tag{3.8}$$

$$\text{Coefficient of transverse mechanical dispersion} = a_j v_i \tag{3.9}$$

where: v_i is the average linear velocity in the i direction [LT^{-1}]; and a_i and a_j are the dispersivity values in the i and j direction [L], respectively.

It should be noted that dispersivity is dependent on the subsurface characteristics only and independent of the fluid or contaminant of interest. However, dispersivity is a scale-dependent property of the subsurface.

Molecular diffusion and mechanical dispersion contribute jointly to the dispersive process of contaminant transport. They are combined together to define a quantity called the *coefficient of hydrodynamic dispersion* as given in the following equations (Fetter 1999):

$$D_L = a_L v_i + D^* \tag{3.10}$$

$$D_T = a_T v_i + D^* \tag{3.11}$$

where: D_L and D_T are the longitudinal and transverse hydrodynamic dispersion coefficients, respectively; and a_L and a_T are the longitudinal and transverse dispersivities, respectively.

Figure 3.3 illustrates the process of hydrodynamic dispersion in two-dimensional groundwater flow with a uniform velocity of V. A tracer was instantaneously injected into the aquifer at $t = 0$. The groundwater flow carries the tracer with it. Due to the effect of dispersion, the solute spreads out such that the maximum concentration decreases with time as shown for times t_1 and t_2 when $t_2 > t_1$. The spread of the tracer results in an elliptically shaped concentration distribution that is normally distributed in both the longitudinal and transverse directions.

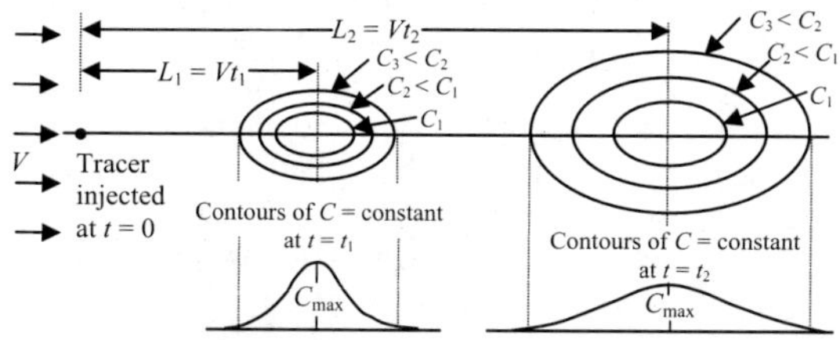

Figure 3.3: Longitudinal and transversal spreading of a tracer of an instantaneous point injection at $t = 0$.

With this distribution, the coefficients of longitudinal and transverse hydrodynamic dispersion can be defined as (Domenico and Schwartz 1990):

$$D_L = \frac{\sigma_L^2}{2t} \tag{3.12}$$

$$D_T = \frac{\sigma_T^2}{2t} \tag{3.13}$$

where: t is time; and σ_L^2 and σ_T^2 are the variances of the longitudinal and transverse spreading of the plume, respectively.

As mentioned earlier, dispersivity is a scale-dependent property that increases with the characteristic transport length. This phenomenon is attributed to the macro-scale effects resulting from heterogeneity of the subsurface at the field scale. Many researchers have developed relationships between longitudinal dispersivity and flow length. Neuman (1990) found that for $L < 3500$ m, $a_L = 0.0175 \times L^{1.46}$ while Xu and Eckstein (1995) showed that $a_L = 0.83 \times (Log\ L)^{2.414}$. Fetter (1999) provides a detailed discussion of these equations and others with the related limitations and assumptions. In his statistical analysis, Saffman (1959) showed that D_L and D_T are related to the mean flow velocity by:

$$D_L = v_x \ln(v_x u) \tag{3.14}$$

where u is a constant dependent on the molecular diffusion coefficient.

In two-dimensional contaminant transport, the ratio of longitudinal to transverse dispersivity α_L/α_T controls the shape of the contaminant plume. As the ratio decreases, the wider the plume will be (Fetter 1999). Bear (1979) indicates that α_L/α_T ranges from 5 to 24. Hydrodynamic dispersion in the longitudinal direction dominates over that in the transverse direction. In cases of very low groundwater flow velocities, the longitudinal and transverse hydrodynamic dispersion coefficients are almost equal. In this case, molecular diffusion is the dominant transport mechanism and the coefficient of hydrodynamic dispersion equals the diffusion coefficient. At high velocities, mechanical dispersion is the dominant dispersive process (Freeze and Cherry 1979). In order to quantify the relative contributions of molecular diffusion and mechanical dispersion to the contaminant transport, a dimensionless parameter expressing advective to diffusive transport called the Peclet number, P_e, is defined (Fetter 1999) as,

$$P_e = \frac{v_x d}{D_d} = \frac{v_x L}{D_L} \tag{3.15}$$

where d is the grain size of the aquifer material and L is a characteristic flow length. As can be concluded from Equations (3.10) and (3.11), when v_x is very small, the dispersion coefficient becomes small and the transport is dominated by diffusion. Likewise, when velocity is high, the impact of diffusion becomes negligible.

Figure 3.4 depicts the behavior of D_L/D_d and D_T/D_d as a function of the Peclet number. Four classes of mixing related to the Peclet number are distinguished. In *Class 1*, molecular diffusion is the dominant process due to the small velocity and the dispersion is said to be diffusion controlled. In *Class 2*, mixing is due to mechanical dispersion and molecular diffusion. With increasing velocity, *Class 3*, mixing is caused mainly by mechanical dispersion and D_L/D_d and D_T/D_d increase with the Peclet number. In *Class 4*, the effect of molecular diffusion is negligible and the mechanical dispersion fully governs the mixing process.

Figure 3.4 shows also that the process of transverse mixing behaves closely to the longitudinal mixing.

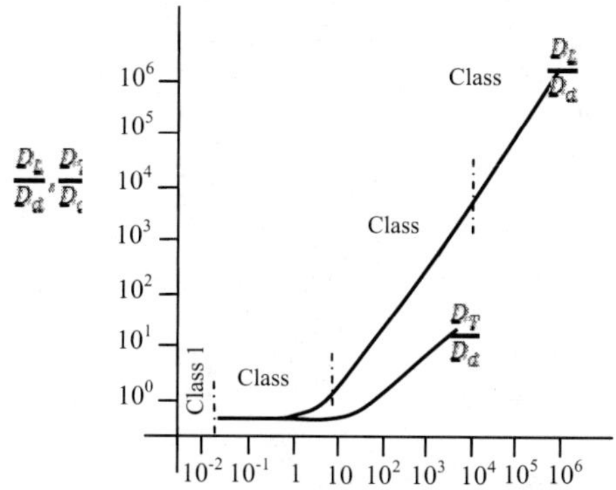

Figure 3.4: Behavior of D_L/D_d and D_T/D_d as a function of the Peclet number. The classes identify regimes where mixing results from the same processes [adapted from Domenico and Schwartz (1990)].

The transport of non-reactive contaminants in three dimensions due to advection and hydrodynamic dispersion can be represented by the following equation in vector form (Zheng and Bennett 1995):

$$\frac{\partial C}{\partial t} = \nabla\left(D\nabla C\right) - \nabla\left(vC\right) \tag{3.16}$$

3.3 Chemical Reactions, Retardation and Decay of Solutes

Generally, the main reactions considered in the fate and transport of contaminants in groundwater are the equilibrium-controlled sorption reactions and the irreversible first-order reactions.

In sorption reactions, contaminants in a solution adhere on or in the solid matrix of the porous medium (Zheng and Bennett 1995; Fetter 1999). Sorption includes both absorption and adsorption. In absorption, chemical species penetrate the solid grains, while in adsorption the species cling to the grain surfaces. As a result of sorption, contaminants will be partitioned between the solution and the solid phase. Sorption causes retardation of the contaminants, which will cause contaminants to move slower than the groundwater flow (Bedient et al. 1999). If the sorption process is fast compared to the velocity of the flow, then the inter-phase mass transfer process can be considered as an equilibrium reaction. When the sorption process is slow compared to the time during which the contaminants and grains are in contact, then the process is described as a non-equilibrium reaction.

The irreversible first-order reactions include mainly radioactive decay and specific forms of biodegradation. Unlike the sorption reactions, the irreversible first-order reactions influence the concentration of the contaminants but most likely have no impact on the contaminant velocity distribution (Bedient et al. 1999). Detailed illustrations of groundwater reactions can be found in Domenico and Schwartz (1990), Deutsch (1997), Fetter (1999), and Bedient et al. (1999).

3.3.1 Equilibrium sorption isotherms

The relationship between the solute concentration and the amount sorbed under a constant temperature is called an isotherm (Appelo and Postma 1996). In reality, isotherms have no prearranged shapes. They might be linear, concave, convex, or a combination of these shapes (Domenico and Schwartz 1990). In the next sections, the linear, Freundlich, and Langmuir sorption isotherms are discussed.

3.3.1.1 *Linear sorption isotherm*

The linear sorption isotherm assumes that mass sorbed on the solid grains is directly proportional to the concentration of the solute in the solution. The linear sorption isotherm is described by the following equation:

$$S = K_d C \tag{3.17}$$

where: S is the mass of solute sorbed per dry unit weight of solid [M/M]; C is the concentration of the solute in the solution in equilibrium with the mass of solute sorbed on the solid [M/L^3]; and K_d is the distribution coefficient [L^3/M]. Figure 3.5 depicts the linear sorption isotherm along with the Freundlich and Langmuir isotherms.

The slope of the linear sorption isotherm equals K_d. A high value of K_d is indicative of a high propensity for sorption. Since sorption causes retardation of the moving contaminants, a retardation factor can be defined as

$$R = 1 + \frac{\rho_b}{n} K_d \tag{3.18}$$

where: R is the retardation factor; ρ_b is the bulk density of the aquifer; and n is the porosity.

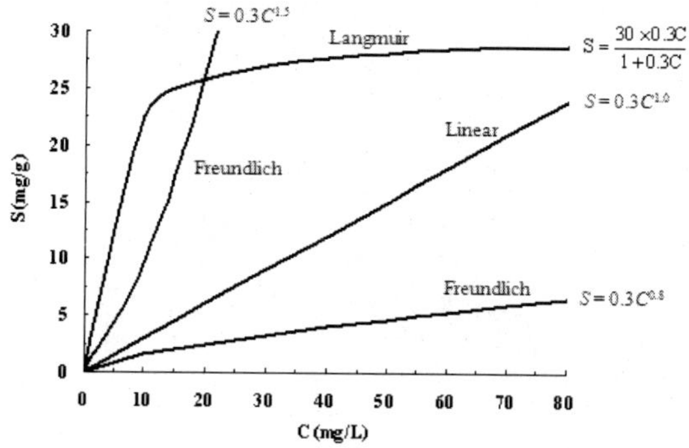

Figure 3.5: Example of linear, Langmuir, and Freundlich sorption isotherms.

The minimum value of R equals one (unity) corresponding to a K_d value of zero. In general, K_d values range from zero to 10^3 mL/g (Bedient et al. 1999). The effect of retardation or sorption is to decrease the apparent values of the hydrodynamic dispersion coefficients and the average linear groundwater flow velocity. Basically, the velocity of the retarded contaminant becomes less than the velocity of the groundwater. The retardation factor relates the two velocities by the retardation equation (Fetter, 1999)

$$v_c = \frac{v_w}{R} = \frac{v_w}{1 + \frac{\rho_b}{n} K_d} \tag{3.19}$$

where v_w is the velocity of the water and v_c is the velocity of the contaminant. This equation predicts the position of the advective front when considering equilibrium sorption using a linear isotherm. The ratio v_w/v_c indicates how much faster the groundwater flow is moving relative to the contaminant being sorbed. When K_d is zero, $v_w = v_c$. Therefore, a retardation factor of R connotes that the contaminant plume moves R times slower than the groundwater velocity (Bedient et al. 1999).

Fetter (1999) studied the effect of equilibrium retardation on solute concentration. Figure 3.6 depicts the effect of retardation on the breakthrough curve of a solute (dashed line) and compares it to the breakthrough with no retardation (solid line). As can be concluded, the retarded substance has a lower maximum concentration than the non-retarded one. Additionally, for retarded substance, it takes more pore volumes to reach the maximum concentration.

Modeling sorption with the linear sorption isotherm is the most widespread approach due to its simplicity. However, as stated in Domenico and Schwartz (1990), the sorption process is far too complex to be modeled correctly by linear isotherms using just K_d, which may be difficult to characterize. K_d values change from site to site as a function of the physical and chemical properties of the porous medium and the chemical characteristics of the contaminant.

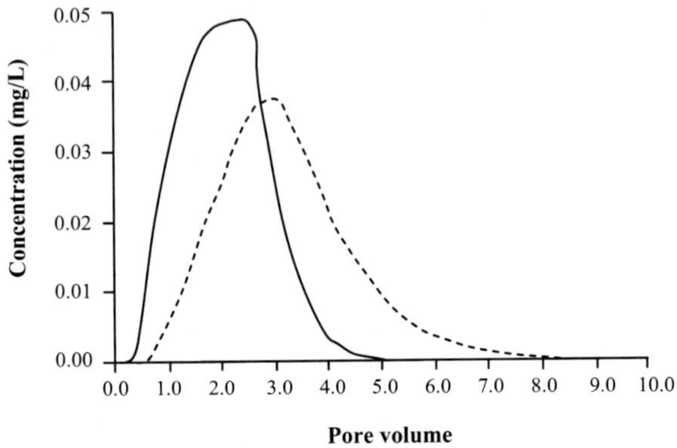

Figure 3.6: Comparison of the breakthrough curves for a non-retarded and a linearly retarded solute. Dashed curve is for the retardation curve.

The main limitation of the linear sorption isotherm is that there is no upper limit of the mass of solute that can be sorbed. Since the linear sorption isotherm may be obtained by fitting few experimental data points, one should be cautious in extrapolating the linear relationship beyond the range of the data points (Fetter 1999).

3.3.1.2 *Freundlich sorption isotherm*

The Freundlich sorption isotherm is defined by the following nonlinear relationship

$$S = KC^N \tag{3.20}$$

where K and N are constants and S and C are as defined earlier.

Figure 3.5 shows the Freundlich sorption isotherms for $N \in \{0.8, 1.5\}$. As can be concluded, when $N > 1$, the sorbed concentration for a given solute concentration is greater as compared to $N < 1$. In addition, for $N < 1$, the ratio of sorbed mass to solute mass decreases with increasing solute concentration solution (Appelo and Postma 1996). If $N = 1$, the Freundlich sorption isotherm becomes the linear sorption isotherm. The retardation coefficient for a Freundlich isotherm, R_F, is (Fetter 1999)

$$R_F = 1 + \frac{\rho_b KNC^{N-1}}{n} \tag{3.21}$$

As in the linear sorption isotherm, there is no upper limit for mass of solute that could be sorbed.

3.3.1.3 *Langmuir sorption isotherm*

The basic premise of the Langmuir sorption isotherm is that there are a finite number of sorption sites on the solid surface that when all filled, sorption will cease. The most common form of the Langmuir sorption isotherm is

$$S = \frac{\alpha \beta C}{(1 + \alpha C)} \tag{3.22}$$

where α is a partition coefficient reflecting the extent of sorption [L^3/M] and β is the maximum sorptive capacity for the solid surface [M/M]. Figure 3.5 depicts the Langmuir sorption isotherm for $\alpha = 30$ L/g and $\beta = 0.3$ mg/g. The retardation factor for the Langmuir sorption isotherm, R_L is (Fetter 1999)

$$R_L = 1 + \frac{\rho_b}{n}\left(\frac{\alpha \beta}{(1 + \alpha C)^2}\right) \tag{3.23}$$

3.3.2 Nonequilibrium sorption models

Equilibrium sorption models assume that the groundwater flow rate is slow relative to the rate of inter-phase mass transfer. When equilibrium conditions are not met, kinetic models are more appropriate than the equilibrium sorption isotherms. There are four main kinetic sorption models: *irreversible first-order kinetic sorption model*, *reversible linear kinetic sorption model*, *reversible nonlinear kinetic sorption model*, and *bilinear adsorption model*.

The irreversible first-order kinetic sorption model is simple and presumes that the rate of sorption is a function of the solute concentration. In this model, the sorbed substance cannot be desorbed but can be attenuated. The irreversible first-order kinetic model can be defined by the following equation:

$$\frac{\partial S}{\partial t} - k_1 C \tag{3.24}$$

where k_1 is a first-order rate constant.

The reversible linear kinetic sorption model assumes that the reaction is reversible and the sorbed substance can be desorbed. This model can be expressed as in the following:

$$\frac{\partial S}{\partial t} = k_2 C - k_3 S \tag{3.25}$$

where k_2 is a forward first-order rate constant and k_3 is a backward first-order rate constant. If sufficient time is given, equilibrium conditions will be attained due to the reversibility of the kinetics. Under such conditions, the rate of change of sorption is zero $(\partial S/\partial t = 0)$, and Equation (3.25) becomes $S = (k_2/k_3)C$ which is equivalent to the linear equilibrium sorption isotherm. Equation (3.25) can be written in the following form:

$$\frac{\partial S}{\partial t} = \gamma(k_4 C - S) \tag{3.26}$$

where γ is the first-order rate coefficient and k_4 is a constant equivalent to K_d.

The reversible nonlinear kinetic sorption model can be defined as

$$\frac{\partial S}{\partial t} = k_5 C^N - k_6 S \tag{3.27}$$

where k_5, k_6, and N are constants. As can be noticed, the forward sorption reaction is nonlinear and the backward sorption is linear. If sufficient time is given to attain equilibrium, it can be proved that $S = (k_5/k_6)C^N$, which is the Freundlich sorption isotherm.

The bilinear adsorption model can be defined as

$$\frac{\partial S}{\partial t} = k_7 C(\beta - S) - k_8 S \tag{3.28}$$

where: β is the maximum amount of solute that can be sorbed; k_7 is the forward rate constant; and k_8 is the backward rate constant.

3.3.3 Hydrophobic sorption of organic compounds

Many organic compounds dissolved in groundwater sorb to the solid phase due to hydrophobic effects. These same effects result in reducing the solubility of these compounds depending on the extent to which they are attracted by the polar water molecule. In general, the more hydrophobic an organic compound is, the greater its propensity to partition into the solid phase (Domenico and Schwartz 1990; Fetter 1999). Karickhoff (1984) provides a detailed description of the underlying physico-chemical processes.

The sorption process of non-polar organic compounds occurs due to the changes in the structure of water that occur when a non-polar molecule dissolves (Domenico and Schwartz 1990). This sorption process is generally modeled using the linear isotherm. The distribution coefficient is proportional to the weight fraction of organic carbon as

$$K_d = K_{oc} f_{oc} \tag{3.29}$$

where K_{oc} is the organic carbon partition coefficient and f_{oc} is the fraction of organic carbon comprising the solid phase. In estimating K_d, a partition coefficient, K_{om}, based on soil or aquifer organic matter can be used as in the following equation:

$$K_{oc} = 1.724 K_{om} \tag{3.30}$$

Many researchers showed the existence of different relationships between the octanol-water partition coefficient, K_{ow}, and the K_{oc} value for different organic compounds. A list of these relationships is provided in Knox et al. (1993), Fetter (1999), and Bedient et al. (1999) among others.

3.3.4 Chemical and biological reactions

Specific chemical reactions, such as hydrolysis, radioactive decay, and some forms of biodegradation, can be characterized as zero- and first-order, irreversible processes.

A general kinetic expression for representing decay as a zero-order, irreversible process can be expressed as:

$$\frac{\partial C}{\partial t} = -\lambda \tag{3.31}$$

where λ is the rate constant of the reaction. In zero-order reactions, the rate of transformation of a substance is unaffected by changes in the solute concentration because the reaction rate is determined by factors other than the constituent concentration. Zero-order decay is simple and easy to use. Nevertheless, as this reaction does not consider changes in the substance concentration, the use of zero-order reaction kinetics is limited.

A first-order kinetic expression for decay, which considers the solute concentration, is

$$\frac{\partial C}{\partial t} = -\gamma C \tag{3.32}$$

The integration of Equation (3.32) yields the following exponential relationship:

$$C(t) = C_o e^{-\lambda t} \tag{3.33}$$

where C_0 indicates the initial concentration. The rate constant, λ, is a function of the substance half-life time, $t_{1/2}$, which is the time required for the mass of reacting substance to decrease to half the original mass. The half-life is related to the rate constant for decay as follows:

$$t_{1/2} = \frac{0.693}{\lambda} \tag{3.34}$$

Equation (3.32) is commonly used to represent the radioactive decay of radionuclides. Half-lives for the principal radionuclides along with their associated decay chains are summarized by Yu et al. (1993).

Because organic chemicals and hydrocarbons are common groundwater contaminants, the biodegradation of the organic contaminants in groundwater is of significant importance. Biodegradation is a biochemical process in which organic contaminants are broken into harmless products, such as CO_2 and water mediated by the microorganisms of the subsurface in the presence of a suitable electron acceptor such as oxygen. There are two types of biodegradation, aerobic and anaerobic. In aerobic biodegradation, oxygen is used as the electron acceptor while in anaerobic biodegradation, alternate electron acceptors, such as nitrate, Mn(IV), Fe(III), sulfate, and CO_2 are used. Chapelle (1993) provides a detailed discussion of the metabolic pathways that are of particular importance to microbial processes in groundwater systems. An example of the aerobic biodegradation reaction for benzene is given (Rifai and Bedient 1994)

$$C_6H_6 + \frac{15}{2}O_2 \rightarrow 6CO_2 + 3H_2O \tag{3.35}$$

This equation is mediated by microorganisms and results in the breakdown of benzene into carbon dioxide and water and the formation of biomass.

There are six basic requirements for biodegradation (Bedient et al. 1999): (i) presence of the appropriate microorganism to degrade the contaminants; (ii) energy source to be used for cell maintenance and growth; (iii) carbon source to be used with energy to generate new cells; (iv) electron acceptor; (v) nutrients; and (vi) acceptable environmental conditions such as temperature, pH, and salinity.

Most biodegradation models fall into two different conceptual approaches. The first approach is known as the biofilm theory where it is assumed that solid particles are uniformly covered by a thin biofilm. This biofilm provides the necessary media for consumption of the substrate in the presence of electron acceptors and nutrients. In addition, diffusion of the substrate through the biofilm occurs. The second approach considers bacterial growth to take place in small discrete microcolonies attached to the particle surface. A detailed illustration of the two approaches is provided in Bedient et al. (1999) along with the related references.

Biodegradation of contaminants in groundwater systems is typically represented using an instantaneous reaction model or a Monod kinetic model. The instantaneous reaction model assumes that the reaction between the contaminant and the electron acceptor mediated by the microorganisms is fast or almost instantaneous, and that the mass of contaminant degraded and electron accepter are related stoichiometrically. In the aerobic biodegradation of benzene, for example, Equation (3.35) indicates that 1 mole of benzene is degraded for every 7.5 moles of oxygen consumed. Alternatively, considering the molecular weights of benzene and oxygen, 1 gram of benzene is degraded for every 3 grams of oxygen utilized. While the instantaneous reaction model has proven to be a useful tool for representing the biodegradation of some contaminants, particularly the aerobic biodegradation of petroleum hydrocarbons, the approach does have its limitations. For example, instantaneous consumption of the contaminant may not be supported by the existing hydrogeological conditions. Also, the instantaneous reaction model does not have a mechanism to account for the differences between the biodegradability potentials of contaminants and may not be applicable to all electron acceptors.

The Monod kinetic formulation is a more flexible conceptualization for predicting the rates of biodegradation in groundwater. The model representing Monod kinetics is (Bedient et al. 1999)

$$\mu = \mu_{\max} \frac{C}{K_c + C} \tag{3.36}$$

and the reduction of contaminant and oxygen concentrations for aerobic biodegradation can be expressed as

$$\Delta C = M_t \mu_{\max} \left(\frac{C}{K_c + C} \right) \left(\frac{O}{K_o + O} \right) \Delta t \tag{3.37}$$

$$\Delta O = M_t \mu_{\max} F \left(\frac{C}{K_c + C} \right) \left(\frac{O}{K_o + O} \right) \Delta t \tag{3.38}$$

where: μ is the growth rate [T^{-1}]; $\mu_{\max}$ is the maximum growth rate [T^{-1}]; C is the concentration of the growth-limiting substrate [M/L^3]; K_c is the contaminant half-saturation constant [M/L^3]; M_t is the total microbial concentration [M/M]; O is the oxygen concentration [M/L^3]; K_o is the oxygen half-saturation constant [M/L^3]; F is the ratio of oxygen to contaminant consumed [M/M]; and Δt is the time interval being considered [T]. Note that the order of reaction rate varies with the based on the magnitude of the solute concentration relative to the half-saturation constant. Consider Equation (3.36) for example. When $C >> K_c$, $\mu \approx \mu_{\max}$ and the reaction rate is zero order. When $C << K_c$, $\mu \approx \mu_{\max} C/K_c$ and the reaction rate is first order.

Incorporating Equations (3.37) and (3.38) into the general form of the transport equation results in a system of partial differential equations as follows (Bedient et al. 1999):

$$\frac{\partial C}{\partial t}=\frac{1}{R_c}\nabla\cdot\left(D\nabla C-vC\right)-M_t\left(\frac{\mu_{\max}}{R_c}\right)\left(\frac{C}{K_c+C}\right)\left(\frac{O}{K_o+O}\right) \tag{3.39}$$

$$\frac{\partial O}{\partial t}=\nabla\cdot\left(D\nabla O-vO\right)-M_t\mu_{\max}F\left(\frac{C}{K_c+C}\right)\left(\frac{O}{K_o+O}\right) \tag{3.40}$$

Bedient et al. (1999) provide a list of models that simulate the biodegradation process with the corresponding description and authors.

3.4 Mathematical Model of Contaminant Transport

Solving the time-dependent partial differential equations governing contaminant transport requires the identification of the initial and boundary conditions. The initial conditions provide the status of the contaminant in the groundwater at the beginning of simulation. The boundary conditions provide the effect of the areas outside the system boundaries on the system being modeled.

A mathematical model of the fate and transport of a contaminant in groundwater involves the governing equations, initial and boundary conditions, information on the sources and sinks in the system, characterization of all the reactions, and determination of the flow and transport parameters. Solving the mathematical model yields the contaminant concentration as a function of space and time.

The approach used in formulating and solving the mathematical model can be classified generally into numerical and analytical approaches. In the analytical methods, an exact solution to the mathematical model is obtained by solving the governing equations analytically, but under simplifying assumptions. Numerical methods provide an approximate solution to the mathematical model by solving a set of algebraic equations that replace the original partial differential equations. Numerical solutions are widely used and preferred because they have the capability to handle more general conditions.

3.4.1 General governing equation

The advection-dispersion-reaction partial differential equation that governs the three-dimensional transport of a single chemical constituent in groundwater, considering advection, dispersion, fluid sinks/sources, equilibrium-controlled sorption, and first-order irreversible rate reactions, is given in the following (Zheng and Bennett 1995):

$$R\frac{\partial C}{\partial t}=\frac{\partial}{\partial x_i}\left(D_{ij}\frac{\partial C}{\partial x_j}\right)-\frac{\partial}{\partial x_i}\left(v_iC\right)+\frac{q_s}{n}C_s-\lambda\left(C+\frac{\rho_b}{n}S\right) \tag{3.41}$$

where the retardation factor is defined as,

$$R = 1 + \frac{\rho_b}{n}\frac{\partial S}{\partial C} \qquad (3.42)$$

and C is the dissolved concentration [ML^{-3}]; S is the sorbed concentration [MM^{-1}] which is a function of the dissolved concentration, C, as defined by a sorption isotherm; t is time [T]; D_{ij} is the dispersion coefficient tensor [L^2T^{-1}]; v_i is the average linear groundwater velocity [LT^{-1}]; q_s is the volumetric flow rate per unit volume of aquifer and represents fluid sources and sinks [T^{-1}]; C_s is the concentration of the fluid source or sink flux [ML^{-3}]; λ is the reaction rate constant [T^{-1}]; R is the retardation factor [L^0]; ρ_b is the bulk density of the porous medium [ML^{-3}]; and n is the porosity [L^0]. A complete derivation of this equation is provided in Freeze and Cherry (1979) and Zheng and Bennett (1995).

3.4.2 Initial and boundary conditions

The initial condition describes the distribution of the contaminant in the groundwater within the domain at the beginning of simulation at time equal zero (Fetter 1999). For a one-dimensional contaminant transport system, a general form of initial condition can be written

$$C(x, t = 0) = f(x) \qquad (3.43)$$

where $f(x)$ is a function defining the one-dimensional variation in concentration in the x-direction at $t = 0$. A common initial condition is $C(x,t) = 0$ or $C(x,t) = C_i$ to provide a constant concentration within the system.

The boundary conditions can be classified into three categories (Fetter 1999): (i) specified concentration boundary, called the Dirichlet condition; (ii) a specified concentration gradient boundary, called the Neumann condition; and. (iii) a variable flux boundary, called the Cauchy condition. Sometimes, these boundaries are referred to as first-, second-, and third-type boundary conditions, respectively. An example of a specified concentration boundary condition for a one-dimensional system is

$$C(x = 0, t) = g_1(t) \qquad (3.44)$$

This equation indicates that at the boundary ($x = 0$), the concentration is given for all times as the function $g_1(t)$. This Dirichlet boundary acts as a source providing contaminant mass to the system, or as a sink removing contaminant mass from the system. An example of the specified concentration gradient normal to the boundary is

$$\frac{\partial C}{\partial x} = g_2(t) \qquad at \qquad x = 0 \qquad (3.45)$$

where $g_2(t)$ is a known function. For the Cauchy boundary condition, the dispersive and advective fluxes across the boundary are specified. An example of the variable flux boundary condition is given as

$$-D\frac{\partial C}{\partial x}+vC=vg_3(t) \qquad at \quad x=0 \tag{3.46}$$

where $g_3(t)$ is a known function. This variable flux boundary condition is in essence a continuity equation equating the mass flux across the boundary to the amount the system is able to transport by advection and dispersion (Domenico and Schwartz 1990).

3.5 Analytical Solutions to the Mass Transport Equation

Over the past few decades, researchers have worked extensively on developing closed-form analytical solutions to the advective-dispersive-reaction equation under a variety of initial and boundary conditions (e.g., Domenico 1987; Domenico and Schwartz 1990). Although these solutions are readily programmable using spreadsheets and can be used to obtain quick solutions under simplified assumptions, the applicability in field-scale scenarios is limited due to the existence subsurface heterogeneity, complex geometry, and complex reaction kinetics. Therefore, most mathematical analyses related to field-scale simulations are carried out using numerical models, especially with the availability of powerful computers. For completeness sake, a few analytical solutions for some simple scenarios are presented here.

3.5.1 Solution to one-dimensional advection-dispersion equation with 1st-order decay

The one-dimensional governing transport equation for a contaminant undergoing 1st-order decay of rate λ in a uniform flow field of v is given as

$$\frac{\partial C}{\partial t}=D_L\frac{\partial^2 C}{\partial x^2}-v\frac{\partial C}{\partial x}-\lambda C \tag{3.47}$$

For initial and boundary conditions given as

$$C(0,t)=C_o \quad \text{and} \quad C(x,0)=0 \tag{3.48}$$

Bear (1979) provides the analytical solution as

$$C(x,t)=\frac{C_o}{2}\exp\left(\frac{vx}{2D_L}\right)\left(\exp(-x\beta)erfc\left(\frac{x-\left(v^2+4\lambda D_L\right)^{1/2}t}{2(D_Lt)^{1/2}}\right)+\exp(x\beta)erfc\left(\frac{x+\left(v^2+4\lambda D_L\right)^{1/2}t}{2(D_Lt)^{1/2}}\right)\right) \tag{3.49}$$

where $\beta^2 = \left(\frac{v}{2D_L}\right)^2 + \frac{\lambda}{D_L}$.

For the case of no decay, $\lambda = 0$, the above solution reduces to

$$C(x,t) = \frac{C_o}{2}\left(erfc\left(\frac{x - vt}{2(D_L t)^{1/2}}\right) + \exp\left(\frac{vx}{D_L}\right) erfc\left(\frac{x + vt}{2(D_L t)^{1/2}}\right)\right) \tag{3.50}$$

The second term of Equation (3.50) can be neglected for large Peclet numbers, and the solution to the one-dimensional dispersion equation can be approximated as,

$$C(x,t) = \frac{C_o}{2}\left(erfc\left(\frac{x - vt}{2(D_L t)^{1/2}}\right)\right) \tag{3.51}$$

Here *erfc* is the complementary error function and can be obtained from any mathematical reference book.

3.5.2 Slug injection into a uniform two-dimensional flow field

If a slug of contaminant is injected in a two-dimensional uniform groundwater flow field over an area A at a point (x_0, y_0) within a short period of time, the concentration at a downgradient location (x, y) at time t after the injection is given as (Fetter 1999):

$$C(x,y,t) = \frac{C_o A}{4\pi t (D_L D_T)^{1/2}} \exp\left(-\frac{((x - x_o) - vt)^2}{4D_L t} - \frac{(y - y_o)^2}{4D_T t}\right) \tag{3.52}$$

The slug will move in the direction of flow and spread with time. Figure 3.7 shows the development of the contamination plume at three different times.

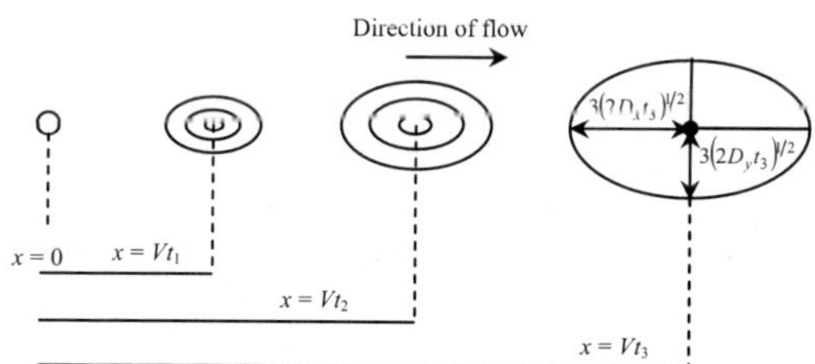

Figure 3.7: Plan view of a contaminant plume developed from the injection of a slug into a uniform flow field and shown at three different times.

The maximum concentration of the contaminant from the slug injection is found at the center of the plume. At any time, the center of the plume is at a location $x = v_x t$ and $y = 0$. If the spill was at a location $x_0 = 0$ and $y_0 = 0$, then the maximum concentration is obtained as follows (Fetter 1999):

$$C_{\max}(t) = \frac{C_o A}{4\pi t (D_L D_T)^{1/2}} \tag{3.53}$$

The dimensions of the contamination plume are such that the total widths along the longitudinal and transverse directions are $3\sigma_x$ and $3\sigma_y$, respectively where $3\sigma_x = 3(2D_L t)^{1/2}$ and $3\sigma_y = 3(2D_T t)^{1/2}$; here σ_x and σ_y are the standard deviations such that about 99.7% of the mass is contained within $3\sigma_x$ and $3\sigma_y$.

3.5.3 Solution to the three-dimensional advection-dispersion equation with 1st-order decay due to a continuous source

Consider a vertical rectangular source of contaminant of vertical height Z and width Y located at $x = 0$. When the source is continuous and the flow is uniform along the x-direction, Domenico (1987) provided the following closed-form solution for three-dimensional mass transport with 1st-order decay:

$$C(x,y,z,t) = \left(\frac{C_o}{8}\right)\exp\left\{\left(\frac{x}{2\alpha_x}\right)\left[1-\left(1+\frac{4\lambda\alpha_x}{v}\right)^{1/2}\right]\right\} erfc\left[\frac{x - vt\left(1+\frac{4\lambda\alpha_x}{v}\right)^{1/2}}{2(\alpha_x vt)^{1/2}}\right]$$

$$\cdot\left\{erf\left[\frac{\left(y+\frac{Y}{2}\right)}{2(\alpha_x x)^{1/2}}\right] - erf\left[\frac{\left(y-\frac{Y}{2}\right)}{2(\alpha_y x)^{1/2}}\right]\right\}\left\{erf\left[\frac{(z+Z)}{2(\alpha_z x)^{1/2}}\right] - erf\left[\frac{(z-Z)}{2(\alpha_z x)^{1/2}}\right]\right\} \tag{3.54}$$

where C_0 is the source concentration.

3.5.4 Solution to the three-dimensional advection-dispersion equation with 1st-order decay due to an instantaneous point source

For an instantaneous contamination point source, Baetslé (1969) presented an analytical model for the concentration distribution as a function of x, y, z, and t as in the following:

$$C(x,y,z,t) = M\exp(-\lambda t)\left[\frac{1}{2\sqrt{\pi D_x t}}\exp\left(\frac{-(x-vt)^2}{4D_x t}\right)\right]\left[\frac{1}{2\sqrt{\pi D_y t}}\exp\left(\frac{-y^2}{4D_y t}\right)\right]\left[\frac{1}{2\sqrt{\pi D_z t}}\exp\left(\frac{-z^2}{4D_z t}\right)\right] \tag{3.55}$$

or

$$C(x,y,z,t)=\left[\frac{M}{8(\pi t)^{3/2}\left(D_xD_yD_z\right)^{1/2}}\right]\exp\left[-\frac{(x-vt)^2}{4D_xt}-\frac{y^2}{4D_yt}-\frac{z^2}{4D_zt}-\lambda t\right] \quad (3.56)$$

where: M is the amount of the contaminant substance deposited in the point source; D_x, D_y, D_z are the coefficients of the hydrodynamic dispersion; x, y, and z are the space coordinates; t is time; v is the velocity of the contaminant; and λ is the decay coefficient for the contaminant substance. It is obvious from the above equation that the maximum concentration is at the center of the plume where $x = vt$ and $y = z = 0$. Maximum concentration is given by the following expression Baetslé (1969):

$$C_{\max}(x,y,z,t)=\frac{Me^{-\lambda t}}{8(\pi t)^{3/2}\sqrt{D_xD_yD_z}} \quad (3.57)$$

3.6 Numerical Solutions to the Mass Transport Equation

Numerical methods for solving the advection-dispersion-reaction equation divide the model domain into small cells, approximate the governing equation over each cell at time t, and then predict a new value at time $t + \Delta t$. This process continues forward in time for small Δt values. The most common numerical methods for solving the advection-dispersion-reaction equation include the finite difference method, the finite element method, the random walk method, and the method of characteristics. A detailed illustration of these methods can be found in Zheng and Bennett (1995) and Sun (1996). Bedient et al. (1999) cited many references that provide detailed descriptions of these numerical methods.

Over the past two decades, many computer codes were developed to solve the advection-dispersion reaction equation. These codes utilize the different numerical techniques to simulate contaminant fate and transport in one, two, or three dimensions. A list of available computer software for simulating contaminant fate and transport in the saturated zone can be found at the International Groundwater Modeling Center (http://igwmc.mines.edu/).

One of the commonly used codes in simulating the fate and transport of contaminants in groundwater is MT3D (Zheng 1990). MT3D is a three-dimensional contaminant transport model for the simulation of advection, dispersion, and chemical reactions of dissolved constituents in groundwater systems. The model uses a modular structure to make it possible to independently simulate advection, dispersion, sink/source mixing, and chemical reactions without using computer memory for unused options.

MT3D uses a mixed Eulerian-Lagrangian approach to solve the three-dimensional advection-dispersion-reaction equation with three options: the method of characteristics (MOC), the modified method of characteristics (MMOC), and a hybrid of these two methods (HMOC). This approach combines the strength of the MOC in eliminating numerical dispersion with the computational efficiency of the MMOC. The availability of both MOC and MMOC options, and their selective use based on an automatic adaptive procedure under the HMOC option, make MT3D uniquely suitable for a wide range of field problems. In addition, another version of MT3D, known as MT3DMS4, includes the third-order total-variation-diminishing (TVD) scheme for solving the advection term.

MT3D has been widely accepted by practitioners and researchers alike and has been applied in numerous field-scale modeling studies. The chemical reactions modeled by MT3D include equilibrium-controlled linear and nonlinear sorption, first-order irreversible decay, and biodegradation. MT3D is based on a modular structure to permit simulation of transport components independently or jointly with finite difference groundwater flow models, such as MODFLOW (Harbaugh and McDonald 1996), which can be accessed at http://water.usgs.gov/software/modflow-96.html. Latest versions of MODFLOW can be downloaded from http://water.usgs.gov/nrp/gwsoftware/modflow.html. MT3D interfaces directly with MODFLOW, retrieves the saturated thickness, fluxes across cell interfaces in all directions, and locates flow rates of the various sources and sinks. MT3D supports all the hydrologic and discretization features of MODFLOW. MT3D is public domain and latest versions of MT3D can be downloaded from http://hydro.geo.ua.edu. Further information on MT3D can be found in Zheng (1990), Zheng and Bennett (1995), and Zheng and Wang (1999).

3.7 Demonstration Example

The following example illustrates the behavior of a contaminant plume resulting from the instantaneous accidental release of a contamination product. The analytical solution of Baetslé (1969) is used to study the sensitivity of the plume development to velocity, distribution coefficient, half-life, longitudinal dispersivity, ratio of transverse to longitudinal dispersivity, and simulation time. The plume originates from an accidental release of 1,000 liters of a waste with a concentration of 600 mg/L at $x = y = z = 0$.

To evaluate the sensitivity of the plume development for the different parameters, a base case example scenario was developed to assist in the evaluation. For the base case scenario, longitudinal dispersivity α_x = 20 m, ratio of transverse dispersivity to longitudinal dispersivity α_x/α_y = 0.1 and α_z/α_x = 0.1, average flow velocity v = 0.2 m/d, and simulation time t = 20 years. In addition, both sorption and decay are neglected. For instance, $\lambda = 0$ (Equation 3.34) and $R = 1$ (Equation 3.18). Figure 3.8, Figure 3.9, and Figure 3.10 depict the horizontal two-dimensional extent of the plume for the different parameters.

Figure 3.8a shows the plume development for $v = 0.1$, 0.2, and 0.3 m/d. Because the center of the plume travels a distance of $v \times t$, then by increasing the groundwater flow velocity, the plume departs more and more from the contamination source. In addition, the longitudinal and lateral extent of the plume increases due to the increase in the dispersion coefficients. Apparently, the maximum concentration decreases by increasing the velocity due to the increase in mixing. Figure 3.8b shows the sensitivity of plume extent to $\alpha_L = 10$, 20, and 30 m. Simulation results show that with increasing the longitudinal dispersivity, the longitudinal and lateral extent of the plume increase without affecting the position of the center of the plume since both v and t were kept constant. Obviously, the maximum concentration decreases with increasing the dispersivity due to the increase in mixing.

The impact of retardation on plume extent is depicted in Figure 3.9a. In this case, Equations (3.18) and (3.19) were used to compute the retardation factor, R, and the contaminant velocity, v_c, respectively. Simulations were performed for $K_d = 0$, 0.1, and 0.4 cm^3/g. By increasing the distribution coefficient, the rate at which the contaminant plume travels is decreased due to retardation (recall that $v_c = v/R$) while the maximum concentration increases due to mixing reduction. Figure 3.9b illustrates the effect of the half-life time on the plume development. In this case, Equation (3.34) was utilized to compute the decay coefficient. Simulations were performed for $t_{1/2} = \infty$, 100, and 10 years. It can be concluded the maximum contaminant concentration decreases with increasing half-life. Clearly, with decreasing half-life, the extent of the plume reduces while the center of the plume remained at the same place since both v and t were kept constant.

Figure 3.10a shows the plume development for different simulation times. As time, t, increases from 10 to 30 years, the plume moves away more and more from the contamination source (distance = $v \times t$) and the extent of the plume increases. In addition, maximum concentration decreases with increasing simulation time. Lastly, Figure 3.10b shows the effect of varying the ratio of transverse to longitudinal dispersivity α_x/α_y on the plume shape. Unsurprisingly, for a low value of α_x/α_y, the transverse extent of the plume reduces while the longitudinal extent increases. By increasing this ratio, the contaminants increasingly spread out in the transverse direction with a decrease in the longitudinal extent.

Finally, Figure 3.11 depicts the sensitivity of the breakthrough curves along a longitudinal cross section taken through the center of the plume from west-east direction for the different simulations considered in Figure 3.8, Figure 3.9, and Figure 3.10.

3.8 On the Use of Artificial Neural Networks in the Analysis of Nitrate Distribution in Groundwater

There are many fundamental difficulties associated with developing distributed models of fate and transport of nitrate for both soil and groundwater. The key difficulties are (Morshed and Kaluarachchi 1998b; McGrail 2001; Almasri and Kaluarachchi 2005a): (i) models need enormous data, which are generally difficult and costly to obtain; (ii) the development of these models require detailed characterization of the study area including the physical, chemical, and biological processes when such processes are not fully known; and (iii) models often use fine spatial and temporal discretization that require substantial computational resources to simulate multiple scenarios.

To overcome these difficulties, many researchers have successfully used artificial neural networks (ANNs). ANNs are interconnections of simple processing elements called neurons that have the ability to identify the relationship between the input-output responses from given patterns (Beale and Jackson 1991; Haykin 1994). In this pattern recognition, ANN overcomes the difficulties of physically-based techniques used in simulating complex features of different relationships. As such, ANN is a powerful tool for representing the underlying relationships governing the input-output response patterns for different physical problems (Morshed and Kaluarachchi 1998b).

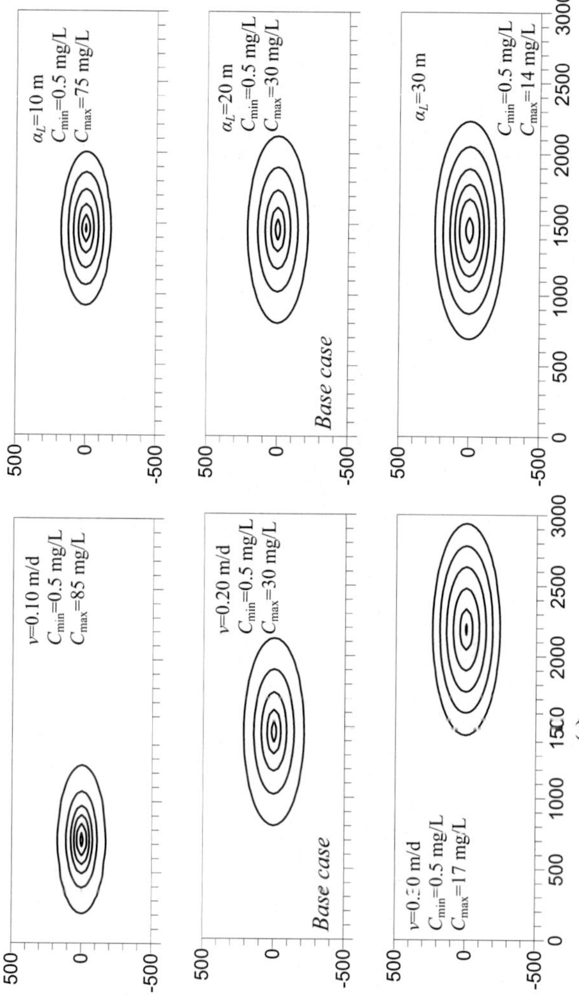

Figure 3.8: The effect of flow velocity (a) and longitudinal dispersivity (b) on the plume development.

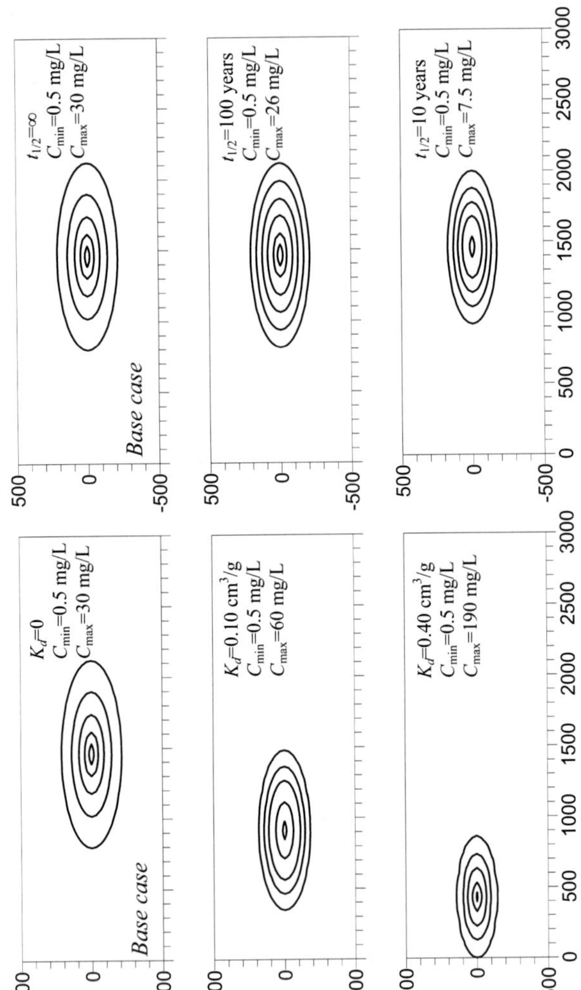

Figure 3.9: The effect of the distribution coefficient (a) and half-life (b) time on the plume development.

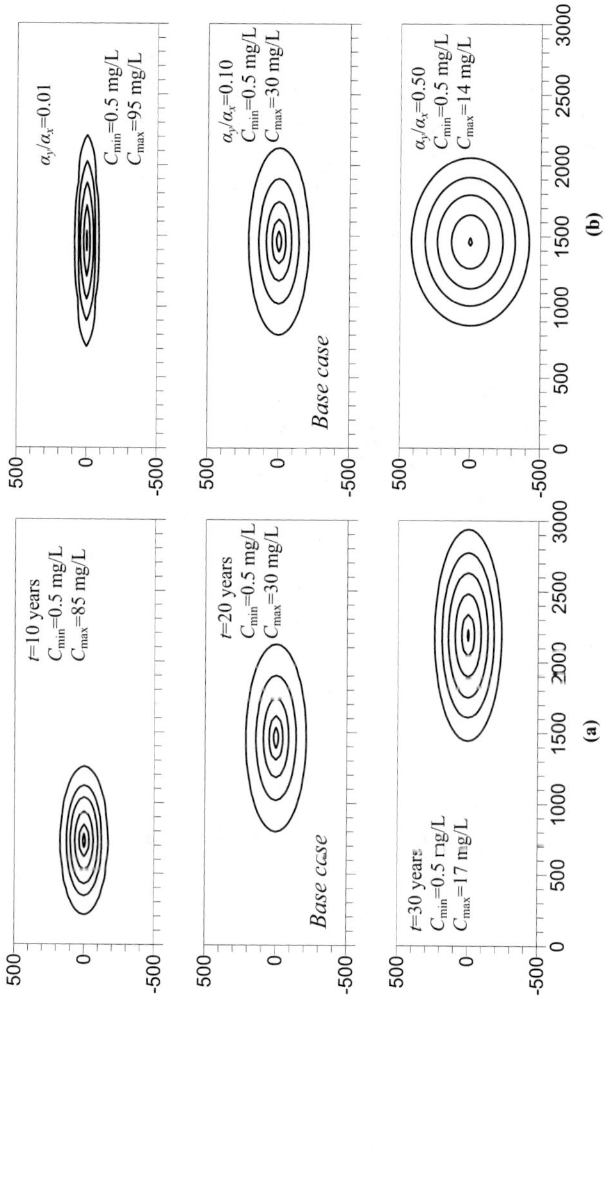

Figure 3.10: The effect of simulation time (a) and the ratio of the transverse to longitudinal dispersivity (b) on the plume development.

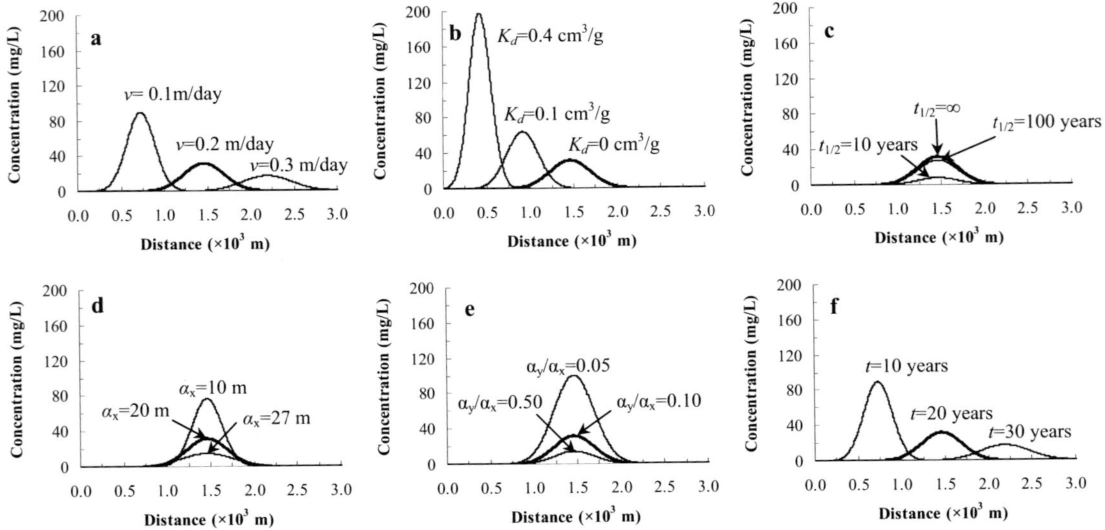

Figure 3.11: Sensitivity of breakthrough curves along a longitudinal cross section taken through the center of the plume from west-east direction for different parameters. The parameters are (a) velocity, (b) distribution coefficient, (c) half-life time, (d) longitudinal dispersivity, (e) ratio of transverse to longitudinal dispersivity, and (f) simulation time.

Due to the capability of ANN in successfully handling large-scale physical problems, ANN has found use in a wide range of hydrologic applications (ASCE Task Committee on Artificial Neural Networks in Hydrology 2000). Ray and Klindworth (2000) used ANN to assess nitrate contamination of rural private wells. Input parameters were the thickness of the aquifer, sampling depth, distance to cropland, streams, disposal sites, and septic systems, topography around wells, season of sample collection, and time between the probable dates of nitrate application and the first flushing storms. The output of ANN was either *none, low, medium,* or *high* when the predicted nitrate concentration is < *10, 10-20, 20-30,* or > *30* mg/L, respectively. Kaluli et al. (1998) developed an ANN model that simulated nitrate leaching to groundwater from a 4.2 ha site. ANN inputs included time of year, daily denitrification rate, cropping systems, water table depth, nitrogen fertilizer application rate, daily antecedent precipitation index, initial nitrate concentration, and the daily drain flow. They used ANN to simulate the impacts of different management options on nitrate leaching. In a different approach, many researchers used ANN as a proxy to a mathematical model of contaminant fate and transport (Ranjithan et al. 1993; Rogers and Dowla 1994; Sawyer et al. 1995; Johnson and Rogers 1995, 2000; Aly and Peralta 1999; Morshed and Kaluarachchi 1998a,b; Gumrah et al. 2000). Besides, ANN was utilized in simulating many environmental and water resources engineering systems (e.g., Bockreis and Jager 1999; Maier and Dandy 2000; Kralisch et al. 2003; Mas et al. 2004; Anctil et al. 2004).

Recently, Almasri and Kaluarachchi (2005a) used a special class of ANN known as modular neural networks (MNNs) for simulating the spatial distribution of *long-term* nitrate concentration in agriculture-dominated aquifers. MNN utilizes high-order computational units to perform multifaceted tasks. These tasks were developed by dividing the physical problem into simpler subtasks via partitioning the input-output response patterns into regions of identical features. As such, MNN has the ability to understand the complex physical problem better than a typical multilayer ANN (Haykin 1994). Nitrate occurrences in groundwater when related to the different on-ground nitrogen sources, land use distribution and practices, and the physical and chemical processes in soil and groundwater do not exhibit well-behaved relationships. Therefore, MNN can be a potential analytical tool to simulate the spatial distribution of long-term nitrate concentrations in agriculture-dominated aquifers where well-behaved relationships do not exist.

The MNN-based approach was applied for Sumas-Blaine aquifer located in the northwest corner of Washington State where the degradation of groundwater quality from nitrate is of great concern to the residents in the area (Blake and Peterson 2001). High concentrations of nitrate have been detected in this aquifer since early 1970's (Kaluarachchi et al. 2002). The overall MNN approach as developed by Almasri and Kaluarachchi (2005a) utilizes the National Land Cover Database (NLCD) grid of the U. S. Geological Survey to determine the spatial sources of nitrogen in the study area based on the land use distribution. A Geographic Information System (GIS) of ESRI (1999) was used to facilitate the MNN input and output processing and was integrated with the NLCD. After conceptualizing the relationship between long-term nitrate concentration in groundwater and the influential parameters and after making the simplifying approximations and assumptions (see Figure 3.12), the long-term steady state nitrate concentration C_{NO3} was reduced to the following simple formulation (Almasri and Kaluarachchi 2005a):

$$C_{NO3} = f\left(\Gamma, \delta, \tau, \Re\right) \tag{3.58}$$

where: C_{NO3} is the corresponding nitrate concentration; Γ is the spatial location of the nitrate receptor, δ is the spatial distribution of the NLCD classes; τ is the on-ground nitrogen loading; and $\Re$ is the groundwater recharge.

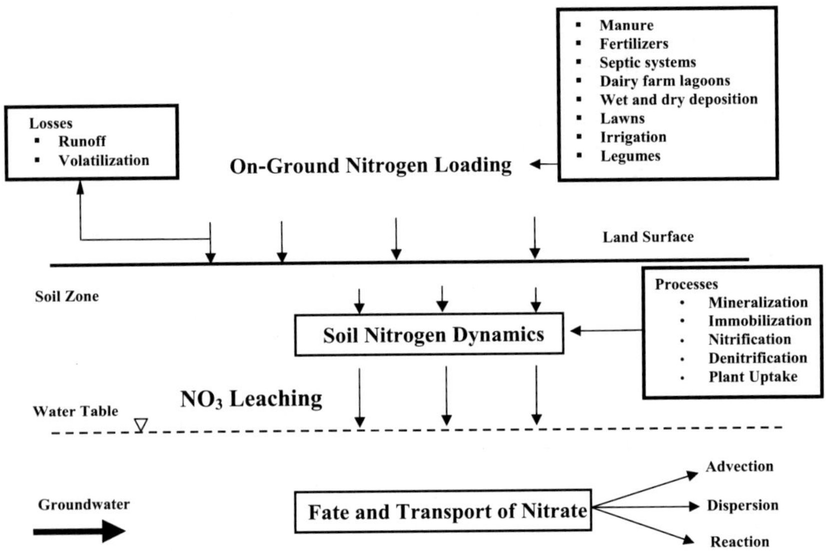

Figure 3.12: A schematic describing the integrated three-process approach to conceptualize the on-ground nitrogen loading, soil nitrogen transformations, and fate and transport of nitrate in groundwater.

The basic premise followed in developing the MNN input-output response patterns was to designate the optimal radius of a specified circular-buffered zone centered by the nitrate receptor so that the input parameters at the upgradient areas correlate with nitrate concentrations in groundwater (see Figure 3.13). The radius of the circular area was determined based on the maximum correlation between the input value and nitrate concentration based on different radii values as shown in Figure 3.14.

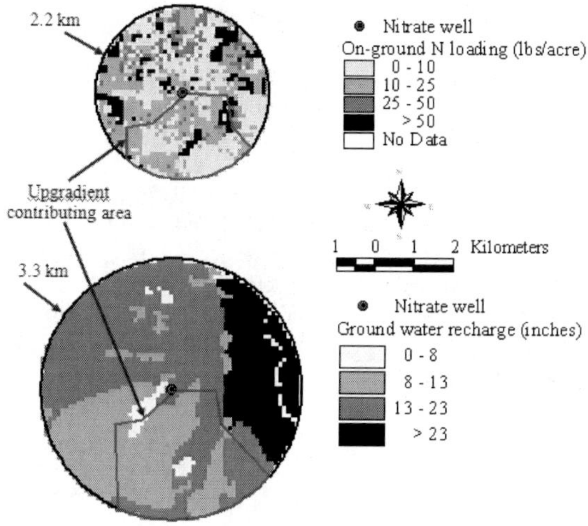

Figure 3.13: A schematic for the upgradient contributing areas for a nitrate well for two different input parameters pertaining to (a) on-ground nitrogen loading, and (b) groundwater recharge. Parameter categories are summarized to four classes for ease of presentation.

After the development (training and testing) of MNN, it was utilized to predict the nitrate distribution in the Sumas-Blaine aquifer. To assess the probable anthropogenic effects on groundwater quality, nitrate concentrations across the Sumas-Blaine aquifer were classified into four groups based on the work of Madison and Brunett (1985) and Cox and Kahle (1999). The four concentration ranges are: 0–1 mg/L to indicate the most likely background concentration, 1–3 mg/L to indicate a possible human influence, 3–10 mg/L to indicate pollution due to human influence, and greater than 10 mg/L to indicate that the MCL was exceeded as a result of extensive human activities.

The study area was divided into a uniform finite difference grid of 100×100 m^2 cell size where MNN predicted the concentration at each cell of 38,695 locations. To check the performance of MNN in prediction, a mathematical fate and transport model for the area was developed using MT3D. Figure 3.15 shows the comparison between the distribution of nitrate concentration predicted by MNN and that simulated by the fate and transport model. It can be seen that MNN overestimated the concentrations in the northernmost area of the aquifer. Nevertheless, the areas predicted to exceed the MCL by the MNN are coinciding to a good extent with those of MT3D predictions. These areas consist of berry plantations with substantial manure applications.

Although there is a discrepancy between the MNN and MT3D predictions, it can be said that MNN captured the general trend of nitrate concentration in most parts of the aquifer. However, nitrate concentrations as simulated by MT3D show higher spatial variability due to model's ability to take into account local effects due to actual processes such as denitrification, dispersion, and advection as well as nitrate leaching to groundwater. It should also be noted that MNN simulations were 180 times faster than MT3D while requiring much less data. Table 3.2 further shows the accuracy of MNN predictions as compared to those of MT3D using different statistical indices. The explanations of these statistical indices can be found in Anderson and Woessner (1992) and Legates and McCabe (1999). On the whole, the statistics show less accuracy in the MNN predictions, especially in the correlation-based measures, whereas the overall RMSE is 4.83 mg/L. Nevertheless, the mean relative error shows an acceptable value of 16%.

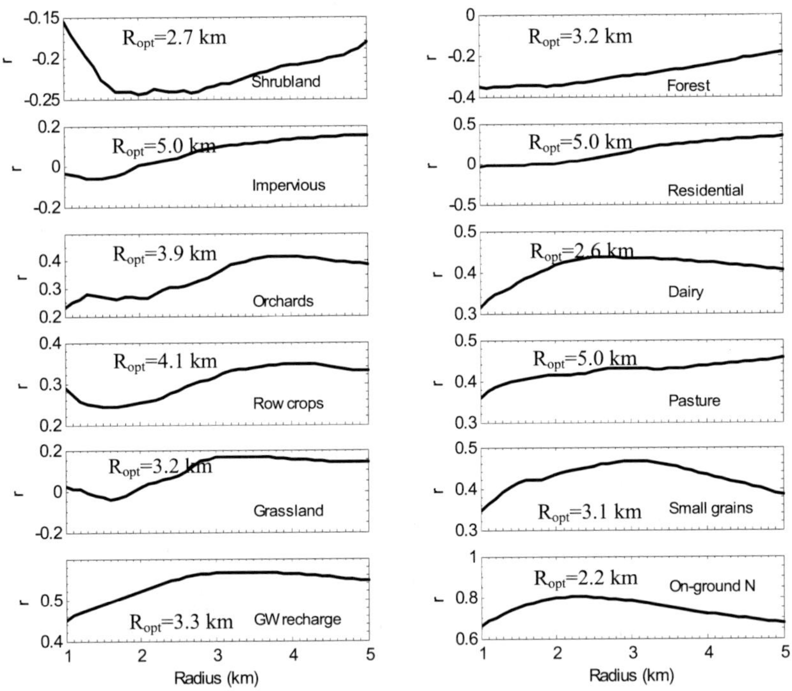

Figure 3.14: Correlation coefficients (r) between the average long-term nitrate concentration from 1990 to 1998 and a selected set of major parameters presented for different radii of the upgradient areas surrounding each sampling well.

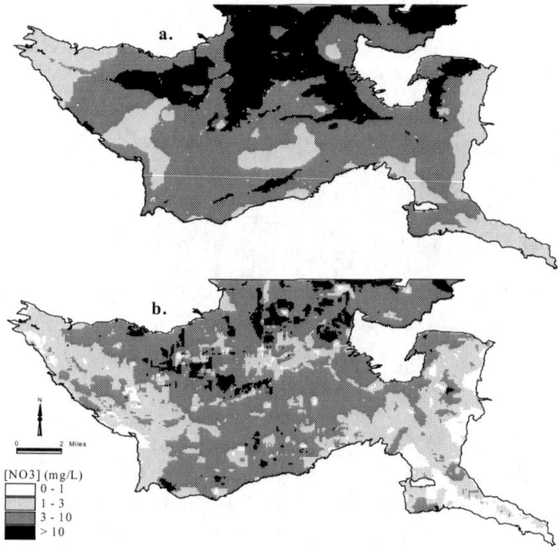

Figure 3.15: Distributions of nitrate concentration for the current land use practices for Sumas-Blaine aquifer; (a) MNN prediction, and (b) MT3D simulation.

Table 3.2: Key statistics for MNN performance in comparison to the MT3D results.

Statistics	MT3D	MNN
Mean (mg/L)	4.51	6.89
Median (mg/L)	3.88	5.20
Standard deviation (mg/L)	3.11	4.45
Range (mg/L)	29.99	17.90
Root mean squared error (mg/L)		4.83
Mean absolute error (mg/L)		3.55
Mean relative error (%)		16.09
Correlation coefficient (-)		0.43
Coefficient of efficiency[§] (-)		-1.41
Index of agreement[§] (-)		0.57
Modified coefficient of efficiency[§] (-)		-0.46
Modified index of agreement[§] (-)		0.41

[§] The coefficient of efficiency ranges from -∞ to 1 and the index of agreement ranges from 0 to 1. Higher values signify better match.

List of Notation

C	Contaminant concentration [ML^{-3}]
C_0	The initial concentration [ML^{-3}]
C_{NO3}	Nitrate concentration [ML^{-3}]
d	The grain size of the aquifer material [L]
D^*	Effective diffusion coefficient [L^2T^{-1}]
dC/dl	The concentration gradient [ML^{-4}]
D_d	The diffusion coefficient [L^2T^{-1}]
dh/dl	Hydraulic gradient [L^0]
D_L	The longitudinal hydrodynamic dispersion coefficient [L^2T^{-1}]
D_T	The transverse hydrodynamic dispersion coefficient [L^2T^{-1}]
F	The diffusive mass flux [ML^2T^{-1}]
F	The ratio of oxygen to contaminant consumed [M/M]
f_{oc}	The fraction of organic carbon comprising the solid phase
$g_1(t)$, $g_2(t)$, and $g_3(t)$	Known functions
J	The advective mass flux [ML^2T^{-1}]
K	the hydraulic conductivity [LT^{-1}]
K	Constant used in the calculation of Freundlich sorption isotherm
K_0	The oxygen half-saturation constant [ML^{-3}]
k_1	The first-order rate constant
k_2	The forward first-order rate constant
k_3	The backward first-order rate constant
k_4	A constant equivalent to K_d
k_5	Constant
k_6	Constant
k_7	The forward rate constant
k_8	The backward rate constant
K_c	The contaminant half-saturation constant [ML^{-3}]
K_d	The distribution coefficient [L^3M^{-1}]
K_{oc}	The organic carbon partition coefficient
K_{om}	A partition coefficient based on soil or aquifer organic matter
L	The characteristic flow length [L]
M	Amount of the contaminant substance deposited in the point source
M_t	The total microbial concentration [M/M]

n	The porosity
N	Constant used in the calculation of Freundlich sorption isotherm
n_e	Effective porosity through which flow occurs
P_e	Peclet number
R	The retardation factor
R_F	The retardation coefficient for a Freundlich sorption isotherm
R_L	The retardation factor for the Langmuir sorption isotherm
S	The mass of solute sorbed per dry unit weight of solid [MM^{-1}]
t	Time [T]
$t_{1/2}$	The half-life time [T]
u	A constant dependent on the molecular diffusion coefficient
v	Average linear velocity of groundwater [LT^{-1}]
v_c	The velocity of contaminant [LT^{-1}]
v_i	The velocity value in the *i* direction [LT^{-1}]
v_j	The velocity value in the *j* direction [LT^{-1}]
v_w	The velocity of water [LT^{-1}]
v_x	The velocity in the *x* direction [LT^{-1}]
x_0, y_0	Point location of slug injection
α	Partition coefficient [L^3M^{-1}]
α_i	The dispersivity value in the *i* direction [L]
α_j	The dispersivity value in the *j* direction [L]
α_L	The longitudinal dispersivity [L]
α_T	The transverse dispersivity [L]
β	The maximum sorptive capacity for the solid surface [MM^{-1}]
γ	The first-order rate coefficient
Γ	The spatial location of the nitrate receptor
δ	The spatial distribution of the NLCD classes
λ	The decay reaction rate constant [T^{-1}]
μ	The growth rate [T^{-1}]
μ_{max}	The maximum growth rate [T^{-1}]
ρ_b	The bulk density of the aquifer [ML^{-3}]
σ_L^2	The variance of the longitudinal spreading of the plume [L^2]
σ_T^2	The variance of the transverse spreading of the plume [L^2]
$\Re$	Ground-water recharge [L^3L^{-2}]

3.9 References

Adriano, D. C., Iskandar, A. K. and Murarka, I. P. (1994). *Contamination of groundwaters*, Science Reviews, London.

Almasri, M. N. (2003). "Optimal management of nitrate contamination of groundwater." Ph.D. Dissertation, Utah State University, Logan, Utah.

Almasri, M. N., and Kaluarachchi, J. J. (2004a). "Implications of on-ground nitrogen loading and soil transformations on groundwater quality management." *Journal of the American Water Resources Association*, 40(1), 165-186.

Almasri, M. N., and Kaluarachchi, J. J. (2004b). "Assessment and management of long-term nitrate pollution of groundwater in agriculture-dominated watersheds." *Journal of Hydrology*, (295), 225-245.

Almasri, M. N., and Kaluarachchi, J. J. (2005a). "Modular neural networks to predict the nitrate distribution in groundwater using the on-ground nitrogen loading and recharge data." *Environmental Modelling and Software*, 20(7), 851-871.

Almasri, M. N., and Kaluarachchi, J. J. (2005b). "Multi-criteria decision analysis for the optimal management of nitrate contamination of aquifers." *Journal of Environmental Management*, In Press.

Aly, A. H., and Peralta, R. C. (1999). "Optimal design of aquifer cleanup systems under uncertainty using a neural network and a genetic algorithm." *Water Resources Research*, 35(8), 2523-2532.

Appelo, C. A., and Postma, D. (1996). *Geochemistry, groundwater and pollution*, A.A. Balkema, Leiden, The Netherlands.

Anctil, F., Perrin, C., and Andréassian, V. (2004). "Impact of the length of observed records on the performance of ANN and of conceptual parsimonious rainfall-runoff forecasting models." *Environmental Modelling and Software*, 19(4), 357-368.

Anderson, M. P., and Woessner, W. W. (1992). *Applied groundwater modeling: Simulation of flow and advective transport*, Academic Press, San Diego.

Arnade, L. J. (1999). "Seasonal correlation of well contamination and septic tank distance." *Groundwater*, 37(6), 920-923.

ASCE Task Committee on application of artificial neural networks in hydrology (2000). "Artificial neural networks in hydrology. I: Preliminary concepts." *Journal of Hydrologic Engineering*, 5(2),115-123.

Baestlé, L. H. (1969). "Migration of radionuclides in porous media." *Progress in nuclear energy Series XII, Health physics*, A. M. F. Duhamel, ed., Pergamon Press, Elmsford, New York, 707-7630.

Barcelona, M., Wehrmann, A., Keely, J. F., and Pettyjohn, W. A. (1990). Contamination of groundwater, prevention, assessment, restoration, Noyes Data Corporation.

Beale, R., and Jackson, T. (1991). *Neural computing: An introduction*, Adam Hilger, Techno House, Bristol.

Bear, J. (1979). *Hydraulics of groundwater*, McGraw-Hill, New York.

Bedient, P. B., Rifai, H. S., and Newell, C. J. (1999). *Groundwater contamination: Transport and remediation*, Prentice Hall, New York.

Blake, S., and Peterson, B. (2001). Summary characterization for Water Resource Inventory Area #1, Water Resources Department, Whatcom County, Bellingham, Washington.

Bockreis, A., and Jager, J. (1999). "Odour monitoring by the combination of sensors and neural networks." *Environmental Modelling and Software*, 14(5), 421-426.

Cantor, L., and Knox, R. C. (1984). Evaluation of septic tank effects on ground-water quality, EPA-600/2-284-107, U. S. Environmental Protection Agency, Washington, DC.

Chapelle, F. H. (1993). *Groundwater microbiology & geochemistry*, John Wiley & Sons, Inc., New York.

Cox, S. E., and Kahle, S. C. (1999). Hydrogeology, ground-water quality, and sources of nitrate in lowland glacial aquifer of Whatcom County, Washington, and British Columbia, Canada, Water Resources Investigation Report 98-4195, U. S. Geological Survey, Tacoma, Washington.

Deutsch, W. J. (1997). *Groundwater geochemistry: Fundamentals and applications to contamination*, Lewis Publishers, Florida.

Domenico, P. A., and Schwartz, F. W. (1990). *Physical and chemical hydrology*, John Wiley & Sons, New York.

Domenico, P. A. (1987). "An analytical model for multidimensional transport of a decaying contaminant species." *Journal of Hydrology*, 91, 49-58.

ESRI (1999). ArcView. Environmental Systems Research Institute, Inc.

Fetter, C. W. (1994). *Applied hydrogeology*, Prentice Hall, New York.

Fetter, C. W. (1999). *Contaminant hydrogeology*, Prentice Hall, New York.

Fitts, C. R. (2002). *Groundwater science*, Academic Press, New York.

Follett, R. F., and Delgado, J. A. (2002). "Nitrogen fate and transport in agricultural systems." *Journal of Soil and Water Conservation*, 57, 402-408.

Freeze, R. A., and Cherry, J. A. (1979). *Groundwater*, Prentice Hall, New York.

Gumrah, F., Oz, B., Guler, B. and Evin, S. (2000). "The application of artificial neural networks for the prediction of water quality of polluted aquifer." *Water Air and Soil Pollution*, 119(1-4), 275-294.

Harbaugh, A. W., and McDonald, M. G. (1996). User's documentation for MODFLOW-96, an update to the U.S. Geological Survey modular finite-difference ground-water flow model, Open-File Report 96-485. U. S. Geological Survey.

Harter, T., Davis, H., Mathews, M., and Meyer, R. (2002). "Shallow groundwater quality on dairy farms with irrigated forage crops." *Journal of Contaminant Hydrology*, 55, 287-315.

Haykin, S. (1994). *Neural networks: A comprehensive foundation*, Maxwell Macmillan International, New York.

Hii, B., Liebscher, H., Mazalek, M., and Tuominen, T. (1999). Groundwater quality and flow rates in the Abbotsford Aquifer, British Columbia, Environment Canada, Vancouver, BC.

Ingebritsen, S. E., and Sanford, W. E. (1998). *Groundwater in geologic processes*, Cambridge University Press.

Johnson, V. M., and Rogers, L. L. (1995). "Location analysis in groundwater remediation using neural networks." *Groundwater*, 33(5), 749-758.

Johnson, V. M., and Rogers, L. L. (2000). "Accuracy of neural network approximators in simulation-optimization." *Journal of Water Resources Planning and Management*, 126(2), 48-56.

Kaluarachchi, J. J., Kra, E., Twarakavi, N., and Almasri, M. N. (2002). Nitrogen and pesticide contamination of groundwater in Water Resource Inventory Area 1, Groundwater quality report for WRIA 1, Phase II Report, Utah State University, Logan, Utah.

Kaluli, J. W., Madramootoo, C. A., and Djebbar, Y. (1998). "Modeling nitrate leaching using neural networks." *Water Science and Technology*, 38(7), 127-134.

Karickhoff, S. W. (1984). "Organic pollutant sorption in aquatic systems." *Journal of Hydraulic Engineering*, 110(6), 707-735.

Katsifarakis, K. L. (2000). *Groundwater pollution control*, WIT Press.

Keeney, D. R. (1986). "Sources of nitrate to groundwater." CRC Crit. Rev. Environmental Control, 16, 257-303.

Knox, R. C., Sabatini, D. A., and Canter, L. W. (1993). *Subsurface transport and fate processes*, Lewis Publishers, Florida.

Kralisch, S., Fink, M., Flügel, W.-A., and Beckstein, C. (2003). "A neural network approach for the optimisation of watershed management." *Environmental Modelling and Software*, 18(8-9), 815-823.

Latinopoulos, P. (2000). "Nitrate contamination of groundwater: Modeling as a tool for risk assessment, management and control." *Groundwater pollution control*, K. L. Katsifarakis, ed., WIT Press.

Lee, Y. W., Dahab, M. F., and Bogardi, I. (1991). "Nitrate risk management under uncertainty." *Journal of Water Resources Planning and Management*, 118(2), 151-165.

Legates, D. R., and McCabe, G. J. (1999). "Evaluating the use of "goodness-of-fit" measures in hydrologic and hydroclimatic model validation." *Water Resources Research*, 35(1), 233-241.

MacQuarrie, K. T. B., Sudicky, E., and Robertson, W. D. (2001)." Numerical simulation of a fine-grained denitrification layer for removing septic system nitrate from shallow groundwater." *Journal of Hydrology*, 52, 29-55.

Madison, R. J., and Brunett, J. O. (1985). Overview of the occurrence of nitrate in groundwater of the United States, National Water Summary, 1984, Water-Supply Paper 2275, U. S. Geological Survey.

Maier, H. R., and Dandy, G. C. (2000). "Neural networks for the prediction and forecasting of water resources variables: A review of modelling issues and applications." *Environmental Modelling and Software*, 15(1), 101-124.

Mas, J. F., Puig, H., Palacio, J. L., and Sosa-López, A. (2004). "Modelling deforestation using GIS and artificial neural networks." *Environmental Modelling and Software*, 19(5), 461-471.

McGrail, B. P. (2001). "Inverse reactive transport simulator (INVERTS): An inverse model for contaminant transport with nonlinear adsorption and source terms." *Environmental Modelling and Software*, 16(8), 711-723.

Mercado, A. (1976). "Nitrate and chloride pollution of aquifers: A regional study with the aid of a single-cell model." *Water Resources Research*, 12(4), 731-747.

Morshed, J., and Kaluarachchi, J. J. (1998a). "Parameter estimation using artificial neural network and genetic algorithm for free-product migration and recovery." *Water Resources Research*, 34(5), 1101-1113.

Morshed, J., and Kaluarachchi, J. J. (1998b). "Application of artificial neural network and genetic algorithm in flow and transport simulations." *Advances in Water Resources*, 22(2), 145-158.

National Research Council (1984). *Groundwater contamination*, National Academy Press.

Neuman, S. P. (1990). "Universal scaling of hydraulic conductivities and dispersivities in geologic media." *Water Resources Research*, 26(8), 1749-1758.

Nolan, B. T., Hitt, K., and Ruddy, B. (2002). "Probability of nitrate contamination of recently recharged groundwaters in the conterminous United States." *Environmental Science and Technology*, 36(10), 2138-2145.

Office of Technology Assessment (OTA) (1984). Protecting the nation's groundwater from contamination, OTA-0-233, Office of Technology Assessment, Washington, DC.

Paul, J. W., and Zebarth, B. J. (1997). "Denitrification and nitrate leaching during the fall and winter following dairy cattle slurry application." *Canadian Journal of Soil Science*, 77, 231-240.

Pye, V. I., and Kelley, J. (1984). "The extent of groundwater contamination in the United States." *Groundwater contamination*, National Academy Press.

Rifai, H. S., and Bedient, P. B. (1994). "Modeling contaminant transport and biodegradation in groundwater." *Contamination of groundwaters*, D. C. Adriano, A. K. Iskandar, and I. P. Murarka, eds., Science Reviews.

Ranjithan, S., Eheart, J. W., and Garrett, J. H. (1993). "Neural network-based screening for groundwater reclamation under uncertainty." *Water Resources Research*, 29(3), 563-574.

Ray, C., and Klindworth, K. (2000). "Neural networks for agrichemical vulnerability assessment of rural private wells." *Journal of Hydrologic Engineering*, 5(2), 162-171.

Rogers, L. L., and Dowla, F. U. (1994). "Optimization of groundwater remediation using artificial neural networks with parallel solute transport modeling." *Water Resources Research*, 30(2), 457-481.

Roy, W. R. (1994). "Groundwater contamination from municipal landfills in the USA." *Contamination of groundwaters*, D. C. Adriano, A. K. Iskandar, and I. P. Murarka, eds., Science Reviews.

Saffman, P. G. (1959). "A theory of dispersion in a porous medium." *Journal of Fluid Mechanics*, 6(3), 321-349.

Sawyer, C. S., Achenie, L. K., and Lieullen, K. K. (1995). "Estimation of aquifer hydraulic conductivities: A neural network approach." *Models for assessing and monitoring groundwater quality* (Proceedings of a Boulder Sym., July 1995), IAHS Publ. No. 227.

Skaggs, T. H., and Kabala, Z. J. (1994). "Recovering the release history of a groundwater contaminant." *Water Resources Research*, 30(1), 71-79.

Solley, W. B., Pierce, R. R., and Perlman, H. A. (1998). Estimated use of water in the United States in 1995, Circular 1200, U. S. Geological Survey.

Sun, N. (1996). *Mathematical modeling of groundwater pollution*, Springer, New York.

Todd, D. K., 1980. *Groundwater hydrology*, John Wiley & Sons.

Wolfe, A. H., and Patz, J. A. (2002). "Reactive nitrogen and human health: Acute and long-term implications." *Ambio*, 31(2), 120-125.

Xu, M., and Eckstein, Y. (1995). "Use of weighted least-squares method in evaluation of the relationship between dispersivity and field scale." *Groundwater*, 33(6), 905-908.

Yu, C., Zielen, A. J., Cheng, J.-J., Yuen, Y. C., Jones, L. G., LePoire, D. J., Wang, Y. Y., Loureiro, C. O., Gnanapragasam, E., Faillace, E., Wallo III, A., Williams, W. A., and Peterson, H. (1993). Manual for implementing residual radioactive material guidelines using RESRAD, Version 5.0, ANL/EAD/LD-2, Argonne National Laboratory, Argonne, Illinois, September.

Zheng, C. (1990). MT3D, A modular three-dimensional transport model for simulation of advection, dispersion and chemical reactions of contaminants in groundwater systems, Report to the U. S. Environmental Protection Agency.

Zheng, C., and Bennett, G. D. (1995). *Applied contaminant transport modeling: Theory and practice*, International Thomson Publishing.

Zheng, C., and Wang, P. P. (1999). MT3DMS, A modular three-dimensional multi-species transport model for simulation of advection, dispersion and chemical reactions of contaminants in groundwater systems; documentation and user's guide, Contract Report SERDP-99-1, U. S. Army Engineer Research and Development Center, Vicksburg, Mississippi.

CHAPTER 4

REVIEW OF ANALYTICAL METHODS OF MODELING CONTAMINANT FATE AND TRANSPORT

Venkataraman Srinivasan[1] and T. Prabhakar Clement[2]
[1]Graduate Student, Department of Civil and Environmental Engineering, University of Illinois at Urbana Champaign, Urbana, IL 61801
[2]Arthur H. Feagin Chair Professor, Department of Civil Engineering, Auburn University, AL 36849

4.1 Introduction

Analytical models are useful tools for predicting the fate and transport of groundwater contaminant plumes. Furthermore, they can be used to test numerical models. In recent years, researchers in the area of contaminant remediation employed analytical models as screening tools to simulate the fate and transport of reactive compounds. Examples of common organic reactive groundwater contaminants include tetrachloroethylene (PCE), trichloroethylene (TCE), and benzene, toluene, ethylbenzene, and xylene (BTEX). Examples of inorganic contaminants include metals such as uranium, chromium, arsenic, and lead. These reactive species are of special interest to researchers because the degradation by-products of some of these compounds are toxic and can cause more harm than the original compound. For example, in the case of PCE, one of its degradation by-products, vinyl chloride, is more toxic than PCE and could pose serious health risks. The objective of this chapter is to provide a historical review of the development of analytical solutions to contaminant transport equations with a major focus on solutions to reactive transport problems.

4.2 Single Species Transport

Since the early 1950s, several researchers have developed analytical solutions to single-species transport problems (Brenner 1962; Lapidus and Amundson 1952; Lindstrom and Narasimham 1973; Ogata and Banks 1961). van Genuchten and Alves (1982) provide a comprehensive review of one-dimensional analytical solutions to advection-dispersion equations for various initial and boundary conditions. Multi-dimensional analytical solutions of the transport equation for a single species are also available in the literature (Cleary and Ungs 1978; Hunt 1973; Wilson and Miller 1978). Wexler (1992) provides a comprehensive review of one-, two- and three-dimensional, single-species reactive transport problems. As the combined work of van Genuchten and Alves (1982) and Wexler (1992) represents a detailed review of analytical solutions to the single-species, advection-dispersion equation, with and without a decay term, in this chapter we present only a cursory overview of these solutions. Interested readers are encouraged to refer to the original works of van Genuchten and Alves (1982) and Wexler (1992).

The general three-dimensional transport problem describing the fate and transport of a decaying single species with a patch source of constant concentration c_0 located at the origin in a clean, semi-infinite aquifer is shown in Figure 4.1. As shown in the figure, the contaminant is subject to advection in the x direction and dispersive mixing in all three directions. It is assumed that the contaminant decays through a first-order process.

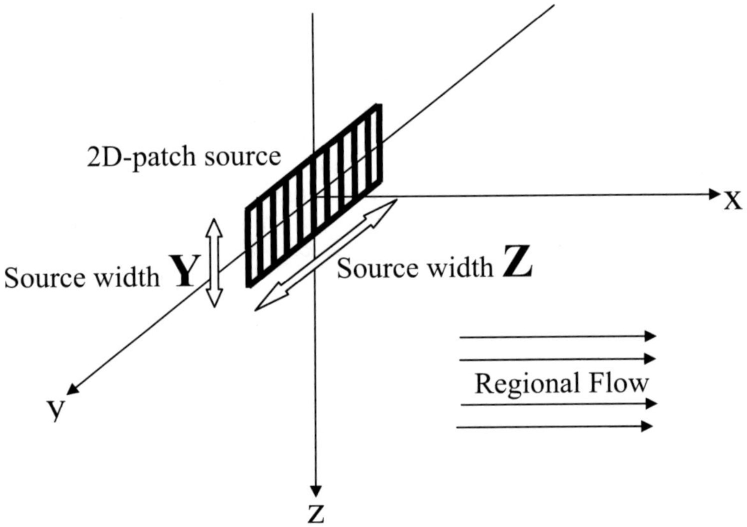

Figure 4.1 Schematic of the three-dimensional transport problem.

The governing transport equation for the three-dimensional system can be written as (Wexler 1992)

$$\frac{\partial c}{\partial t} = -v\frac{\partial c}{\partial x} + D_x\frac{\partial^2 c}{\partial x^2} + D_y\frac{\partial^2 c}{\partial y^2} + D_z\frac{\partial^2 c}{\partial z^2} - kc \tag{4.1}$$

where: c is the concentration of the contaminant [ML^{-3}]; k is the first-order destruction rate constant of the contaminant [T^{-1}]; v is the transport velocity [LT^{-1}]; and D_x, D_y, D_z are the dispersion coefficients in the x, y and z dimensions [L^2T^{-1}]. The initial and boundary conditions are given by

$$
\begin{aligned}
&c(x,y,z,t)=0 \qquad \forall \quad 0<x<\infty, \quad -\infty<y<\infty, \quad -\infty<z<\infty \\
&c(x,y,z,t)=c_0 \qquad -\frac{Z}{2}<z<\frac{Z}{2}, \quad -\frac{Y}{2}<y<\frac{Y}{2}, \quad \forall \quad t>0 \\
&\qquad\qquad\quad =0 \qquad otherwise, \quad \forall \quad t>0 \\
&\lim_{x\to\infty}\frac{\partial c(x,y,z,t)}{\partial x}=0 \\
&\lim_{y\to\pm\infty}\frac{\partial c(x,y,z,t)}{\partial y}=0 \\
&\lim_{z\to\pm\infty}\frac{\partial c(x,y,z,t)}{\partial z}=0
\end{aligned}
\tag{4.2}
$$

where Y and Z are the source widths in the y and z dimension. The solution to the above set of governing equations can be expressed as (Wexler 1992)

$$
c(x,y,z,t)=\frac{c_o}{8}\int_{\tau=0}^{\tau=t} f'_x(x,\tau)\ f'_y(y,\tau)\ f'_z(z,\tau)\ d\tau \tag{4.3}
$$

The $f'_x(x,\tau)$, $f'_y(y,\tau)$ and $f'_z(z,\tau)$ terms are given as

$$
\begin{aligned}
f'_x(x,\tau)&=\frac{x}{\sqrt{\pi D_x}}\exp\left(\frac{vx}{2D_x}\right)\frac{\exp\left(\dfrac{-v^2}{4D_x}\tau-k\tau+\dfrac{-x^2}{4D_x\tau}\right)}{\tau^{3/2}} \\
f'_y(y,\tau)&=\left[erf\left\{\frac{y+\frac{Y}{2}}{2(D_y\tau)^{1/2}}\right\}-erf\left\{\frac{y-\frac{Y}{2}}{2(D_y\tau)^{1/2}}\right\}\right] \\
f'_z(z,\tau)&=\left[erf\left\{\frac{z+\frac{Z}{2}}{2(D_z\tau)^{1/2}}\right\}-erf\left\{\frac{z-\frac{Z}{2}}{2(D_z\tau)^{1/2}}\right\}\right]
\end{aligned}
\tag{4.4}
$$

Reducing the problem to one dimension by dropping all the dependencies in the y and z directions, we get the classic analytical solution given by Ogata and Banks (1961), i.e.,

$$
c(x,t)=\frac{c_o}{2}f_x(x,t) \tag{4.5}
$$

where the $f_x(x,t)$ term is given by

$$f_x(x,t) = \left(\exp\left\{ \frac{x}{2\alpha_x}\left[1 - \left(1 + \frac{4k\alpha_x}{v} \right)^{1/2} \right] \right\} erfc\left\{ \frac{x - vt\left(1 + \frac{4k\alpha_x}{v}\right)^{1/2}}{2(\alpha_x vt)^{1/2}} \right\} \right) + \left(\exp\left\{ \frac{x}{2\alpha_x}\left[1 + \left(1 + \frac{4k\alpha_x}{v} \right)^{1/2} \right] \right\} erfc\left\{ \frac{x + vt\left(1 + \frac{4k\alpha_x}{v}\right)^{1/2}}{2(\alpha_x vt)^{1/2}} \right\} \right) \quad (4.6)$$

Note that $\alpha_x = D_x/v$, $\alpha_y = D_y/v$ and $\alpha_z = D_z/v$ are the dispersivities in the x, y and z directions, respectively [L].

One of the main drawbacks of the multi-dimensional analytical solution presented in Equation (4.3) is that it is a non-closed form solution and requires numerical evaluation of the indefinite integral. An attractive alternative to this would be the use of the approximate closed-form expression given by Domenico and Robbins (1985) and Martyn-Hayden and Robbins (1997). The Domenico solution is an approximate three-dimensional, closed-form expression that describes the fate and transport of a decaying contaminant plume evolving from a finite patch source. It must be stressed that this is not an exact analytical solution, but an approximate closed-form expression that does not involve numerical procedures to compute the final results. The approximate expression given by Domenico and Robbins (1985) and Martyn-Hayden and Robbins (1997) is

$$c(x,y,z,t) = \frac{c_o}{8}\, f_x(x,t)\, f_y(x,y)\, f_z(x,z) \quad (4.7)$$

The $f_x(x,t)$, in Equation (4.7) is given in Equation (4.6). The $f_y(x,y)$ and $f_z(x,z)$ terms in Equation (4.7) are given as

$$f_y(y,x) = \left[erf\left\{ \frac{y + \frac{Y}{2}}{2(\alpha_y x)^{1/2}} \right\} - erf\left\{ \frac{y - \frac{Y}{2}}{2(\alpha_y x)^{1/2}} \right\} \right]$$
$$f_z(z,x) = \left[erf\left\{ \frac{z + \frac{Z}{2}}{2(\alpha_z x)^{1/2}} \right\} - erf\left\{ \frac{z - \frac{Z}{2}}{2(\alpha_z x)^{1/2}} \right\} \right] \quad (4.8)$$

More recently, Srinivasan et al. (2007) presented a detailed analysis of the limitations and errors associated with this approximate solution including qualitative insights into the nature of these errors. They conclude that the Domenico solution will be more accurate when the transport problem involves a small value of longitudinal dispersivity. Furthermore, they conclude that the approximate solutions will have minimal errors when used under high advection velocities and large simulation times.

It is useful to note that in the absence of longitudinal dispersion a closed-form expression can be derived for the three-dimensional transport problem (Srinivasan et al. 2007). The corresponding solution is

$$c(x,y,z,t) = \frac{C_o}{8}\; f_x^o(x,t)\; f_y(x,y)\; f_z(x,z) \tag{4.9}$$

The $f_x^o(x,t)$ term in Equation (4.9) is defined as

$$f_x^o(x,t) = 2\exp\left(-\frac{k\,x}{v}\right)u\left\{t-\frac{x}{v}\right\}$$

$$where;\quad u\left\{t-\frac{x}{v}\right\} = \begin{cases} 0 & if\ t \le \dfrac{x}{v} \\ 1 & if\ t > \dfrac{x}{v} \end{cases} \tag{4.10}$$

The $f_y(x,y)$ and $f_z(x,z)$ terms in Equation (4.9) are identical to the ones given in Equation (4.8).

The Ogata and Banks (1961) one-dimensional solution and the Domenico and Robbins (1985) three-dimensional approximate solution are the two most widely used analytical solutions in the groundwater literature. The U. S. Environmental Protection Agency screening tools BIOSCREEN (Newell et al. 1996) and BIOCHLOR (Aziz et al. 2000) use modified forms of these solutions, which consider multi-species transport, to solve practical bioremediation problems. In the sections below, we provide a detailed review of analytical solutions to coupled, multi-species reactive transport problems.

4.3 Multi-species Reactive Transport in One-Dimensional Systems

This section considers solutions to a set of partial differential equations that describe the transport of multiple reactive species in saturated porous media systems. The kinetics of the reactions are assumed to be first order. The general governing set of differential equations that describe the transport of multiple species in a saturated three-dimensional porous medium with uniform steady flow in the x direction are given as (Clement 2001)

$$R_i \frac{\partial c_i}{\partial t} + v \frac{\partial c_i}{\partial x} - D_x \frac{\partial^2 c_i}{\partial x^2} - D_y \frac{\partial^2 c_i}{\partial y^2} - D_z \frac{\partial^2 c_i}{\partial z^2} = \sum_{j=1}^{i-1} y_{i/j} k_j c_j - k_i c_i + \sum_{j=i+1}^{n} y_{i/j} k_j c_j \;;\; \forall\, i = 1,2,\ldots n \tag{4.11}$$

where: c_i is the concentration of species i [ML^{-3}]; R_i is the retardation factor of species i [dimensionless]; $y_{i/j}$ is the effective yield factor that describes the mass of a species i produced from another species j [MM^{-1}]; k_i is the first-order destruction rate constant of species i [T^{-1}]; v is the transport velocity [LT^{-1}]; D_x, D_y, D_z are the dispersion coefficients [L^2T^{-1}]; and n is the total number of species in the reaction network. Note that the reaction network represented in Equation (4.11) can be used to represent serial, parallel, diverging and/or converging reaction networks. A conceptual diagram of the different types of reaction networks is given in Figure 4.2. Equation (4.11) assumes that the degradation reactions occur only in the liquid phase.

If we assume degradation reactions to occur in both solid and liquid phases, then Equation (4.11) should be modified as

$$R_i \frac{\partial c_i}{\partial t} + \mathrm{v} \frac{\partial c_i}{\partial \mathrm{x}} - D_x \frac{\partial^2 c_i}{\partial x^2} - D_y \frac{\partial^2 c_i}{\partial x^2} - D_z \frac{\partial^2 c_i}{\partial x^2} = \sum_{j=1}^{i-1} R_j y_{i/j} k_j c_j - R_i k_i c_i + \sum_{j=i+1}^{n} R_j y_{i/j} k_j c_j,\ \forall\, i = 1,2,\ldots,n \tag{4.12}$$

Equations of the form (4.11) and (4.12) can be used to model several types of environmental transport problems. For example, Cho (1971) uses it to model the fate and transport of nitrate species in soil-water systems, van Genuchten (1985) notes its applicability for modeling radionuclide migration, and Clement et al. (1998, 2000, 2002) use similar forms of the equation to model the fate and transport of mixed chlorinated solvent plumes. A historical perspective of the development of analytical solutions to these equations is presented in the section below.

4.3.1 Pure Advection Problems (Zero Dispersion)

The analytical solution given by McLaren (1969, 1970), which describes the steady-state, one-dimensional transport of a five species nitrogen chain in a saturated porous medium, is one of the first multi-species solution derived for solving sequentially-coupled, reactive transport problems. This work assumes that transport is governed by advection, and hence the effects of dispersion and sorption can be ignored. The simplified governing equation solved by McLaren can be expressed as

$$\begin{aligned} v \frac{\partial c_i}{\partial x} &= k_{i-1} c_{i-1} - k_i c_i \;;\; \forall\, i = 2,3\ldots 5 \\ &= -k_i c_i \;;\; i = 1 \end{aligned} \tag{4.13}$$

Equation (4.13) was solved sequentially using a Dirichlet boundary condition, i.e.,

$$c_i(x=0)=c_i^o \tag{4.14}$$

Solutions to this coupled set of Equations (4.13) and (4.14) were obtained sequentially using a zero initial condition.

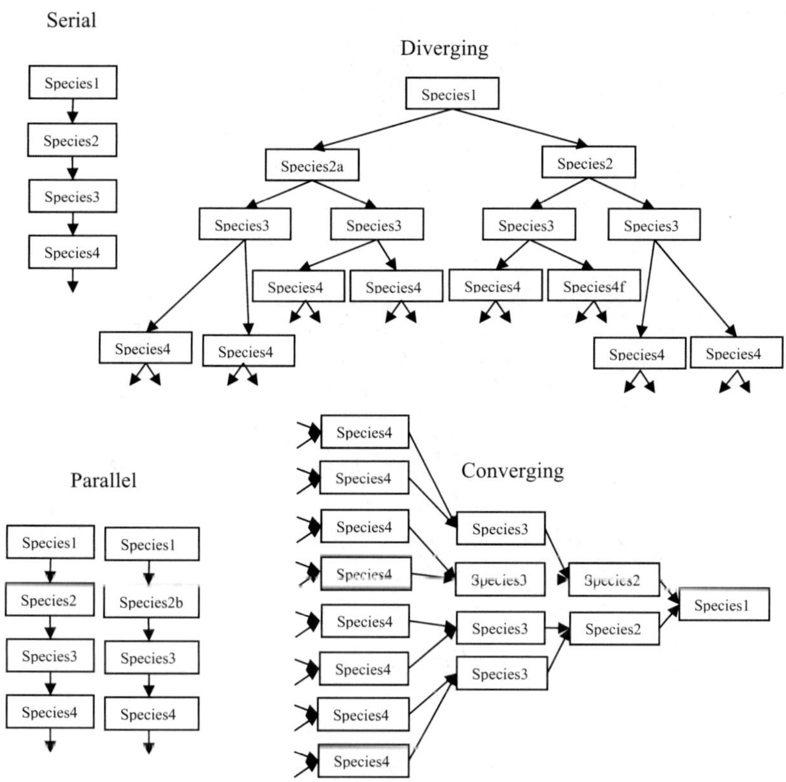

Figure 4.2 Conceptual diagrams of the different types of reaction networks.

Burkholder and Rosinger (1980) and Lester et al. (1975) developed solutions for the advective transport of radionuclide chains subject to linear equilibrium sorption. Explicit analytical solutions are presented for a three-species problem involving distinct retardation factors for each species, for both impulse and decaying band release boundary conditions. In addition, they also provide solutions for the case of identical retardation factors. The governing equations adopted by Lester et al. (1975)are expressed as

$$\begin{aligned} R_i \frac{\partial c_i}{\partial t} + v\frac{\partial c_i}{\partial x} &= y_{i-1}k_{i-1}c_{i-1} - k_i c_i \ ; \ \forall\, i = 2, 3, \ldots n \\ &= -k_i c_i\left(x,t\right) \ ; \ i = 1 \end{aligned} \tag{4.15}$$

The boundary condition for the decaying band release boundary is given by:

$$c_i\left(0,t\right) = \begin{cases} \sum_{i_1=1}^{i} B_i^{i_1} e^{-k_{i_1} t}, \ 0 < t \le t_o \\ 0, \ t > t_o \end{cases} \ ; \ \forall\, i = 1, 2, 3 \tag{4.16}$$

where $B_i^{i_1}$ is a constant.

Harada et al. (1980) and Higashi and Pigford (1980) provided explicit closed-form solutions for a set of purely advective (no dispersion) transport problems for several types of boundary conditions including preferential release, exponential release, step release, band release and impulse release. The governing equation for this problem is identical to Equation (4.15). The solution to Equation (4.15) subject to the boundary condition given by Equation (4.16) for a zero initial condition is given as

$$c_i\left(x,t\right) = \sum_{i_1=1}^{i}\left[\left(\prod_{i_2=i_1+1}^{i} y_{i_2} k_{i_2-1}\right)\sum_{i_2=i_1}^{i}\sum_{i_3=1}^{i_1}\left\{G_1^1 + h\left(G_1^1\right)\ G_2^1\right\}\right] \ ; \ \forall\, i = 1, 2, \ldots n$$
$$where \ ; \ h\left(M\right) = \begin{cases} 1 \ if \ \text{M} \ loop \ is \ not \ executed \\ 0 \ if \ \text{M} \ loop \ is \ executed \end{cases} \tag{4.17}$$

The G_1^1 and G_2^1 terms in Equation (4.17) are defined as

$$G_1^1 = \sum_{i_4=i_1,\left(i_4 \ne i_2, R_{i_4} \ne R_{i_2}\right)}^{i} \frac{B_{i_1}^{i_3}\left\langle \begin{array}{l} F_{i_2,i_3,0}\left[x,t\right] - e^{-t_o \lambda_{i_3}} F_{i_1,i_2,i_3}\left[x,(t-t_o)\right] \\ -F_{i_2,i_2,i_4}\left[x,t\right] + e^{-t_o \lambda_{i_3}} F_{i_2,i_2,i_4}\left[x,(t-t_o)\right] \end{array}\right\rangle}{\left(a_{i_2,i_4} - \lambda_{i_3}\right)\left(\prod_{i_5=i_1,\left(i_5 \ne i_2, R_{i_5} = R_{i_2}\right)}^{i} -k_{i_2,i_5}\right)\left(-R_{i_2,i_4}\right)\prod_{i_5=i_1,\left(i_5 \ne i_2, i_5 \ne i_4, R_{i_5} \ne R_{i_2}\right)}^{i} -R_{i_2,i_5}\left(a_{i_2,i_5} - a_{i_2,i_4}\right)} \tag{4.18}$$

$$G_2^1 = \frac{B_{i_1}^{i_3}\left\langle F_{i_2,i_3,0}\left[x,t\right] - e^{-t_o \lambda_{i_3}} F_{i_1,i_2,i_3}\left[x,(t-t_o)\right]\right\rangle}{\prod_{i_4=i_1,\left(i_4 \ne i_2, R_{i_4} = R_{i_2}\right)}^{i} -k_{i_2,i_4}}$$

The F_{i_1,i_2,i_3} term in Equation (4.18) is given by

$$F_{i_1,i_2,i_3}[x,t] = u\left(t - \frac{R_{i_1} x}{v}\right) e^{-a_{i_2,i_3}\left(t - \frac{R_{i_1} x}{v}\right)} \tag{4.19}$$

We later show that the solutions presented in this section are special case solutions of the general solution obtained by Srinivasan and Clement (2008a, 2008b).

4.3.2 Short Chain Advection-Dispersion Problems

One of the pioneering works in obtaining analytical solutions to multi-species, coupled reactive transport problems was completed by Cho (1971). Cho developed an explicit analytical solution to a three-species transport problem that is subject to advection, dispersion, linear equilibrium sorption, coupled through sequential, first-order reactions. The governing equation used by Cho (1971) can be expressed as

$$\begin{aligned} R_i \frac{\partial c_i}{\partial t} + v \frac{\partial c_i}{\partial x} - D_x \frac{\partial^2 c_i}{\partial x^2} &= k_{i-1} c_{i-1} - k_i c_i \ ; \ \forall\, i = 2,3 \\ &= -k_i c_i \ ; \ i = 1 \end{aligned} \tag{4.20}$$

Explicit closed-form analytical solutions were obtained for the Dirichlet boundary condition and a zero initial condition. The initial and boundary conditions used by Cho (1971) are given by

$$\begin{aligned} & c_i(x,0) = 0,\ 0 < x < \infty \ ; \ \forall\, i = 1,2,3 \\ & \frac{\partial c_i(\infty,t)}{\partial x} = 0,\ t > 0 \ ; \ \forall\, i = 1,2,3 \\ & c_i(0,t) = B_i \ ; \ \forall\, i = 1,2,3 \end{aligned} \tag{4.21}$$

The procedure adopted by Cho involves the use of Laplace transform methods. Equation (4.20) is transformed in the Laplace domain along with the corresponding boundary conditions. This results in a set of coupled ordinary differential equations that is solved by successive substitution of the solutions to higher reaction orders. Finally, the solutions in the Laplace domain are retransformed in the time domain using inverse transform procedures (look up tables for Laplace transforms). The final solution given by Cho (1971) can be expressed compactly as

$$\begin{aligned} & c_i(x,t) = \sum_{i_1=1}^{i} \left[\left(\prod_{i_2=i_1+1}^{i} k_{i_2-1} \right) \sum_{i_2=i_1}^{i} \left\{ G_1^1 + h\left(G_1^1\right)\ G_2^1 \right\} \right] \ ; \ \forall\, i = 1,2,3 \\ & where \ ; \ h(M) = \begin{cases} 1 \ \mathit{if}\ \ \mathrm{M}\ \ \mathit{loop\ is\ not\ executed} \\ 0 \ \mathit{if}\ \ \mathrm{M}\ \ \mathit{loop\ is\ executed} \end{cases} \end{aligned} \tag{4.22}$$

The G_1^1 and G_2^1 terms in Equation (4.22) are given by

$$G_1^1 = \sum_{i_4=i_1,(i_4\neq i_2, R_{i_4}\neq R_{i_2})}^{i} \frac{B_{i_1}\left\langle F_{i_2,i_1,0}[x,t] - F_{i_2,i_2,i_4}[x,t]\right\rangle}{-k_{i_2,i_4}\left(\prod_{i_5=i_1,(i_5\neq i_2, R_{i_5}=R_{i_2})}^{i} -k_{i_2,i_5}\right)\prod_{i_5=i_1,(i_5\neq i_2, i_5\neq i_4, R_{i_5}\neq R_{i_2})}^{i} -R_{i_2,i_5}\left(a_{i_2,i_5}-a_{i_2,i_4}\right)} \tag{4.23}$$

$$G_2^1 = \frac{B_{i_1}\left\langle F_{i_2,i_1,0}[x,t]\right\rangle}{\prod_{i_4=i_1,(i_4\neq i_2, R_{i_4}=R_{i_2})}^{i} -k_{i_2,i_4}}$$

Also note that

$$\begin{aligned} k_{i_1,i_2} &= k_{i_1} - k_{i_2} \\ R_{i_1,i_2} &= R_{i_1} - R_{i_2} \\ a_{i_1,i_2} &= \frac{k_{i_1,i_2}}{R_{i_1,i_2}} \end{aligned} \tag{4.24}$$

The F_{i_1,i_2,i_3} term in Equation (4.23) is defined as

$$F_{i_1,i_2,i_3}[x,t] = \frac{e^{-a_{i_2,i_3}t}e^{\frac{xv}{2D_x}}}{2}\left[\begin{aligned} & e^{\frac{-x\omega_{i_1,i_2,i_3}}{2D_x}} erfc\left\{\frac{R_{i_1}x - \omega_{i_1,i_2,i_3}t}{2\sqrt{R_{i_1}D_x t}}\right\} \\ & +e^{\frac{x\omega_{i_1,i_2,i_3}}{2D_x}} erfc\left\{\frac{R_{i_1}x + \omega_{i_1,i_2,i_3}t}{2\sqrt{R_{i_1}D_x t}}\right\} \end{aligned}\right] \tag{4.25}$$

$$where\ ;\ \omega_{i_1,i_2,i_3} = \sqrt{v^2 + 4R_{i_1}D_x\left(\frac{k_{i_1}}{R_{i_1}} - a_{i_2,i_3}\right)}$$

It must be noted that the solution method proposed by Cho (1971) is feasible only for problems involving a small number of species. When the number of species increases, the terms involved in successive substitutions grows rapidly, making the solution process cumbersome. Also, one of the limitations of Cho's (1971) solution is that only the first species in the chain is subject to sorption and hence the retardation factors of the second and the third species are assumed to be one. Furthermore, the above solution imposes several parameter limitations that are not acknowledged by Cho (1971), including $k_{i_2,i_4} \neq 0$, $k_{i_2,i_5} \neq 0$ and $a_{i_2,i_5} \neq a_{i_2,i_4}$ in Equation (4.23). The above solution is valid only for real values of ω_{i_1,i_2,i_3} and Cho (1971) ignores the possibility of ω_{i_1,i_2,i_3} being complex. For the sake of completeness, we provide the solution expressions for problems involving complex values of ω_{i_1,i_2,i_3}. For this case, the F_{i_1,i_2,i_3} terms given in Equation (4.25) are modified as

$$F_{i_1,i_2,i_3}\left[x,t\right]=e^{-a_{i_2,i_3}t}e^{\frac{xv}{2D_x}}\left[ACos\left(\frac{x\omega^{*}_{i_1,i_2,i_3}}{2D_x}\right)-BSin\left(\frac{x\omega^{*}_{i_1,i_2,i_3}}{2D_x}\right)\right]$$

$$where\ ;\ \omega^{*}_{i_1,i_2,i_3}=\sqrt{\left|v^2+4R_{i_1}D_x\left(\frac{k_{i_1}}{R_{i_1}}-a_{i_2,i_3}\right)\right|} \tag{4.26}$$

$$and\ \left(A+iB\right)=erfc\left\{\frac{R_{i_1}x+i\omega^{*}_{i_1,i_2,i_3}t}{2\sqrt{R_{i_1}D_xt}}\right\}$$

Misra et al. (1974) derived semi-analytical solutions to a problem similar to Cho (1971) using a pulse source boundary. The solution methodology involves the use of Laplace transforms and sequential substitution. However, their solution was not a major improvement to the solution given by Cho (1971) due to the semi-analytical nature of the final expression presented by the authors.

Burkholder and Rosinger (1980) and Lester et al. (1975) developed solutions for the advective-dispersive transport of radionuclide chains subject to linear equilibrium sorption. Explicit analytical solutions are presented for a three-species problem involving distinct retardation factors for each species, for both impulse and decaying-band release boundary conditions. In addition, they also provided solutions for the case of identical retardation factors. The work by Burkholder and Rosinger (1980) and Lester et al. (1975) was more comprehensive compared to Cho (1971), although they fail to mention the problems involving complex values of ω_{i_1,i_2,i_3}. The governing equation adopted by Lester et al. (1975) is identical to Equation (4.20); however, the boundary condition for the decaying band release boundary is given by Equation (4.16). The solution to the problem solved by Lester et al. (1975) can be expressed as

$$c_i\left(x,t\right)=\sum_{i_1=1}^{i}\left[\left(\prod_{i_2=i_1+1}^{i}y_{i_2}k_{i_2-1}\right)\sum_{i_2=i_1}^{i}\sum_{i_3=1}^{i_1}\left\{G_1^1+h\left(G_1^1\right)\,G_2^1\right\}\right]\ ;\ \forall\, i=1,2,3 \tag{4.27}$$

The G_1^1 and G_2^1 terms in Equation (4.27) are defined as

$$G_1^1 = \sum_{i_4=i_1,(i_4\neq i_2, R_{i_4}\neq R_{i_2})}^{i} \frac{B_{i_1}^{i_3}\left\langle \begin{array}{l} F_{i_2,i_3,0}[x,t]-u(t-t_o)e^{(-\lambda_{i_3}t_o)}F_{i_2,i_3,0}[x,(t-t_o)] \\ -F_{i_2,i_2,i_4}[x,t]+u(t-t_o)e^{(-\lambda_{i_3}t_o)}F_{i_2,i_2,i_4}[x,(t-t_o)] \end{array}\right\rangle}{\left(a_{i_2,i_4}-\lambda_{i_3}\right)\left(\prod_{i_5=i_1,(i_5\neq i_2, R_{i_5}=R_{i_2})}^{i} -k_{i_2,i_5}\right)\left(-R_{i_2,i_4}\right)\prod_{i_5=i_1,(i_5\neq i_2, i_5\neq i_4, R_{i_5}\neq R_{i_2})}^{i} -R_{i_2,i_5}\left(a_{i_2,i_5}-a_{i_2,i_4}\right)} \tag{4.28}$$

$$G_2^1 = \frac{B_{i_1}^{i_3}\left\langle F_{i_2,i_3,0}[x,t]-u(t-t_o)e^{(-\lambda_{i_3}t_o)}F_{i_2,i_3,0}[x,(t-t_o)]\right\rangle}{\prod_{i_4=i_1,(i_4\neq i_2, R_{i_4}=R_{i_2})}^{i} -k_{i_2,i_4}}$$

The definitions of other terms in Equation (4.28) remain the same as before except that the a_{i_1,i_2} term is modified to be

$$a_{i_1,i_2} = \begin{cases} \dfrac{k_{i_1,i_2}}{R_{i_1,i_2}} & ;\ when\ i_2 > 0 \\ \lambda_{i_1} & ;\ when\ i_2 = 0 \end{cases} \tag{4.29}$$

Another of the pioneering works in analytical solutions to multi-species reactive transport problems is attributed to van Genuchten (1985). van Genuchten developed explicit analytical solutions to model a sequentially-coupled, four-species transport problem governed by advection, dispersion and linear equilibrium sorption processes involving first-order reactions. It was assumed that all the species had distinct retardation factors. Solutions are presented for both generic Bateman-type Dirichlet and Cauchy boundary conditions that could take into account the effect of a pulse source input. One of the key contributions of van Genuchten (1985) was that he developed a robust computer code (CHAIN) for implementing his analytical solution. It must be noted that several key techniques and analytical manipulations are required to formulate a computer code for this problem; van Genuchten (1985) provides a fairly detailed and extremely useful analysis of the computational scheme. The solutions provided by van Genuchten look similar in structure to the solutions provided by Cho (1971) and Lester et al. (1975). We later show that all these are simpler cases of a more general solution provided by Srinivasan and Clement (2008a, 2008b).

4.3.3 Short Chain Advection-Dispersion Problems Including Non-Zero Initial Conditions

Lunn et al. (1996) solved a three-species transport problem, which was similar to the Cho (1971) problem, using the Fourier transform method. The authors demonstrated that the use of Fourier transforms enabled them to solve problems having non-zero initial conditions by solving two special problems with a constant Dirichlet boundary condition. Their key contribution involved the development of a closed-form solution for the case of an exponentially decaying initial condition for the first species. This initial condition can be expressed as

$$c_i(x,0)=\begin{cases} c_i^o e^{-\mu_i x} \ ; \ i=1 \\ 0 \ ; \ \forall\, i=2,3 \end{cases}, \quad 0<x<\infty \tag{4.30}$$

We later show that the solution presented by Lunn et al. (1996) is also a special case of a more general solution given by Srinivasan and Clement (2008a, 2008b).

4.3.4 Long Chain Advection-Dispersion Problems Including Non-Zero Initial Conditions

Harada et al. (1980) published a research report presenting general semi-analytical solutions to sequentially-coupled, one-dimensional reactive transport problems of arbitrary chain length subject to arbitrary release modes. The corresponding governing equation is

$$\begin{aligned} R_i\frac{\partial c_i}{\partial t}+v\frac{\partial c_i}{\partial x}-D_x\frac{\partial^2 c_i}{\partial x^2} &= y_i k_{i-1}c_{i-1}-k_i c_i \ ; \ \forall\, i=2,3,\dots n \\ &= -k_i c_i \ ; \ i=1 \end{aligned} \tag{4.31}$$

The solution method involves the use of Green's function and Fourier integrals to obtain generic semi-analytical (non-closed form) solution expressions for arbitrary boundary conditions. The solutions for the predecessor species are substituted back into the governing equation for the successor species. Although the solution (not presented here) seems very powerful at the outset, it must be realized that one of the major limitations of this solution strategy is that the semi-analytical solution for a given species in the chain requires the computation of its entire predecessor species. This results in computationally-inefficient algorithms especially when analyzing transport problems involving long reactive chains. Furthermore, the errors due to numerical evaluation (approximation) of the semi-analytical solution accumulate with increasing chain length, making the solution less accurate.

Prior to Sun et al. (1999a), most analytical solutions in the literature of multi-species reactive transport problems were restricted to small chain lengths or involved semi-analytical solutions having complex sequential integrals. Sun et al. (1999a) developed a method that could solve multi-species, advective-dispersive transport equations coupled with sequential first-order reactions involving an arbitrary number of species for different types of initial and boundary conditions. Their method uses a transformation procedure to uncouple the system of equations, which can then be solved analytically in the transformed domain. The final solutions are obtained by retransforming the solutions to the original domain. Later, Sun et al. (1999b) extended the transformation format to solve problems involving a combination of serial and parallel reactions. Clement (2001) presented a more general and fundamental approach to derive the Sun et al. (1999b) solution by employing the similarity transformation method. This approach can also be used to solve problems involving serial, parallel, converging, diverging and/or reversible first-order reaction networks. However, all of these methods are only applicable for solving problems involving identical retardation factors. As shown by Clement (2001) the underlying concept employed by all of these solutions is the use of a similarity transformation technique that can uncouple the set of coupled equations. A brief overview of the solution methodology, as adopted by Clement (2001), is reviewed here.

The general governing equation considered by Clement (2001) is

$$\begin{aligned}\frac{\partial c_i}{\partial t} + \mathrm{v}\frac{\partial c_i}{\partial \mathrm{x}} - D_x\frac{\partial^2 c_i}{\partial x^2} - D_y\frac{\partial^2 c_i}{\partial y^2} - D_z\frac{\partial^2 c_i}{\partial z^2} = \\ \sum_{j=1}^{i-1} y_{i/j}k_j c_j - k_i c_i + \sum_{j=i+1}^{n} y_{i/j}k_j c_j,\ \forall i = 1,2,\cdot\cdot,n\end{aligned} \tag{4.32}$$

The system of governing equations given by Equation (4.32) can be conveniently expressed in a matrix format as

$$\frac{\partial}{\partial t}\{c\} + v\frac{\partial}{\partial x}\{c\} - D_x\frac{\partial^2}{\partial x^2}\{c\} - D_y\frac{\partial^2}{\partial y^2}\{c\} - D_z\frac{\partial^2}{\partial z^2}\{c\} = [YK]\{c\} \tag{4.33}$$

Now let us assume an arbitrary $n \times n$ matrix $[A]$ and use its inverse to perform the following matrix operation:

$$\{b\} = [A]^{-1}\{c\} \tag{4.34}$$

Conversely, we can express $\{c\} = [A]\{b\}$. Substituting $\{c\} = [A]\{b\}$ in Equation (4.33) we get

$$[A]\frac{\partial}{\partial t}\{b\} + v[A]\frac{\partial}{\partial x}\{b\} - D_x[A]\frac{\partial^2}{\partial x^2}\{b\} - D_y[A]\frac{\partial^2}{\partial y^2}\{b\} - D_z[A]\frac{\partial^2}{\partial z^2}\{b\} = [YK][A]\{b\} \tag{4.35}$$

Pre-multiplying Equation (4.35) by $[A]^{-1}$ we get

$$\frac{\partial}{\partial t}\{b\} + v\frac{\partial}{\partial x}\{b\} - D_x\frac{\partial^2}{\partial x^2}\{b\} - D_y\frac{\partial^2}{\partial y^2}\{b\} - D_z\frac{\partial^2}{\partial z^2}\{b\} = \left\langle [A]^{-1}[YK][A]\right\rangle\{b\} \tag{4.36}$$

If the columns of the transformation matrix $[A]$ are assigned as the eigenvectors of the combined reaction coefficient matrix $[YK]$, then the matrix $\left\langle [A]^{-1}[YK][A]\right\rangle$ will have a diagonal form. Under this condition, Equation (4.36) reduces to a system of n uncoupled differential equations. Note that the diagonalization process describe above has essentially helped us transform a system of n coupled equations in the c-domain into a set of n uncoupled equation in the linearly transformed b-domain. The uncoupled set of advection-disperision-reaction equations can now be independently solved using an elementary solution with transformed initial and boundary conditions. Once we obtain the independent solutions in the b-domain, we perform the transformation $\{c\}=[A]\{b\}$ to obtain the solution in the original c-domain. The elegant formulation of this method allows the generalization of the problem to any arbitrary chain length.

Quezada et al. (2004) extended the Clement (2001) approach and developed a method that can solve multi-species transport equations coupled with a network of first-order reactions involving distinct retardation factors. This method involves transforming the system of governing equations to a Laplace domain and then solving the transformed system of equations using the Clement (2001) approach. The solutions in the Laplace domain are then retransformed to the time domain using an inverse Laplace transform procedure. The general solution strategy provided by Quezada et al. (2004) is described below.

The governing system of equations under consideration is

$$R_i\frac{\partial c_i}{\partial t}+v\frac{\partial c_i}{\partial x}-D_x\frac{\partial^2 c_i}{\partial x^2}-D_y\frac{\partial^2 c_i}{\partial y^2}-D_z\frac{\partial^2 c_i}{\partial z^2}= \sum_{j=1}^{i-1} y_{i/j}k_jc_j-k_ic_i+\sum_{j=i+1}^{n} y_{i/j}k_jc_j \ ; \ \forall\, i=1,2,\ldots n \tag{4.37}$$

In matrix notation, Equation (4.37) can be expressed as

$$[R]\frac{\partial}{\partial t}\{c\}+v\frac{\partial}{\partial x}\{c\}-D_x\frac{\partial^2}{\partial x^2}\{c\}-D_y\frac{\partial^2}{\partial y^2}\{c\} \quad D_z\frac{\partial^2}{\partial z^2}\{c\}=[YK]\{c\} \tag{4.38}$$

Laplace transform of the system of equations given by Equation (4.38) yields

$$s[R]\{p\}+v\frac{\partial}{\partial x}\{p\}-D_x\frac{\partial^2}{\partial x^2}\{p\}-D_y\frac{\partial^2}{\partial y^2}\{p\}-D_z\frac{\partial^2}{\partial z^2}\{p\}=[YK]\{p\} \tag{4.39}$$

where s is the Laplace variable and $\{p\}$ is the Laplace-transformed concentration vector. Rearranging Equation (4.39), we get

$$v\frac{\partial}{\partial x}\{p\}-D_x\frac{\partial^2}{\partial x^2}\{p\}-D_y\frac{\partial^2}{\partial y^2}\{p\}-D_z\frac{\partial^2}{\partial z^2}\{p\}=\langle [YK]-s[R]\rangle\{p\} \tag{4.40}$$

The elements of $\{p\}$ are defined by the standard Laplace transformation relation:

$$p_i(x,y,z,s) = \int_0^\infty \exp(-st)c_i(x,y,z,t)dt \tag{4.41}$$

We now use the uncoupling strategy proposed by Clement (2001) by performing the following matrix operation $\{p\}=[A]\{b\}$. Substituting this into Equation (4.40) we get

$$v[A]\frac{\partial}{\partial x}\{b\} - D_x[A]\frac{\partial^2}{\partial x^2}\{b\} - D_y[A]\frac{\partial^2}{\partial y^2}\{b\} - D_z[A]\frac{\partial^2}{\partial z^2}\{b\} = \langle [YK]-s[R]\rangle [A]\{b\} \tag{4.42}$$

Pre-multiplying Equation (4.42) by $[A]^{-1}$ and simplifying results in

$$v\frac{\partial}{\partial x}\{b\} - D_x\frac{\partial^2}{\partial x^2}\{b\} - D_y\frac{\partial^2}{\partial y^2}\{b\} - D_z\frac{\partial^2}{\partial z^2}\{b\} = [A]^{-1}\langle [YK]-s[R]\rangle [A]\{b\} \tag{4.43}$$

If the columns of the transformation matrix $[A]$ are assigned as the eigenvectors of the combined reaction coefficient matrix $\langle [YK]-s[R]\rangle$ then the $[A]^{-1}\langle [YK]-s[R]\rangle [A]$ matrix will have a diagonal form. Under this condition, Equation (4.43) reduces to a system of n uncoupled differential equations. Note the diagonalization process describe above has essentially helped us transform a system of n coupled equations in the p-domain into a set of n uncoupled equation in the linearly transformed b-domain. The uncoupled set of advection-diffusion-reaction equations can now be independently solved using an elementary solution with transformed initial and boundary conditions to obtain the solution in the b-domain. Then the following matrix operation must be performed on the $\{b\}$ vector to deduce the solution in the p-domain:

$$\{p\} = [A]\{b\} \tag{4.44}$$

Finally, the solution is transformed to the c-domain by applying the Laplace inversion operation on the $\{p\}$ vector. Quezada et al. (2004) state that the Laplace inversion step can be completed by either using an analytical procedure or by using a numerical procedure. Figure 4.3 provides a summary of the procedure described by Quezada et al. (2004). They demonstrate their strategy by providing a closed-form solution to a one-dimensional, two-species reactive transport problem. Although robust in its methodology, their framework has some important drawbacks. One of its key limitations is that, except for a simple two-species transport problem, the solutions require a numerical inverse Laplace transform routine to evaluate the expressions.

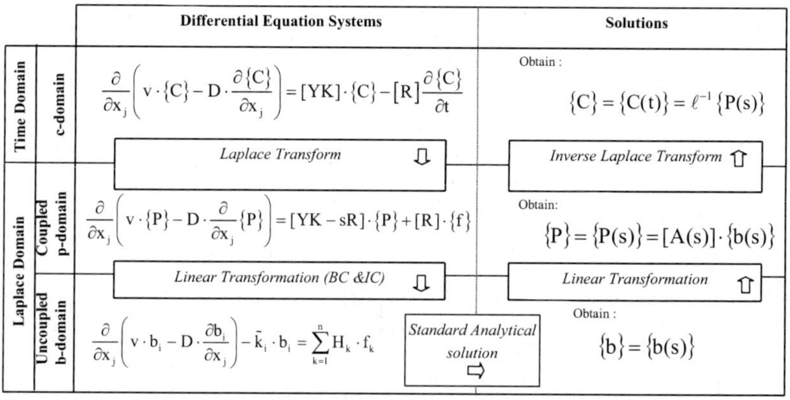

Figure 4.3 Summary of various transformation steps used in the solution procedure (Quezada, et al., 2004).

Srinivasan and Clement (2008a, 2008b) presented a series of closed-form analytical solutions to a generalized set of sequentially-coupled, reactive transport problems for arbitrary chain length involving different initial and boundary conditions. This work provides a first set of closed-form solutions to series of advective-dispersive, sequentially-coupled, reactive transport problems of arbitrary chain length. They adopt a flexible Bateman-type boundary condition that includes a generic exponentially-decaying pulse source of either Dirichlet or Cauchy type. A spatially exponentially-varying initial condition is also incorporated in the solution, and all these resulting in a compact, closed-form expression that is suitable for implementing in a computational platform. Their solution framework is based on that of Quezada et al. (2004), but unlike Quezada et al. (2004), Srinivasan and Clement (2008a, 2008b) use a fully analytical method for computing the Laplace and linear transforms and associated inverse transforms. Details of their solution strategy can be obtained from Srinivasan and Clement (2008a, 2008b); an overview is provided below.

The governing equations solved by Srinivasan and Clement (2008a, 2008b) are

$$
\begin{aligned}
R_i\frac{\partial c_i(x,t)}{\partial t}+v\frac{\partial c_i(x,t)}{\partial x}-D_x\frac{\partial^2 c_i(x,t)}{\partial x^2} &= y_i k_{i-1}c_{i-1}(x,t)-k_i c_i(x,t)\ ;\ \forall\, i=2,3,\ldots n \\
&= -k_i c_i(x,t)\ ;\ i=1 \qquad (4.45)\\
&\ ;\ \forall\, t>0 \ \ and\ \ 0<x<\infty
\end{aligned}
$$

Note that the definition of the yield factor y_i is slightly modified such that y_i is the effective yield factor that describes the mass of a species i produced from species i-1 [MM^{-1}]. Equation (4.45) is solved for a generic exponentially-distributed initial condition given by

$$c_i(x,0) = c_i^o e^{-\mu_i x},\ 0 < x < \infty\ ;\ \forall\, i = 1, 2, \ldots n \tag{4.46}$$

where 'c_i^o' is the initial source concentration of species i [ML^{-3}] and 'μ_i' is the first-order decay parameter of the initial distribution of species i [L^{-1}]. The boundary condition at '∞' is given as

$$\frac{\partial c_i(\infty,t)}{\partial x} = 0,\ t > 0\ ;\ \forall\, i = 1, 2, \ldots n \tag{4.47}$$

Explicit solutions are provided for both Dirichlet and Cauchy inlet (source) boundary conditions. In the Dirichlet case, the boundary condition at the source is given as

$$c_i(0,t) = \begin{cases} \sum_{i_1=1}^{i} B_i^{i_1} e^{-\lambda_{i_1} t}, & 0 < t \le t_o \\ 0, & t > t_o \end{cases}\ ;\ \forall\, i = 1, 2, \ldots n \tag{4.48}$$

where '$B_i^{i_1}$' is the source boundary concentration of specie i_1 that contributes to species i [ML^{-3}] and 'λ_{i_1}' is the first-order decay of the corresponding '$B_i^{i_1}$' term [T^{-1}]. The final solution is expressed as

$$\begin{aligned} c_i(x,t) = {} & \sum_{i_1=1}^{i}\left[\left(\prod_{i_2=i_1+1}^{i} y_{i_2} k_{i_2-1}\right)\sum_{i_2=i_1}^{i}\sum_{i_3=1}^{i_1}\left\{G_1^1 + h\left(G_1^1\right)\, G_2^1\right\}\right] \\ & + \sum_{i_1=1}^{i}\left[R_{i_1} c_{i_1}^o\left(\prod_{i_2=i_1+1}^{i} y_{i_2} k_{i_2-1}\right)\sum_{i_2=i_1}^{i}\left\{G_1^2 + h\left(G_1^2\right) G_2^2\right\}\right] \\ & ;\ \forall\, i = 1, 2, \ldots n \end{aligned} \tag{4.49}$$

The G_1^1, G_2^1, G_1^2 and G_2^2 terms in Equation (4.49) are given by

$$G_1^1 = \sum_{i_4=i_1,(i_4\neq i_2,R_{i_4}\neq R_{i_2})}^{i} \frac{B_{i_1}^{i_3}\left\langle \begin{matrix} F_{i_2,i_3,0}[x,t]-u(t-t_o)e^{(-\lambda_{i_3}t_o)}F_{i_2,i_3,0}[x,(t-t_o)] \\ -F_{i_2,i_2,i_4}[x,t]+u(t-t_o)e^{(-\lambda_{i_3}t_o)}F_{i_2,i_2,i_4}[x,(t-t_o)] \end{matrix}\right\rangle}{(a_{i_2,i_4}-\lambda_{i_3})\left(\prod_{i_5=i_1,(i_5\neq i_2,R_{i_5}=R_{i_2})}^{i} -k_{i_2,i_5}\right)(-R_{i_2,i_4})\prod_{i_5=i_1,(i_5\neq i_2,i_5\neq i_4,R_{i_5}\neq R_{i_2})}^{i} -R_{i_2,i_5}(a_{i_2,i_5}-a_{i_2,i_4})}$$

$$G_2^1 = \frac{B_{i_1}^{i_3}\left\langle F_{i_2,i_3,0}[x,t]-u(t-t_o)e^{(-\lambda_{i_3}t_o)}F_{i_2,i_3,0}[x,(t-t_o)]\right\rangle}{\prod_{i_4=i_1,(i_4\neq i_2,R_{i_4}=R_{i_2})}^{i} -k_{i_2,i_4}}$$

$$G_1^2 = \sum_{i_3=i_1,(i_3\neq i_2,R_{i_3}\neq R_{i_2})}^{i} \frac{\left\langle F_{i_2,i_1,-i_2}[x,t]-e^{(-\mu_{i_1}x-a_{i_1,-i_2}t)}-F_{i_2,i_2,i_3}[x,t]+e^{(-\mu_{i_1}x-a_{i_2,i_3}t)}\right\rangle}{(a_{i_2,i_3}-a_{i_1,-i_2})R_{i_2}\left(\prod_{i_4=i_1,(i_4\neq i_2,R_{i_4}=R_{i_2})}^{i} -k_{i_2,i_4}\right)R_{i_2,i_3}\prod_{i_4=i_1,(i_4\neq i_2,i_4\neq i_3,R_{i_4}\neq R_{i_2})}^{i} -R_{i_2,i_4}(a_{i_2,i_4}-a_{i_2,i_3})}$$

$$G_2^2 = \frac{-\left\langle F_{i_2,i_1,-i_2}[x,t]-e^{(-\mu_{i_1}x-a_{i_1,-i_2}t)}\right\rangle}{R_{i_2}\prod_{i_3=i_1,(i_3\neq i_2,R_{i_3}=R_{i_2})}^{i} -k_{i_2,i_3}} \qquad (4.50)$$

The F_{i_1,i_2,i_3} terms are given by either Equation (4.25) or Equation (4.26) depending on whether ω_{i_1,i_2,i_3} is real or complex. Note that the definition of the term a_{i_2,i_3} is modified to be

$$a_{i_1,i_2} = \begin{cases} \dfrac{k_{i_1,i_2}}{R_{i_1,i_2}} \ ; \ when\ i_2 > 0 \\ \lambda_{i_1} \ ; \ when\ i_2 = 0 \\ \dfrac{-\mu_{i_1}^2 D_x - \mu_{i_1} v + k_{i_2}}{R_{i_2}} \ ; \ when\ i_2 < 0 \end{cases} \qquad (4.51)$$

For the Cauchy case, the boundary conditions are given by

$$-D_x \frac{\partial c_i(0,t)}{\partial x} + v c_i(0,t) = \begin{cases} \sum_{i_1=1}^{i} B_i^{i_1} v e^{-\lambda_{i_1} t}, \ 0 < t \le t_o \\ 0, \ t > t_o \end{cases} ; \ \forall\, i = 1, 2, \ldots n \qquad (4.52)$$

The solution for the Cauchy boundary condition is also given by Equation (4.49); however, the G_1^1, G_2^1, G_1^2 and G_2^2 terms in Equation (4.49) are modified as follows:

$$G_1^1 = \sum_{i_4=i_1,\left(i_4 \neq i_2, R_{i_4} \neq R_{i_2}\right)}^{i} \frac{B_{i_1}^{i_3} \left\langle \begin{matrix} F_{i_2,i_3,0}\left[x,t\right] - u\left(t-t_o\right) e^{\left(-\lambda_{i_3} t_o\right)} F_{i_2,i_3,0}\left[x,\left(t-t_o\right)\right] \\ -F_{i_2,i_2,i_4}\left[x,t\right] + u\left(t-t_o\right) e^{\left(-\lambda_{i_3} t_o\right)} F_{i_2,i_2,i_4}\left[x,\left(t-t_o\right)\right] \end{matrix} \right\rangle}{\left(a_{i_2,i_4} - \lambda_{i_3}\right)\left(\prod_{i_5=i_1,\left(i_5 \neq i_2, R_{i_5} = R_{i_2}\right)}^{i} -k_{i_2,i_5}\right)\left(-R_{i_2,i_4}\right) \prod_{i_5=i_1,\left(i_5 \neq i_2, i_5 \neq i_4, R_{i_5} \neq R_{i_2}\right)}^{i} -R_{i_2,i_5}\left(a_{i_2,i_5} - a_{i_2,i_4}\right)}$$

$$G_2^1 = \frac{B_{i_1}^{i_3} \left\langle F_{i_2,i_3,0}\left[x,t\right] - u\left(t-t_o\right) e^{\left(-\lambda_{i_3} t_o\right)} F_{i_2,i_3,0}\left[x,\left(t-t_o\right)\right] \right\rangle}{\prod_{i_4=i_1,\left(i_4 \neq i_2, R_{i_4} = R_{i_2}\right)}^{i} -k_{i_2,i_4}}$$

$$G_1^2 = \sum_{i_3=i_1,\left(i_3 \neq i_2, R_{i_3} \neq R_{i_2}\right)}^{i} \frac{\left\langle \left(1+\frac{\mu_{i_1} D_x}{v}\right) F_{i_2,i_1,-i_2}\left[x,t\right] - e^{\left(-\mu_{i_1} x - a_{i_1,-i_2} t\right)} - \left(1+\frac{\mu_{i_1} D_x}{v}\right) F_{i_2,i_2,i_3}\left[x,t\right] + e^{\left(-\mu_{i_1} x - a_{i_2,i_3} t\right)} \right\rangle}{\left(a_{i_2,i_3} - a_{i_1,-i_2}\right) R_{i_2} \left(\prod_{i_4=i_1,\left(i_4 \neq i_2, R_{i_4} = R_{i_2}\right)}^{i} -k_{i_2,i_4}\right) R_{i_2,i_3} \prod_{i_4=i_1,\left(i_4 \neq i_2, i_4 \neq i_3, R_{i_4} \neq R_{i_2}\right)}^{i} -R_{i_2,i_4}\left(a_{i_2,i_4} - a_{i_2,i_3}\right)} \tag{4.53}$$

$$G_2^2 = \frac{-\left\langle \left(1+\frac{\mu_{i_1} D_x}{v}\right) F_{i_2,i_1,-i_2}\left[x,t\right] - e^{\left(-\mu_{i_1} x - a_{i_1,-i_2} t\right)} \right\rangle}{R_{i_2} \prod_{i_3=i_1,\left(i_3 \neq i_2, R_{i_3} = R_{i_2}\right)}^{i} -k_{i_2,i_3}}$$

The F_{i_1,i_2,i_3} terms in Equation (4.53) are given by

$$F_{i_1,i_2,i_3}\left[x,t\right] = v e^{-a_{i_2,i_3} t} e^{\frac{xv}{2D_x}} \left[\begin{matrix} \dfrac{e^{\frac{-x\omega_{i_1,i_2,i_3}}{2D_x}}}{\left(v+\omega_{i_1,i_2,i_3}\right)} erfc\left\{\dfrac{R_{i_1} x - \omega_{i_1,i_2,i_3} t}{2\sqrt{R_{i_1} D_x t}}\right\} + \\ \dfrac{e^{\frac{x\omega_{i_1,i_2,i_3}}{2D_x}}}{\left(v-\omega_{i_1,i_2,i_3}\right)} erfc\left\{\dfrac{R_{i_1} x + \omega_{i_1,i_2,i_3} t}{2\sqrt{R_{i_1} D_x t}}\right\} \end{matrix} \right]$$

$$+ \frac{2v^2}{\left(\omega_{i_1,i_2,i_3}{}^2 - v^2\right)} e^{\left[\frac{xv}{D_x} - \frac{k_{i_1} t}{R_{i_1}}\right]} erfc\left\{\frac{R_{i_1} x + vt}{2\sqrt{R_{i_1} D_x t}}\right\} \; ; \; when \; \frac{k_{i_1}}{R_{i_1}} \neq a_{i_2,i_3}, \; and$$

$$= e^{\frac{-k_{i_1} t}{R_{i_1}}} \left[\begin{matrix} \dfrac{1}{2} erfc\left(\dfrac{R_{i_1} x - vt}{2\sqrt{R_{i_1} D_x t}}\right) + \sqrt{\dfrac{v^2 t}{\pi R_{i_1} D_x}} e^{\frac{-\left(R_{i_1} x - vt\right)^2}{4 R_{i_1} D_x t}} \\ -\dfrac{1}{2}\left(1 + \dfrac{xv}{D_x} + \dfrac{v^2 t}{R_{i_1} D_x}\right) e^{\frac{xv}{D_x}} erfc\left(\dfrac{R_{i_1} x + vt}{2\sqrt{R_{i_1} D_x t}}\right) \end{matrix} \right] \; ; \; when \; \frac{k_{i_1}}{R_{i_1}} = a_{i_2,i_3} \tag{4.54}$$

The above solution is valid only for real values of ω_{i_1,i_2,i_3}. For problems involving complex values for ω_{i_1,i_2,i_3} in the case when $\frac{k_{i_1}}{R_{i_1}} \neq a_{i_2,i_3}$ the F_{i_1,i_2,i_3} terms are defined as

$$F_{i_1,i_2,i_3}[x,t]=\frac{2v}{\left(v^2+\omega^{*\,2}_{i_1,i_2,i_3}\right)}e^{-a_{i_2,i_3}t}e^{\frac{xv}{2D_x}}\left[\begin{array}{l}\left(Av-B\omega^*_{i_1,i_2,i_3}\right)Cos\left(\frac{x\omega^*_{i_1,i_2,i_3}}{2D_x}\right)\\-\left(A\omega^*_{i_1,i_2,i_3}+Bv\right)Sin\left(\frac{x\omega^*_{i_1,i_2,i_3}}{2D_x}\right)\end{array}\right]$$
$$+\frac{v^2}{2R_{i_1}D_x\left(\frac{k_{i_1}}{R_{i_1}}-a_{i_2,i_3}\right)}e^{\left[\frac{xv}{D_x}-\frac{k_{i_1}t}{R_{i_1}}\right]}erfc\left\{\frac{R_{i_1}x+vt}{2\sqrt{R_{i_1}D_xt}}\right\} \quad (4.55)$$

Note that when $\frac{k_{i_1}}{R_{i_1}}=a_{i_2,i_3}$ the F_{i_1,i_2,i_3} terms are unchanged.

The solution provided by Srinivasan and Clement (2008a, 2008b) is subject to the restrictions listed in Table 4.1.

Table 4.1 Summary of parameter limitations, (Srinivasan and Clement, 2008b).

$k_{i_1,i_2}\neq 0\ ;\ \forall\, i_1,i_2=1,2,\ldots n$ $where\ ;\ i_1\neq i_2$
$a_{i_1,i_2}\neq a_{i_1,i_3}\ ;\ \forall\, i_1,i_2,i_3=1,2,\ldots n$ $where\ ;\ i_1\neq i_2\ ,\ i_1\neq i_3\ ,\ i_2\neq i_3\ ,\ R_{i_1}\neq R_{i_2}\ and\ R_{i_1}\neq R_{i_3}$
$a_{i_1,i_2}\neq \lambda_{i_3}\ ;\ i\leq i_1,i_2\leq n\ ;\ 1\leq i_3\leq i\ ;\ \forall\, i=1,2,\ldots n$ $where\ ;\ i_1\neq i_2\ and\ R_{i_1}\neq R_{i_2}$
$a_{i_1,i_2}\neq a_{i,-i_1}\ ;\ i\leq i_1,i_2\leq n\ ;\ \forall\, i=1,2,\ldots n$ $where\ ;\ i_1\neq i_2\ and\ R_{i_1}\neq R_{i_2}$

The implementation of the solution presented by Srinivasan and Clement (2008a) using a computer code can be challenging. This is because round-off errors and underflow/overflow errors that occur during the computations of the 'G', 'F_{i_1,i_2,i_3}' and the complementary error function (Erfc) terms can cause severe stability problems even for short chain lengths involving four species (van Genuchten 1985). These computational errors are further aggravated when solving problems involving the Cauchy boundary condition for long chain lengths and/or large simulation times. Srinivasan and Clement (2008b) provide recommendations for improving the stability of the computer code. Typically, the complementary error function is evaluated by approximating it either as a closed-form analytical expression (Gautschi 1964) or as an infinite series (Gautschi 1970). van Genuchten (1985) suggests that this round-off error can be controlled by first substituting the approximate expressions for the Erfc terms into the 'F_{i_1,i_2,i_3}' terms, and then combining and simplifying the results. Furthermore, using a more accurate approximation of the complementary error function, such as Cody's (1969) Chebyshev approximation, can help reduce round-off errors. Alternatively, Srinivasan and Clement (2008b) propose a log-based formulation that addresses the twin problems of round-off and underflow/overflow errors, especially in transport situations involving long chain lengths and large simulation times. The log-based formulation involves transforming the arguments whose products have to be evaluated into log space. Within log space the products are evaluated by performing a set of additions. Finally, the log inverse of the evaluated product is obtained. By employing the proposed log-based formulation, the underflow and/or overflow errors related to evaluating the product terms can be virtually eliminated. However, the log-based formulation can only be applied to compute product terms and cannot be used to compute summation terms. To address the problem of overflow errors while evaluating summation terms, one can make use of the symmetrical property of the 'F_{i_1,i_2,i_3}' terms (Srinivasan and Clement 2008b). This involves combining the 'F_{i_1,i_2,i_3}' terms that have identical 'i_2' and 'i_3' values and then evaluating the sum of these combination terms separately and then substituting them in the main solution (van Genuchten 1985).

Finally, Srinivasan and Clement (2008b) demonstrate that their solution to the sequential decay chain problem can be readily extended to a general diverging reaction network. A similar strategy was described by Sun et al. (1999b). By conceptualizing a diverging network as a superposition of a series of parallel reaction networks, one can split the network into a set of parallel sequential reaction networks each of which can be solved independently. It must be noted that one diverging species will have no effect on the other diverging species and hence they can be considered independent of each other. Furthermore, the authors deduce several closed-form solutions to simpler problems that are summarized below:

Zero Initial Condition

The general solution given by Equation (4.49) can be readily simplified to solve transport scenarios where the initial concentrations of all contaminants are zero. This is done by substituting the value of 'c^o' as zero. For this case, the general solution simplifies to

$$c_i(x,t) = \sum_{i_1=1}^{i} \left[\left(\prod_{i_2=i_1+1}^{i} y_{i_2} k_{i_2-1} \right) \sum_{i_2=i_1}^{i} \sum_{i_3=1}^{i_1} \left\{ G_1^1 + h\left(G_1^1\right) G_2^1 \right\} \right] \tag{4.56}$$

$$; \forall\, i = 1, 2, \ldots n$$

Note that the 'G_1^1' and 'G_2^1' terms in Equation (4.56), defined in Equation (4.53), remain unchanged. It can be shown that the expressions for the first four species of this special case solution match the solutions given by van Genuchten (1985). Furthermore, from Equation (4.56) one can observe that, for the zero initial condition case, the second term of the general solution given by Equation (4.49) is absent. This not only relaxes some of the restrictions on the transport parameter values but also directly helps in improving the round-off errors, especially during large simulation times.

Identical Retardation Factors

Special cases arise when the retardation factors of all species in the transport problem are identical (e.g., the transport of non-sorbing, sequentially-decaying contaminants for which the retardation factor for all species is unity). One can obtain the solution for this problem from the general solution by simple substitution. The modified solution for this special case problem for both types of boundary conditions is given as

$$
\begin{aligned}
c_i(x,t) = & \sum_{i_1=1}^{i}\left[\left(\prod_{i_2=i_1+1}^{i} y_{i_2} k_{i_2-1}\right)\sum_{i_2=i_1}^{i}\sum_{i_3=1}^{i_1}\left\{G_2^1\right\}\right] \\
& + \sum_{i_1=1}^{i}\left[R_{i_1} c_{i_1}^{o}\left(\prod_{i_2=i_1+1}^{i} y_{i_2} k_{i_2-1}\right)\sum_{i_2=i_1}^{i}\left\{G_2^2\right\}\right] \qquad (4.58) \\
& ;\ \forall\, i = 1, 2, \ldots n
\end{aligned}
$$

Note that the 'G_2^1' and 'G_2^2' terms defined earlier remain unchanged. The exclusion of the 'G_1^1' and 'G_1^2' terms helps in obtaining a computationally stable solution for longer chain lengths. Additionally, this solution imposes fewer restrictions on transport parameter values.

Zero Advection Velocity

In some situations the transport of contaminants is governed by hydrodynamic dispersion. The solution for this condition is identical to the general solution given by Equation (4.49), except that the 'F_{i_1,i_2,i_3}' terms are modified. For the Dirichlet boundary condition, the 'F_{i_1,i_2,i_3}' term becomes

$$F_{i_1,i_2,i_3}[x,t] = \frac{e^{-a_{i_2,i_3}t}}{2}\left[\begin{array}{l} e^{\frac{-x\omega_{i_1,i_2,i_3}}{2D_x}} erfc\left\{\dfrac{R_{i_1}x - \omega_{i_1,i_2,i_3}t}{2\sqrt{R_{i_1}D_x t}}\right\} \\ +e^{\frac{x\omega_{i_1,i_2,i_3}}{2D_x}} erfc\left\{\dfrac{R_{i_1}x - \omega_{i_1,i_2,i_3}t}{2\sqrt{R_{i_1}D_x t}}\right\} \end{array}\right] \tag{4.59}$$

$$where \; ; \; \omega_{i_1,i_2,i_3} = \sqrt{4R_{i_1}D_x\left(\frac{k_{i_1}}{R_{i_1}} - a_{i_2,i_3}\right)}$$

The above expression is valid only for real values of 'ω_{i_1,i_2,i_3}'. If 'ω_{i_1,i_2,i_3}' is complex, the 'F_{i_1,i_2,i_3}' term becomes

$$F_{i_1,i_2,i_3}[x,t] = e^{-a_{i_2,i_3}t}\left[ACos\left(\frac{x\omega^*_{i_1,i_2,i_3}}{2D_x}\right) - BSin\left(\frac{x\omega^*_{i_1,i_2,i_3}}{2D_x}\right)\right] \tag{4.60}$$

$$where \; ; \; \omega^*_{i_1,i_2,i_3} = \sqrt{\left|4R_{i_1}D_x\left(\frac{k_{i_1}}{R_{i_1}} - a_{i_2,i_3}\right)\right|} \quad and \; ; \; (A+iB) = erfc\left\{\frac{R_{i_1}x + i\omega^*_{i_1,i_2,i_3}t}{2\sqrt{R_{i_1}D_x t}}\right\}$$

For the case of the Cauchy boundary condition, forcing the advection velocity to be zero would result in a zero flux boundary condition. To avoid this, it is assumed that the boundary condition at the source does not involve the 'v' term so that the modified Cauchy source boundary condition becomes

$$-D_x\frac{\partial c_i(0,t)}{\partial x} = \begin{cases}\sum_{i_1=1}^{i} B_i^{i_1} e^{-\lambda_{i_1}t}, \; 0 < t \le t_o \\ 0, \; t > t_o\end{cases} \; ; \; \forall\, i = 1, 2, \ldots n \tag{4.61}$$

Under this condition, the 'F_{i_1,i_2,i_3}' term is modified as

$$F_{i_1,i_2,i_3}[x,t] = e^{-a_{i_2,i_3}t}\left[\begin{array}{l}\dfrac{e^{\frac{-x\omega_{i_1,i_2,i_3}}{2D_x}}}{\omega_{i_1,i_2,i_3}} erfc\left\{\dfrac{R_{i_1}x - \omega_{i_1,i_2,i_3}t}{2\sqrt{R_{i_1}D_x t}}\right\} + \\ \dfrac{e^{\frac{x\omega_{i_1,i_2,i_3}}{2D_x}}}{-\omega_{i_1,i_2,i_3}} erfc\left\{\dfrac{R_{i_1}x + \omega_{i_1,i_2,i_3}t}{2\sqrt{R_{i_1}D_x t}}\right\}\end{array}\right] ;\ when\ \frac{k_{i_1}}{R_{i_1}} \neq a_{i_2,i_3},\ and$$

$$= e^{\frac{-k_{i_1}t}{R_{i_1}}}\left[\sqrt{\frac{t}{\pi R_{i_1}D_x}}\, e^{\frac{-(R_{i_1}x - vt)^2}{4R_{i_1}D_x t}} - \frac{x}{2D_x} erfc\left(\frac{R_{i_1}x}{2\sqrt{R_{i_1}D_x t}}\right)\right] ;\ when\ \frac{k_{i_1}}{R_{i_1}} = a_{i_2,i_3} \tag{4.62}$$

$$where\ ;\ \omega_{i_1,i_2,i_3} = \sqrt{4R_{i_1}D_x\left(\frac{k_{i_1}}{R_{i_1}} - a_{i_2,i_3}\right)}$$

The above expression is valid only for real values of 'ω_{i_1,i_2,i_3}'. If 'ω_{i_1,i_2,i_3}' is complex, the 'F_{i_1,i_2,i_3}' term is modified as:

$$F_{i_1,i_2,i_3}[x,t] = \frac{2}{\omega^{*\,2}_{i_1,i_2,i_3}} e^{-a_{i_2,i_3}t} e^{\frac{xv}{2D_x}}\left[\begin{array}{l} -B\omega^*_{i_1,i_2,i_3} Cos\left(\dfrac{x\omega^*_{i_1,i_2,i_3}}{2D_x}\right) \\ -A\omega^*_{i_1,i_2,i_3} Sin\left(\dfrac{x\omega^*_{i_1,i_2,i_3}}{2D_x}\right)\end{array}\right] \tag{4.63}$$

$$where\ ;\ \omega^*_{i_1,i_2,i_3} = \sqrt{\left|4R_{i_1}D_x\left(\frac{k_{i_1}}{R_{i_1}} - a_{i_2,i_3}\right)\right|}\ \ and\ (A + iB) = erfc\left\{\frac{R_{i_1}x + i\omega^*_{i_1,i_2,i_3}t}{2\sqrt{R_{i_1}D_x t}}\right\}$$

The solutions obtained for the zero advection velocity case are specialized solutions that can be applied directly to model reactive transport scenarios involving chemical and nuclear repository leakage through molecular diffusion.

Steady State

The behavior of a contaminant plume under steady-state conditions is of special interest, especially in analyzing monitored natural attenuation (MNA) problems. Steady-state solutions avoid problems related to overflow errors that occur when applying the general solution for large simulation times. Although this problem can be tackled by using certain algebraic manipulation techniques, it can be a tedious effort. Furthermore, numerical codes, such as RT3D (Clement et al. 1998), and semi-analytical solutions, such as those proposed by Quezada et al. (2004), require large amounts of computational resources to accurately model steady-state solutions. For these reasons, it is very attractive to obtain explicit solutions for steady-state conditions. It must be noted that unlike previous special cases, it is not possible to deduce the steady-state solution by direct substitution. Therefore, the steady-state solution is derived by solving the steady-state governing equations assuming a constant source for the Dirichlet and Cauchy boundaries conditions. Because the time term is absent in the governing equations, Laplace transform techniques are not involved, and one can directly use the linear transform method given by Clement (2001) to uncouple the system of equations. The steady-state solution for the Dirichlet boundary is obtained as

$$c_i(x)=\sum_{i_1=1}^{i}\left[\left(\prod_{i_2=i_1+1}^{i} y_{i_2}k_{i_2-1}\right)\left(\sum_{i_2=1}^{i_1} B_{i_1}^{i_2}\right)\sum_{i_2=i_1}^{i}\frac{e^{\left[\frac{x}{2D_x}\left\{v-\sqrt{v^2+4D_xk_{i_2}}\right\}\right]}}{\prod_{i_3=i_1,(i_3\neq i_2)}^{i}-\left(k_{i_2}-k_{i_3}\right)}\right] \quad ;\ \forall\, i=1,2,\ldots n \tag{4.64}$$

The corresponding steady-state solution for the Cauchy boundary becomes

$$c_i(x)=\sum_{i_1=1}^{i}\left[\left(\prod_{i_2=i_1+1}^{i} y_{i_2}k_{i_2-1}\right)\left(\sum_{i_2=1}^{i_1} B_{i_1}^{i_2}\right)\sum_{i_2=i_1}^{i}\frac{ve^{\left[\frac{x}{2D_x}\left\{v-\sqrt{v^2+4D_xk_{i_2}}\right\}\right]}}{\left\{\frac{v}{2}+\frac{1}{2}\sqrt{v^2+4D_xk_{i_2}}\right\}\prod_{i_3=i_1,(i_3\neq i_2)}^{i}-\left(k_{i_2}-k_{i_3}\right)}\right] \quad ;\ \forall\, i=1,2,\ldots n \tag{4.65}$$

From Equations (4.64) and (4.65) it can be observed that steady-state solutions are less complex compared to the generic transient solutions as they do not involve the Erfc terms.

Zero Dispersion Coefficient

If we ignore the effects of dispersion, the governing transport simplifies to

$$\begin{aligned} R_i\frac{\partial c_i(x,t)}{\partial t}+v\frac{\partial c_i(x,t)}{\partial x} &= y_{i-1}k_{i-1}c_{i-1}(x,t)-k_ic_i(x,t)\ ;\ \forall\, i=2,3,\ldots n \\ &= -k_ic_i(x,t)\ ;\ i=1 \\ &;\ \forall\, t>0 \ \textit{and}\ 0<x<\infty \end{aligned} \tag{4.66}$$

Note that in the absence of dispersion, the Cauchy boundary condition will be identical to the Dirichlet boundary condition. The above problem is solved using the generic exponential initial condition given by Equation (4.46). The solution for the zero dispersion condition can also be represented by Equation (4.49). However, the 'G' terms associated with this solution are redefined as

$$G_1^1 = \sum_{i_4=i_1,(i_4 \neq i_2, R_{i_4} \neq R_{i_2})}^{i} \frac{B_{i_1}^{i_3} \left\langle \begin{matrix} F_{i_2,i_3,0}[x,t] - e^{-t_o \lambda_{i_3}} F_{i_1,i_2,i_3}[x,(t-t_o)] \\ -F_{i_2,i_2,i_4}[x,t] + e^{-t_o \lambda_{i_3}} F_{i_2,i_2,i_4}[x,(t-t_o)] \end{matrix} \right\rangle}{(a_{i_2,i_4} - \lambda_{i_3}) \left(\prod_{i_5=i_1,(i_5 \neq i_2, R_{i_5} = R_{i_2})}^{i} -k_{i_2,i_5} \right) (-R_{i_2,i_4}) \prod_{i_5=i_1,(i_5 \neq i_2, i_5 \neq i_4, R_{i_5} \neq R_{i_2})}^{i} -R_{i_2,i_5} (a_{i_2,i_5} - a_{i_2,i_4})}$$

$$G_2^1 = \frac{B_{i_1}^{i_3} \left\langle F_{i_2,i_3,0}[x,t] - e^{-t_o \lambda_{i_3}} F_{i_1,i_2,i_3}[x,(t-t_o)] \right\rangle}{\prod_{i_4=i_1,(i_4 \neq i_2, R_{i_4} = R_{i_2})}^{i} -k_{i_2,i_4}}$$

$$G_1^2 = \sum_{i_3=i_1,(i_3 \neq i_2, R_{i_3} \neq R_{i_2})}^{i} \frac{\left\langle F_{i_2,i_1,-i_2}[x,t] - e^{(-\mu_{i_1} x - a_{i_1,-i_2} t)} - F_{i_2,i_2,i_3}[x,t] + e^{(-\mu_{i_1} x - a_{i_2,i_3} t)} \right\rangle}{(a_{i_2,i_3} - a_{i_1,-i_2}) R_{i_2} \left(\prod_{i_4=i_1,(i_4 \neq i_2, R_{i_4} = R_{i_2})}^{i} -k_{i_2,i_4} \right) R_{i_2,i_3} \prod_{i_4=i_1,(i_4 \neq i_2, i_4 \neq i_3, R_{i_4} \neq R_{i_2})}^{i} -R_{i_2,i_4} (a_{i_2,i_4} - a_{i_2,i_3})}$$

$$G_2^2 = \frac{-\left\langle F_{i_2,i_1,-i_2}[x,t] - e^{(-\mu_{i_1} x - a_{i_1,-i_2} t)} \right\rangle}{R_{i_2} \prod_{i_3=i_1,(i_3 \neq i_2, R_{i_3} = R_{i_2})}^{i} -k_{i_2,i_3}} \tag{4.67}$$

where

$$F_{i_1,i_2,i_3}[x,t] = u\left(t - \frac{R_{i_1} x}{v}\right) e^{-a_{i_2,i_3}\left(t - \frac{R_{i_1} x}{v}\right)} \tag{4.68}$$

It can be shown that for a zero initial condition the solutions derived in this section are identical to those published by Harada et al. (1980) and Higashi and Pigford (1980).

4.3.8 Other Special Case Problems and Alternative Solution Methodologies

Gureghian and Jansen (1985) presented an analytical solution to a transport problem involving a three-member, first-order decay chain in a multi-layered system, subject to advection, dispersion and linear equilibrium sorption processes for both continuous and band source release conditions. Gureghian and Jansen (1985) apply a combination of convolution theorems and Laplace transform techniques to solve their problem. Semi-analytical solutions are presented for the case involving both advective and dispersive transport, and explicit closed-form analytical solutions are presented for the case involving non-dispersive transport. This work is one of the few that provide solutions to coupled reactive transport problems in a heterogeneous porous medium. However, the generality of the model is greatly reduced due to the inherent heterogeneity and hence restricted to short chain lengths.

Angelakis et al. (1987) developed a semi-analytical solution to a sequentially-coupled, two-species transport problem governed by advection, dispersion and linear equilibrium sorption subject to Dirichlet boundary conditions. The transport problem assumes that reactions are first-order with each species having unique dispersion coefficients and distinct retardation factors. The authors also demonstrate that when the dispersion coefficients of both species are equal, their solution reduces to the closed-form solution similar to those presented by Cho (1971). Furthermore, the authors also provide solutions for the no dispersion (pure advection) case. Angelakis et al. (1993) developed an interesting semi-analytical solution for a problem involving the coupled transport of two solutes and a gaseous product in soils. The solute migration is governed by advection, dispersion, linear equilibrium sorption and sequential first-order reaction, whereas the gas migration is governed by diffusive transport coupled with reversible, linear equilibrium dissolution. These solutions represent a special case of problems involving short chain lengths with unique input parameter sets. Although very useful in the particular context, the applicability of these solutions is limited due to the loss of generality.

Khandelwal and Rabideau (1999) developed semi-analytical solutions for a three-species, sequentially-coupled, first-order reactive transport problem. The key contribution of this work is the consideration of non-equilibrium sorption mechanisms. However, their solutions are semi-analytical solutions and are limited to three species.

Eykholt and Li (2000) developed a solution method based on kinetic response functions to solve a linearly-coupled, non-sequential reactive transport problem having different retardation factors. Although, there is no restriction on the number of species in the system, this method requires numerical procedures to evaluate the final solution. Furthermore, for the case of the non-ideal plug flow scenarios (i.e., advective-dispersive transport), the accuracy of this method appears to decrease with decreasing Peclet number.

4.4. Multispecies Transport in Multi-dimensional Systems

Multi-dimensional reactive transport equations have greater applicability compared to one-dimensional reactive transport equations; hence, they appear more frequently in the literature. In this section, we review solutions to some special cases of multi-species, multi-dimensional, coupled reactive transport problems.

4.4.1 Pure Advection Problems (Zero Dispersion)

Srinivasan et al. (2007) derived a three-dimensional, closed-form analytical solution that describes the fate and transport of a decaying contaminant plume evolving from a finite patch source for a single species. The solution for this problem is given by Equation (4.9). Note that the $f_y(x,y)$ and $f_z(x,z)$ terms in Equation (4.9) do not involve the time term. Hence, extending this single-species solution to a coupled, multi-species solution using the approach given by Quezada et al. (2004) is straightforward. This is because the transverse dispersal terms $f_y(x,y)$ and $f_z(x,z)$ do not participate in the Laplace transform process. It can be shown that in the absence of initial contaminants (zero initial conditions) the solution to the three-dimensional, coupled reactive transport problem can be obtained by multiplying the solution to the one-dimensional, coupled reactive transport problem with the transverse dispersal terms and a factor of ¼. The governing transport equation can be expressed as

$$R_i \frac{\partial c_i}{\partial t} + v\frac{\partial c_i}{\partial x} - D_y \frac{\partial^2 c_i}{\partial y^2} - D_z \frac{\partial^2 c_i}{\partial z^2} = y_{i-1}k_{i-1}c_{i-1} - k_i c_i \ ; \ \forall\, i = 2,3,\ldots n$$
$$= -k_i c_i \ ; \ i = 1 \qquad (4.69)$$

with the corresponding initial and boundary conditions given by

$$\begin{aligned}
&c_i(x,y,z,0) = 0, \ ; \ \forall\, i = 1,2,\ldots n\\
&\frac{\partial c_i(x,\infty,z,t)}{\partial y} = 0, \ t>0 \ ; \ \forall\, i = 1,2,\ldots n\\
&\frac{\partial c_i(x,y,\infty,t)}{\partial z} = 0, \ t>0 \ ; \ \forall\, i = 1,2,\ldots n\\
&c_i(0,y,z,t) = \begin{cases} \sum_{i_1=1}^{i} B_i^{i_1} e^{-\lambda_{i_1} t}, & 0 < t \le t_o \\ 0, & t > t_o \end{cases} \ ; \ \forall\, i = 1,2,\ldots n \ ; \ -\frac{Y}{2} < y < \frac{Y}{2} \ ; \ -\frac{Z}{2} < z < \frac{Z}{2}
\end{aligned} \qquad (4.70)$$

The solution to the one-dimensional version of the above problem is given by Equation (4.56) where the 'G_1^1' and 'G_2^1' terms are given in Equation (4.67) and the 'F_{i_1,i_2,i_3}' term in Equation (4.67) is given by Equation (4.68). The solution to the three-dimensional version of the problem can be readily obtained by multiplying Equation (4.56) with the transverse dispersal terms $f_y(x,y)$ and $f_z(x,z)$ and a factor of ¼. This is expressed as

$$c_i(x,y,z,t) = \frac{1}{4} f_y(x,y) f_z(x,z) \sum_{i_1=1}^{i} \left[\left(\prod_{i_2=i_1+1}^{i} y_{i_2} k_{i_2-1} \right) \sum_{i_2=i_1}^{i} \sum_{i_3=1}^{i_1} \left\{ G_1^1 + h\left(G_1^1\right) G_2^1 \right\} \right] \qquad (4.71)$$
$$; \ \forall\, i = 1,2,\ldots n$$

where the terms $f_y(x,y)$ and $f_z(x,z)$ are given in Equation (4.8). Note that this is an exact closed-form analytical solution to the above transport problem.

4.4.2 Short Chain Advection-Dispersion Problems

Bauer et al. (2001) presented a method to solve one-, two-, and three-dimensional, sequentially-coupled reactive transport problems with distinct retardation factors. This method is based on transforming the system of equations to a Laplace domain and then obtaining a set of fundamental solutions to each of the equations in the transformed domain. The specific solutions in the Laplace domain can then be obtained through a linear combination of the fundamental solutions, provided that the fundamental solutions are linearly independent. Finally, the Laplace domain solutions are transformed back to the time domain using the inverse Laplace transform procedure, which can be accomplished either analytically or numerically. Although this method can be applied to solve different types of boundary conditions, the solution procedure is mathematically tedious; obtaining specific analytical inverse transform expressions for long chain lengths can be a challenge.

Montas (2003) developed an analytical procedure to solve a three-species, multi-dimensional transport problem coupled by a first-order, non-sequential reaction network subject to a pulse-type boundary condition. This procedure involves obtaining a basis solution of a convoluted form for the transport equation and then evaluating the basis solution using Laplace transforms. One of the advantages of this procedure is that it can model transport problems with distinct retardation factors. However, as mentioned earlier, this solution is limited to a three-species system.

4.4.3 Long Chain Advection-Dispersion Problems

It must be noted that the solution strategy provided by Quezada et al. (2004) is generic and can be used to solve long-chain, multi-dimensional problems..However, one of the main challenges in multi-dimensional reactive transport modeling is the absence of closed-form analytical solutions even for the single-species transport problem (see Section 4.2). Hence, the analytical solutions in the uncoupled, doubly-transformed 'p' domain would be of non-closed form that one has to evaluate numerically. Furthermore, inverting the solution from the 'p' domain back to the 'c' domain would involve further numerical procedures. These methods make the use of the solution cumbersome and unattractive. Quezada (2004) has provided detailed numerical techniques and algorithms to address the problems associated with the numerical inversion of the solutions.

To tackle this problem, we make use of the approximate closed-form expression provided by Domenico and Robbins (1985). It has been shown that the Domenico solution has minimal errors and performs reasonably well when applied to transport problems involving small longitudinal dispersivities, high advection velocities, and large simulation times (Srinivasan et al. 2007). Using the approach developed by Quezada (2004), described in Section 4.4.1, one can obtain an elegant approximation to multi-species, three-dimensional transport problems by multiplying the solutions of the one-dimensional, multi-species transport problem with the transverse dispersal terms $f_y(x,y)$ and $f_z(x,z)$ of the Domenico solution given in Equation (4.8). Note that this method is valid for problems having zero initial conditions only. The governing equation for the problem considered can be expressed as

$$R_i \frac{\partial c_i}{\partial t} + v\frac{\partial c_i}{\partial x} - D_x \frac{\partial^2 c_i}{\partial x^2} - D_y \frac{\partial^2 c_i}{\partial y^2} - D_y \frac{\partial^2 c_i}{\partial y^2} = y_i k_{i-1} c_{i-1} - k_i c_i \;;\; \forall\, i = 2,3,\ldots n \qquad (4.72)$$

$$= -k_i c_i \;;\; i = 1$$

The corresponding initial and boundary conditions are given by

$$
\begin{aligned}
& c_i(x,y,z,0)=0, \; ; \; \forall\, i=1,2,\ldots n \\
& \frac{\partial c_i(\infty,y,z,t)}{\partial x}=0, \; t>0 \; ; \; \forall\, i=1,2,\ldots n \\
& \frac{\partial c_i(x,\infty,z,t)}{\partial y}=0, \; t>0 \; ; \; \forall\, i=1,2,\ldots n \\
& \frac{\partial c_i(x,y,\infty,t)}{\partial z}=0, \; t>0 \; ; \; \forall\, i=1,2,\ldots n \\
& c_i(0,y,z,t)=\begin{cases} \sum_{i_1=1}^{i} B_i^{i_1} e^{-\lambda_{i_1} t}, & 0<t\le t_o \\ 0, & t>t_o \end{cases} \; ; \; \forall\, i=1,2,\ldots n \; ; \; -\frac{Y}{2}<y<\frac{Y}{2} \; ; \; -\frac{Z}{2}<z<\frac{Z}{2}
\end{aligned}
\tag{4.73}
$$

The exact solution to the one dimensional version of the above problem is given by Equation (4.56) where the 'G_1^1' and 'G_2^1' terms are given in Equation (4.53) and the 'F_{i_1,i_2,i_3}' term in Equation (4.53) is given by Equation (4.54) or Equation (4.55). The approximate closed-form solution expression to the sequentially-coupled, three-dimensional reactive transport problem can also be expressed by Equation (4.71). However, we need to substitute the correct expressions for the 'G_1^1', 'G_2^1' and 'F_{i_1,i_2,i_3}' terms as described in this section. It must be reiterated that this method does not yield provide an exact solution but provides an approximate closed-form analytical expression to the problem.

4.5 Conclusions

This report provides a review of analytical solutions to multi-species, reactively-coupled transport problems. Detailed reviews for single species reactive transport problems are presented by van Genuchten and Alves (1982) and Wexler (1992); readers interested in these solutions are encouraged to consult these original manuscripts. The solutions presented in this chapter can be used as powerful tools for predicting the fate and transport of reactive contaminants. Of particular interest is the application of these solutions to develop screening level tools such as BIOCHLOR (Aziz et al. 2000) and BIOSCREEN (Newell et al. 1996). Analytical models offer a cost-effective alternative for rapidly simulating various types of groundwater contamination problems.

4.6 References

Angelakis, A. N., Kadir, T. N., and Rolston, D. E. (1987). "Solutions for transport of 2 sorbed solutes with differing dispersion coefficients in soil." *Soil Science Society of America Journal*, 51, 1428-1434.

Angelakis, A. N., Kadir, T. N., and Rolston, D. E. (1993). "Analytical solutions for equations describing coupled transport of two solutes and a gaseous product in soil." *Water Resources Research*, 29(4), 945-956.

Aziz, C. E., Newell, C. J., Gonzales, J. R., Hass, P., Clement, T. P., and Sun, Y. (2000). BIOCHLOR: Natural attenuation decision support system, User's manual, Version 1.0, U. S. Environmental Protection Agency, EPA/600/R-00/008.

Bauer, P., Attinger, S., and Kinzelbach, W. (2001). "Transport of a decay chain in homogenous porous media: analytical solutions." *Journal of Contaminant Hydrology*, 49(3-4), 217-239.

Brenner, H. (1962). "The diffusion model of longitudinal mixing in beds of finite length. Numerical values." *Chemical Engineering Science*, 17, 229-243.

Burkholder, H. C., and Rosinger, E. L. J. (1980). "A model for the transport of radionuclides and their decay products through geologic media." *Nuclear Technology*, 49(1), 150-158.

Cho, C. M. (1971). "Convective transport of ammonium with nitrification in soil." *Canadian Journal of Soil Science*, 51, 339-350.

Cleary, R., and Ungs, M. J. (1978). Analytical models for groundwater pollution and hydrology, Report No. 78-WR-15. Water Resources Program, Princeton University, Princeton.

Clement, T. P. (2001). "Generalized solution to multispecies transport equations coupled with a first-order reaction network." *Water Resources Research*, 37(1), 157-163.

Clement, T. P., Johnson, C. D., Sun, Y. W., Klecka, G. M., and Bartlett, C. (2000). "Natural attenuation of chlorinated ethene compounds: Model development and field-scale application at the Dover site." *Journal of Contaminant Hydrology*, 42(2-4), 113-140.

Clement, T. P., Sun, Y., Hooker, B. S., and Petersen, J. N. (1998). "Modeling multispecies reactive transport in ground water." *Ground Water Monitoring and Remediation*, 18(2), 79-92.

Clement, T. P., Truex, M. J., and Lee, P. (2002). "A case study for demonstrating the application of U.S. EPA's monitored natural attenuation screening protocol at a hazardous waste site." *Journal of Contaminant Hydrology*, 59(1-2), 133-162.

Cody, W. J. (1969). "Rational Chebyshev approximations for the error function." *Mathematics of Computation*, 23, 631-638.

Domenico, P. A., and Robbins, G. A. (1985). "A new method of contaminant plume analysis." *Ground Water*, 23(4), 476-485.

Eykholt, G. R., and Li, L. (2000). "Fate and transport of species in a linear reaction network with different retardation coefficients." *Journal of Contaminant Hydrology*, 46(1-2), 163-185.

Gautschi, W. (1964). Error function and fresnel integrals, U. S. Government Printing Office, Washington D.C.

Gautschi, W. (1970). "Efficient computation of the complex error function." *SIAM Journal on Numerical Analysis*, 7, 187-198.

Gureghian, A. B., and Jansen, G. (1985). "One-dimensional analytical solutions for the migration of a three-member radionuclide decay chain in a multilayered geologic medium." *Water Resources Research*, 21(5), 733-742.

Harada, M., Chambre, P. L., Foglia, M., Higashi, K., Iwamoto, F., Leung, D., Pigford, T. H., and Ting, D. (1980). Migrations of radionuclides through sorbing media, analytical solutions – 1, Report No. LBL-10500 (UC-70), Lawrence Berkeley Laboratory, University of California, Berkeley.

Higashi, K., and Pigford, T. H. (1980). "Analytical models for migration of radionuclides in geologic sorbing media." *Journal of Nuclear Science and Technology*, 17, 700-709.

Hunt, B. W. (1973). "Dispersion from pit in uniform seepage." *Journal of the Hydraulics Division, ASCE*, 99(2), 13-21.

Khandelwal, A., and Rabideau, A. J. (1999). "Transport of sequentially decaying reaction products influenced by linear nonequilibrium sorption." *Water Resources Research*, 35(6), 1939-1945.

Lapidus, L., and Amundson, N. R. (1952). "Mathematics of adsorption in beds. VI. The effect of longitudinal diffusion in ion exchange and chromatographic columns." *Journal of Physical Chemistry*, 56, 984-988.

Lester, D. H., Jansen, G., and Burkholder, H. C. (1975). "Migration of radionuclide chains through an adsorbing medium." *Adsorption and Ion Exchange, AICHE Symposium Series 152,* American Institute of Chemical Engineers, New York, 71.

Lindstrom, F. T., and Narasimham, M. N. L. (1973). "Mathematical theory of a kinetic model for dispersion of previously distributed chemicals in a sorbing medium." *SIAM Journal on Applied Mathematics*, 24, 496-510.

Lunn, M., Lunn, R. J., and Mackay, R. (1996). "Determining analytic solutions of multiple species contaminant transport, with sorption and decay." *Journal of Hydrology*, 180(1-4), 195-210.

Martyn-Hayden, J., and Robbins, G. A. (1997). "Plume distortion and apparent attenuation due to concentration averaging in monitoring wells." *Ground Water*, 35(2), 339-346.

McLaren, A. D. (1969). "Nitrification in soil - Systems approaching a steady state." *Soil Science Society of America Proceedings*, 33, 551-&.

McLaren, A. D. (1970). "Temporal and vectorial reactions of nitrogen in soil - A review." *Canadian Journal of Soil Science*, 50, 97-&.

Misra, C., Nielsen, D. R., and Biggar, J. W. (1974). "Nitrogen transformations in soil during leaching: 1. Theoretical considerations." *Soil Science Society of America Journal*, 38, 289-293.

Montas, H. J. (2003). "An analytical solution of the three-component transport equation with application to third-order transport." *Water Resources Research*, 39(2).

Newell, C. J., McLeod, R. K., and Gonzales, J. R. (1996). BIOSCREEN: Natural attenuation decision support system, User's manual, Version 1.3, U. S. Environmental Protection Agency, EPA/600/R-96/087.

Ogata, A., and Banks, R. B. (1961). A solution of the differential equation of longitudinal dispersion in porous media, U. S. Geological Survey, Professional Paper 411-A.

Quezada, C. R. (2004). Generalized solution to multi-species transport equations coupled with a first-order reaction network with distinct retardation factors, Auburn University, Auburn, Alabama.

Quezada, C. R., Clement, T. P., and Lee, K. K. (2004). "Generalized solution to multi-dimensional multi-species transport equations coupled with a first-order reaction network involving distinct retardation factors." *Advances in Water Resources*, 27(5), 507-520.

Srinivasan, V., Clement, T. P., and Lee, K. K. (2007). "Domenico solution - Is it valid?" *Ground Water*, 45(2), 136-146.

Srinivasan, V., and Clement, T. P. (2008a). "Analytical solutions for sequentially coupled one-dimensional reactive transport problems - Part I: Mathematical derivations." *Advances in Water Resources*, 31(2), 203-218.

Srinivasan, V., and Clement, T. P. (2008b). "Analytical solutions for sequentially coupled one-dimensional reactive transport problems - Part II: Special cases, implementation and testing." *Advances in Water Resources*, 31(2), 219-232.

Sun, Y., Petersen, J. N., and Clement, T. P. (1999a). "Analytical solutions for multiple species reactive transport in multiple dimensions." *Journal of Contaminant Hydrology*, 35(4), 429-440.

Sun, Y., Petersen, J. N., Clement, T. P., and Skeen, R. S. (1999b). "Development of analytical solutions for multispecies transport with serial and parallel reactions." *Water Resources Research*, 35(1), 185-190.

van Genuchten, M. Th. (1985). "Convective-dispersive transport of solutes involved in sequential 1st-order decay reactions." *Computers and Geosciences*, 11(2), 129-147.

van Genuchten, M. Th., and Alves, W. J. (1982). Analytical solutions of the one-dimensional convective-dispersive solute transport equation, U. S. Department of Agriculture, Technical Bulletin No. 1661.

Wexler, E. J. (1992). Analytical solutions for one-, two-, and three-dimensional solute transport in ground-water systems with uniform flow, U. S. Geological Survey, Techniques of Water-Resources Investigations, Book 3, Chapter B7.

Wilson, J. L., and Miller, P. J. (1978). "Two-dimensional plume in uniform ground-water flow." *Journal of the Hydraulics Division, ASCE*, 104(4), 503-514.

CHAPTER 5

PHYSICAL AND CHEMICAL CHARACTERIZATION OF GROUNDWATER SYSTEMS

Randall W. Gentry
The University of Tennessee

5.1 Introduction

Characterizing the physical parameters that govern groundwater flow dynamics is a problem that is always plagued with observational data that is traditionally a small dataset statistically and insular with respect to the overall domain of interest. Tools have been developed to measure or infer parameter values based upon physical measurement, geochemically derived information, and inversely through parameter estimation techniques. The purpose of this chapter is to integrate historical information with novel advances in the scientific and engineering literature for the physical characterization of aquifer parameters. The goal is to provide information that will be useful to the practitioner and researcher alike.

5.1.1 Characterization Goals

Comprehensive water-supply or remediation projects require a detailed aquifer characterization. The goal of these characterization efforts is to gain a better understanding of the distribution and magnitude of physical parameters that impact flow and transport in the subsurface system or measurements that are indicative of the flow regime or mixing regimes. Typically, such variables may include hydraulic conductivity, aquitard/aquifer thickness, and effective porosity, depending upon the goals of the characterization effort. The basic theory of fluid flow in porous media governs the variables necessary for basic aquifer characterization. For a heterogeneous, anisotropic, consolidating porous media the constitutive equation for mass conservation can be stated as follows, in modified form (Bear 1972):

$$\frac{\partial}{\partial x}\left(K_x \frac{\partial h}{\partial x}\right)+\frac{\partial}{\partial y}\left(K_y \frac{\partial h}{\partial y}\right)+\frac{\partial}{\partial z}\left(K_z \frac{\partial h}{\partial z}\right)=-S_s \frac{\partial h}{\partial t} \tag{5.1}$$

Where, h is the hydraulic head or potential, K_x, K_y, K_z are the directional hydraulic conductivities along cartesian planes, S_s is the specific mass storativity or specific storage, and t is time. In this formulation, the state variable, head, is impacted by the directional hydraulic conductivities and the specific storage of the aquifer. In order to better determine the flow characteristics of the aquifer, one must better characterize these variables spatially throughout the aquifer.

The specific techniques used to characterize the system range from discrete hydraulic property measurements to broad scale geophysical measurements. These parameters may vary significantly across differing spatial scales, and the need to characterize heterogeneity for better understanding of aquifer behavior has been made in the literature (Wood 2000). New technologies are currently being explored for improving aquifer characterization efforts and will be explored in the sections that follow.

5.1.2 Discrete Measurements and Spatial Interpretation

Another implicit goal of aquifer characterization is to gain a spatially robust representation of the system. However, this goal is complicated by the fact that typically only discrete measurements can be made and then some form of interpretation or model must be used to form a more complete realization of the system. Primarily for sand matrices, an examination of the particle size attributes and distribution from a physical sample will provide an indication of the hydraulic conductivity from the Hazen Method (Hazen 1911):

$$K = C\left(d_{10}\right)^2 \tag{5.2}$$

where K is the hydraulic conductivity (cm/s), d_{10} (cm) is the effective grain size, and C is a constant based upon the complexity of sand matrix, as summarized in Table 5.1.

Table 5.1: Coefficients for Hazen Method

Matrix	C
Very fine sand, poorly sorted	40-80
Find sand with appreciable fines	40-80
Medium sand, well sorted	80-120
Coarse sand, poorly sorted	80-120
Coarse sand, well sorted, clean	120-150

Pneumatic push techniques or drilling of boreholes and emplacement of monitoring wells is the primary tool used for the collection of subsurface discrete data. Generally, the pneumatic-push or direct-push techniques are a less expensive way, yet limited in terms of depth profiling, to obtain preliminary subsurface data, and an entire support industry (e.g., Geoprobe®) has grown and developed a suite of down-hole tools for aquifer characterization (Bjerg et al. 1992; Butler et al. 2002; Cho et al. 2000; Gribb et al. 1998; Henebry and Robbins, 2000; Hinsby et al. 1992; Kodesova et al. 1999). In recent years vibratory drilling has also allowed access to continuous core data. An example of the types of samples from vibratory core sampling is shown in Figure 5.1, where Gentry et al. (2006) used vibratory core data to investigate the distribution of hydraulic conductivities near an aquifer leakage site.

a. b.

Figure 5.1: Photographs showing (a) continuous core section split for examination and grain-size sampling; and (b) rig using vibratory drilling method for continuous core collection.

5.1.3 Applications

The purpose for which aquifer characterization is performed is an important factor in determining the type and degree of characterization tools to be used. The Kücük Menderes River Basin aquifer in western Turkey is an example where many differing tools (modeling, geological, geophysical, etc.) are being used to address aquifer management strategies for sustainable yield, which require detailed conceptual models and aquifer characterization (Sakiyan and Yazicigil 2004). A second level analysis in aquifer characterization involves the development of a conceptual model for combining discrete data into a more broad spatial realization. This analysis may involve some apriori knowledge regarding the stratigraphic layers of the local or regional geology. At a study site scale it may be more appropriate to use some form of interpolative mathematics to determine a continuous realization from discrete data. These techniques may involve a form of linear interpolation, or may involve more complex interpolation algorithms like kriging (Boman et al. 1995; Desbarats and Bachu 1994; Rizzo and Dougherty 1994; Yeh et al. 1983). Many algorithms are now for use in an easily available geostatistics library, known as GSLIB (Deutsch and Journel 1998). As with all techniques, the user should evaluate the interpolative technique and resultant realization on the basis of what is known about the system and the conceptualized behavior between observations. This was recently demonstrated by the use of downhole flowmeter and ordinary kriging methods used to interpret hydraulic conductivity distributions at the Oak Ridge National Lab site (Fienen et al. 2004).

Another common application of characterization is for source water protection, which is defined either from the viewpoint of the wellhead or the recharge source to an aquifer. The viewpoint taken in characterization efforts may depend upon the average residence time in the aquifer system. Regulators are now considering time of travel concepts in defining source contributions and groundwater under the direct influence (GWUDI) of surface water. Recent research has proposed a technique for assessing absolute risk for GWUDI systems using information about certain aquifer and transport characteristics (Chin and Qi 2000). Chin and Qi (2000) develop a theoretical expression for the mass loading of pathogens in pumped groundwater that is characterized as GWUDI. Evaluating the basic relationship presented by Chin and Qi (2000) requires some detail about aquifer characterization and is summarized as follows:

$$\dot{m} = \int_0^\infty \int_A f\left(\tau \middle| x\right) q_o\left(x\right) c_o \, dx d\tau \tag{5.3}$$

Where, $\dot{m}$ is the mass flux of pathogen in the pumped water, A is the surface area of the aquifer in contact with the water, $f(\tau|x)$ is the probability that a pathogen originating at the ground-water/surface-water interface location x arrives at the well at time τ later, $q_o(x)$ is the specific discharge at the groundwater/surface water interface, and c_o is the concentration of pathogens at the groundwater/surface water interface. Each of these parameters requires detailed characterization in order to determine the most reliable mass flux in this example relating to GWUDI and source water protection. Chin and Qi (2000) use this basic expression to develop further the probability functions relating the occurrence of possible pathogens at a wellhead.

Mapping of aquifer properties over broad geographic regions for the purpose of characterizing aquifer vulnerability has been explored through the use of methodologies like DRASTIC (Aller 1987). Other characterization efforts may use broad scale mapping of certain aquifer properties by geophysical means as surrogates for potential risk to aquifers (Thomsen et al. 2004).

5.2 Reconnaissance-Level Investigations

Water-supply studies require knowledge about the sustainable withdrawal from an aquifer system. The sustainable withdrawal may be defined by pumping that will avoid unnecessary groundwater mining and subsidence or may be related to flows that will avoid negative impacts to ecological interfaces in hydrologically connected systems.

In order to characterize an aquifer for water supply, it is important to have a good understanding of all of the hydrologic inputs to the system in addition to the physical characteristics governing flow in the system. The placement of multiple pumping wells in a wellfield requires detailed knowledge about how long-term withdrawals will affect drawdowns throughout the regional area.

Remedial investigation and aquifer characterization with the intent of mitigation will require detailed description of hydraulic and transport properties alike. Aquifer testing may be geared toward better characterization of organic carbon in the geologic matrix for the purpose of understanding possible contaminant partitioning. Other factors, such as porosity in the matrix may govern conceptual understanding of matrix diffusion in a specific system.

5.3 Geophysical Investigations

Geophysical investigations are used to create a more detailed conceptual model of site given certain observational tools and responses. These types of investigations can be conducted using broad scale signal initiation and response interpretations, which are known as surface geophysical methods or downhole techniques to investigate stratigraphy at a single location, also known as borehole or well geophysics.

5.3.1 Surface Geophysical Methods

Typical surface geophysical methods are based upon either seismic reflection, electrical resistivity or gravity responses. These methods may be used to determine the presence and location of groundwater, bedrock depth, and dip, etc. Typically the observational responses from these techniques are calibrated to known locations, where geologic subsurface samples are available for comparison. These techniques can provide subsurface information on a larger spatial scale which may be useful in hydrologic investigations. However, the signals are not always easily acquired and may require some subjective interpretation. For the purposes of description and discussion, two techniques will be presented: 1) electrical resistivity; and 2) seismic reflection, which are two very commonly used surface geophysical techniques for aquifer characterization.

Electrical Resistivity Methods The electrical resistivity method, or its inverse electrical conductivity, uses the surface placement of electrodes to determine the change in specific electrical resistance due to the diverse geologic media which are located in the subsurface environment below the electrodes (Kelley 1962; McGinnis and Kempton 1961; Meidav 1960). The technique is useful as a tool for aquifer characterization because of the factors that determine the electrical resistivity of geologic media: 1) porosity; 2) pore space water salinity; 3) flow geometry of the interconnected pore space; 4) presence of solid conductive materials; and 5) temperature (ASCE 1987). Archie (1941) proposed that a formation factor, F, could be used for characterization purposes. The empirical relationship has been demonstrated to show that a constant formation factor exists for a given resistvity of a geologic media, ρ_o, relative to the resistivity of the saturating fluid, ρ_w, given in the following equation (Archie 1941).

$$F = \frac{\rho_o}{\rho_w} \tag{5.4}$$

The formation factor has been further extended in terms of the fractional porosity, φ, and the cementation, m, shown in the following equation (ASCE 1987).

$$F = \phi^{-m} \tag{5.5}$$

It has been reported that the value of m ranges from 1.2 to 1.5 in well-sorted, noncemented sediments to 1.9 to 2.2 in geologically-mature, well-cemented sediments or igneous rock matrices (ASCE 1987). Collecting measurements is implemented by emplacing two electrically active electrodes into the ground surface and measuring the resulting voltage at two other metal stakes. The interpretation of the data is governed by the geometry of the four points, which typically has a symmetric design. The accuracy of the technique for determining individual stratigraphic layers has been reported to be +15% to +20% (ASCE 1987).

Seismic Reflection Methods – Seismic reflection methodologies have been used for characterizing aquifer profiles (aquifer-aquitard interfaces). For shallow characterization, shear or secondary waves (S-wave), are mapped along transects or cross-section. In contrast, compressional or primary waves (P-waves) are much faster waves and travel in the direction of the wave. The reflection of these waves from interfaces and associated travel time through the geologic media provide the capability to map the depth profile between clay, bedrock and/or unconsolidated media. High-resolution P-wave (compressional wave) seismic reflection investigations targeting bedrock paleovalleys (Pullan and Hunter 1990; Wolfe and Richard 1996) and near-surface faults (Sexton and Jones 1986) have been successful in areas where geological conditions are favorable for the collection of high-quality data (e.g., fine-grained material, shallow water table, and large seismic velocity contrasts). However, there are situations where shallow S-wave (shear wave) reflection methods are more effective. Unconsolidated, water-saturated sediment sequences are ideal for the use of S-waves. In such environments, P-waves travel with the velocity of the fluid and are often less sensitive to lithologic changes with depth. Because S-waves travel with the velocity of the sediment framework, they are not greatly influenced by the degree of saturation and often provide a more accurate image of the subsurface geology. For example, S-wave techniques have been used successfully for near-surface hydrogeologic (Langston et al. 1998; Miles et al. 2000) and earthquake hazard (Harris and Street 1997; Harris et al. 1998; Woolery et al. 1999) studies in the Mississippi embayment.

5.3.2 Downhole Geophysical Methods

Downhole techniques are probably the most commonly used of the geophysical techniques. Many downhole techniques exist, but the most commonly used measurements are the natural-gamma, resistivity and spontaneous potential. The natural-gamma method is used to identify clay layers, which have more naturally-occurring gamma radiation than other geologic materials. Conversely, the resistivity takes advantage of the higher electrical conductivity of more permeable media.

Typically both measurements are used to identify aquifer and aquitard layers throughout a borehole sequence.

More recently researchers have applied magnetic resonance sounding (MRS) for the purpose of aquifer characterization (Legchenko et al. 2004). The MRS technique was originally developed in Russia in the 1980s and further refined in recent research in France. MRS offers a non-invasive means of directly detecting subsurface water (Legchenko et al. 2004).

Aquifer compaction and storativity parameters are often difficult to access. Recently researchers used one-dimensional modeling of aquifer compaction with extensometer data to better characterize aquifer hydraulic conductivities (Pope and Burbey 2004).

5.4 Test Drilling and Well Installation

The most utilized method for aquifer characterization remains to be test hole drilling and the emplacement of observation or pumping wells into the hydrologic system of interest. Various methods exist for the both the drilling and well installation and these data remain the most useful for characterization of stratigraphy and hydraulic properties. At a minimum, test hole drilling should include the following data (ASCE 1987):

i. Identification, geographic coordinates, and elevation of the site of each test hole;
ii. A lithologic and/or geophysical logs of the entire drilling sequence;
iii. Samples of the geologic media representing the various stratigraphic layers;
iv. Static water level associated with aquifer layers; and
v. Water quality samples and hydraulic data collection when possible.

5.4.1 Drilling Methods and Borehole Logging

Various techniques exit for the drilling of boreholes for the purpose of aquifer characterization. A good presentation of the diverse drilling methods has been described by Driscoll (1986). Mud rotary, air rotary or hollow stem auger are the most common types of drilling for relatively shallow aquifer characterization purposes. In all of these techniques a cutting bit is advanced with mechanical rotation. The cuttings are then advanced back up the borehole for removal by air, mud or auger flights. The formation material dictates the best method to use with the goal being to keep the borehole open while the borehole is advanced. This is important for specimen collection and logging of the borehole for characterization purposes. More recently, Rotasonic drilling has been used due to the capability of retrieving continuous core during the drilling operation. The technique and guidance are described by the ASTM International guidance document, "D6914-04e1 Standard Practice for Sonic Drilling for Site Characterization and the Installation of Subsurface Monitoring Devices." Other than with the Rotasonic technique, sampling of the subsurface is limited to the collection of drive tube samples or inspection of cuttings as the cutting head is advanced. It is for this reason, that borehole drilling should also include downhole geophysics when possible.

5.4.2 Well Installation (Observation and Pumping)

The use of a particular well will govern its design and the method of installation. For example, if environmental tracers such as helium-3 or krypton-85 are to be sampled from the well, then no method should be used that would entrain air into the formation during drilling. One of the most effective ways to develop a well and remove drilling contamination is air sparging. However, once air sparging has been used, the well may not be used to sample for environmental tracers such as helium-3 or krypton-85. As much as possible, the influences from the drilling fluid should also be minimized by using an appropriate, well-characterized water source for drilling operations. An important aspect of the well design is the well screen itself. Many observation or monitoring wells have multiple uses for the purpose of characterization. The observation well may be used to perform hydraulic testing, water quality characterization and transient head observations. By far, the most common type of observation well material used is PVC, low carbon steel, or stainless steel. Stainless steel is not used as often due to the higher cost of the material and is reserved for somewhat specialized observation well design. Predominately, 2-, 4-, or 6-inch wells are installed for multi-purpose hydraulic testing or observation purposes. Six-inch and larger wells are more ideally suited for large-scale pumping or aquifer testing. A series of criteria have presented and adapted by Driscoll (1986) for consideration in observation well design:

i. The screens should be constructed from an inert material;
ii. Consider maximizing the open area to facilitate sampling or observation equipment needs;
iii. The slot size of the screen should retain the filter pack or formation and the overall design should not inhibit the development of the well; and
iv. Ensure that the screen is non-plugging.

Specific design criteria are usually governed by the purpose of the well, and are typically established by the regulatory agency with primacy over the work being performed. Due diligence requires that the engineer investigates and uses the appropriate federal, state or local guidelines. Drilling of monitoring wells is still considered to be one of the more expensive activities associated with aquifer characterization, but improper well design can significantly increase costs if replacement wells must be drilled or newly installed wells abandoned.

The proper development of an installed well is necessary to ensure that the data collected from the well are adequate for characterization purposes. Development of the well is a relatively high energy input activity that is meant to surge drilling fluid and smaller fraction fines from the filter pack. This is typically accomplished by the use of surge blocks or cycling of the development pump. This hydraulic interaction tends to rapidly change the velocity profile through the screen and the adjacent filter pack to dislodge fines in the pore space of the filter pack. For long-term pumping wells, or production wells, water is pumped to waste for a long period of time prior to use to accomplish the same process. In the end, the development process will ensure that the water coming from the well is representative of the formation, and not induced by the drilling and installation process. Smaller diameter wells are easiest to develop, while larger diameter wells require much more infrastructure and are thus more difficult to develop.

5.4.3 Hydraulic Testing

Aquifer characterization often involves the collection of hydraulic response data from boreholes or monitoring wells. Hydraulic testing, in combination with geophysical and lithologic logs, can be used for large-scale, robust characterization of aquifers (e.g., Paillet and Reese 2000). The hydraulic interpretations are based upon the fundamental understanding of flow in porous media and well hydraulics. The application of well hydraulics theory has been used to identify hydraulic properties of aquifer layers. These solutions are based upon analytical solutions for radial flow to a well screen, typically considered fully penetrating, in a homogenous and isotropic aquifer system. Two traditional analytical solutions are the steady state Thiem solution and the transient Theis solution. Recently, the United States Geological Survey (USGS) published several spreadsheet programs to aid in the analyses of well hydraulic data (Halford and Kuniansky 2002). Analytical solutions were later developed for partially penetrating wells in single or multilayer aquifer systems (Bouwer and Rice 1976; Cooper et al. 1967; Javandel and Witherspoon 1983; Kabala et al. 1985). Improvements have been suggested to slug test techniques (McElwee 2002). Also, in more recent years, more robust well hydraulics testing for aquifer characterization have advanced to include flowmeter testing (Kabala 1993; Molz et al. 1989) and offer a more robust means of characterizing aquifer hydraulic conductivity via a completed well. In many cases, monitoring or production wells are not completed in an open borehole. Therefore, analytical techniques have been developed to estimate hydraulic properties from boreholes. Research has shown that relatively fine resolution data collection can be accomplished in a borehole, during drilling, and prior to the installation of the well structure (Zlotnik and McGuire 1998).

5.4.3.1 *Slug Tests*

Slug tests use traditional well hydraulics interpretations to determine the hydraulic conductivity of the coarse-grained materials (Cooper et al. 1967; Bouwer and Rice 1976). More recently, researchers have developed smaller scale, in situ "mini-slug" tests (Hinsby et al. 1992). The goal of the more recent research is to provide characterization tools at a finer resolution than has been available. All of these techniques are based upon the instantaneous displacement of a specific volume of water within the well and an analysis of the rising or falling head associated with the appropriate test. The volume of water displace is referred to as a "slug" and may be comprised of a solid matrix or a slug of water added or removed from the well. In recent research, Loáiciga et al. (2006) have provided some guidance and insights into evaluating slug-test data for hydraulic conductivity statistical characterization, and the appropriate probability density function determination, using 201 slug tests from the Mexico City area.

Because slug tests affect a relatively small volume of the formation being tested, the hydraulic conductivity values derived from such tests generally represent localized values.

5.4.3.2 *Pumping Tests*
Pumping tests are designed to stress an aquifer in order to determine hydraulic parameters that are not easily evaluated by single well tests. Typically a pumping well is emplaced in the aquifer with sentinel monitoring well(s) located peripheral to that well. Pumping of the aquifer is then implemented at a constant or stepped fashion with drawdowns measured in the pumped and sentinel wells. These methods are used to determine the effective hydraulic conductivity and storativity of the aquifer media. The tests may also be used to characterize the overall conceptual model of the aquifer system, such as, leaky versus well confined. As pumping tests stress an aquifer over a much larger volume than do slug tests, the parameters estimated from the tests represent volume averages over a correspondingly larger volume.

5.5 Water Quality Characterization

The water quality, chemistry, and radiochemistry can be thought of as a unique characteristic of a localized water source if the dataset is robust enough. In recent years, one of the grand challenges for groundwater engineers and scientists has been the quantification and understanding of groundwater fluxes across interfaces (NRC 2004).

5.5.1 Geochemical Characterization vs Contaminant Characterization

An important aspect in aquifer characterization is the identification of flow paths within and between aquifer units. These interactions will typically result in a unique geochemical signature that can be interpreted based upon a conceptual model. Typically, geologic facies are used to define aquifer thickness and lateral extents. However, geochemical facies can be used to define mixing zones and as a qualitative means of assessing interaquifer hydraulics.

Predicting the fate of contaminants in the environment and designing remediation strategies can be complicated by complex heterogeneities that exist in particular hydrogeologic settings. The complicated role of subsurface heterogeneity and variable hydraulic conductivity has been well demonstrated in the literature (Gerber et al. 2001; Fogg et al. 1999; Martin and Frind 1998; NRC 1993). A lucid and concise editorial addressing the need to better understand the nature of complicated subsurface environments through new and innovative technology to resolve issues of heterogeneity was described by Wood (2000).

Recently Gentry et al. (2006) demonstrated the use of uranium and thorium decay series coupled with aquifer geochemistry to better characterize a localized leakage feature in the upper Claiborne confining unit of the Memphis aquifer in the Sheahan Wellfield. Figure 5.2 shows the conceptual model derived for interpreting the radiochemical (uranium and thorium decay series) and geochemical signatures for the Memphis aquifer leakage site. This interpretive framework allowed for a broader spatial characterization of the confining layer within the local wellfield, shown in Figure 5.3. These types of specialized sample collection and radiochemical analyses are not available from most commercial analytical laboratories, and are primarily used in a research context. However, as aquifer characterization techniques using these specialized techniques enter the mainstream, especially for complex systems, they may be more broadly available. As an example of the collection process, a photograph of the sample collection process is shown in Figure 5.4.

5.5.2 Well Purging and Sampling Methods

Prior to sampling a well must be purged to ensure that the water being collected is representative of the pore water and environmental condition of the formation being characterized. Purging is a less energetic process than the development process described in Section 5.4. Purging is primarily used to remove the stagnant water in the casing or borehole that is not representative of the chemistry. Driscoll (1986) presents a summary of reasons that groundwater samples may not be representative of formation conditions:

i. The sample was collected from the stagnant water in well casing [purging of 3-10 well volumes is typically required (Gibb 1983)];
ii. The samples were collected at inappropriate intervals, given the hydraulics of the well and formation;
iii. Entrained sediment (usually from a poorly developed well) contaminated the sample;
iv. The accuracy of the sample was impacted by the hydraulics of the well and formation (i.e. contaminant dilution from long well screens or near-screen, high hydraulic conductivity zones);
v. Samples taken after an extended period of pumping are not representative, due to the extended volume of aquifer water drawn to the well, and possible non-equilibrium conditions;
vi. Events during pumping that cause changes of pH or redox potential, such as releases of carbon dioxide or introduction of oxygenated water, that may alter metal speciation;
vii. Entraining air into the water sample, which will induce oxidation losses, volatilization or speciation changes;
viii. Cross contamination of the sample by the pump or sample collection apparatus;
ix. Appropriate preservation techniques were not used when transporting samples;
x. Holding times were not met by the laboratory performing the analyses; and
xi. The testing laboratory or field analyses were not set-up properly or inappropriate methods were used to perform the analyses.

To ensure that proper purging has occurred, typically field parameters are measured at steady time intervals (ensuring the time interval is appropriate). Traditionally, pH, temperature and conductivity are measured to ensure that the water coming from the well is not experiencing redox shifts and is most likely representative of formation conditions. Recent technical literature and regional sampling experience indicate that use of low stress purging and sampling procedure improves data quality for all inorganic and organic groundwater sample results, with the water quality data generated using this procedure being more reproducible and representative of actual groundwater conditions (USEPA 1996).

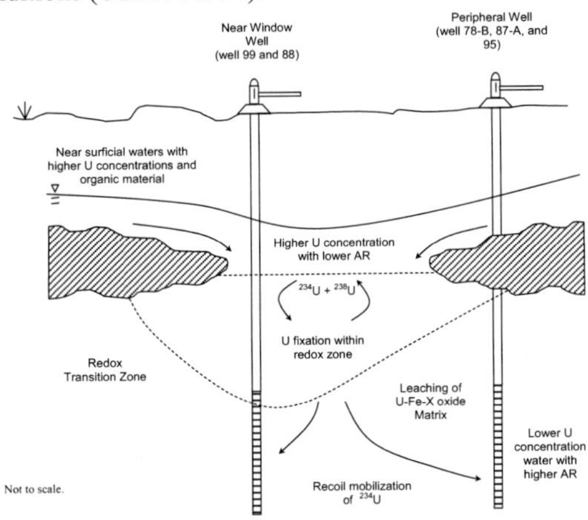

Figure 5.2: Conceptual model for a localized leakage feature in the Memphis Aquifer based upon radiochemistry (Gentry et al. 2006).

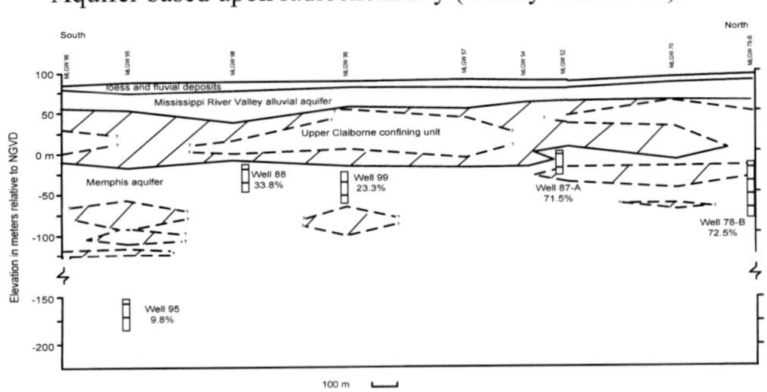

Note: Percentages are based upon the number of hours that the well was active during the 2003 calendar year. Hatched areas in the figure represent clay lenses.

Figure 5.3: North-South stratigraphic cross-section of the Sheahan Wellfield in Memphis, Tennessee (Gentry et al. 2006).

Figure 5.4: Photograph of uranium and thorium radioisotope filter equipment installation at a production well in the Sheahan Wellfield, Memphis, Tennessee. Shown are Dr. Teh-Lung Ku (left), Dr. John McCarthy (center) and Dr. Shangde Luo (right).

5.5.3 Analytical Methods

Guidance on the appropriate analytical techniques used in geochemical or contaminant characterization in an aquifer investigation is driven by the agency that has ultimate primacy over the activity. For most wet chemical characterization analyses, the primary guideline for acceptable chemical analyses is Standard Methods 20th edition (1998) jointly published by the American Public Health Association, American Water Works Association, and Water Environment Federation. For regulatory environmental characterization, the U.S. Environmental Protection Agency (USEPA) provides an acceptable list of methodologies by print or alternatively is available on the USEPA's website [http://www.epa.gov/OGWDW/methods/methods.html]. In some cases an alternative source of guidance may be acceptable, but due dilligence requires approval by the agency who has primacy of the characterization effort. Another listing that may include acceptable methodologies is the National Environmental Methods Index (NEMI) which is also searchable by website [http://www.nemi.gov].

5.6 Characterization Guidance and Specifications

Many organizations have begun to make guidelines for characterization efforts available via the Internet. Two of the primary organizations for this activity, as described in some previous sections, are ASCE and ASTM International [www.astm.org]. As of the date of this publication many guidelines and standards have been prepared for the purposes the physical and chemical characterization of aquifers. Although, somewhat general in some cases, these standards set a baseline for discussion and framework for the methodologies used in characterization efforts. The goal of the publication of some of these standards is to provide a quick reference for practitioners to use that is generally agreed upon by most of those involved in this field of study and practice. Although not exhaustive, a summary of some of the most pertinent standards are as follows:

ASCE

ASCE 50-08 Standard Guideline for Fitting Saturated Hydraulic Conductivity Using Probability Density Functions.

ASCE 51-08 Standard Guideline for Calculating the Effective Saturated Hydraulic Conductivity.

ASTM International

D4043 Standard Guide for Selection of Aquifer-Test Method in Determining of Hydraulic Properties by Well Techniques.

D4044 Standard Test Method (Field Procedure) for Instantaneous Change in Head (Slug Tests) for Determining Hydraulic Properties of Aquifers.

D4050 Standard Test Method (Field Procedure) for Withdrawal and Injection Well Tests for Determining Hydraulic Properties of Aquifer Systems.

D4448 Standard Guide for Sampling Ground-Water Monitoring Wells.

D5092 Standard Practice for Design and Installation of Ground Water Monitoring Wells in Aquifers.

D5254 Standard Practice for Minimum Set of Data Elements to Identify a Ground-Water Site.

5.7 References

Aller, L. (1987). *DRASTIC: A standardized system for evaluating ground water pollution potential using hydrogeologic settings*, Robert S. Kerr Environmental Research Laboratory, Office of Research and Development, U. S. Environmental Protection Agency, Ada, Okla.

American Society of Civil Engineers (ASCE) (1987). *Ground Water Management, 3rd Edition*, ASCE Manuals and Reports on Engineering Practice No. 40, New York.

Archie, G. E. (1941). "The electrical resistivity log as an aid in determining some reservoir characteristics." *Well Logging*, T.P. 25H, ASME, New York.

Bear, J. (1972). *Dynamics of fluids in porous media*, American Elsevier Pub. Co., New York.

Bjerg, P. L., Hinsby, K., Christensen, T. H., and Gravesen, P. (1992). "Spatial variability of hydraulic conductivity of an unconfined sandy aquifer determined by a mini slug test." *Journal of Hydrology*, 136(1-4), 107-122.

Boman, G. K., Molz, F. J., and Guven, O. (1995). "An evaluation of interpolation methodologies for generating 3-dimensional hydraulic property distributions from measured data." *Ground Water*, 33(2), 247-258.

Bouwer, H., and Rice, R. C. (1976). "Slug test for determining hydraulic conductivity of unconfined aquifers with completely or partially penetrating wells." *Water Resources Research*, 12(3), 423-428.

Butler, J. J., Healey, J. M., McCall, G. W., Garnett, E. J., and Loheide, S. P. (2002). "Hydraulic tests with direct-push equipment." *Ground Water*, 40(1), 25-36.

Chin, D. A., and Qi, X. (2000). "Ground water under direct influence of surface water." *Journal of Environmental Engineering*, 126(6), 501-508.

Cho, J. S., Wilson, J. T., and Beck, F. P. (2000). "Measuring vertical profiles of hydraulic conductivity with in situ direct-push methods." *Journal of Environmental Engineering*, 126(8), 775-777.

Cooper, H., Bredehoeft, J., and Papadopulos, I. (1967). "Response to a finite diameter well to an instantaneous charge of water." *Water Resources Research*, 3, 263-269.

Desbarats, A. J., and Bachu, S. (1994). "Geostatistical analysis of aquifer heterogeneity from the core scale to the basin-scale: A case-study." *Water Resources Research*, 30(3), 673-684.

Deutsch, C. V., and Journel, A. G. (1998). *GSLIB: Geostatistical software library and user's guide*, Oxford, New York.

Driscoll, F. G. (1986). *Groundwater and Wells*, Johnson Screens, St. Paul, Minn.

Fienen, M. N., Kitanidis, P. K., Watson, D., and Jardine, P. (2004). "An application of Bayesian inverse methods to vertical deconvolution of hydraulic conductivity in a heterogeneous aquifer at Oak Ridge National Laboratory." *Mathematical Geology*, 36(1), 101-126.

Fogg, G. E., LaBolle, E. M., and Weissmann, G. S. (1999). "Groundwater vulnerability assessment: Hydrogeologic perspective and example from Salinas Valley, California." *Assessment of non-point source pollution in the vadose zone*, D. L. Corwin, K. Loague, and T. R. Ellsworth, T.R., eds., Geophysical Monograph 108, AGU Press, Washington, D.C., Monograph 108, 45-61.

Gentry, R. W., McKay, L., Thonnard, N., Anderson, J. L, Larsen, D., Carmichael, J. K., and Solomon, K. (2006). *Novel techniques for investigating recharge to the Memphis aquifer*, American Water Works Association, 97 pp.

Gerber, R. E., Boyce, J. I., and Howard, K. (2001). "Evaluation of heterogeneity and field-scale groundwater flow regime in a leaky till aquitard." *Hydrogeology Journal*, 9(1), 60-78.

Gibb, J. P. (1983). "Sampling procedures for monitoring wells." *Groundwater Monitoring Technology Conference*, University of Wisconsin-Extension, Philadelphia, Pa.

Gribb, M. M., Simunek, J., and Leonard, M. F. (1998). "Development of cone penetrometer method to determine soil hydraulic properties." *Journal of Geotechnical and Geoenvironmental Engineering*, 124(9), 820-829.

Halford, K. J., and Kuniansky, E. L. (2002). *Documentation of spreadsheets for the analysis of aquifer-test and slug-test data*, U. S. Geological Survey, Open-File Report 02-197.

Harris, J. B., Berman, S. A., Beard, W. C., Street, R. L., and Cox, R. T. (1998). "Shallow seismic reflection investigations of neotectonic activity in the Lower Mississippi Valley." *68th Annual International Meeting, Society of Exploration Geophysicists*, Expanded Abstracts, 848-851.

Harris, J. B., and Street, R. L. (1997). "Seismic investigation of near-surface geological structure in the Paducah, Kentucky, area: Application to earthquake hazard evaluation." *Engineering Geology*, 46(3-4), 369-383.

Hazen, A. (1911). "Discussion of 'Dams on sand foundations' by A. C. Koenig." *Trans. Am. Soc. Civ. Eng.*, 73, 199–203.

Henebry, B. J., and Robbins, G. A. (2000). "Reducing the influence of skin effects on hydraulic conductivity determinations in multilevel samplers installed with direct push methods." *Ground Water*, 38(6), 882-886.

Hinsby, K., Bjerg, P. L., Andersen, L. J., Skov, B., and Clausen, E. V. (1992). "A mini slug test method for determination of a local hydraulic conductivity of an unconfined sandy aquifer." *Journal of Hydrology*, 136(1-4), 87-106.

Javandel, I., and Witherspoon, P. A. (1983). "Analytical solution of a partially penetrating well in a 2-layer aquifer." *Water Resources Research*, 19(2), 567-578.

Kabala, Z. J. (1993). "The dipole flow test - A new single-borehole test for aquifer characterization." *Water Resources Research*, 29(1), 99-107.

Kabala, Z. J., Pinder, G. F., and Milly, P. C. D. (1985). "Analysis of well-aquifer response to a slug test." *Water Resources Research*, 21(9), 1433-1436.

Kelley, S. F. (1962). "Geophysical exploration for water by electrical resistivity." *Journal, New England Water Works Association*, 76, 118-189.

Kodesova, R., Ordway, S. E., Gribb, M. M., and Simunek, J. (1999). "Estimation of soil hydraulic properties with the cone permeameter: Field studies." *Soil Science*, 164(8), 527-541.

Langston, C., McIntyre, J., Street, R., and Harris, J. (1998). "Investigation of the shallow subsurface near the Paducah Gaseous Diffusion Plant using SH-wave seismic methods." *68th Annual International Meeting, Society of Exploration Geophysicists*, Expanded Abstracts, 878-880.

Legchenko, A., Baltassat, J. M., Bobachev, A., Martin, C., Robain, H., and Vouillamoz, J. M. (2004). "Magnetic resonance sounding applied to aquifer characterization." *Ground Water*, 42(3), 363-373.

Loáiciga, H. A., W. W.-G. Yeh, and M. A. Ortega-Guerrero (2006), "Probability density functions in the analysis of hydraulic conductivity data." *Journal of Hydrologic Engineering*, 1a(5), 442-450.

Martin, P. J., and Frind, E. O. (1998). "Modeling a complex multi-aquifer system: The Waterloo moraine." *Ground Water*, 36(4), 679-690.

McElwee, C. D. (2002). "Improving the analysis of slug tests." *Journal of Hydrology*, 269(3-4), 122-133.

McGinnis, L. D., and Kempton, J. P. (1961). *Integrated seismic, resistivity, and geologic studies of glacial deposits*, Circular 323, Illinois Geological Survey.

Meidav. T. (1960). "An electrical resistivity survey for groundwater." *Geophysics*, 25, 1077-1093.

Miles, J. E., Galicki, S. J., and Harris, J. B. (2000). "A hydrogeological and geophysical study of an unconfined aquifer on Millsaps College campus, Jackson, Mississippi." *Journal of the Mississippi Academy of Sciences*, 45(1), 47-48.

Molz, F. J., Morin, R. H., Hess, A. E., Melville, J. G., and Guven, O. (1989). "The impeller meter for measuring aquifer permeability variations - Evaluation and comparison with other tests." *Water Resources Research*, 25(7), 1677-1683.

National Research Council (NRC) (1993). *Ground water vulnerability assessment: Contamination potential under conditions of uncertainty*, National Academy Press, Washington, D.C., 204 pp.

National Research Council (NRC) (2004). *Groundwater fluxes across interfaces.* National Academy Press, Washington, D.C., 85 pp.

Paillet, F. L., and Reese, R. S. (2000). "Integrating borehole logs and aquifer tests in aquifer characterization." *Ground Water*, 38(5), 713-725.

Pope, J. P., and Burbey, T. J. (2004). "Multiple-aquifer characterization from single borehole extensometer records." *Ground Water*, 42(1), 45-58.

Pullan, S. E., and Hunter, J. A. (1990). "Delineation of buried bedrock valleys using the optimum offset shallow seismic reflection technique." *Geotechnical and Environmental Geophysics, Volume III*, S. H. Ward, ed., Society of Exploration Geophysicists, 75-87.

Rizzo, D. M., and Dougherty, D. E. (1994). "Characterization of aquifer properties using artificial neural networks - Neural kriging." *Water Resources Research*, 30(2), 483-497.

Sakiyan, J., and Yazicigil, H. (2004). "Sustainable development and management of an aquifer system in western Turkey." *Hydrogeology Journal*, 12(1), 66-80.

Sexton, J. L., and Jones, P. B. (1986). "Evidence for recurrent faulting in the New Madrid seismic zone from Mini-Sosie high-resolution reflection data." *Geophysics*, 51, 1760-1788.

Thomsen, R., Sondergaard, V. H., and Sorensen, K. I. (2004). "Hydrogeological mapping as a basis for establishing site-specific groundwater protection zones in Denmark." *Hydrogeology Journal*, 12(5), 550-562.

U. S. Environmental Protection Agency (USEPA) (1996). *Low stress (low flow) purging and sampling procedure for the collection of ground water samples from monitoring wells, Revision 2*, U. S. Environmental Protection Agency, Region I, July 30.

Wolfe, P. J., and Richard, B. H. (1996). "Integrated geophysical studies of buried valley aquifers." *Journal of Environmental and Engineering Geophysics*, 1, 75-84.

Wood, W. W. (2000). "It's the heterogeneity!" *Ground Water*, 38(1), 1-1.

Woolery, E. W., Street, R. L., Wang, Z., Harris, J. B., and McIntyre, J. (1999). "Neotectonic structure in the central New Madrid seismic zone: Evidence from multimode seismic-reflection data." *Seismological Research Letters*, 70, 554-576.

Yeh, W. W.-G., Yoon, Y. S., and Lee, K. S. (1983). "Aquifer parameter-identification with kriging and optimum parameterization." *Water Resources Research*, 19(1), 225-233.

Zlotnik, V. A., and McGuire, V. L. (1998). "Multi-level slug tests in highly permeable formations: 2. Hydraulic conductivity identification, method verification, and field applications." *Journal of Hydrology*, 204(1-4), 283-296.

CHAPTER 6

QUANTITATIVE ANALYSIS OF GROUNDWATER SYSTEMS

David Ahlfeld

Department of Civil and Environmental Engineering
University of Massachusetts Amherst
Massachusetts, USA, 01003

Abstract

The governing equations for groundwater flow and contaminant transport introduced in preceding Chapters can be applied to individual groundwater sites to yield useful tools for analysis, prediction and design. A wide variety of groundwater problems can be addressed with a properly constructed groundwater simulation model. When combined with optimization methods, the design capabilities of simulation models can be enhanced. In this chapter, the basic steps in model building are presented, the various methods for solving the governing equations are reviewed, optimization methods for groundwater management are introduced and readily available computer programs for simulation and optimization are discussed.

6.1 Use of Models in Groundwater Management

The general equations that describe groundwater flow and contaminant transport in groundwater systems can be applied to individual sites if sufficient data are available to characterize the system. Once a properly constructed and calibrated simulation model has been created, it can be used for site specific analysis, parameter estimation, design and operation. For complex groundwater problems, a simulation model may be the only practical way to obtain reasonably accurate quantitative answers to questions posed. A sampling of several common applications of models is presented here.

Systems for groundwater pumping for municipal or industrial use are often managed using models. For a single well or clustered well field, models can be used to predict the available yields. Models can also be used to assess the impact of production pumping on the surrounding hydrologic system, including other wells and ecological systems such as streams and wetlands. In many locales, models are the preferred method for determining the recharge zone of a well or well field. This zone is used to define well-head protection zones where certain classes of development that have the potential for releasing contaminants are limited. Models can also be used to design operational strategies for well field use. For example, to minimize the adverse impacts of groundwater withdrawal on ecological systems, or to determine pumping locations, either from different strata or different areal locations that may shift with the season.

For problems in groundwater remediation, models are used to predict the fate and transport of contaminants in the subsurface. Models are accepted as an essential tool for assessing the potential for natural attenuation at contamination sites. Models are used to design long-term monitoring systems to insure that adverse contaminant migration is detected. Engineered remediation strategies are often designed with models. For pump-and-treat strategies, models are used to determine capture zones obtained using a specific well and pump rate arrangement and to estimate the time required for clean-up to be accomplished. For engineered bioremediation systems, models are used for analysis of fate and transport of contaminants, microorganisms and associated constituents. Models are available for analysis of other remediation approaches such as those that simulate air sparging, oil/water contamination and other special cases.

At the regional scale, where an aquifer may be subject to multiple pumping and artificial recharge systems, models are used to assess long-term sustainability issues. For systems in coastal areas, large-scale pumping may induce migration of saline waters from the sea into the freshwater aquifer diminishing the availability and quality of groundwater. Large-scale development tends to diminish the permeable land surface area available for infiltration. Artificial recharge may be necessary to replenish the aquifer. By considering all pumping and recharge activity in a region models are used to analyze the interactions between various components of the system and make predictions about long-term water availability and water quality.

6.2 Constructing Site-Specific Simulation Models

A site-specific groundwater simulation model is intended to approximate the true behavior of groundwater at a site. Constructing a site-specific simulation model requires assembling appropriate site data, selecting the model form and parameter values based on the specific hydrogeologic and chemical characteristics of the site and testing the model for its ability to reproduce observed site conditions. If the site is relatively simple and the questions posed to the model are limited, a model can often be constructed quickly and efficiently. However, for large, complex sites where the model is to be used for sophisticated analysis, model construction can be expensive and time-consuming.

The success of model construction hinges on several decisions. Among these are the selection of an appropriate domain and associated boundary and initial conditions. The domain is the volume of aquifer over which the governing equations will be solved. Boundary conditions describe the state variable at the boundary of the domain. Initial conditions are required for transient problems and consist of the value of the state variable throughout the domain at the beginning of the simulated time period.

A second set of decisions are the specific governing equations to be used and the specific method for solving the governing equations. A wide variety of methods are available for solving groundwater flow and contaminant transport problems. Analytical methods provide closed-form solutions to the governing equations, based on simplifying assumptions, that can often be evaluated by hand calculation. Numerical methods provide more general solutions but require use of computer software.

The domain, boundary and initial conditions and specific parameters needed to solve a groundwater problem depend on the type of governing equation under consideration and the assumptions implied by the solution method used. In turn, the governing equations selected and solution method used may be selected based on the data available to identify parameters and boundary conditions. The interdependence of these decisions complicates the steps in model development, as described below.

In this section, the wide range of specific solution techniques that are available are considered first. Next, the steps in model construction are considered and an example application is described.

6.2.1 Analytical Solution Methods

Analytical solution of the governing equations for groundwater flow and solute transport generally requires use of a simplified domain with simple boundary conditions and uniform parameter values. The modeler should be aware of the assumptions invoked to obtain the analytical solution used. Analytical solutions are most widely available for solution of the groundwater flow equation, especially, for problems in well hydraulics. Other solutions exist for certain problems in solute transport and simple multiphase flow problems. For complex domain geometries or variable parameter values it is generally necessary to use numerical solution approaches.

Groundwater Flow Solutions: Analytical solutions to the groundwater flow equation can be obtained for a variety of settings. Generally, aquifer parameters must be assumed uniform to achieve these solutions. In addition, simplified geometries must be assumed. Two common classes of geometries are one-dimensional flow and radial flow to a well or away from a point recharge. Solutions for well hydraulics are useful both for predicting the response of the aquifer to pumping and for inferring values of hydraulic parameters from field data.

The simplest well hydraulics solutions are based on assuming that the aquifer is horizontal, confined with uniform thickness and has infinite areal extent. The aquifer properties are assumed to be homogeneous and isotropic. The well is assumed to fully penetrate the aquifer and pump at a constant rate. Taken together these assumptions imply that the impact of pumping on heads will be radially symmetric around the well. The Theim equation (Theim 1906) predicts the difference in heads at radii r_1 and

r_2 from a pumping well. The solution uses the assumptions above and further assumes that steady-state, confined flow is present. The solution is:

$$h(r_2) - h(r_1) = \frac{Q}{2\pi T} \ln\left(\frac{r_2}{r_1}\right) \tag{6.1}$$

where Q is the steady pumping rate, T is the aquifer transmissivity and h is the hydraulic head. Given a known head at a known radius from the well, this equation can be used to predict head at a second location. Alternately, if the difference in heads at two observation wells is determined and the pumping rate is known then the transmissivity of an aquifer can be inferred.

The Theim equation requires that head be known at two locations for determination of aquifer transmissivity. An alternate approach is to use transient head data in a single well to infer aquifer properties. This requires that a solution to the transient groundwater flow equation be obtained. The Theis equation (Theis 1935) is the most widely used for this purpose. It solves the equations invoking the assumptions above except that flow is transient. Initial conditions are required and these are assumed to be that the piezometric surface is horizontal at the initiation of pumping. The Theis solution predicts the head as a function of radial distance from the well, r, and time since the beginning of pumping, t, as:

$$h_0 - h(r,t) = \frac{Q}{4\pi T} W\left(\frac{r^2 S}{4Tt}\right) \tag{6.2}$$

where h_0 is the head before pumping begins, S is the aquifer storage coefficient and $W(\cdot)$ is the Theis well function which takes the form of the exponential integral.

The Theis solution can be used, in conjunction with measurements taken in a pumping test, to determine aquifer properties. Typically, measurements of drawdown are collected over time at an observation well located at a distance from the pumping well. Depending on the properties of the aquifer and the distance between the pumping well and the observation well, the pumping test may continue for hours or days. With a time history of drawdown in hand, methods exist to infer values for both T and S in the Theis equation (Batu 1998). Aquifer properties can also be inferred from slug tests. These tests either add or withdrawal water from a single well and observe the recovery of the water level in the same well as it returns to static conditions. Analytical solutions exist that can be used to infer aquifer properties in the immediate vicinity of the well from the transient recovery data.

Numerous alternate well hydraulics solutions have been developed for both steady and transient cases. The solutions obtained, accommodate such conditions as unconfined aquifers (Boulton 1954; Neuman 1972), partially penetrating wells (Hantush 1964), large diameter wells (Papadopulos and Cooper 1967), anisotropic aquifers (Hantush 1966) and leaky aquifers (Hantush and Jacob 1955; Neuman and

Witherspoon 1968). These analytical solutions can be used for prediction of drawdown or, in combination with steady or transient drawdown data, for inference of aquifer parameters. See Batu (1998) for detailed presentation of these methods and references to the many contributions to this literature.

Analytical solutions to groundwater multiphase flow problems are very limited. For example, several analytical solutions are available for problems of one-dimensional infiltration in unsaturated soils (Segol 1994). Multi-phase flow problems are more commonly solved with numerical methods.

Contaminant Transport Solutions: Solutions to the solute transport equations generally require assumptions of a uniform velocity field and uniform properties of dispersion, decay and retardation. The source strength and geometry must also take a relatively simple form. One of the simplest solutions is that for the one-dimensional transport of a solute by advection and dispersion processes. The solute originates from a continuous source with concentration magnitude, C_0. The resulting concentration, C, as a function of distance, x, from the source and time, t, since the release of solute began is given by:

$$C(x,t)=\frac{C_0}{2}\left\{\operatorname{erfc}\left[\frac{(x-Vt)}{2\sqrt{Dt}}\right]+\exp\left(\frac{Vx}{D}\right)\operatorname{erfc}\left[\frac{(x+Vt)}{2\sqrt{Dt}}\right]\right\} \tag{6.3}$$

where V is the uniform groundwater velocity, D is the dispersion coefficient and erfc(.) is the complimentary error function (Ogata and Banks 1961).

More sophisticated analytical solutions are available for cases of two and three dimensions, temporally varying source strengths, source areas of finite dimension, inclusion of decay and retardation processes and radial transport of solute from a well (Domenico and Schwartz 1990; Fitts 2002). See also Chapters 3 and 4 for additional analytical solutions.

As with analytical solutions to the groundwater flow equation, solutions to the solute transport equation can be used both for prediction and for parameter estimation. Use of these models for prediction requires that the assumptions implied by the models be reasonably accurate. The primary use of these models for parameter evaluation is for the estimation of the dispersion coefficient. A number of methods are available for determining the dispersion coefficient when the velocity is known and a time history of concentration is available. This is commonly done at the laboratory scale in a soil column or in the field over relatively short distances in a two-well tracer test. Care must be taken in applying estimated dispersion coefficients. It has been found that the value of the dispersion for a particular site is dependent upon the scale at which the measurement is taken.

6.2.2 Numerical Solution Methods

Using numerical solution methods, with associated computer programs, provides the means to obtain high quality, approximate solutions to the governing equations for groundwater flow and contaminant transport. The methods substitute a system of algebraic equations for the governing differential equation. The algebraic equations are then solved to determine the value of the state variable at discrete points on a grid. The accuracy of the numerical solution is related to the density of the grid and other factors. In general, no significant assumptions on the domain, boundary conditions or parameters are required for use of numerical solution methods. However, the specific computer programs that implement numerical methods may invoke certain assumptions.

The discretization of the spatial and temporal domain is common to all numerical methods that are applied to groundwater problems. Commonly-used numerical methods include the finite difference and finite element methods. In each case, the domain is divided into many small volumes. These are called cells, elements or grid blocks depending on the method used. These volumes may be restricted to rectilinear, in the case of the finite difference method, or may take tetrahedral or other complex shapes for other methods. Using these small volumes, complex domain geometries can be represented.

Grid points or nodes are associated with the small volumes. Depending on the method used, the nodes may be centered in each volume or at the vertices of each volume. For most numerical methods, uniform parameter values are assigned to each of the individual nodes or individual volumes. Hence, the resolution with which spatial variability in the parameters can be represented is limited by the resolution of the grid. Similarly, boundary condition and initial condition values are assigned for each node or volume limiting the resolution that can be represented.

Definition of the numerical grid is a key step in building a numerical model. A grid with a large number of grid points will be able to capture spatial variability of properties with greater resolution and produce a better approximation of the governing equations. However, computational time increases with the number of grid points. Present-day computing power makes it possible to routinely solve very large problems. As computing speed continues to increase, computational limitations on typical grid sizes will diminish further.

In the 1980's and 1990's there was an enormous growth in the availability of high quality software for solving problems in groundwater flow and contaminant transport. Starting from some the earliest numerical solution programs (Pinder and Bredehoeft 1968; Prickett and Lonnquist 1971; Pinder 1973; Konikow and Bredehoeft 1974; Trescott et al. 1976; Pinder and Gray 1977), this growth was driven by both the demand for good software and the availability of inexpensive high-speed computing facilities.

Hundreds of different computer programs are currently available for solution of groundwater-related problems. These programs range from those designed to solve highly specific problems, such as capture-zone analysis, to those that are capable of modeling general multi-phase and multi-species flow and transport. Some programs are public domain while others are commercially available. The user-friendliness of programs also varies. Some programs use sophisticated graphical user interfaces making it relatively easy for the new user to organize input data and display model results. Other programs are drawn from academic research programs and have less sophisticated user support available and require a high level of user training. A sampling of major software packages are described below with an emphasis on those that are easily available and user-friendly. References to commercial software products are provided solely for the convenience of the reader. Such references do not imply any endorsement by the authors or by ASCE.

A number of internet-based catalogs of models are available. Programs can also be downloaded or purchased from these sites. The U.S. Environmental Protection Agency (USEPA) supports the Center for Subsurface Modeling Support (http://www.epa.gov/ada/csmos.html). The U.S. Geological Survey maintains a listing of programs that they support (http://water.usgs.gov/software/lists/groundwater/). The International Ground Water Modeling Center, located at the Colorado School of Mines, provides listings of a range of programs (http://igwmc.mines.edu/).

The most widely used computer program for groundwater flow analysis is MODFLOW. The program solves transient, three dimensional groundwater flow using the finite difference method. This program was developed by the U.S. Geological Survey (Harbaugh and MacDonald 1996; Harbaugh et al. 2000; Harbaugh 2005). The MODFLOW computer program is structured so that new groundwater features can be easily added. These added packages represent processes such as aquifer-connected streamflow, recharge and evapotranspiration. MODFLOW continues to evolve with added packages from many researchers. Recent versions of MODFLOW include processes for sensitivity analysis and automated parameter estimation (Hill et al. 2000) and for local grid refinement (Mehl and Hill 2005). MODFLOW is available from the U.S. Geological Survey and is in the public domain.

A number of computer programs are designed to be integrated with MODFLOW output for subsequent solute transport simulations. MODPATH (Pollock 1994) computes three-dimensional flow paths using output from steady-state or transient ground-water flow simulations by MODFLOW using a particle-tracking method. MOC and MOC3D (Heberton et al. 2000) uses a MODFLOW defined flow field to solve three-dimensional solute transport equations using the method of characteristics. MT3DMS (Zheng and Wang 1999) is designed to be used with MODFLOW and can model single or multiple solute species including species interaction reactions. Two programs couple MODFLOW with stream-routing programs to obtain improved models of stream/aquifer interactions. These programs are MODBRNCH (Swain and Wexler 1996) and MODFLOW/DAFLOW (Jobson and Harbaugh 1999).

The U.S. Geological Survey supports and distributes a number of computer programs for solute transport that are independent of MODFLOW. To obtain the groundwater velocities needed to solve these problems these programs incorporate or are coupled to methods for solving the groundwater flow equations. SUTRA (Voss 1984) solves for saturated and unsaturated, constant or variable-density fluid flow, and solute or energy transport in two or three dimensions using the finite element method. HST3D (Kipp 1997) simulates groundwater flow, heat and solute transport in three dimensions. The processes are coupled through the dependence of density on temperature and solute concentration. PHREEQC (Parkhurst and Appelo 1999) is designed to perform aqueous geochemical calculations. It handles speciation, batch-reaction, one-dimensional transport, and inverse geochemical calculations. SHARP (Essaid 1990) is a quasi-three-dimensional model to simulate freshwater and saltwater flow separated by a sharp interface in layered coastal aquifer systems

The USEPA also supports several computer programs for contaminant transport. 3DFEMWATER models flow and transport in three-dimensional, variably-saturated porous media under transient conditions using the finite element method. PRZM3 models one-dimensional flow in the unsaturated zone and transport of pesticide and nitrogen in the crop root zone. It includes modeling capabilities for such phenomena as soil temperature simulation, volatilization and vapor phase transport in soils, irrigation simulation and microbial transformation. WhAEM2000 is distributed by the USEPA and is designed to facilitate capture zone delineation and protection area mapping. It uses the analytical element method.

The U.S. Department of Agriculture Salinity Laboratory has produced the HYDRUS program (Šimůnek et al. 2008) that uses the finite element method to simulate movement of water, heat, and multiple solutes in variably saturated media. Flow and transport can occur in the vertical plane, the horizontal plane, or in a three-dimensional region exhibiting radial symmetry about the vertical axis. The flow equation can account for water uptake by plant roots.

A number of other programs have been produced by researchers for advanced applications. BIOPLUME III (Rifai et al. 1996) is a two-dimensional, finite difference model for simulating the natural attenuation of organic contaminants in groundwater due to the processes of advection, dispersion, sorption, and biodegradation. It is based on the U.S. Geological Survey solute transport code MOC. UTCHEM is a general purpose simulator for multiphase flow and transport including temperature-dependent geochemical reactions and microbiological transformations. It is particularly focused on non-aqueous phase liquids (NAPL) that might occur at a groundwater contamination site. HYDROGEOCHEM is a coupled model of transient hydrologic transport and geochemical reaction in saturated-unsaturated media. It is designed to simulate multiple components and multiple complexed, adsorbed, ion-exchanged, and potentially precipitated species. FEFLOW is a program for three-dimensional, transient saturated and unsaturated flow and contaminant transport. It uses the finite element method and provides a graphically-based groundwater modeling environment

A number of proprietary software programs are available that provide easy-to-use graphical user interfaces for pre- and post-processing for many of the computer programs described above. These pre- and post-processors typically bundle together a suite of programs for both groundwater flow and contaminant transport. Also included may be database software for efficient organization, statistical analysis and display of field information. Assembly of these data into appropriate input formats for the simulation model is typically available. Once the simulation models are executed, these programs provide means to graphically display the results, often using advanced visualization and animation techniques. Popular versions of these programs include Visual MODFLOW, GMS (Groundwater Modeling System), Groundwater Vistas and Processing MODFLOW.

6.2.3 Steps in Model Construction

Constructing a model for a specific site includes a number of steps that are required regardless of the physical processes that are modeled (Anderson and Woessner 1992; Pinder 2002). The problem must be explicitly defined. The specific governing equations and solution method must be identified. A spatial domain must be chosen. Boundary conditions must be identified. Initial conditions are required if the problem is transient. Parameters relevant to the particular physical process studied must be determined. Finally, steps must be taken to calibrate and test the model before use.

Problem Definition: Definition of the problem includes identifying the purpose of the model and determining the specific questions that will be addressed using the model. Definition of the problem impacts all the subsequent steps in model construction.

For a relatively simple task a simple model may be adequate. For example, to estimate the average direction and rate of groundwater flow, a two-dimensional, steady-state model with homogeneous aquifer properties may be sufficient. Complex questions may require complex models. For example, predicting the fate of contaminants in an aquifer over extended time periods, may require a three-dimensional, transient, heterogeneous model that simulates both groundwater flow and solute transport.

Overextending the capabilities of a model should be avoided. Using a simple model to answer complex questions can lead to poor results. If the problem definition evolves during the life of a project, it may be necessary to construct new models.

Domain Definition: The domain is the volume of the subsurface over which the governing equations will be solved. The domain is defined by specifying the location of domain boundaries, both horizontally and vertically. Even if the governing equation is not three-dimensional, the full domain must be considered, since one- or two-dimensional models still imply assumptions about subsurface behavior over a domain volume. For example, a vertically-averaged groundwater flow model requires specification of an aquifer thickness (implying vertical location of the domain) and predicts vertically-averaged flow behavior.

Selection of the domain is critical to successful model construction. The domain defines the needs for boundary and initial conditions and for parameters. Boundary conditions must be imposed on the exterior of the domain. Parameters must be defined within the domain. If the problem is transient, initial conditions must also be defined within the domain.

The domain may be limited to just the saturated zone or just the unsaturated zone or may encompass both zones. When the unsaturated zone is excluded from the domain but unconfined conditions are present, the water table defines the top of the domain. Determination of the water table location, and hence, the location of the top boundary, becomes part of the solution process.

For models that encompass the saturated zone, the bottom of the domain is often located at a significant geologic transition, such as a bedrock interface or low-permeability layer. This can be a good choice of domain location if it can be assumed that aquifer behavior beneath this boundary has insignificant effect on aquifer behavior within the domain. The domain may extend horizontally far beyond the vicinity of the area of interest. For example, a model to predict the change in head from pumping at a well may have an areal domain extending several kilometers from the well. This may be necessary to capture all significant hydrologic elements, such as, nearby recharge areas and surface water bodies.

Generally, the domain should encompass all portions of the subsurface that will have significant interaction with the behaviors that will be modeled. In addition, the domain should be selected so that boundary conditions can be readily identified. The size of the domain will determine the amount of data required to specify boundary and initial conditions and parameters. For this reason, the domain should be no larger than needed. However, if the domain is too small then the computed solution may not accurately reflect aquifer behavior.

Selection of Governing Equation: Selection of the governing equation includes both selection of the physical processes to be represented and the dimensionality of the problem. Solution of the saturated groundwater flow equation is common to most groundwater models. For saturated flow problems, choices are needed concerning the dimensionality of the problem, the importance of representing transient phenomena and the use of the confined assumption. When multiple fluid phases are present, such as air and water or water and oil, the modeler must decide if these phenomena are sufficiently important to include in the model.

When solute transport is included in a model then a wide variety of additional choices are needed. Generally, advection and dispersion are important transport processes.

This implies that a model of groundwater flow is needed. The choices made in modeling flow, such as dimensionality, will follow through to the transport simulation. The most significant choice made in transport simulation is the selection of the specific transport processes to be included in the model, the species to be represented and the phases which the species may occupy. Modeling all the species present in a system and all their possible phases can quickly produce an intractable problem. Clear understanding of the dominant biological and geochemical processes is needed along with consideration of data availability and computational tractability.

Selection of Solution Methodology: The specific method used to solve the governing equations depends on the complexity of the equations, characteristics of the domain and boundary conditions and variability of the parameters and initial conditions. Both analytical and numerical solution methods are available, as described earlier in this chapter. Generally, to reduce the potential for error, the simplest model available should be used. For example, if a problem has appropriate parameters and boundary conditions so that it can be solved analytically, then this would be the preferred approach.

Selection of Boundary and Initial Conditions: Boundary conditions describe the state variable at the boundary of the domain. Depending on the governing equation, this description can take one of several forms. For example, the boundary condition may be the value of the state variable at the boundary or the value the derivative of the state variable at the boundary. Boundary conditions must be specified on the entire surface of the volume defined by the domain. Many solution methods, both analytical and numerical, assume certain boundary conditions on certain boundaries. For example, some groundwater flow solutions may assume that no flow occurs across the bottom boundary of the domain. The modeler should be aware of all boundary conditions implied by the solution method.

Previous chapters describe the general mathematical conditions that are available for the flow and contaminant transport equations at the boundaries of the modeled domain. At a specific site these general boundary conditions are appropriately quantified to match physical conditions at the domain boundary. For example, at an impermeable boundary no flow or solute transport can occur across the boundary. The no-flow condition can be imposed by requiring that the derivative of head normal to the boundary be zero (a Neumann condition, as described in Chapter 2). The absence of solute transport can be imposed by requiring that the total flux normal to the boundary (Cauchy condition in Chapter 3) be zero. Similar usage of the general conditions can represent such physical phenomena as contaminant sources, surface water bodies and recharge areas.

Initial conditions are required for transient problems. These conditions consist of the value of the state variable throughout the domain at the beginning of the simulated time period. If the model is used to simulate past behavior of the aquifer then the beginning of the simulated time period will be a specified point in the past. Alternately, the simulated time period may begin at some time in the future. Finally, simulated time may begin in the present. Several methods are used to obtain initial conditions depending on the availability of data. If past or present data are available, they may be used to construct initial conditions. Care should be taken that constructed initial conditions not have anomalous features that might cause spurious results during early iterations of the simulation. Alternately, a preliminary simulation may be used to produce values of the state variable that can in turn be used as initial conditions for subsequent simulations.

Selection of Parameters: The specific parameters needed to solve a groundwater problem depend on the type of governing equation under consideration and the assumptions implied by the solution method used. The primary parameter of the flow equation is the hydraulic conductivity. Solute transport depends on wide variety of parameters, depending on the processes modeled, such as, dispersivity, sorption parameters and decay parameters. Parameter values may be determined from a combination of field observation, literature values, model calibration and professional judgment. Chapters 2, 3 and 4 describe in more detail the parameters required for different forms of the flow and transport equations and describe some methods for estimating parameter values. Chapter 5 discusses field and laboratory methods for characterizing aquifer parameters.

Calibration: Model calibration is the process of adjusting model elements to obtain a model that matches observed subsurface conditions. Because of the difficulty of accurately determining parameter values and other model elements from field measurements alone, model calibration is usually necessary. The observed conditions used in model calibration typically include hydraulic heads measured throughout the aquifer. For solute transport problems the concentrations of solutes in the groundwater are also considered. Other measured properties, such as streamflow that results from groundwater discharge, may be considered as the observed condition. The ability of the model to reproduce the observed conditions becomes the primary test of success in model construction.

Figure 6.1 depicts the model building process including calibration. After all model elements; domain, boundary conditions and parameters, are set the model is solved. If the aquifer conditions predicted by the model (e.g., heads, concentrations) do not match the observed conditions then the model needs adjustment. Adjustment continues until the match between observed and modeled conditions is satisfactory. It is rare for this match to be exact. In practice, a reasonable degree of error is acceptable, where the allowed error depends upon data reliability, intended use of the model and other site-specific conditions.

In general, any model element can be adjusted during the calibration process, including the governing equation, domain and boundary conditions. This is suggested in Figure 6.1, where the return from a failed test goes to any one of the model elements depending on the type of failure and the judgment of the modeler. For example, it may be determined that modeled flow near a boundary is different than observed flow direction suggesting that a change in boundary condition is needed; or results from a two-dimensional model may suggest that a three-dimensional representation is needed to fully capture the observed flow behavior. While all model elements are subject to change during calibration, it is most common to limit calibration to adjustment of the parameters. The modeler should be careful that parameter values selected are consistent with available field measurements and with literature values obtained for similar sites. Computer programs are available that automate the process of parameter estimation by choosing parameter values that minimize the deviation of the simulated state variable from observed values of the state variable (Hill and Tiedeman 2007).

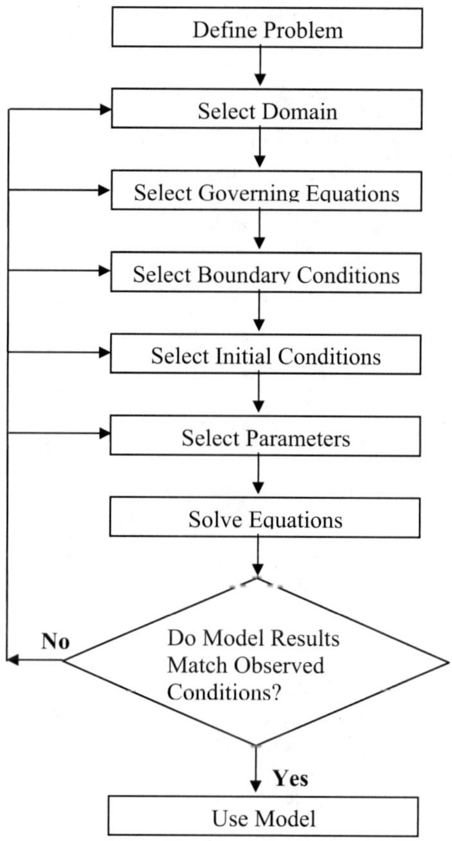

Figure 6.1: Steps in model construction and the model calibration process.

The choice of observations to use is important in the model construction process. There should be sufficient data available to test model reliability for the range of conditions over which the model will be used. For example, if the model is to be used to study transient flow over different hydrologic seasons, then it is advisable to calibrate the model using hydraulic head data over several seasons. It is often the case that insufficient data are available to fully test model reliability. For example, the model may be used to simulate the fate of contaminants based on future conditions but concentration data may only be available for present conditions. The impact of insufficient data on model reliability is site-specific and must always be considered in assessing the credibility of model results.

For hydrogeologically complex sites, it is common to use all available data for calibration purposes so that subsequent validation is not possible without collecting additional data.

6.2.4 Example of Model Construction

To demonstrate the concepts in model construction a small example is presented. The system under study consists of an aquifer that is hydraulically connected to a stream. Consideration is being given to the placement of new pumping wells near the stream. The pumping rates for these wells are limited, in part, by the impact of the wells on the streamflow. A model is needed to predict how heads near the stream, and the related streamflow will be affected by pumping.

A model domain is selected as shown in Figure 6.2 with a horizontal extent measuring 6,000 meters by 5,000 meters. Four wells are to be considered in the analysis as shown on Figure 6.2. The domain is approximately centered around these wells. It is believed that on the east and west boundaries ambient groundwater flow is towards the stream and that this behavior will not be significantly affected by pumping at the four wells. On the north and south boundaries, it is presumed that the head is controlled by recharge outside the domain and will remain at a constant level for the range of pumping rates to be used in the analysis.

It is determined that the system can be adequately represented by modeling just the saturated zone which is relatively thin and unconfined. It is assumed that the wells will pump at steady rates for a long period and that the recharge and streamflow have insignificant seasonal variation. Based on these considerations, the two-dimensional, steady-state, unconfined groundwater flow equation is selected for this problem. Because of the irregular geometry of the stream, analytical solutions are not readily available. Instead a numerical solution method is selected that requires discretizing the domain into rectangular grid cells. For graphical clarity, Figure 6.2 depicts a coarse discretization, although typical models would have many times the number of cells shown here.

Boundary conditions are required for all boundaries of the domain. Along the east and west boundaries, no-flow conditions are imposed. This is consistent with the expectation that flow along these boundaries will be parallel to the boundary so that no flow crosses the boundary. Along the north and south boundaries, specified head (Dirichlet conditions) are imposed. This is consistent with the expectation that heads along these boundaries will remain fixed. The values used at these boundaries are determined from field data. Along the bottom of the domain a no-flow condition is used while on the top of the domain a specified flow is used which is related to the annual precipitation rate for the site. Along the length of the stream a Robbins condition is used to represent the flow that might occur between the stream and the aquifer. Since the problem is steady state, no initial conditions are needed. Model construction proceeds with identification of parameters including hydraulic conductivity, aquifer thickness and stream properties.

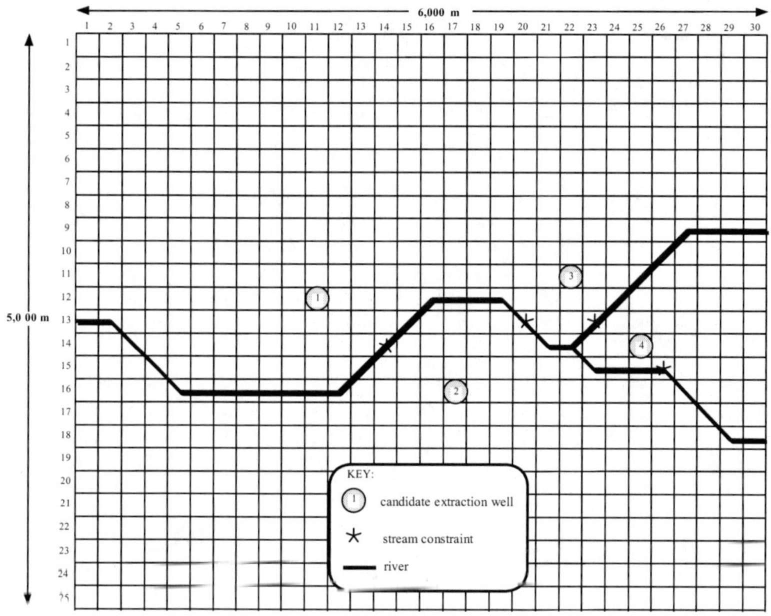

Figure 6.2: Example problem showing model domain, numerical discretization, location of stream through the domain, location of candidate extraction wells and location of stream constraint locations.

6.3 Optimization Methods for Groundwater Management

Simulation models can be used to make predictions of future behavior of the groundwater system under different management scenarios. This "what-if" approach is relatively easy to use and understand, although its breadth is limited by the number of different management scenarios that are tested. If the range of different scenarios is too large to be practically tested by repeated simulation, then the optimization approach may be appropriate.

The optimization approach depends upon a coupling of the simulation model with an optimization algorithm. In the direct simulation approach, each management scenario is selected, a simulation is run and the simulated system state is compared to the desired state of the system. In contrast, the optimization approach requires that the desired state of the system be explicitly defined at the beginning of the analysis. Through the coupled simulation-optimization program, a search is conducted of all possible management scenarios to find the best one that achieves the desired system state. Taken together, the combined simulation and optimization programs and the formulation of the problem are called the management model.

6.3.1 Components of the Management Model Formulation

The management model formulation defines the problem that is to be solved. It has three main components. These are the decision variables, the constraints and the objective function.

The decision variables are the properties of the management system that are to be decided in the design process. These variables are typically any elements of the system that can be controlled by human intervention. Examples of decision variables include the pumping rates to be used at specified wells, the location of wells or recharge facilities, the stage to be maintained in a recharge basin, the location and orientation of a subsurface drain and the configuration of a slurry wall.

Constraints are requirements on the state of the system that must be met when the management scenario is in place and operating. Constraints are often placed both on the state of the system and on the values of the decision variables. Typical constraints on the state of the system include requirements that simulated hydraulic head must be above specified elevations at specified locations and that simulated solute concentrations must be below specified level at specified locations. Typical constraints on the values of decision variables might be that the pumping rate at any well can not exceed a specified rate and that the total pumping rate for the system must be greater than a specified level.

Once constraints are defined two possibilities exist. The first is that no management scenario exists that satisfies all the constraints. A problem with these conditions is referred to as infeasible. An infeasible problem results when the constraints impose requirements that are physically impossible to achieve (where physical possibility is assessed using the simulation model). The second possibility is that management scenarios do exist that satisfy all the constraints. If the problem is feasible, there will generally be a large number of management scenarios that satisfy the constraints. To choose from among them, one more component of the management model is required: the objective function.

The objective function defines the criteria used to determine the best solution. The objective function is typically stated as a function of decision variables and system state variables. Once defined it is either minimized or maximized. Examples of objective functions include those that minimize cost, maximize water supply withdrawals or minimize deviations from a target system state.

The range of possible management model formulations is enormous. Different choices of decision variables and system variables, different constraints and objective functions and different simulation models lead to a wide range of possible uses of the optimization approach. Optimization has been coupled with groundwater flow models for drawdown control, head gradient control, and velocity control (Maddock 1973; Molz and Bell 1977; Aguado and Remson 1980; Ahlfeld et al. 1995; Barlow et al. 1995; Reichard 1995; Nishikawa 1998). Optimization has been coupled with solute transport models for remediation design and related applications (Gorelick et al. 1984; Wagner and Gorelick 1987; Culver and Shoemaker 1992; Minsker and Shoemaker 1996). Many other applications of groundwater management models are possible (Gorelick 1983; Willis and Yeh 1987; Wagner 1995; Ahlfeld and Mulligan 2000).

6.3.2 Management Model Example

To demonstrate the concepts involved in creating a management model, a simple hypothetical example follows. The example is applied to the system described in Section 6.2. The objective of the management problem is to maximize steady-state groundwater extraction rates while limiting drawdown at select locations and maintaining a specified a minimum flow in the stream.

The optimization formulation for this problem can be stated as:

$$\text{maximize} \quad z = Q_1 + Q_2 + Q_3 + Q_4 \tag{6.4}$$

such that

$$Q_j \leq Q_j^u \qquad j = 1,\ldots,4 \tag{6.5}$$

$$S_k\left(Q_1, Q_2, Q_3, Q_4\right) \geq S_k^l \qquad k = 1,\ldots,4 \tag{6.6}$$

$$h_i\left(Q_1, Q_2, Q_3, Q_4\right) \geq h_i^l \qquad i = 1,\ldots,6 \tag{6.7}$$

The decision variables are the pumping rates at each of four candidate extraction well locations, numbered 1 through 4 on Figure 6.2 and notated as Q_1, Q_2, Q_3, Q_4. Equation (6.4) represents the objective function and serves to maximize the sum of pumping rates. The problem contains a total of 14 constraint equations which are clustered into three groups. The first group is represented by Equation (6.5) which states that each of the four pumping rates must be below its specified upper bound, Q_j^u. This bound might be selected based on well yield considerations. The second group is represented by Equation (6.6) and states that the streamflow at location k, S_k, must be above a specified lower bound, S_k^l. This lower bound might be established by regulatory requirements on streamflow. For this example, the streamflow constraints are imposed on the four locations marked as stream constraints on Figure 6.2. The last group of constraints consists of lower bounds on hydraulic head, as expressed in Equation (6.7) where a specified lower bound, h_i^l, is provided for each of six grid locations (not shown).

The streamflow and head at each constraint location depends on the pumping rates, as indicated in Equations (6.6) and (6.7). This relationship is determined by the simulation model. Evaluating the constraints at a particular set of pumping rates requires running the simulation model. The optimization formulation seeks the maximum pump rates that can be achieved while obeying the limitations on simulated head and streamflow.

6.3.3 Software for Groundwater Management Modeling

Several computer programs are available for solving optimization problems. These programs couple specific simulation models with one or more optimization algorithm and make it possible to solve optimization formulations with a variety of objective function and constraint types. Two programs are coupled with MODFLOW and are intended for problems in which the state variables include head and velocity and the primary decision variables are pumping rates at specified locations. A MODFLOW process called GWM (Ground Water Management) contains an internal optimization solver that makes it possible to solve mildly nonlinear problems and mixed binary problems (Ahlfeld et al. 2005). MODMAN (Greenwald 1998) produces a linear program from the formulation components defined by the user and is coupled with a commercial optimization program to obtain a solution. MGO (Modular Groundwater Optimizer) (Zheng and Wang 2002) is a general simulation-optimization program which incorporates MODFLOW and MT3DMS for groundwater resource and quality management.

6.4 References

Aguado, E., and Remson, I. (1980). "Groundwater management with fixed charges." *Journal of Water Resources Planning and Management Division, ASCE,* 106(2), 375-382.

Ahlfeld, D. P., Barlow, P. M., and Mulligan, A. E. (2005). *GWM – A ground-water management process for the U.S. Geological Survey modular ground-water model (MODFLOW-2000),* U.S. Geological Survey Open File Report 2005-1072.

Ahlfeld, D. P., and Mulligan, A.E. (2000). *Optimal management of flow in groundwater systems*, Academic Press, San Diego, Calif..

Ahlfeld, D. P., Page, R. H., and Pinder, G. F. (1995). "Optimal ground-water remediation methods applied to a Superfund site: From formulation to implementation." *Ground Water*, 33(1), 58-70.

Anderson, M. P., and Woessner, W. W. (1992). *Applied groundwater modeling*, Academic Press, San Diego, Calif.

Barlow, P. M., Wagner, B. J., and Belitz, K. (1995). "Pumping strategies for management of a shallow water table: The value of the simulation-optimization approach." *Ground Water*, 34(2), 305-317.

Batu, V. (1998). *Aquifer hydraulics,* John Wiley & Sons, New York.

Boulton, N. S. (1954). "Unsteady radial flow to a pumped well allowing for delayed yield from storage." *International Association of Scientific Hydrology*, Vol. II, 472-477.

Culver, T. B., and Shoemaker, C. A. (1992). "Dynamic optimal control for groundwater remediation with flexible management periods." *Water Resources Research*, 28(3), 629-641.

Domenico, P. A., and Schwartz, F. W. (1990). *Physical and chemical hydrogeology*, John Wiley & Sons, New York, 1990.

Essaid, H. I. (1990). *The computer model SHARP, a quasi-three-dimensional finite-difference model to simulate freshwater and saltwater flow in layered coastal aquifer systems*, U.S. Geological Survey Water-Resources Investigations Report 90-4130.

Fitts, C. R. (2002). *Groundwater science,* Academic Press, Amsterdam, 2002.

Gorelick, S. M. (1983). "A review of distributed parameter ground-water management modeling methods." *Water Resources Research*, 19(2), 305-319.

Gorelick, S. M., Voss, C. I., Gill, P. E., Murray, W., Saunders, M. A., and Wright, M. H. (1984). "Aquifer reclamation design: The use of contaminant transport simulation combined with nonlinear programming." *Water Resources Research*, 20(4), 415-427.

Greenwald, R. M. (1998). *Documentation and user's guide: MODMAN, an optimization module for MODFLOW, version 4.0*, HSI GeoTrans, Freehold, N.J.

Hantush, M. S. (1964). "Hydraulics of Wells." *Advances in Hydroscience*, Ven Te Chow, ed., Academic Press, New York, 281-442.

Hantush, M. S. (1966). "Wells in homogeneous anisotropic aquifers." *Water Resources Research*, 2(2), 273-279.

Hantush, M. S., and Jacob, C. E. (1955). "Non-steady radial flow in an infinite leaky aquifer." *Transactions, American Geophysical Union*, 36(1), 95-100.

Harbaugh, A. W., and McDonald, M. G. (1996). *User's documentation for MODFLOW-96 – An update to the U.S. Geological Survey modular finite-difference ground-water flow model*, U.S. Geological Survey Open-File Report 96-485.

Harbaugh, A. W., Banta, E. R., Hill, M. C., and McDonald, M. G. (2000). *MODFLOW-2000, the U.S. Geological Survey modular ground-water model – User guide to modularization concepts and the groundwater glow process*, U.S. Geological Survey Open-File Report 00-92.

Harbaugh, A. W. (2005). *MODFLOW-2005, the U.S. Geological Survey modular ground-water model – The ground-water flow process*, U.S. Geological Survey Techniques and Methods 6-A16.

Heberton, C. I., Russel, T. F., Konikow, L. F., and Hornberger, G. Z. (2000). *A three-dimensional finite volume eulerian-lagrangian localized adjoint method (ELLAM) for solute-transport modeling*, U.S. Geological Survey Water-Resources Investigations Report 00-4087.

Hill, M. C., Banta, E. R., Harbaugh, A. W., and Anderman, E. R. (2000). *MODFLOW-2000, the U.S. Geological Survey modular ground-water model – User guide to the observation, sensitivity, and parameter estimation processes and three post-processing programs*, U.S. Geological Survey Open-File Report 00-184.

Hill, M. C., and Tiedeman, C. R. (2007). *Effective groundwater model calibration with analysis of data, sensitivities, prediction and uncertainty*, Wiley & Sons, Hoboken, N.J.

Jobson, H. E., and Harbaugh, A. W. (1999). *Modifications to the diffusion analogy surface-water flow model (DAFLOW) for coupling to the modular finite-difference ground-water flow model (MODFLOW)*, U.S. Geological Survey Open-File Report 99-217.

Kipp, K. L. (1997). *Guide to the revised heat and solute transport simulator: HST3D - version 2*, U.S. Geological Survey Water-Resources Investigations Report 97-4157.

Konikow, L. F., and Bredehoeft, J. D. (1974). "Modeling flow and chemical quality changes in an irrigated stream-aquifer system." *Water Resources Research*, 10(3), 546-562.

Maddock, T., III (1973). "Management model as a tool for studying the worth of data." *Water Resources Research*, 9(3), 270-280.

Mehl, S. W., and Hill, M. C. (2005). *MODFLOW-2005, the U.S. Geological Survey modular ground-water model - Documentation of shared node local grid refinement (LGR) and the boundary flow and head (BFH) package*, U.S. Geological Survey Techniques and Methods 6-A12.

Minsker, B. S., and Shoemaker, C. A. (1996). "Differentiating a finite element biodegradation simulation model for optimal control." *Water Resources Research*, 32(1), 187-192.

Molz, F. J., and Bell, L. C. (1977). "Head gradient control in aquifers used for fluid storage." *Water Resources Research*, 13(4), 795-798.

Neuman, S. P. (1972). "Theory of flow in unconfined aquifers considering delayed response of the water table." *Water Resources Research*, 8(4), 1031-1045.

Neuman, S. P., and Witherspoon, P. A. (1968). "Theory of flow in aquicludes adjacent to slightly leaky aquifers." *Water Resources Research*, 4(1), 103-112.

Nishikawa, T. (1998). "Water resources optimization model for Santa Barbara, California." *Journal of Water Resources Planning and Management*, 124(5), 252-263.

Ogata, A., and Banks, R. B. (1961). *A solution of the partial differential equation of longitudinal dispersion in porous media*, U.S. Geological Survey Professional Paper 411-A.

Papadopulos, I. S., and Cooper, H. H., Jr. (1967). "Drawdown in a well of large diameter." *Water Resources Research*, 3(1), 241-244.

Parkhurst, D. L., and Appelo, C. A. J. (1999). *User's guide to PHREEQC (version 2) - A computer program for speciation, batch-reaction, one-dimensional transport, and inverse geochemical calculations*, U.S. Geological Survey Water-Resources Investigations Report 99-4259.

Pinder, G. F. (1973). "A galerkin finite element simulation of groundwater contamination on Long Island, N.Y." *Water Resources Research*, 9(6), 1657-1669.

Pinder, G. F. (2002). *Groundwater modeling using geographical information systems*, John Wiley & Sons, New York.

Pinder, G. F., and Bredehoeft, J. D. (1968). "Application of the digital computer for aquifer evaluation." *Water Resources Research*, 4(5), pp 1069-1093.

Pinder, G. F., and Gray, W. G. (1977). *Finite element simulation in surface and subsurface hydrology*, Academic Press, New York.

Prickett, T. A., and Lonnquist, C. G. (1971). *Selected digital computer techniques for groundwater evaluation*, Illinois State Water Survey Bulletin 55.

Pollock, D. W. (1994). *User's guide for MODPATH/MODPATH-PLOT, version 3: A particle tracking post-processing package for MODFLOW, the U.S. Geological Survey finite-difference ground-water flow model*, U.S. Geological Survey Open-File Report 94-464.

Reichard, E. G. (1995). "Groundwater-surface water management with stochastic surface water supplies: A simulation optimization approach." *Water Resources Research*, 31(11), 2845-2865.

Rifai, H. S., Newell, C. J., Gonzales, J. R., Dendrou, S., Dendrou, B., Kennedy, L., and Wilson, J. (1996). *BIOPLUME III – Natural attenuation decision support system, user's manual, version 1.0*, Air Force Center For Environmental Excellence, San Antonio, Tex.

Segol, G. (1994). *Classic groundwater simulations,* PRT Prentice Hall, Englewood Cliffs, N.J.

Šimůnek, J., van Genuchten, M. Th., and Šejna, M. (2008). "Development and applications of the HYDRUS and STANMOD software packages and related codes." *Vadose Zone Journal*, 7(2), 587-600.

Swain, E. D., and Wexler, E. J. (1996). *A coupled surface-water and ground-water flow model (MODBRNCH) for simulation of stream-aquifer interaction*, U.S. Geological Survey Techniques of Water-Resources Investigations, Book 6, Chap. A6.

Theis, C. V. (1935). "The relation between the lowering of the piezometric surface and the rate and duration of discharge of a well using ground-water storage." *Transactions, American Geophysical Union*, 16, 519-524.

Thiem, G. (1906). *Hydrologische methoden* (in German), J. M. Gebhardt, Leipzig, Germany.

Trescott, P. C., Pinder, G. F., and Larson, S. P. (1976). *Finite difference model for aquifer simulation in two dimensions with results of numerical experiments*, Techniques of Water Resources Investigations of the U.S. Geological Survey Book 7, Chap. C1.

Voss, C. I. (1984). *A finite-element simulation model for saturated-unsaturated, fluid-density-dependent ground-water flow with energy transport or chemically-reactive single-species solute transport,* U.S. Geological Survey Water-Resources Investigations Report 84-4369.

Wagner, B. J. (1995). "Recent advances in simulation-optimization groundwater management modeling." *Reviews of Geophysics, Supplement*, 33, 1021-1028.

Wagner, B. J., and Gorelick, S. M. (1987). "Optimal groundwater quality management under parameter uncertainty." *Water Resources Research*, 23(7), 1162-1174.

Willis, R., and Yeh, W. W.-G. (1987). *Groundwater systems planning and management*, Prentice-Hall, Englewood Cliffs, N.J.

Zheng, C., and Wang, P. P. (1999). *MT3DMS: A modular three-dimensional multispecies transport model for simulation of advection, dispersion and chemical reactions of contaminants in groundwater systems; Documentation and user's guide*, Contract Report SERDP-99-1, U.S. Army Engineer Research and Development Center, Vicksburg, Miss.

Zheng, C., and Wang, P. P. (2002). *MGO: A modular groundwater optimizer incorporating MODFLOW and MT3DMS; Documentation and user's guide*, The University of Alabama and Groundwater Systems Research Ltd.

CHAPTER 7

MODEL CALIBRATION AND PARAMETER STRUCTURE IDENTIFICATION IN CHARACTERIZATION OF GROUNDWATER SYSTEMS

Frank T.-C. Tsai[1] and William W.-G. Yeh[2]
[1]Department of Civil & Environmental Engineering, Louisiana State University
Baton Rouge, LA 70803-6405
[2]Department of Civil & Environmental Engineering, University of California
Los Angeles, CA 90095-1593

Abstract

This chapter summarizes recent progress in model calibration and model uncertainty assessment in groundwater modeling. The main focus is on parameter structure identification of hydraulic conductivity using the Bayesian maximum likelihood approach. In general, parameter structure identification seeks to identify the parameter dimension, parameter pattern, and associated parameter values. We discuss the importance of parameterization, which is used to define parameter structure and to formulate the inverse problem of parameter structure identification. We propose a generalized parameterization (GP) method for parameter structure identification. GP integrates the traditional zonation and interpolation methods (including kriging) and avoids the shortcomings of each individual method. It captures aquifer heterogeneity by a combination of zonal structure and continuous distribution. An indicator generalized parameterization (IGP) is further introduced to address the interpolation point selection problem, which is an important issue in parameterization. We introduce model selection, model discrimination, and statistical information criteria to select the best parameterization method. To resolve the non-uniqueness problem in parameterization due to lack of data and data uncertainty, we employ Bayesian model averaging (BMA) to consider multiple parameterization methods for parameter structure identification. Finally, based on several optimality criteria we offer the experimental design as the ultimate goal of collecting additional informative data to reduce estimation uncertainty.

7.1 Introduction

Model calibration in groundwater modeling is a difficult inverse problem because of the lack of knowledge about the real aquifer system and the scarcity of observations. A more complex model requires more data for calibration. Furthermore, there are difficulties in identifying aquifer structure due to the heterogeneity and anisotropy of the physical parameters embedded in the governing equation. Other than physical parameters, insufficient information regarding the sink/source term, initial condition, as well as the location and types of boundary conditions, presents additional obstacles.

7.1.1 Model Structure Identification

Model structure identification is a critical component in model calibration for several reasons. Principally, an incorrect model structure may lead to failure in model applications and, consequently, erroneous policies. Model structure identification plays an important role not only in parameter identification but also in model structure distinguishability and experimental design. The traditional assumption that model structure is known is unrealistic and impractical.

Model structure incorporates a broad spectrum of elements from physical laws to conceptual models. For example, in order to simulate groundwater flow, physical laws and assumptions are used to represent an aquifer as an electronic model, a lumped parameter model or a distributed parameter model. From a macroscopic point of view, we consider groundwater flow based on Darcy's law. However, at the pore scale, microscopic velocity is governed by the Navier-Stokes equations. Different governing equations entail different model structures. For instance, a two-dimensional model has a different model structure than a three-dimensional model. In the conceptualization of hydrogeologic activity, model structures also vary according to initial and boundary conditions. In a distributed parameter model, there are a number of ways to describe the heterogeneity and anisotropy of hydraulic conductivity that will result in different model structures.

Model structure error results when an inexact model structure is used to simulate a real aquifer system. Model structure error also can represent the difference between different model structures. In groundwater modeling, model structure error is impossible to avoid because the conceptual model always differs from the real subsurface system and never captures the real aquifer. According to the concept of model structure, it is easy to recognize the sources of model structure error. For example, we may approximate a distributed system by a lumped system. We may simplify the heterogeneous and anisotropic aquifer to a homogeneous and isotropic one. We may simplify a three-dimensional flow field to a two-dimensional model. Finally, the specified initial and boundary conditions may be incorrect.

In most instances, model structure error dominates numerical and measurement errors and is likely to cause failures in model applications. Therefore, it is highly desirable to reduce model structure error in model calibration. However, the inverse problem of model structure identification is extremely difficult to solve and usually requires a large amount of data from geological and hydrogeological investigations. Even though we recognize the importance of reducing model structure error by properly specifying the initial condition, boundary conditions, and pumping/recharging rates, a full implementation of model structure identification is generally infeasible. With a limited amount of data in quantity and quality, most published studies have focused on reducing the parameter structure error in connection with the identification of distributed parameters, such as hydraulic conductivity. This chapter only considers parameter structure identification.

7.1.2 Parameter Structure Identification

The inverse problem of parameter structure identification in a distributed parameter system seeks to identify the parameter structure (pattern) and the corresponding parameter values of a distributed parameter. The first crucial task is to characterize the distributing manner of a parameter for parameter structure identification. The second crucial task is to identify the parameter structure.

A standard way of characterizing a distributed parameter is by means of parameterization. For model simulation and application, each distributed parameter must be represented by a selected parameterization scheme. For example, a distributed hydraulic conductivity field can be parameterized into zones or a continuous distribution. Early literature dealing with the determination of a parameterization pattern for the hydraulic conductivity field includes Emsellem and de Marsily (1971), Yeh and Yoon (1976), Yeh and Yoon (1981), Yeh et al. (1983), and Sun and Yeh (1985). Eppstein and Dougherty (1996) used an extended Kalman filter to identify transmissivity and zonation pattern simultaneously. Mayer and Huang (1999) compared genetic algorithm (GA) and quasi-Newton methods in a three-dimensional, coupled inverse problem. In their study, the distributed hydraulic conductivity was identified using deterministic (zonation) and stochastic distributions. They concluded that the GA method provided a robust inverse solution, but computationally was less efficient than the gradient-based method. Moreover, the identified stochastic hydraulic conductivity field showed good agreement with observations. Karpouzos et al. (2001) developed a multi-population GA to identify distributed transmissivity in a two-dimensional steady-state groundwater flow. Gangnis and Smith (2001) proposed a Bayesian approach to quantify the model error in groundwater prediction. Recently, Sun et al. (1998), Sun and Sun (2002a), and Tsai et al. (2003b) introduced the concept of the extended inverse problem (EIP) and the generalized inverse problem (GIP) for parameter structure identification. Many studies have investigated the parameter structure identification problem by examining the data configuration and basis functions in interpolation and zonation methods (Yeh and Yoon 1976; Yeh and Yoon 1981; Sun and Yeh 1985; Zheng and Wang 1996; Sun et al. 1998; Tsai et al. 2003a; Tsai et al. 2003b.; Tsai et al. 2005). Parameter structure identification in connection with kriging mainly has been concerned with the estimation of the structural parameters embedded in spatial correlation functions (Kitanidis and Vomvoris 1983; Wagner and Gorelick 1987; Cassiani and Medina 1997). Parameter structure identification also has been studied using the pilot point method under the kriging paradigm (Certes and de Marsily 1991; LaVenue et al. 1995; RamaRao et al. 1995; Gomez-Hernanez et al. 1997; Doherty 2003).

Moreover, geological structure information obtained from well logs and seismic measurements can be incorporated directly into the inversion (Rubin and Hubbard

2005; Sun et al. 1995; Hyndman and Gorelick 1996; Koltermann and Gorelick 1996). To reduce uncertainty, Poeter and McKenna (1995) used hard data (e.g., observation of lithology), soft data (e.g., expert opinion) and stochastic simulations to interpret the hydraulic conductivity distribution in three dimensions. However, geological information is often insufficient to support certain parameter structures. Statistical model selection criteria frequently are used to identify the best model among a set of alternative models with different structures (Carrera and Neuman 1986a; Carrera and Neuman 1986c; Hyun and Lee 1998). Aschenbrenner and Ostin (1995) used conditional standard deviation and Jiao and Lerner (1996) used sensitivity analysis to determine the best zonation model.

We note that parameter structure is not limited to physical parameters only. It also can be used for other kinds of distributed parameters in a conceptual model, such as the initial conditions, locations and types of boundary conditions, and sink/source terms.

7.1.3 Non-Uniqueness of Parameterization Methods

Groundwater inverse modeling has a long history of using just one parameterization method along with a conceptualized simulation model for hydraulic conductivity estimation (Yeh 1986; McLaughlin and Townley 1996) and model error analysis (Gaganis and Smith 2001). Even in the case of single parameterization, there are several inherent difficulties. For example, inverse modeling requires preparing various types of geological and hydrological data for the development of a simulation model, as well as running the computationally-intensive simulation model when it is developed. Due to data limitations in quantity and quality, uncertainties in selecting suitable parameterization methods and developing simulation models are unavoidable. Model selection, model elimination, model reduction, and model discrimination techniques have been employed to resolve these difficulties in inverse modeling (Carrera and Neuman 1986a; Carrera and Neuman 1986b; Knopman and Voss 1988; Usunoff et al. 1992; Aschenbrenner and Ostin 1995; Hyun and Lee 1998; Jiao and Lerner 1996; Hill et al. 1998). Most studies have aimed at selecting a single parameterization method among the candidates.

However, it has been recognized that, for a given set of data, many independent parameterization methods are acceptable under a certain criterion. By selecting a single parameterization method, we are more likely to ignore parameterization uncertainty and underestimate prediction uncertainty. Moreover, parameterization schemes published in the literature, including interpolation, zonation, and geostatistical methods, have been criticized as very rigid because they generate either a completely smooth or completely piecewise-homogeneous distributions. Adoption of either type of distribution is unrealistic and will result in inaccurate estimation. To better characterize heterogeneity, it is necessary to increase the flexibility of parameterization.

In this chapter, we limit our discussion to parameter structure identification. In Section 7.2, we introduce a generalized parameterization (GP) method that integrates zonation and interpolation. In Section 7.3, we introduce the extended inverse problem (EIT) and the generalized inverse problem (GIP) for parameter structure identification. In Section 7.4, we discuss the interpolation point selection problem, which is also an important issue in parameter structure identification. Section 7.5 discusses the model selection problem in identifying the best parameterization method. Section 7.6 introduces the Bayesian model averaging method, which integrates all candidate parameterization methods for inverse groundwater modeling. Section 7.7 reviews optimal experimental design for parameter structure identification. Section 7.8 summarizes and concludes this chapter.

7.2 Parameterization of Heterogeneity

The key to parameter structure identification in inverse groundwater modeling is parameterization, which reduces a continuously distributed parameter to a finite and small number of parameters. Zonation and interpolation methods commonly are used for parameterization.

7.2.1 Zonation

The zonation method, a piece-wise constant method, is easy to implement (Yeh and Yoon 1976; Cooley et al. 1986; Jiao and Lerner 1996; Eppstein and Dougherty 1996; Sun et al. 1998; Hyun and Lee 1998; Tsai et al. 2003a; Tsai et al. 2003b). Using this method, the field of a distributed parameter (e.g., hydraulic conductivity) is partitioned into a number of zones. Each zone is characterized by a constant parameter value. The zonation method is easy to implement; however, difficulty arises with the determination of the number and shape of zones to form a zonal structure.

To illustrate the zonation method, consider p, which represents any form of hydraulic conductivity field (e.g., logarithmic values of hydraulic conductivity) in a region (Ω). Let the region be partitioned into m zones, $\Omega_j, j=1,2,\cdots,m$. The zones do not overlap each other, i.e. $\{\Omega_i \cap \Omega_j = 0 \text{ if } i \neq j\}$, and all zones cover the entire region, i.e. $\Omega = \Omega_1 + \Omega_2 + \cdots \Omega_m$. The zonal value is denoted by p_j. Consequently, the parameter value at site $\mathbf{x}_0$ is determined by the zonal value

$$p_{\text{zonation}}(\mathbf{x}_0) = \{p_j \lfloor \mathbf{x}_0 \in \Omega_j\}. \tag{7.1}$$

One approach to partitioning the domain (Ω) is to sequentially place a set of basis points into the domain, where the zonal values are determined by the values of the basis points. As shown in Figure 7.1(a)-(e), the process starts from two basis points, and a bisector divides the domain by separating the two points. Add one basis point at a time into the domain to divide a zone, where the new basis point resides, into two zones. This can be seen as a tree-regression procedure (Sun and Sun 2002), which creates nested zonal structures. As a result, each zone (Ω_j) contains only one basis point located at $\mathbf{x}_j \in \Omega_j$. The domain is the sum of all zones, i.e. $\Omega = \sum_j \Omega_j$. In the case where two adjacent basis points have the same values, two zones still result.

In this approach, the zonal structure depends on the ordering of the basis points to be added. As shown in Figure 7.1 (f)-(j), using the same basis point locations, the final zonal structure changes when the ordering changes.

Voronoi Tessellation (VT): Voronoi tessellation (or Dirichlet tessellation), originally developed by mathematicians (Dirichlet 1850; Voronoi 1908), is a commonly used zonation method and has been applied to meteorology and hydrology (Thiessen 1911), computational geometry (Green and Sibson 1978; Watson 1981; Bowyer 1981), geography (Boots 1986), and hydrogeology (Tsai et al. 2003a; Tsai et al. 2003b). Readers should refer to Okabe et al. (2000) for various applications of VT.

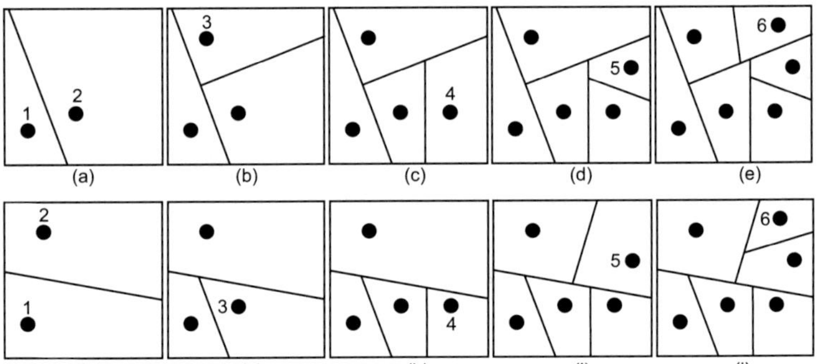

Figure 7.1: A zonation method based on tree-regression procedure.

To tessellate a region (Ω) into m Voronoi zones, m distinct basis points are placed at locations $\{\mathbf{x}_j, j = 1,2,\cdots,m\}$ in the region. The Voronoi zone, Ω_j, associated with a basis point $\mathbf{x}_j$ is defined by,

$$\Omega_j = \left\{\mathbf{x} \mid d(\mathbf{x},\mathbf{x}_j) < d(\mathbf{x},\mathbf{x}_\ell),\ \ \forall \ell \neq j, \ell = 1,2,\cdots,m\right\}, \tag{7.2}$$

where $d(\mathbf{x},\mathbf{x}_j) = \sqrt{(\mathbf{x}-\mathbf{x}_j)^T(\mathbf{x}-\mathbf{x}_j)}$ is the Euclidean distance in any dimension between two points $\mathbf{x}$ and $\mathbf{x}_j$. Figure 7.2 shows the Voronoi zones based on a sequence of placing basis points into the domain.

Each Voronoi zone includes only one basis point. A Voronoi zone is a line segment in a one-dimensional region, an area in a two-dimensional region, and a volume in a three-dimensional region. A generalized Voronoi diagram can be obtained by considering the weighted Euclidean distance in Equation (7.2) (Okabe et al. 2000; Tan et al. 2008).

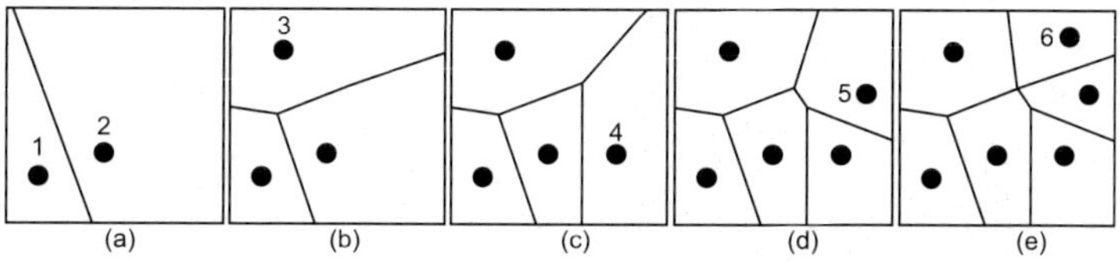

Figure 7.2: A Voronoi tessellation method.

The zonal structure made by VT is unique regardless of the ordering of the addition of the basis points. VT does not create nested zonal structures by adding or deleting basis points. Two basis points are considered "natural neighbors" when their Voronoi zones have a common edge. For example, in Figure 7.2(e), point 5 and point 2 are natural neighbors, but point 5 and point 1 are not. In general, three Voronoi edges join to form a vertex, which is the center of a circle passing three natural neighbors. A non-degenerate Voronoi diagram is formed if each vertex has only three Voronoi edges; otherwise, a degenerate Voronoi diagram results. For example, Figure 7.2(e) shows a degenerate Voronoi diagram because points 2, 3, 5 and 6 join at one vertex. In other words, a circle centered at the vertex passes points 2, 3, 5 and 6.

7.2.2 Interpolation

Considering parameters are continuously distributed in a region, interpolation methods are often used to obtain smooth distributions, e.g. the finite element interpolation methods (Distefano and Rath 1975, Yeh and Yoon 1981) and the natural neighbor interpolation method (Sambridge et al. 1995; Tsai et al. 2005). Sun and Yeh (1985) suggested an adjustable inverse distance-based parameterization method for parameter structure identification. Their method built a bridge between zonation and the inverse distance-based method by tuning a set of shape parameters. The kriging method provides another parameterization scheme that incorporates the geostatistical properties from sample data (Delhomme 1979; Neuman and Yakowitz 1979; Clifton and Neuman 1982; de Marsily 1986; Kitanidis 1997; Sun et al. 1995). Sufficient sample data are needed to generate a reliable semivariogram.

Using an interpolation method, unknown parameter values are interpolated using a linear combination of basis functions and corresponding values of basis points. To estimate the hydraulic conductivity value $p(\mathbf{x}_0)$ at an unsampled site $\mathbf{x}_0$, a generic form of interpolation with m basis points with values $\{p_j\}$ is expressed as

$$p_{\text{interpolation}}(\mathbf{x}_0) = \sum_{j=1}^{m} \phi_j(\mathbf{x};\mathbf{v}) p_j , \tag{7.3}$$

where $\phi_j(\mathbf{x};\mathbf{v})$ is the basis function and $\mathbf{v}$ represents a set of parameters governing the basis functions. An unbiased estimation requires $\sum_{j=1}^{m} \phi_j(\mathbf{x};\mathbf{v}) = 1$. This chapter illustrates the natural neighbor interpolation method, inverse distance interpolation method, and ordinary kriging method.

Natural Neighbor Interpolation (NN) Method: This method is based on the first-order and second-order Voronoi zones shown in Figure 7.3. The first-order Voronoi cells (Ω_j) originally are created by the sample data. An unsampled location $\mathbf{x}_0$ creates another first-order Voronoi cell (Ω_0) that overlaps parts of the first-order Voronoi cells of the sample data. A second-order Voronoi cell Ω_{0j} is defined as the overlapping area of Ω_0 to Ω_j, represented as

$$\Omega_{0j} = \left\{ \mathbf{x} \mid d(\mathbf{x},\mathbf{x}_0) < d(\mathbf{x},\mathbf{x}_j) < d(\mathbf{x},\mathbf{x}_\ell), \ \forall \ell \neq j, 0, \ \ell = 1, \cdots, m \right\}. \tag{7.4}$$

The sampled location $\mathbf{x}_j$ is a natural neighbor of $\mathbf{x}_0$ if the second-order Voronoi cell Ω_{0j} (the overlapping area) exists. Sibson (1981) introduced an NN method with the basis functions based on the natural neighbor coordinates defined as

$$\phi_j^{NN}(\mathbf{x}_0) = \frac{\Omega_{0j}}{\Omega_0}. \tag{7.5}$$

For those sampled sites that are not natural neighbors of $\mathbf{x}_0$, the basis function is $\phi_j = 0$.

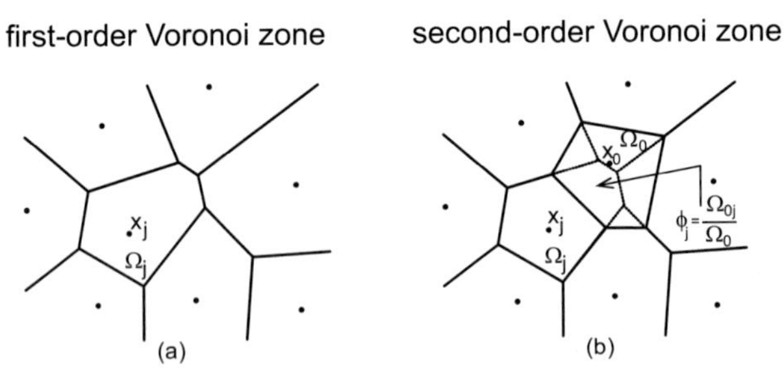

Figure 7.3: The first and second order Voronoi zones.

The method uses local sample data for parameterization based on the weighted average of its natural neighboring basis points. It is different from the global methods (e.g., the inverse distance interpolation and kriging) as both use information from all basis points for the estimator.

Inverse Distance Interpolation (ID) Method: The inverse distance interpolation (ID) method assumes that the weight of a sampled site relative to an unsampled site is inversely proportional to the distance. The basis function is

$$\phi_j^{ID}(\mathbf{x}_0) = \frac{d^{-\alpha}(\mathbf{x}_0, \mathbf{x}_j)}{\sum_{i=1}^{m} d^{-\alpha}(\mathbf{x}_0, \mathbf{x}_i)} \tag{7.6}$$

where α is the weighting power. The influence of the nearest sample point will dominate when the value of the weighting power is large. A significant high α value will form a Voronoi zonation (Gotway et al. 1996). In general, an inverse squared distance method (i.e., $\alpha = 2$) commonly is used.

Ordinary Kriging (OK) Method: The ordinary kriging (OK) method considers the spatial correlation structure of the sampled data to determine the best basis function values such that the conditional variance is minimized (Cressie and Wikle 1998; Olea 1999). The OK method is a stochastic interpolation method, which provides the conditional mean and conditional variance of the estimate. In the method, ϕ_j represents the kriging weight and is determined by

$$\sum_{j=1}^{m} \phi_j^{OK} \gamma(\mathbf{x}_i, \mathbf{x}_j) - \mu = \gamma(\mathbf{x}_i, \mathbf{x}_0) \quad , i = 1, 2, \cdots, m \tag{7.7}$$

where γ is the semivariogram. The Lagrange multiplier μ ensures the unbiasedness constraint.

7.2.3 Generalized Parameterization (GP) Method

Presumption of either a pure zonation structure or a pure continuous distribution is unrealistic and impractical and can lead to poor model predictions. Moreover, the traditional interpolation and zonation methods are generally applicable to a deterministic field. Tsai and Yeh (2004) developed a generalized parameterization (GP) method to increase parameterization flexibility by combining zonation with interpolation. Tsai (2006) further advanced GP under a geostatistical framework and extended geostatistical theory to non-kriging interpolation.

This section introduces the GP method, which combines the zonal conditional estimate (Equation (7.1) and the interpolation estimate Equation (7.3)) through a set of weighting coefficients $\boldsymbol{\beta}_m = \{\beta_1, \beta_2, \cdots, \beta_m\}$. The conditional estimate for the value at an unsampled site $\mathbf{x}_0$ from the m sample sites is

$$p_{\mathrm{GP}}(\mathbf{x}_0) = \sum_{\substack{j=1 \\ j \neq k(\mathbf{x}_0)}}^{m} \phi_j \left(p_j - p_{k(\mathbf{x}_0)}\right)\beta_j + p_{k(\mathbf{x}_0)} \tag{7.8}$$

where β_j are the weighting coefficients bounded between 0 and 1; $p_{k(\mathbf{x}_0)}$ is the kth sample value when location $\mathbf{x}_0$ is in the kth zone; and $k(\mathbf{x}_0)$ represents the kth sample data index for the unsampled site $\mathbf{x}_0$ in the kth zone.

The GP method splits the ultimate weights of the sample data to an unsampled site in two parts. One part is $\sum_{\substack{j=1 \\ j \neq k(\mathbf{x}_0)}}^{m} \phi_j \beta_j p_j$, which contains a set of sample data with zones that do not contain the unsampled site $\mathbf{x}_0$. The other part is $\left(1 - \sum_{\substack{j=1 \\ j \neq k(\mathbf{x}_0)}}^{m} \phi_j \beta_j\right) p_{k(\mathbf{x}_0)}$, where the weight is solely from the sample point $p_{k(\mathbf{x}_0)}$. The sum of these two parts constitutes the GP method.

Let p be an intrinsic field. Then Equation (7.8) represents a conditional estimation (or conditional mean) for the random field of p, which depends on a set of sample data. The calculated estimation error is

$$\mathrm{E}[p - p_{\mathrm{GP}}] = \mathrm{E}[p] - \left\{ \sum_{\substack{j=1 \\ j \neq k(\mathbf{x}_0)}}^{m} \phi_j \mathrm{E}\left[p_j - p_{k(\mathbf{x}_0)}\right]\beta_j + \mathrm{E}\left[p_{k(\mathbf{x}_0)}\right] \right\} = 0 \tag{7.9}$$

which is an unbiased estimation regardless of the weighting coefficient values.

The GP method provides greater flexibility for parameterizing a spatially-correlated random field. If the weighting coefficients are all zeros, a zonal structure, which is a random field due to the geostatistical properties of the sample data, results. With the same data set, if all weighting coefficients are unity, a smooth interpolated field, which is also a conditional estimate for a random field, results. This indicates that the kriging and non-kriging interpolation methods are applicable to stochastic estimations. For an arbitrary combination of weighting coefficients varying from 0 to 1, a mixed distribution of zonal structure and interpolated distribution results. Moreover, the GP method is applicable to interpolation methods with a global as well as local neighborhood (e.g., the natural neighbor interpolation).

Setting $0 \leq \beta_j \leq 1$ ensures that p_{GP} is between p_{zonation} and $p_{\text{interpolation}}$. If β_j takes on arbitrary values, Equation (7.8) is still an unbiased estimator, but may provide unrealistic parameter values outside the range between the zonal values and interpolated values. With $0 \le \beta_j \le 1$, the GP method ensures a conditional estimate bounded between the zonal structure and the interpolated field.

The GP method extends the concept of conditional estimation by incorporating a zonal structure in the stochastic estimation problem. It can be expected that the GP method has more flexibility and potential to characterize the non-smoothness in the stationary and non-stationary fields. Non-smoothness estimation cannot be done solely by the interpolation method or the zonation method. In the next section, we extend geostatistical theory beyond kriging.

7.2.3.1 *Covariance using Generalized Parameterization for Intrinsic Field*

For an intrinsic random field, the covariance of the estimation error for a pair of estimates at two locations $\mathbf{x}_p$ and $\mathbf{x}_q$ using the GP method is (Tsai 2006)

$$\begin{aligned} C_{\mathrm{GP}}\left(\mathbf{x}_p,\mathbf{x}_q\right) = & \sum_{\substack{i=1 \\ i\neq k(\mathbf{x}_p)}}^{m} \sum_{\substack{j=1 \\ j\neq k(\mathbf{x}_q)}}^{m} \beta_i^p \beta_j^q \phi_i^p \phi_j^q R\left(\mathbf{x}_i,\mathbf{x}_j\right) \\ & - \sum_{\substack{i=1 \\ i\neq k(\mathbf{x}_p)}}^{m} \beta_i^p \phi_i^p R\left(\mathbf{x}_i,\mathbf{x}_q\right) - \sum_{\substack{j=1 \\ j\neq k(\mathbf{x}_q)}}^{m} \beta_j^q \phi_j^q R\left(\mathbf{x}_p,\mathbf{x}_j\right) + R\left(\mathbf{x}_p,\mathbf{x}_q\right) \end{aligned} \tag{7.10}$$

where

$$R\left(\mathbf{x}_a,\mathbf{x}_b\right) = \gamma\left(\mathbf{x}_a,\mathbf{x}_{k(\mathbf{x}_q)}\right) + \gamma\left(\mathbf{x}_b,\mathbf{x}_{k(\mathbf{x}_p)}\right) - \gamma\left(\mathbf{x}_a,\mathbf{x}_b\right) - \gamma\left(\mathbf{x}_{k(\mathbf{x}_p)},\mathbf{x}_{k(\mathbf{x}_q)}\right). \tag{7.11}$$

The subscript (a,b) in Equation (7.11) represents the subscripts (i,j), (i,q), (p,j), and (p,q) in the functions $R\left(\mathbf{x}_i,\mathbf{x}_j\right)$, $R\left(\mathbf{x}_i,\mathbf{x}_q\right)$, $R\left(\mathbf{x}_p,\mathbf{x}_j\right)$, and $R\left(\mathbf{x}_p,\mathbf{x}_q\right)$, respectively, in Equation (7.10). The GP covariance is symmetric, i.e., $C_{\mathrm{GP}}\left(\mathbf{x}_p,\mathbf{x}_q\right) = C_{\mathrm{GP}}\left(\mathbf{x}_q,\mathbf{x}_p\right)$, but the function R in Equation (7.11) is not symmetric, i.e., $R\left(\mathbf{x}_p,\mathbf{x}_q\right) \neq R\left(\mathbf{x}_q,\mathbf{x}_p\right)$, except for the case of the same location $\mathbf{x}_p = \mathbf{x}_q$. The GP covariance is zero when the location $\mathbf{x}_p$ or $\mathbf{x}_q$ is at a sampled location.

Equation (7.10) represents a general form of assessing the covariance of estimation error for any combination of zonation and interpolation methods. The variance of estimation error at $\mathbf{x}_0$ is obtained by setting $\mathbf{x}_p = \mathbf{x}_q = \mathbf{x}_0$ in Equation (7.10):

$$\sigma_{\mathrm{GP}}^2(\mathbf{x}_0)=\sum_{\substack{i=1\\ i\neq k(\mathbf{x}_0)}}^{m}\sum_{\substack{j=1\\ j\neq k(\mathbf{x}_0)}}^{m}\beta_i\beta_j\phi_i\phi_j R(\mathbf{x}_i,\mathbf{x}_j)-2\sum_{\substack{i=1\\ i\neq k(\mathbf{x}_0)}}^{m}\beta_i\phi_i R(\mathbf{x}_i,\mathbf{x}_0) +2\gamma\left(\mathbf{x}_0,\mathbf{x}_{k(\mathbf{x}_0)}\right) \quad (7.12)$$

For the zonation case ($\boldsymbol{\beta}_m=0$), the variance of the zonal structure is the variogram ($2g$), measured according to the distance between $\mathbf{x}_0$ and $\mathbf{x}_{k(\mathbf{x}_0)}$ within the kth zone:

$$\sigma_{\text{zonation}}^2(\mathbf{x}_0)=2\gamma\left(\mathbf{x}_0,\mathbf{x}_{k(\mathbf{x}_0)}\right). \quad (7.13)$$

For the interpolation case ($\boldsymbol{\beta}_m=1$), the variance of the interpolated field is

$$\sigma_{\text{interpolation}}^2(\mathbf{x}_0)=\sum_{\substack{i=1\\ i\neq k(\mathbf{x}_0)}}^{m}\sum_{\substack{j=1\\ j\neq k(\mathbf{x}_0)}}^{m}\beta_i\beta_j\phi_i\phi_j R(\mathbf{x}_i,\mathbf{x}_j)-2\sum_{\substack{i=1\\ i\neq k(\mathbf{x}_0)}}^{m}\beta_i\phi_i R(\mathbf{x}_i,\mathbf{x}_0) +2\gamma\left(\mathbf{x}_0,\mathbf{x}_{k(\mathbf{x}_0)}\right) \quad (7.14)$$

Using the OK weights, the OK conditional covariance is

$$\mathrm{C}_{OK}(\mathbf{x}_p,\mathbf{x}_q)=-\sum_{\substack{j=1\\ j\neq k(\mathbf{x}_q)}}^{m}\phi_j^{(OK)}R(\mathbf{x}_p,\mathbf{x}_j)+R(\mathbf{x}_p,\mathbf{x}_q). \quad (7.15)$$

The variance of the OK is

$$\sigma_{OK}^2(\mathbf{x}_0)=-\sum_{\substack{j=1\\ j\neq k(\mathbf{x}_0)}}^{m}\phi_j^{OK}R(\mathbf{x}_j,\mathbf{x}_0)+2\gamma\left(\mathbf{x}_0,\mathbf{x}_{k(\mathbf{x}_0)}\right). \quad (7.16)$$

The zonal structure has the largest GP variances because only one sample datum is used for the estimation. The interpolation method reduces the GP variances by incorporating more sample points. The kriging method results in the minimum conditional variance (kriging variance) because the kriging is known as the "best" linear unbiased estimator (BLUE).

7.2.3.2 *GP Covariance with OK Weights for Second-order Stationary Field*

For a second-order stationary random field, the following relationship can be established:

$$\mathrm{cov}_u(\mathbf{x}_p,\mathbf{x}_q)=\sigma_u^2-\gamma(\mathbf{x}_p,\mathbf{x}_q) \quad (7.17)$$

where $\mathrm{cov}_u(\mathbf{x}_p, \mathbf{x}_q)$ is the unconditional covariance between a pair of estimates at locations $(\mathbf{x}_p, \mathbf{x}_q)$; and $\sigma_u^2 = \mathrm{cov}_u(0)$ is the unconditional variance. Consequently, the function $R(\mathbf{x}_a, \mathbf{x}_b)$ becomes

$$R(\mathbf{x}_a, \mathbf{x}_b) = \mathrm{cov}_u(\mathbf{x}_a, \mathbf{x}_b) + \mathrm{cov}_u\left(\mathbf{x}_{k(\mathbf{x}_p)}, \mathbf{x}_{k(\mathbf{x}_q)}\right) - \mathrm{cov}_u\left(\mathbf{x}_a, \mathbf{x}_{k(\mathbf{x}_q)}\right) - \mathrm{cov}_u\left(\mathbf{x}_b, \mathbf{x}_{k(\mathbf{x}_p)}\right). \quad (7.18)$$

Using the OK weights in the GP method, the GP covariance can be derived using Equation (7.12) in terms of the unconditional covariances as

$$\begin{aligned}\sigma_{\mathrm{GP}}^2(\mathbf{x}_0) = {} & \sigma_u^2 - \sum_{i=1}^{m} \phi_i^{OK} \mathrm{cov}_u(\mathbf{x}_i, \mathbf{x}_0) + \sum_{i=1}^{m} \phi_i^{OK} \mathrm{cov}_u\left(\mathbf{x}_i, \mathbf{x}_{k(\mathbf{x}_0)}\right) \\ & - \mathrm{cov}_u\left(\mathbf{x}_0, \mathbf{x}_{k(\mathbf{x}_0)}\right) + \sum_{\substack{i=1 \\ i \neq k(\mathbf{x}_0)}}^{m} \sum_{\substack{j=1 \\ j \neq k(\mathbf{x}_0)}}^{m} (\beta_i - 1)(\beta_j - 1) \phi_i^{OK} \phi_j^{OK} R(\mathbf{x}_i, \mathbf{x}_j)\end{aligned} \quad (7.19)$$

In Equation (7.19), the first four terms on the equation's right side represent the OK conditional variance:

$$\sigma_{OK}^2(\mathbf{x}_0) = \sigma_u^2 - \sum_{i=1}^{m} \phi_i^{OK} \mathrm{cov}_u(\mathbf{x}_i, \mathbf{x}_0) + \sum_{i=1}^{m} \phi_i^{OK} \mathrm{cov}_u\left(\mathbf{x}_i, \mathbf{x}_{k(\mathbf{x}_0)}\right) - \mathrm{cov}_u\left(\mathbf{x}_0, \mathbf{x}_{k(\mathbf{x}_0)}\right). \quad (7.20)$$

If referring the OK method in Equation (7.7), the Lagrange multiplier is

$$\mu = \mathrm{cov}_u\left(\mathbf{x}_0, \mathbf{x}_{k(\mathbf{x}_0)}\right) - \sum_{i=1}^{m} \phi_i^{OK} \mathrm{cov}_u\left(\mathbf{x}_i, \mathbf{x}_{k(\mathbf{x}_0)}\right). \quad (7.21)$$

Therefore, the conditional variance of the GP can be rewritten as

$$\sigma_{\mathrm{GP}}^2(\mathbf{x}_0) = \sigma_{OK}^2 + \sum_{\substack{i=1 \\ i \neq k(\mathbf{x}_0)}}^{m} \sum_{\substack{j=1 \\ j \neq k(\mathbf{x}_0)}}^{m} (\beta_i - 1)(\beta_j - 1) \phi_i^{OK} \phi_j^{OK} R(\mathbf{x}_i, \mathbf{x}_j). \quad (7.22)$$

In summary, above we introduce a broad aspect of geostatistics through the GP method, where the conditional estimates of zonation and interpolation methods are combined. The extended geostatistical theory is applicable to the intrinsic field and second-order stationary field. The GP covariance and variance are generalized for any interpolation method. We also show the relationship between the GP variance and OK variance.

7.3 Parameter Structure Identification

The parameter structure identification problem discusses the estimation of the distribution of a parameter in the distributed parametric system. For running a numerical model, each distributed parameter must be expressed by a selected parameterization scheme. The ultimate task of parameter structure identification is to choose the best parameterization scheme that approximates the true distribution as closely as possible. We note that solving the parameter structure identification problem is expensive and the best parameterization scheme is always case-dependent. To reduce the problem complexity, we choose to identify the components of a parameterization method that include the parameter dimension, parameter values, and basis functions.

The complexity level of the parameter structure depends on available information from the sample data as well as on the chosen parameterization method. If there are no hydraulic conductivity measurements available, the parameter structure identification procedure needs to start from scratch, including the systematic search for the best number of basis points (Sun et al. 1998). If the sample data are sufficient, one merely can identify the weighting coefficients in the GP method. If the sample data are not sufficient, the concept of the pilot point method (Certes and de Marsily 1991; LaVenue et al. 1995; RamaRao et al. 1995) along with a chosen parameterization method can be used to help identify parameter structure.

Using the GP, a matrix-vector form of Equation (7.8) for n unsampled locations has the following form:

$$\mathbf{p}_n = \boldsymbol{\theta}_m \mathbf{p}_m \tag{7.23}$$

where $\mathbf{p}_n$ is a vector of estimates at n nodes; $\boldsymbol{\theta}_m = \boldsymbol{\theta}_m(\mathbf{x}_m, \boldsymbol{\beta}_m)$ is an $(n \times m)$ structure matrix; $\mathbf{p}_m$ is a vector of m basis point values; and $\mathbf{x}_m$ is a vector of m basis point locations. For an over-determined inverse problem, we have $n > m$. Therefore, parameter structure identification is reduced to the task of identifying all components $(m, \boldsymbol{\theta}_m, \mathbf{p}_m)$ associated with the chosen parameterization scheme.

In model calibration, a general form that represents the fitting residual in describing the discrepancy between the conceptual model with a parameter structure $(\boldsymbol{\theta}_m, \mathbf{p}_m)$ and the true model can be expressed by the following generalized least squares criterion:

$$E_m = \left(\mathbf{u}_D(\boldsymbol{\theta}_m \mathbf{p}_m) - \mathbf{u}_D^{obs}\right)^T V_u^{-1} \left(\mathbf{u}_D(\boldsymbol{\theta}_m \mathbf{p}_m) - \mathbf{u}_D^{obs}\right) + \eta \left(\boldsymbol{\theta}_m \mathbf{p}_m - \mathbf{p}_n^0\right)^T V_p^{-1} \left(\boldsymbol{\theta}_m \mathbf{p}_m - \mathbf{p}_n^0\right) \tag{7.24}$$

where $\mathbf{u}_D^{obs}$ is a vector of state variables observed at pointwise space and time; $\mathbf{u}_D$ is a vector of state variables reproduced by modeling at the same pointwise space and time as $\mathbf{u}_D^{obs}$; $\mathbf{p}_n^0$ is the prior model parameter values at all nodes; V_u is the error covariance matrices of the state variables; V_p is the error covariance matrices of prior model parameters; and η is a weighting factor.

Equation (7.24) contains two parts. The first part is the fitting residual between the reproduced states and the observed states at a certain number of points in space and time. This part takes into account the model outputs at the observation space. The second part is the fitting residual between the estimated parameter distribution and the prior parameter distribution. It counts the parameter distinction at each single point in the domain. This part takes into account the model inputs at the parameter space.

When the parameter dimension m and the structure matrix $\boldsymbol{\theta}_m$ are fixed, minimizing Equation (7.24) reduces to the *parameter identification* problem, in which the optimal parameter values are sought. In this case, m and $\boldsymbol{\theta}_m$ do not necessarily show up in model calibration. The fitting residual in the parameter space is merely the difference of m parameter values. The parameter identification problem is

$$RE = \min_{\mathbf{p}_m} \left(\mathbf{u}_D(\mathbf{p}_m) - \mathbf{u}_D^{obs}\right)^T V_u^{-1} \left(\mathbf{u}_D(\mathbf{p}_m) - \mathbf{u}_D^{obs}\right) + \eta \left(\mathbf{p}_m - \mathbf{p}_m^0\right)^T V_p^{-1} \left(\mathbf{p}_m - \mathbf{p}_m^0\right) \tag{7.25}$$

where $\mathbf{p}_m \in R^m$ is a vector of the unknown model parameters; and $\mathbf{p}_m^0 \in R^m$ is the prior information of the unknown model parameters.

This traditional inverse problem assumes that the parameter dimension and parameter structure pattern of parameterization are known and only the parameter values need to be identified; however, this is generally unrealistic and impractical. Instead, the parameter structure identification based on Equation (7.24) has to be considered in order to identify all of the components $\{m, \boldsymbol{\theta}_m, \mathbf{p}_m\}$ in the parameterization scheme, including the determination of the optimal parameter dimension m.

To launch the parameter structure identification procedure, we use a regression procedure to systematically increase the number of basis points to the field. To do this, we denote $(\boldsymbol{\theta}_m, \mathbf{p}_m)$ as the m-level parameter structure. The procedure systematically identifies a series of parameter structures with increasing parameter dimensions, starting from a homogeneous structure:

$$(\boldsymbol{\theta}_1, \mathbf{p}_1) \rightarrow (\boldsymbol{\theta}_2, \mathbf{p}_2) \rightarrow \cdots \rightarrow (\boldsymbol{\theta}_m, \mathbf{p}_m) \rightarrow \cdots \tag{7.26}$$

The identification procedure starts with the simplest structure and systematically increases structure complexity and parameter heterogeneity until certain stopping criteria are met. The stopping criteria used include the parameter uncertainty (Tsai and Yeh 2004; Tsai et al. 2005), parameter structure error (Sun et al. 1998; Tsai et al. 2003b), or parameter structure discrimination (Tsai et al. 2003a).

7.3.1 The Extended Inverse Problem (EIT)

The EIP is predicated on available observations and prior information. At each level of complexity of parameter structure (a given parameter dimension), the optimal structure pattern $\boldsymbol{\theta}_m$ and the corresponding parameter values $\mathbf{p}_m$ are obtained by minimizing the fitting residual:

$$RE_m = \min_{\boldsymbol{\theta}_m, \mathbf{p}_m} \left(\mathbf{u}_D(\boldsymbol{\theta}_m \mathbf{p}_m) - \mathbf{u}_D^{obs}\right)^T V_u^{-1} \left(\mathbf{u}_D(\boldsymbol{\theta}_m \mathbf{p}_m) - \mathbf{u}_D^{obs}\right) + \eta \left(\boldsymbol{\theta}_m \mathbf{p}_m - \mathbf{p}_n^0\right)^T V_p^{-1} \left(\boldsymbol{\theta}_m \mathbf{p}_m - \mathbf{p}_n^0\right). \tag{7.27}$$

We note that the EIT does not consider model prediction or model management in model calibration. Therefore, observation data and prior information are the only information available for the determination of the stopping criteria. For example, if the minimal fitting residual is less than the observation error, it is not necessary to identify higher parameter dimensions. If the minimization cannot significantly decrease the fitting residual or if the difference between RE_m and RE_{m-1} becomes small, we stop the procedure. Over-parameterization may result if we continue to increase the parameter dimension.

7.3.2 Parameter Uncertainty as a Stopping Criterion

A norm of the error covariance matrix of the estimated parameters can be utilized to assess the parameter uncertainty and used as a stopping criterion. An increase in parameter dimension generally is accompanied by a decrease in reliability of the identified parameters. It is intuitively apparent that a higher-level parameter structure will capture more heterogeneity. However, beyond a certain complexity level, a further increase in parameter dimension will result in a drastic decrease in reliability of the estimated parameters.

For the parameter estimation problem in Equation (7.25), the covariance matrix of the estimated parameter on $\mathbf{p}_m \in R^m$ is

$$\mathrm{Cov}(\mathbf{p}_m) = (J_D^T V_u^{-1} J_D + V_p^{-1})^{-1} \tag{7.28}$$

where $J_D = \partial \mathbf{u}_D / \partial \mathbf{p}_m$ is the Jacobian matrix. This form has been introduced to the groundwater problem (Carrera and Neuman 1986b) as well as coupled groundwater and mass transport problems (Wagner and Gorelick 1987). However, we cannot see the structure matrix in this form. To compare the parameter uncertainty among different parameter structures, Tsai et al. (2003a) extended the work of Shah et al. (1978) and derived a general form of error covariance matrix with respect to the estimated parameter values at n nodes:

$$\mathrm{Cov}(\mathbf{p}_n) = \boldsymbol{\theta}_m \left(J_D^T V_u^{-1} J_D + \boldsymbol{\theta}_m^T V_p^{-1} \boldsymbol{\theta}_m \right)^{-1} \boldsymbol{\theta}_m^T . \tag{7.29}$$

A specific case of Equation (7.29) applied to the groundwater problem can be found in Yeh and Yoon (1981).

Overall parameter structure uncertainty can be assessed in terms of the trace $tr\left[\mathrm{Cov}(\mathbf{p}_n)\right]$ or a general norm $\left\|\mathrm{Cov}(\mathbf{p}_n)\right\|$. Accordingly, if the sensitivities of basis points become small and if the overall parameter uncertainty becomes too high, over-parameterization and incorrect parameter structure may occur. If these signals occur, the identification procedure must be stopped. As a result, the EIP maximizes the identifiable parameter dimension, avoids over-parameterization, and guarantees that the identified parameters are sensitive to observations.

7.3.3 The Generalized Inverse Problem (GIP)

A new concept to the inverse problem is the incorporation of an accuracy requirement of model application in model calibration. Model application includes model prediction and model management. The inverse problem that incorporates model application is called the generalized inverse problem (GIP). The goal of studying the GIP is to quantitatively link the model calibration procedure to parameter identifiability, parameter structure complexity, and model application reliability.

7.3.3.1 *Parameter Structure Error*

Let J be the objective in model application. The difference between two objectives of immediately adjacent parameter dimensions is defined by

$$d_J(\mathbf{p}_m, \mathbf{p}_{m-1}) = \| J(\boldsymbol{\theta}_m, \mathbf{p}_m; \mathbf{q}_E) - J(\boldsymbol{\theta}_{m-1}, \mathbf{p}_{m-1}; \mathbf{q}_E) \|_J \tag{7.30}$$

where $\mathbf{q}_E$ is a vector of control variables (planned actions); and $\| \cdot \|_J$ is a norm defined in the model application space. In a situation where a simpler structure $(\boldsymbol{\theta}_{m-1}, \mathbf{p}_{m-1})$ is used to replace a more complex structure $(\boldsymbol{\theta}_m, \mathbf{p}_m)$ in connection with model application, Sun et al. (1998) introduced the following parameter structure error:

$$SE_m = \max_{\mathbf{p}_m} \min_{\mathbf{p}_{m-1}} \| J(\boldsymbol{\theta}_m, \mathbf{p}_m; \mathbf{q}_E) - J(\boldsymbol{\theta}_{m-1}, \mathbf{p}_{m-1}; \mathbf{q}_E) \|_J . \tag{7.31}$$

The minimization part seeks the minimum difference over the proper choice of $\mathbf{p}_{m-1}$ given $\mathbf{p}_m$. The maximization part seeks the maximum distinction in model application if $\boldsymbol{\theta}_m$ is replaced by $\boldsymbol{\theta}_{m-1}$.

For each complexity level m, we can make the following assessment: (i) if RE_m is larger than the measurement error and SE_m does not satisfy the accuracy requirement in model application, a more complex structure should be explored; (ii) if SE_m satisfies the accuracy requirement in model application (without considering RE_m), we conclude that the identified parameter structure is sufficient in terms of model application; or (iii) if SE_m is large but RE_m is smaller than the measurement error, we conclude that existing data are insufficient to identify a parameter structure that satisfies the accuracy requirement in model application.

7.3.3.2 *Parameter Structure Discrimination*

The goal of parameter structure discrimination is to detect the distinction among the identified parameter structures. The distance between two model outputs is defined as

$$\mathrm{d_m} = \left\| \mathbf{u}\left(\boldsymbol{\theta}_{m-1}\mathbf{p}_{m-1}\right) - \mathbf{u}\left(\boldsymbol{\theta}_{m}\mathbf{p}_{m}\right) \right\| + \mu \left\| \boldsymbol{\theta}_{m-1}\mathbf{p}_{m-1} - \boldsymbol{\theta}_{m}\mathbf{p}_{m} \right\| \tag{7.32}$$

where μ is a weighting factor. Parameter structure discrimination is defined as the minimum distance between two structures $\left(\boldsymbol{\theta}_{m-1}, \boldsymbol{\theta}_{m}\right)$ such that

$$SD_m = \min_{\substack{\mathbf{p}_{m-1} \in Ad_{m-1} \\ \mathbf{p}_m \in Ad_m}} \left\| \mathbf{u}\left(\boldsymbol{\theta}_{m-1}\mathbf{p}_{m-1}\right) - \mathbf{u}\left(\boldsymbol{\theta}_{m}\mathbf{p}_{m}\right) \right\| + \mu \left\| \boldsymbol{\theta}_{m-1}\mathbf{p}_{m-1} - \boldsymbol{\theta}_{m}\mathbf{p}_{m} \right\|. \tag{7.33}$$

The admissible sets Ad_{m-1} and Ad_m for Equation (7.33) are much narrower than those for the parameter structure error since Ad_{m-1} and Ad_m are determined posteriorly once the corresponding optimal parameter values are identified. In minimizing the fitting residual and calculating the parameter structure error, a wide range usually is assigned according to prior information. After parameter values are identified, the reliability of each parameter value is estimated posteriorly according to the Gaussian distribution assumption and linear statistical theory, which provide the covariance matrix of estimated parameters (Equation (7.28)). In general, there are many ways to determine the new parameter admissible set. When the admissible sets Ad_{m-1} and Ad_m are small, the identified structures have high reliability and are discriminated. Consequently, the SD_m value is large. However, wide ranges of admissible sets are obtained when the identified structures contain high uncertainty that causes the SD_m value to become smaller.

Parameter structure discrimination provides another stopping criterion for the parameter structure identification procedure. According to both minimal fitting residual and parameter structure discrimination, if both SD_m and RE_m are large, the current observation data allow for exploring a more complex parameter structure. If RE_m is less than the observation errors, there is no additional information which can be extracted from the observation data. Therefore, it is not necessary to identify the parameter dimensions higher than m. If SD_m is less than the accuracy requirement of structure discrimination, $\boldsymbol{\theta}_{m-1}$ and $\boldsymbol{\theta}_m$ are viewed as not discriminated from each other. The procedure should be stopped to avoid over-parameterization and high parameter structure uncertainty. The parameter structure discrimination links with the requirement of structure discrimination, linear statistical analysis, and parameter uncertainty analysis.

7.4 Interpolation Point Selection

Given the sample data points, the manner in which we select the interpolation points has a significant influence on the results of parameter structure identification. Isaaks and Srivastava (1989, p. 347-348) noted that not all sample data used in the interpolation equation are relevant to unsampled locations. What works in certain areas of a particular data point may not work in others. One task in parameter structure identification is to examine the configuration of the nearby data points and decide which sample data should be included in the estimation. The problem of interpolation point selection is a challenging one because it deals not only with which sample data should be used, but also when and how to use sample data in the parameterization procedure. Most studies simplify this problem by defining local neighborhoods centered on unsampled locations. The local neighborhoods are based on circles (ellipses or squares), natural neighbors, or the integral distance in semivariograms. However, the local neighborhood approach merely reduces the number of interpolation points. It does not select the points according to the role of data points in relation to the ones to be included or discarded for the unsampled locations.

This section introduces binary data indicators to address the interpolation point selection problem (Tsai and Li 2008a; Tsai 2009). This approach is similar to the variable selection concept in the linear regression context where (binary) latent variables were introduced to determine which variables should be selected to obtain the best linear model (George and McCulloch 1993; Kohn et al. 2001; Kuo and Mallick 1997). We consider in particular that data points must contribute to their surrounding unsampled locations. However, beyond a certain range, the data points might suddenly be irrelevant to remote unsampled locations. Moreover, we focus on the neighborhoods centered on data points to quantify the relevance of sample data to unsampled locations, as opposed to focusing on local neighborhoods centered on unsampled locations.

7.4.1 Neighborhoods for Interpolation Point Selection

To ensure that a data point must have relevance to its surrounding unsampled locations, the interpolation estimator is split into two parts:

$$p(\mathbf{x}) = \theta_k p_{k(\mathbf{x})} + \sum_{\substack{j=1 \\ j \neq k(\mathbf{x})}}^{m} \theta_j p_j \,. \tag{7.34}$$

The first term on the right side of Equation (7.34) is the contribution solely from the sample data p_k. The second term on the right side of Equation (7.34) is the contribution from all other data points. The data index $k(\mathbf{x})$ defines a close relationship between the unsampled location $\mathbf{x}$ and the data point $\mathbf{x}_k$. Equation (7.34) implies an assured contribution from data point $p_{k(\mathbf{x})}$ to location $\mathbf{x}$ as long as θ_k is non-zero. Equation (7.34) requires m distinct neighborhoods, denoted *d-neighborhoods* and formed by the unsampled locations centered on data points to determine the data index k for all unsampled locations.

The d-neighborhoods play a significant role in determining the influential boundaries of data points. Each d-neighborhood contains only one data point. No intersection is allowed between the d-neighborhoods. For example, one can use the tree-regression procedure (Figure 7.1) or Voronoi tessellation (Figure 7.2) to determine the d-neighborhoods. Then the data index $k(\mathbf{x})$ refers to the data point $\mathbf{x}_k$ because the unsampled location $\mathbf{x}$ is inside the d-neighborhood of the data point $\mathbf{x}_k$. In this way, the data point $\mathbf{x}_k$ makes an assured contribution to unsampled locations inside its d-neighborhood.

7.4.2 Indicator Generalized Parameterization

Once the d-neighborhoods are determined, one can introduces m binary data indicators $\{I_j, j = 1, 2, \cdots, m\}$ to the data points to determine the ultimate contributing weights from the sample data as follows:

$$\begin{aligned} \theta_j &= \phi_j I_j \qquad , \forall j \neq k \\ \theta_k &= 1 - \sum_{\substack{j=1 \\ j \neq k}}^{m} \phi_j I_j \end{aligned} \tag{7.35}$$

where ϕ_j are the basis functions of an interpolation method and $I_j \in \{0,1\}$ are the binary data indicators.

Equation (7.35) addresses the interpolation point selection problem centered on the sample data. For sample points not including the unsampled location $\mathbf{x}$ in their d-neighborhoods, $I_j = 1$ indicates the selection of sample point j for interpolation. Then p_i contributes $p_j \phi_j$ to the unsampled locations outside its d-neighborhood. By contrast, $I_j = 0$ indicates the non-selection of sample point j for interpolation for unsampled location $\mathbf{x}$ outside its d-neighborhood. The data point p_k always contributes $\left(1-\sum_{\substack{j=1 \\ j\neq k}}^{m} \phi_j I_j\right) p_k$ to the unsampled locations inside its d-neighborhood.

Substituting Equation (7.35) into Equation (7.34), an indicator generalized parameterization (IGP) method is formulated:

$$p_{\text{IGP}}(\mathbf{x}) = \sum_{\substack{j=1 \\ j\neq k(\mathbf{x})}}^{m} \phi_j \left(p_j - p_{k(\mathbf{x})}\right) I_j + p_{k(\mathbf{x})} . \tag{7.36}$$

The IGP method possesses similar advantages to the generalized parameterization (GP) method. If all indicator values are unity, then all data points are selected and Equation (7.36) represents traditional interpolation. If all indicator values are zero, then all data points only contribute their values to unsampled locations inside their own d-neighborhoods. Thus, Equation (7.36) creates a conditional zonal distribution, whose zonal values are conditioned on the sample data. The data indicators in the IGP are used specifically for the interpolation point selection problem.

7.4.3 Estimation Covariance using IGP

Similar to the GP method, Tsai (2009) derived the covariance of using the IGP method for a pair of estimates at location $(\mathbf{x}, \mathbf{x}')$ for an intrinsic field as follows

$$\begin{aligned} \mathrm{C}_{\text{IGP}}(\mathbf{x},\mathbf{x}') = & \sum_{\substack{i=1 \\ i\neq k(\mathbf{x})}}^{m} \sum_{\substack{j=1 \\ j\neq k(\mathbf{x}')}}^{m} \phi_i \phi_j R(\mathbf{x}_i,\mathbf{x}_j) I_i I_j - \sum_{\substack{i=1 \\ i\neq k(\mathbf{x})}}^{m} \phi_i R(\mathbf{x}_i,\mathbf{x}') I_i \\ & - \sum_{\substack{i=1 \\ i\neq k(\mathbf{x})}}^{m} \phi_j R(\mathbf{x},\mathbf{x}_j) I_j + R(\mathbf{x},\mathbf{x}') \end{aligned} \tag{7.37}$$

The function $R(\mathbf{x}_a, \mathbf{x}_b)$ is defined in Equation (7.11). The conditional variance at location $\mathbf{x} = \mathbf{x}'$ is

$$\begin{aligned} \sigma^2_{\text{IGP}}(\mathbf{x}) = & \sum_{\substack{i=1 \\ i\neq k(\mathbf{x})}}^{m} \sum_{\substack{j=1 \\ j\neq k(\mathbf{x})}}^{m} \phi_i \phi_j R(\mathbf{x}_i,\mathbf{x}_j) I_i I_j - 2\sum_{\substack{i=1 \\ i\neq k(\mathbf{x})}}^{m} \phi_i R(\mathbf{x}_i,\mathbf{x}) I_i \\ & + 2\gamma\left(\mathbf{x},\mathbf{x}_{k(\mathbf{x})}\right) \end{aligned} \tag{7.38}$$

The relationship between C_{IGP} and the covariance of the ordinary kriging (OK), C_{OK}, is

$$C_{IGP}(\mathbf{x},\mathbf{x}') = C_{OK}(\mathbf{x},\mathbf{x}') + \sum_{\substack{i=1 \\ i\neq k(\mathbf{x})}}^{m} \sum_{\substack{j=1 \\ j\neq k(\mathbf{x}')}}^{m} \left(\phi_i I_i - \phi_i^{(OK)}\right) R\left(\mathbf{x}_i,\mathbf{x}_j\right)\left(\phi_j I_j - \phi_j^{(OK)}\right). \quad (7.39)$$

Tsai and Li (2008a) used a genetic algorithm to identify the binary values of the data indicators. Tsai (2009) further studied the probabilistic data indicators in the IGP.

7.5 Model Selection

Data uncertainty and insufficient information motivate the use of many parameterization methods and groundwater models. Model selection involves using data to find the best model from a set of candidate models. Model selection methods developed in the literature to evaluate the relative importance among candidate models include the statistical hypothesis test (Luis and McLaughlin 1992), likelihood ratio tests (Suchard et al. 2003), decision theory (Bernardo and Smith 1994), Occam's window (Madigan and Raftery 1994; Hoeting et al. 1999), Bayes factors (Usunoff et al. 1992; Kass and Raftery 1995), cross-validation (Stone 1974), and information-theoretic criteria (Akaike 1973; Mallows 1973; Akaike 1974; Sugiura 1978; Schwarz 1978; Hannan and Quinn 1979; Akaike 1981; Chow 1981; Atkinson 1981; Kashyap 1982; Hurvich and Tsai 1989; Foster and George 1994; Claeskens and Hjort 2003; Seghouane and Bekara 2004). In the field of groundwater modeling, statistical information criteria (Carrera and Neuman 1986c; Hyun and Lee 1998) and sensitivity analysis (Jiao and Lerner 1996) have been used to identify the best groundwater model. The trade-off between bias (goodness-of-fit) and variance (parameter estimation uncertainty) in light of the parsimony principle have been applied to determine the best parameterization dimension for hydraulic conductivity estimation (Yeh and Yoon 1981; Sun et al. 1998; Tsai et al. 2003a; Tsai et al. 2003b). Model selection also includes model discrimination, used to discriminate among rival groundwater models through experimental designs (Knopman and Voss 1988; Knopman and Voss 1989; Knopman et al. 1991; Usunoff et al. 1992).

The purpose of model selection is to determine the best model(s) among competing candidate models. It is important that each model be developed and calibrated carefully before it is judged for model selection suitability. It would be premature to select the "best" model without thorough study of each candidate model (Poeter and Anderson 2005).

7.5.1 Bayes factor and Occam's Window

We first discuss the Bayes factor for model selection. Through Bayes' rule, the posterior model probabilities, $\Pr\left(M^{(p)} \mid \mathbf{D}\right)$, for two rival models $M^{(0)}$ and M_1 are proportional to their likelihoods, $\Pr\left(\mathbf{D} \mid M^{(p)}\right)$, and their prior model probability, $\Pr\left(M^{(p)}\right)$:

$$\Pr\left(M^{(p)} \mid \mathbf{D}\right) \propto \Pr\left(\mathbf{D} \mid M^{(p)}\right)\Pr\left(M^{(p)}\right) \quad , p = 0,1 \tag{7.40}$$

where $\mathbf{D}$ is the data. The Bayes factor B_{01} is the likelihood ratio of M_0 to M_1 as follows (Congdon 2001):

$$B_{01} = \frac{\Pr\left(\mathbf{D} \mid M^{(0)}\right)}{\Pr\left(\mathbf{D} \mid M^{(1)}\right)} = \frac{\Pr\left(M^{(0)} \mid \mathbf{D}\right)}{\Pr\left(M^{(1)} \mid \mathbf{D}\right)} \Bigg/ \frac{\Pr\left(M^{(0)}\right)}{\Pr\left(M^{(1)}\right)}, \tag{7.41}$$

which also represents the ratio of posterior odds to prior odds of two rival models. The Bayes ratio is used as a measure of strength of evidence for and against various hypotheses for model selection. The scale of evidence can be used to interpret the strong, moderate, or weak evidence favoring either of two rival models (Jeffreys 1961; Kass and Wasserman 1995; Wasserman 2000). The model M_0 is judged better than model M_1 if the Bayes factor presents strong evidence that is much larger than unity.

For equal prior model probabilities, the Bayes factor is the posterior odd, which develops Occam's window as shown in Figure 7.4. A threshold c determines the size of Occam's window, $\overline{O_L O_R} = 2\ln c$, that rejects model M_1 when $\Pr(M_0 \mid \mathbf{D})/\Pr(M_1 \mid \mathbf{D}) > c$. A smaller threshold value renders M_1 likely for rejection. A threshold value of 10 may be considered sufficient to determine that model M_0 is significantly better than model M_1 (Hoeting et al. 1999).

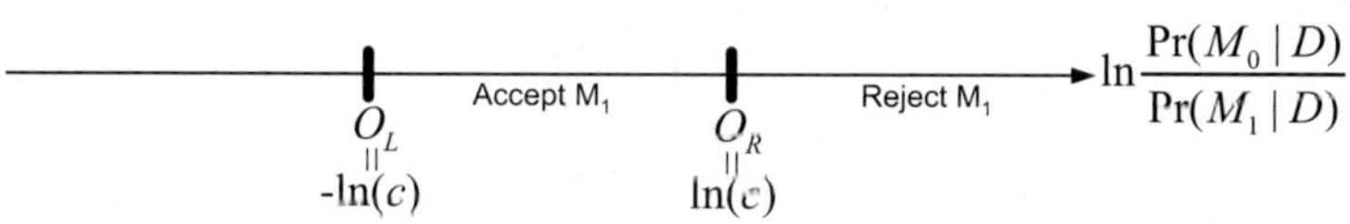

Figure 7.4: Occam's Window

The Bayes factor is associated with likelihood ratio tests and is used to perform hypothesis tests on pairwise likelihood ratios in a specific sequence until a final model is found. However, it has been argued that hypothesis testing and model selection are two distinct issues (Burnham and Anderson 2002; Posada and Buckley 2004). Using the Bayes factor with hypothesis tests strongly depends on the selection of P-values and sequence of model pairs in hypothesis testing and does not guarantee finding the optimal model.

7.5.2 Statistical Information Criteria

Statistical information criteria also can be used to evaluate model likelihood for parametric models. The criteria include the Akaike information criterion (AIC) (Akaike 1973; Akaike 1974), corrected Akaike information criterion (AICc) (Sugiura 1978; Hurvich and Tsai 1989), Mallows' criterion (Mallows 1973; Hansen 2007), Bayesian information criterion (BIC) (Schwarz 1978), Hannan and Quinn's information criterion (HQ) (Hannan and Quinn 1979), Kashyap information criterion (KIC) (Kashyap 1982), Kullback information criterion (Seghouane and Bekara 2004), and focused information criterion (FIC) (Claeskens and Hjort 2003). HQ and Kullback information criterion have been used for model selection purposes. AIC, AICc, BIC, and KIC often are used for model selection and model averaging purposes. The derivation of the AIC and its related criteria is based on the Kullback-Leibler (K-L) distance, itself used to obtain an information complexity criterion by Bozdogan (2000). AIC aims at finding a model that produces a density function to be, on average, close to the true density function. However, KIC and BIC are obtained from the Bayesian perspective and are designed for finding the most probable model using observation data. We note that the information criteria are closely related to the Bayes factor with regard to model selection and model averaging. In what follows, we focus our discussions on the KIC and BIC.

7.5.2.1 *Kashyap Information Criterion (KIC)*

Consider a model $M^{(p)}$ and a set of model parameters, $\boldsymbol{\beta}^{(p)}$. We obtain the marginal likelihood of model $M^{(p)}$ by considering all possible values of model parameters that are embedded in model $M^{(p)}$:

$$\Pr\left(\mathbf{D} \mid M^{(p)}\right) = \int_{\boldsymbol{\beta}^{(p)}} \Pr\left(\mathbf{D} \mid M^{(p)}, \boldsymbol{\beta}^{(p)}\right) \Pr\left(\boldsymbol{\beta}^{(p)} \mid M^{(p)}\right) d\boldsymbol{\beta}^{(p)} . \tag{7.42}$$

The marginal likelihood has an analytic form if data can be approximated as a linear function of parameters and the Gaussian probability distribution is considered for the likelihood and prior. Kitanidis (1996) suggested this to calculate the marginal distribution using a linearization approach, and it is suitable for linear regression models in statistics. For given density distributions other than the Gaussian distribution, a common approach is to use the Markov Chain Monte Carlo (MCMC) method to estimate the marginal likelihood (George and McCulloch 1993; Madigan and Raftery 1994; Madigan and York 1995; Raftery et al. 1997). Using Laplace approximation along with the maximum likelihood of parameters is another common approach to calculate the marginal likelihood. This approach requires the evaluation of the Hessian matrix or Fisher information matrix (Tierney and Kadane 1986; Gelfand and Dey 1994; Raftery 1995; Draper 1995; Raftery 1996).

To make the calculation more efficient, a Laplace approximation based on a Taylor series expansion can be applied to approximate the integration with the accuracy of $O\left(n^{-1}\right)$ (Tierney and Kadane 1986; Raftery 1995; Draper 1995):

$$\Pr\left(\mathbf{D} \mid M^{(p)}\right) = \exp\left[g_D^{(p)}\right]\left(2\pi\right)^{m^{(p)}/2}\left|\mathbf{A}^{(p)}\right|^{-1/2} + O\left(n^{-1}\right) \tag{7.43}$$

where $g_D^{(p)} = \ln\left[\Pr\left(\mathbf{D} \mid M^{(p)}, \hat{\boldsymbol{\beta}}^{(p)}\right)\Pr\left(\hat{\boldsymbol{\beta}}^{(p)} \mid M^{(p)}\right)\right]$; n is the number of data D; $m^{(p)}$ is the number of unknown parameters $\boldsymbol{\beta}^{(p)}$; and $\hat{\boldsymbol{\beta}}^{(p)}$ are the maximum likelihood estimators (MLE) that give $g_D^{(p)}$ the maximum value. For a large number of data, the matrix $[\mathbf{A}_{ij}^{(p)}] = -\left[\partial^2 g_D^{(p)} / \partial\beta_i^{(p)} \partial\beta_j^{(p)}\right]$ can be approximated to the Fisher information matrix (FIM), $\left[\mathbf{F}_D^{(p)}\right]_{ij} = -\mathrm{E}\left[\partial^2 g_D^{(p)} / \partial\beta_i^{(p)} \partial\beta_j^{(p)}\right]$, the expectation taken over the data. The Kashyap information criterion (KIC) (Kashyap 1982) then is obtained as

$$\mathrm{KIC}_D^{(p)} = -2 g_D^{(p)} + m^{(p)} \ln\frac{n}{2\pi} + \ln\left|\tfrac{1}{n}\mathbf{F}_D^{(p)}\right| \tag{7.44}$$

to approximate the marginal likelihood, i.e., $\mathrm{KIC}^{(p)} \approx -2\ln\Pr\left(\mathbf{D} \mid M^{(p)}\right)$. The use of the averaged FIM in Equation (7.44) is similar to the consideration of one observation with maximum information (Raftery 1995; Draper 1995). The advantage of using the FIM is to avoid a Hessian matrix calculation in the matrix $\mathbf{A}^{(p)}$ and to reduce computation demands.

Substituting the KIC into Equation (7.41), the Bayes factor becomes $B_{01} = e^{\frac{1}{2}\Delta_{01}}$, where $\Delta_{01} = \left(\mathrm{KIC}^{(1)} - \mathrm{KIC}^{(0)}\right)$.

7.5.2.2 *Bayesian Information Criterion (BIC)*

When the size of the observation data increases, the prior probability of model parameters in $g_D^{(p)}$ asymptotically approaches a multivariate normal distribution, $\Pr\left(\boldsymbol{\beta}^{(p)} \mid M^{(p)}\right) \sim N(\hat{\boldsymbol{\beta}}^{(p)}, \left[\tfrac{1}{n}\mathbf{F}_D^{(p)}\right]^{-1})$. As a consequence, the KIC can be approximated by the BIC for large amounts of data:

$$\mathrm{BIC}_D^{(p)} = -2\ln\Pr\left(\mathbf{D} \mid M^{(p)}, \hat{\boldsymbol{\beta}}^{(p)}\right) + m^{(p)} \ln n . \tag{7.45}$$

This implies that prior parameters contain the same amounts of information as data (Raftery 1995). We note further that the error in BIC is $O(1)$ and does not reduce to zero when data increases; however, the relative error of BIC with respect to the likelihood value $\ln \Pr\left(\mathbf{D} \mid M^{(p)}\right)$ does reduce to zero (Kass and Wasserman 1995).

Similarly, we can use BIC in the Bayes factor and $\Delta_{01} = \left(\mathrm{BIC}^{(1)} - \mathrm{BIC}^{(0)}\right)$.

Several information criteria other than BIC and KIC, e.g., the Akaike information criterion (AIC), corrected Akaike information criterion (AICc), and Hannan and Quinn's information criterion (HQ), also have been derived to approximate the marginal likelihood of a model:

$$
\begin{aligned}
\mathrm{AIC}_D^{(p)} &= -2\ln \Pr\left(\mathbf{D} \mid M^{(p)}, \hat{\boldsymbol{\beta}}^{(p)}\right) + 2m^{(p)} \\
\mathrm{AIC}_{cD}^{(p)} &= -2\ln \Pr\left(\mathbf{D} \mid M^{(p)}, \hat{\boldsymbol{\beta}}^{(p)}\right) + 2m^{(p)} + \frac{2m^{(p)}(m^{(p)}+1)}{n - m^{(p)} - 1}. \\
\mathrm{HQ}_D^{(p)} &= -2\ln \Pr\left(\mathbf{D} \mid M^{(p)}, \hat{\boldsymbol{\beta}}^{(p)}\right) + 2m^{(p)} \ln[\ln n]
\end{aligned}
\tag{7.46}
$$

The major difference among BIC, AIC, AICc, and HQ is the penalty term (the second term in Equation (7.45)), which penalizes the number of unknown parameters in light of the parsimony principle. Additionally, the KIC considers the FIM, which is related to parameter uncertainty. As a result, using AIC, AICc, and HQ will result in the same form of the Bayes factor, but different Δ_{01} values. For the purpose of model selection, the best model is the one selected with the minimum values of information criteria.

7.6 Bayesian Model Averaging

Issues regarding the use of a single model for parameter estimation have been raised. This is because, for a given set of data, many models are acceptable under a certain criterion. By selecting a single model, we are more likely to ignore model uncertainty and underestimate prediction uncertainty. Hence, it may be desirable to integrate multiple models for parameter estimation.

Model averaging is a process of integrating candidate models and, in many instances, is more robust than model selection because model averaging does not exclude good models. In model averaging, model weights usually are calculated based on posterior model probabilities that provide the strength of evidence to rank models. Using posterior model probabilities and the law of total probability (Leamer 1978), Bayesian model selection methods (Kass and Raftery 1995; Raftery 1995; Gelman et al. 1996) and Bayesian model averaging (BMA) methods (Draper 1995; Madigan and Raftery 1994; Raftery 1995; Hoeting et al. 1999; Brown et al. 2002) can be developed and linked (Wasserman 2000).

To deal with multiple models for parameter identification and model prediction, the Bayesian modeling averaging method provides an effective means of integrating different models (Neuman 2003; Ye et al. 2004; Poeter and Anderson 2005; Tsai and

Li 2008b; Tsai and Li 2008c). The following section discusses the use of the BMA to integrate multiple parameterization methods.

7.6.1 Statistical Inference through Bayesian Model Averaging (BMA)

We consider a set of parameterization methods $\Theta = \left\{ M^{(p)}; p = 1, 2, \cdots \right\}$ in order to estimate the hydraulic conductivity distribution over the same region. Given available data ($\mathbf{D}$) and multiple parameterization methods (Θ), the conditional probability of a predicted quantity of interest (Δ) can be obtained through Bayesian model averaging (BMA), which is based on the law of total probability (Leamer 1978):

$$\Pr(\Delta \mid \mathbf{D}) = \sum_p \Pr\left(\Delta \mid \mathbf{D}, M^{(p)}\right) \Pr\left(M^{(p)} \mid \mathbf{D}\right) \tag{7.47}$$

where $\Pr(\Delta \mid \mathbf{D})$ is the conditional probability of the predicted quantity given data $\mathbf{D}$; and $\Pr\left(\Delta \mid \mathbf{D}, M^{(p)}\right)$ is the conditional probability of the predicted quantity given the data $\mathbf{D}$ and a parameterization method $M^{(p)}$. In addition, $\Pr\left(M^{(p)} \mid \mathbf{D}\right)$ is the posterior probability of a parameterization method given data $\mathbf{D}$, which represents posterior model weights. The models in the BMA refer to the parameterization methods.

According to Bayesian decision theory (Berger 1985), the posterior model probability of a parameterization method is

$$\Pr\left(M^{(p)} \mid \mathbf{D}\right) = \frac{\Pr\left(\mathbf{D} \mid M^{(p)}\right) \Pr\left(M^{(p)}\right)}{\sum_j \Pr\left(\mathbf{D} \mid M^{(j)}\right) \Pr\left(M^{(j)}\right)} \tag{7.48}$$

where $\sum_p \Pr\left(M^{(p)} \mid \mathbf{D}\right) = 1$. The prior probabilities of parameterization methods also represent prior model weights and $\sum_p \Pr\left(M^{(p)}\right) = 1$.

The prior model probability is a subjective value and is based on the analysts' prior information and their philosophical beliefs. Choosing proper prior probabilities is a challenging practical issue, especially in the absence of substantial prior knowledge, and usually engenders philosophical debates. It could be argued that in one hand the posterior model probability is rather sensitive to the specification of the prior. On the other hand, the prior model probability should not dominate the likelihood as supported by data. At any rate, it is imperative to avoid using improper prior that substantially affects the analysis. A Kullback-Leibler (K-L) prior was suggested by Burnham and Anderson (2004) in the prior BIC weights, which lead to the AIC weights. Ye et al. (2005) suggested the identification of prior model probabilities through the entropy maximization method. If there is no informational support, it is reasonable to have equal prior probabilities as a neutral choice (Hoeting et al. 1999; Wasserman 2000).

Considering equal prior probabilities and the KIC, we can approximate the posterior model probability as

$$\Pr\left(M^{(p)} \mid \mathbf{D}\right) \approx \frac{\exp\left(-\frac{1}{2}\mathrm{KIC}_D^{(p)}\right)}{\sum_j \exp\left(-\frac{1}{2}\mathrm{KIC}_D^{(j)}\right)}. \tag{7.49}$$

Equation (7.49) can be rewritten by considering the KIC difference to the minimum KIC value:

$$\Pr\left(M^{(p)} \mid \mathbf{D}\right) \approx \frac{\exp\left(-\frac{1}{2}\Delta\mathrm{KIC}_D^{(p)}\right)}{\sum_j \exp\left(-\frac{1}{2}\Delta\mathrm{KIC}_D^{(j)}\right)} \tag{7.50}$$

where $\Delta\mathrm{KIC}_D^{(p)} = \mathrm{KIC}_D^{(p)} - \mathrm{KIC}_{D,\min}^{(p)}$. Again, the KIC can be replaced by other informational criteria as needed.

According to the law of total expectation, the expectation of the predicted quantity is

$$\mathrm{E}[\Delta \mid \mathbf{D}] = \sum_p \mathrm{E}\left[\Delta \mid \mathbf{D}, \theta^{(p)}\right] \Pr\left(\theta^{(p)} \mid \mathbf{D}\right). \tag{7.51}$$

Similarly, the law of total (co)variance obtains the covariance matrix of the predicted quantity as

$$\begin{aligned}\mathrm{Cov}[\Delta \mid \mathbf{D}] = &\sum_p \mathrm{Cov}\left[\Delta \mid \mathbf{D}, M^{(p)}\right] \Pr\left(M^{(p)} \mid \mathbf{D}\right) \\ &+\sum_p \left(\mathrm{E}\left[\Delta \mid \mathbf{D}, M^{(p)}\right] - \mathrm{E}[\Delta \mid \mathbf{D}]\right)\left(\mathrm{E}\left[\Delta \mid \mathbf{D}, M^{(p)}\right] - \mathrm{E}[\Delta \mid \mathbf{D}]\right)^T \Pr\left(M^{(p)} \mid \mathbf{D}\right)\end{aligned} \tag{7.52}$$

The first term on the right side of Equation (7.52) represents the covariance for individual parameterization methods (within covariance). The second term represents the covariance between different parameterization methods (between covariance).

We note that, in practice, the true groundwater model is impossible to obtain. All the results from model selection and model averaging are based on models that are imperfect. A 100% model weight in the BMA does not imply a 100% correct model, but merely indicates the single best model among candidate models. To be sure, the ideal situation is to obtain an exhaustive set of models. However, this is very expensive in groundwater modeling. Using a limited number of models is inevitable, but with an understanding that model weights calculated by posterior model probabilities are the relative model weights among the selected models, not the true model weights over the model space.

7.6.2 Multigeneralized Parameterization (MultiGP)

Applications of the BMA depend on the data $\mathbf{D}$ and the predicted quantity Δ to be specified. In this section, we consider multiple generalized parameterization (GP) methods to estimate hydraulic conductivity. The data is the sample data $\mathbf{D} = \mathbf{p}^{data}$ and the predicted quantity is $\Delta = \mathbf{p}$.

In the multiGP method, we consider a set of zonal distributions $Z = \{\Omega^{(1)}, \Omega^{(2)}, \cdots\}$ and a set of interpolation methods $\Phi = \{\phi^{(1)}, \phi^{(2)}, \cdots\}$ to parameterize the same hydraulic conductivity field (Ω) using the same hydraulic conductivity data. Combinations of these zonal distributions and interpolation methods pose a multiGP scheme that involves many GP methods $\Theta = Z \times \Phi = \{M^{(p)}; p = 1, 2, \cdots\}$. Each GP method has its own weighting coefficients $\boldsymbol{\beta}^{(p)} = \left\{\beta_j^{(p)}, j = 1, 2, \cdots, m\right\}$. The weighting coefficients $\boldsymbol{\beta}^{(p)}$ are the model parameters embedded in the posterior probability $\Pr\left(M^{(p)} \mid \mathbf{D}\right)$. The mean and covariance of predicted hydraulic conductivity using a GP method are

$$\mathrm{E}\left[\Delta \mid \mathbf{D}, M^{(p)}\right] = \mathrm{E}\left[\mathbf{p} \mid \mathbf{p}^{data}, M^{(p)}\right] = \mathbf{p}_{\mathrm{GP}}^{(p)} \tag{7.53}$$

$$\mathrm{Cov}\left[\Delta \mid \mathbf{D}, M^{(p)}\right] = \mathrm{Cov}\left[\mathbf{p} \mid \mathbf{p}^{data}, M^{(p)}\right] = \mathrm{C}_{\mathrm{GP}}^{(p)}. \tag{7.54}$$

The expectation of the BMA hydraulic conductivity is

$$\mathrm{E}\left[\mathbf{p} \mid \mathbf{p}^{data}\right] = \overline{\mathbf{p}}_{GP} = \frac{\sum_p \mathbf{p}_{GP}^{(p)} \exp\left(-\tfrac{1}{2}\Delta\mathrm{KIC}_D^{(p)}\right)}{\sum_j \exp\left(-\tfrac{1}{2}\Delta\mathrm{KIC}_D^{(j)}\right)}. \tag{7.55}$$

The covariance of the BMA hydraulic conductivity is

$$\mathrm{Cov}\left[\mathbf{p} \mid \mathbf{p}^{data}\right] = \frac{\sum_p \left[\mathrm{Cov}_{GP}^{(p)} + \left(\mathbf{p}_{GP}^{(p)} - \overline{\mathbf{p}}_{GP}\right)\left(\mathbf{p}_{GP}^{(p)} - \overline{\mathbf{p}}_{GP}\right)^T\right] \exp\left(-\tfrac{1}{2}\Delta\mathrm{KIC}_D^{(p)}\right)}{\sum_j \exp\left(-\tfrac{1}{2}\Delta\mathrm{KIC}_D^{(j)}\right)}. \tag{7.56}$$

7.6.3 BMA Groundwater Inverse Modeling

In BMA groundwater inverse modeling, the optimal weighting coefficients $\hat{\boldsymbol{\beta}}^{(p)}$ of individual GP methods need to be estimated using observation data. In this case, the data $\mathbf{D} = \mathbf{h}^{data}$ would be the observed groundwater heads and the predicted quantity $\Delta = \mathbf{h}$ would be the predicted groundwater heads at the observation space. One cannot use the maximum likelihood of the posterior probability $\Pr(\Delta | \mathbf{D})$ in Equation (7.47) to estimate $\hat{\boldsymbol{\beta}}^{(p)}$ values because optimization will force 100% posterior probability to the best GP method that has the minimum KIC (or BIC) value and zero posterior probability for rest of the GP methods. Instead, one can implement maximum likelihood Bayesian model averaging (MLBMA) (Neuman 2003; Ye et al. 2004) to obtain optimal $\hat{\boldsymbol{\beta}}^{(p)}$ in $\Pr\left(\Delta = \mathbf{h} | \mathbf{D} = \mathbf{h}^{data}, M^{(p)}\right)$ or use the inverse methods presented in Section 7.4. for individual GP methods. Then, $\Pr\left(\Delta | \mathbf{D}, M^{(p)}\right)$ can be approximated by $\Pr\left(\Delta | \mathbf{D}, M^{(p)}, \hat{\boldsymbol{\beta}}^{(p)}\right)$ to avoid calculation burden (Draper 1995) for BMA predictions.

Another approach that considers the maximum likelihood estimation in $\Pr\left(\Delta | \mathbf{D}, M^{(p)}\right)$ is the following integration form:

$$\Pr\left(\Delta | \mathbf{D}, M^{(p)}\right) = \int_{\boldsymbol{\beta}} \Pr\left(\Delta | \mathbf{D}, M^{(p)}, \boldsymbol{\beta}^{(p)}\right) \Pr\left(\boldsymbol{\beta}^{(p)} | \mathbf{D}, M^{(p)}\right) d\boldsymbol{\beta}^{(p)} . \tag{7.57}$$

Because Δ is the predicted quantity at the observation space, $\Pr\left(\Delta | \mathbf{D}, M^{(p)}\right)$ can be seen as a likelihood function. Using the Laplace approximation, one can obtain

$$\Pr\left(\Delta | \mathbf{D}, M^{(p)}\right) \approx \exp(-\frac{1}{2}\mathrm{KIC}_{\Delta}^{(p)}) \tag{7.58}$$

where $\mathrm{KIC}_{\Delta}^{(p)}$ is the KIC with respect to the predicted quantity Δ:

$$\mathrm{KIC}_{\Delta}^{(p)} = -2g_{\Delta}^{(p)} - m^{(p)} \ln 2\pi + \ln\left|\mathbf{F}_{\Delta}^{(p)}\right| . \tag{7.59}$$

The Fisher information matrix is $\left[\mathbf{F}_{\Delta}^{(p)}\right]_{ij} = -\mathrm{E}\left[\partial^2 g_{\Delta}^{(p)} / \partial\beta_i^{(p)} \partial\beta_j^{(p)}\right]$, where $g_{\Delta}^{(p)} = \ln\left[\Pr\left(\Delta | \mathbf{D}, M^{(p)}, \hat{\boldsymbol{\beta}}^{(p)}\right) \Pr\left(\hat{\boldsymbol{\beta}}^{(p)} | \mathbf{D}, M^{(p)}\right)\right]$.

Then, the optimal $\hat{\boldsymbol{\beta}}^{(p)}$ can be obtained by minimizing the KIC:

$$\min_{0 \le \boldsymbol{\beta}^{(p)} \le 1} \quad \mathrm{KIC}_{\Delta}^{(p)} \quad , p = 1, 2, \cdots \tag{7.60}$$

7.6.4 $\mathrm{KIC}_{D}^{(p)}$ vs. $\mathrm{KIC}_{\Delta}^{(p)}$

In the general case, $\mathrm{KIC}_{D}^{(p)}$ and $\mathrm{KIC}_{\Delta}^{(p)}$ are different because the data in $\Pr(\Delta | \mathbf{D}, M^{(p)}, \hat{\boldsymbol{\beta}}^{(p)})$ and the data in $\Pr(\mathbf{D} | M^{(p)}, \boldsymbol{\beta}^{(p)})$ can differ. For example, one merely can use measured hydraulic conductivity data through the cross-validation

(CV) approach (Stone 1974) to estimate the $\boldsymbol{\beta}$ values. Using CV, $\mathbf{D}=\mathbf{p}^{data}$ are the measured hydraulic conductivity values and Δ are the cross-validated hydraulic conductivity values at the sample locations. This approach has been suggested in Ye et al. (2004). Both $\mathrm{KIC}_{\Delta}^{(p)}$ and $\mathrm{KIC}_{D}^{(p)}$ can be evaluated by the CV method. In groundwater inverse modeling, we still can choose $\mathbf{D}=\mathbf{p}^{data}$ to calculate $\mathrm{KIC}_{D}^{(p)}$ to obtain the GP weights in $\Pr\left(M^{(p)} \mid \mathbf{D}\right)$ through the CV method. However, $\mathrm{KIC}_{\Delta}^{(p)}$ is based on the predicted groundwater heads $\Delta=\mathbf{h}^{cal}$ at the head observation space. In a case where the same data are used in both $\Pr(\Delta \mid \mathbf{D}, M^{(p)}, \hat{\boldsymbol{\beta}}^{(p)})$ and $\Pr(\mathbf{D} \mid M^{(p)}, \boldsymbol{\beta}^{(p)})$, $\mathrm{KIC}_{D}^{(p)}$ and $\mathrm{KIC}_{\Delta}^{(p)}$ have the same forms for the inverse problem. Moreover, $\mathrm{KIC}_{D}^{(p)}$ and $\mathrm{KIC}_{\Delta}^{(p)}$ are different when the predicted quantity Δ is at the prediction space, not at the observation space. Furthermore, even though $\mathrm{KIC}_{D}^{(p)}$ and $\mathrm{KIC}_{\Delta}^{(p)}$ are the same in the inverse problem, $\mathrm{KIC}_{D}^{(p)}$ is mainly used to calculate the importance (weights) of individual GP methods in BMA. From a practical viewpoint, $\mathrm{KIC}_{D}^{(p)}$ can be modified or scaled as the empirical Bayesian inference in order to obtain reasonable GP weights. Recently, Tsai and Li (2008b) and Tsai and Li (2008c) introduced a variance window to scale $\mathrm{KIC}_{D}^{(p)}$ to cope with the narrow window size of Occam's window. However, scaling $\mathrm{KIC}_{\Delta}^{(p)}$ should not be considered.

As a consequence, the data $\mathbf{D}$ in BMA in Equation (7.47) actually represent two data sets: one data set $\mathbf{D}_{\Delta} \in \mathbf{D}$ for calculating $\mathrm{KIC}_{\Delta}^{(p)}$ and the other data set $\mathbf{D}_{D} \in \mathbf{D}$ for calculating $\mathrm{KIC}_{D}^{(p)}$. For the KIC case in the following analysis, we consider the prior $\boldsymbol{\beta}$ distribution in g_{Δ} and g_{D} to be independent within and between individual GP methods and to be uniformly distributed between [0,1].

7.6.5 Multi-Gaussian Distributions

Consider a subset of data $\mathbf{D}_{\Delta} \in \mathbf{D}$ is used to calculate $\Pr\left(\Delta \mid \mathbf{D}, M^{(p)}\right)$ and the prediction $\Delta(\boldsymbol{\beta}^{(p)})$ is at the $\mathbf{D}_{\Delta}$ space. For groundwater inverse modeling purposes, $\Pr\left(\Delta \mid \mathbf{D}_{\Delta}, M^{(p)}, \hat{\boldsymbol{\beta}}^{(p)}\right)$ describes the probability of errors between the predicted quantity and observation data, and is assumed to be a multi-Gaussian distribution with zero mean and covariance matrix $\mathbf{C}_{\Delta}$. Therefore, the $\mathrm{KIC}_{\Delta}^{(p)}$ is

$$\mathrm{KIC}_{\Delta}^{(p)} = Q_{\Delta}^{(p)} + \left(n_{\Delta} - m^{(p)}\right)\ln 2\pi + \ln\left|\mathbf{C}_{\Delta}\right| + \ln\left|\mathbf{F}_{\Delta}^{(p)}\right| \tag{7.61}$$

where n_{Δ} is the number of $\mathbf{D}_{\Delta}$ and

$$Q_{\Delta}^{(p)} = \left(\Delta(\boldsymbol{\beta}^{(p)}) - \mathbf{D}_{\Delta}\right)^{T} \mathbf{C}_{\Delta}^{-1} \left(\Delta(\boldsymbol{\beta}^{(p)}) - \mathbf{D}_{\Delta}\right) \tag{7.62}$$

is the fitting residual with respect to data, $\mathbf{D}_\Delta$. The Fisher information matrix, $\mathbf{F}_\Delta^{(p)}$, is (Kitanidis and Lane 1985)

$$\left[\mathbf{F}_\Delta^{(p)}\right]_{ij} = \frac{1}{2}\mathrm{Tr}\left[\mathbf{C}_\Delta^{-1}\frac{\partial \mathbf{C}_\Delta}{\partial \beta_i^{(p)}}\mathbf{C}_\Delta^{-1}\frac{\partial \mathbf{C}_\Delta}{\partial \beta_j^{(p)}}\right] + \frac{\partial \Delta^T}{\partial \beta_i^{(p)}}\mathbf{C}_\Delta^{-1}\frac{\partial \Delta}{\partial \beta_j^{(p)}}. \tag{7.63}$$

If the covariance matrix $\mathbf{D}_\Delta$ is constant or the trace is relatively small, the Fisher information matrix can be approximated to the inverse covariance matrix of the estimated weighting coefficients:

$$\left[\mathbf{F}_\Delta^{(p)}\right]_{ij} = \frac{\partial \Delta^T}{\partial \beta_i^{(p)}}\mathbf{C}_\Delta^{-1}\frac{\partial \Delta}{\partial \beta_j^{(p)}} \approx \mathrm{Cov}\left(\boldsymbol{\beta}^{(p)}\right)^{-1}. \tag{7.64}$$

Yeh and Yoon (1981) considered the groundwater heads $\Delta = \mathbf{h}$ in Equation (7.64).

Similarly, the subset of data $\mathbf{D}_D \in \mathbf{D}$ is used to calculate model weights $\Pr\left(\mathbf{D} \mid M^{(p)}\right)$, and $\mathbf{D}^{cal}(\boldsymbol{\beta}^{(p)})$ is the calculated data corresponding to the $\mathbf{D}_D$ space. The error between the calculated data and observation data is assumed to be multi-Gaussian with zero mean and covariance matrix $\mathbf{C}_D$ in the likelihood function $\Pr\left(\mathbf{D}_D \mid M^{(p)}, \hat{\boldsymbol{\beta}}^{(p)}\right)$. Accordingly, the $\mathrm{KIC}_D^{(p)}$ is

$$\mathrm{KIC}_D^{(p)} = Q_D^{(p)} + \left(n_D - m^{(p)}\right)\ln 2\pi + \ln\left|\mathbf{C}_D\right| + \ln\left|\mathbf{F}_D^{(p)}\right| \tag{7.65}$$

where n_D is the number of $\mathbf{D}_D$ and

$$Q_D^{(p)} = \left(\mathbf{D}^{cal} - \mathbf{D}_D\right)^T \mathbf{C}_D^{-1}\left(\mathbf{D}^{cal} - \mathbf{D}_D\right) \tag{7.66}$$

is the fitting residual with respect to data, $\mathbf{D}_D$. The Fisher information matrix, $\mathbf{F}_D^{(p)}$, is

$$\left[\mathbf{F}_D^{(p)}\right]_{ij} = \frac{1}{2}\mathrm{Tr}\left[\mathbf{C}_D^{-1}\frac{\partial \mathbf{C}_D}{\partial \beta_i^{(p)}}\mathbf{C}_D^{-1}\frac{\partial \mathbf{C}_D}{\partial \beta_j^{(p)}}\right] + \frac{\partial \mathbf{D}^{cal^T}}{\partial \beta_i^{(p)}}\mathbf{C}_D^{-1}\frac{\partial \mathbf{D}^{cal}}{\partial \beta_j^{(p)}}. \tag{7.67}$$

The multi-standard normal distribution is also applicable to BIC as follows:

$$\begin{aligned}\mathrm{BIC}_\Delta^{(p)} &= Q_\Delta^{(p)} + n_\Delta \ln 2\pi + \ln\left|\mathbf{C}_\Delta\right| + m^{(p)}\ln n_\Delta \\ \mathrm{BIC}_D^{(p)} &= Q_D^{(p)} + n_D \ln 2\pi + \ln\left|\mathbf{C}_D\right| + m^{(p)}\ln n_D\end{aligned} \tag{7.68}$$

7.7 Experimental Design

7.7.1 Experimental Design for Parameter Estimation

Experimental design deals with the selection of experimental conditions such that a specified criterion is optimized. The experimental conditions also are referred to as the decision variables. The decision variables of a groundwater experimental design generally consist of two groups: the excitation group and the observation group (Hsu and Yeh 1989; Sun 1994). In the excitation group, the decision variables may include the number and locations of pumping and injection wells, the pumping and injection rates, stress periods of pumping and injection, and artificial changes of boundary conditions. In the observation group, the decision variables may include the state variables to be observed, the number and locations of observation wells, and the observation frequency. When the decision variables in the excitation group are fixed, experimental design simplifies to observation network design. The optimization of experimental design is generally subject to a set of constraints. The constraints frequently encountered include: budget, allowable drawdown at selected locations, maximum pumping/recharge rates, duration of the experiment, allowable time interval between consecutive measurements, and reliability of the estimated parameters. To apply the optimal experimental design in practice requires that we: 1) establish a criterion (performance measure) so that different experimental designs can be compared, and 2) develop an algorithm so that the established criterion can be optimized over the proper choice of the decision variables. The formulated optimal experimental problem invaluably lends itself to a combinatorial optimization problem which, in principle, can be solved by a mixed integer nonlinear programming algorithm (Yeh 1992).

Steinberg and Hunter (1984) presented an extensive review of the classical criteria derived for linear statistical models. The most popular optimality criteria are:
(i) *D*-Optimality: A design is said to be *D*-optimal if it minimizes the determinant of the covariance matrix of the estimated parameters.
(ii) *A*-Optimality. A design is said to be *A*-optimal if it minimizes the trace of the covariance matrix of the estimated parameters.
(iii) *E*-Optimality: A design is said to be *E*-optimal if it minimizes the maximal eigenvalue of the covariance matrix of the estimated parameters.
(iv) *G*-Optimality: A design is said to be *G*-optimal if it minimizes max $d(\mathbf{x})$, where $d(\mathbf{x})$ is the variance of the estimated response at $\mathbf{x}$, and the maximum is taken over all possible vectors $\mathbf{x}$ of predictor variables.
(v) I_λ-Optimality: A design is said to be I_λ-optimal if it minimizes $\int d(\mathbf{x})\lambda(d\mathbf{x})$, where λ is a probability measure on the space of predictor variables. This criterion, also called average integrated variance, belongs to a more general class of L-optimality criterion (Fedorov 1972). Among the proposed design criteria, the *A*-Optimality, *D*-Optimality and *E*-Optimality are the most widely used.

The inherent difficulty in experiment design is that the designs are predicated on unknown parameters which themselves are to be estimated. However, in practice, initial estimates of parameters are determined from prior information. Using the initial estimates, an optimal experimental design is carried out and data are collected accordingly. With the collected data, the inverse problem of parameter identification is solved to update the parameter values. If necessary, another round of experimental design can be carried out with the updated parameter values. This procedure is called sequential design. Its convergence property has been investigated by Nishikawa and Yeh (1989) and Cleveland and Yeh (1990).

7.7.2 Experimental Design for Parameter Structure Identification

Traditional experimental design methods, such as *D*-optimal design or the *A*-optimal design, do not consider whether the information provided by a design is sufficient. As a result, under-sampling or over-sampling frequently occurs. Chang et al. (2005) utilized the GIP and presented a methodology for observation network design aimed at finding a minimum cost design that provides sufficient information for identifying both the parameter structure and parameter values. Sequential Gaussian simulation was used to generate different realizations for the unknown distributed parameter. For each possible realization, the number of observation wells was increased gradually during the design process until the information provided by the design was sufficient. The selection criterion for locating a new observation well was the maximization of the information content for parameter identification. The overall sufficiency of a design was assessed by Monte Carlo simulation.

To circumvent the need for initial estimates of parameters and a large number of Monte Carlo simulations, Sun and Yeh (2007b) proposed a robust experimental design procedure based on the worst-case parameter in the parameter admissible region. The worst-case parameter (WCP) is defined as one that requires the most information for its identification (Sun and Yeh 2007a). Therefore, if the data provided by a design are sufficient for identifying the WCP, then the design must be sufficient for identifying any other parameters with the same structure or a simplified structure. Sun and Yeh (2007b) presented the following procedure for constructing a sufficiency and robustness design D:

Step 1. Compile all available prior information including the objectives of model application and their accuracy requirements.

Step 2. Based on the prior information, estimate a structure $\boldsymbol{\theta}_A$ as the true structure and find its WCP, $\tilde{\mathbf{p}}_A$.

Step 3. Run the simulation model to generate a set of "observation data" $\mathbf{u}_D(\boldsymbol{\theta}_A, \tilde{\mathbf{p}}_A)$ according to the designed excitation strengths, observation locations and times.

Step 4. Run the application model to generate a set of "application results" $\mathbf{g}_E(\boldsymbol{\theta}_A, \tilde{\mathbf{p}}_A)$.

Step 5. Generate a series of structures one by one and calculate the following fitting residual RE_m and model application error AE_m for each structure $\boldsymbol{\theta}_m$:

$$RE_m = \min_{\boldsymbol{\theta}_m, \mathbf{p}_m} \left\| \mathbf{u}_D(\boldsymbol{\theta}_A, \tilde{\mathbf{p}}_A) - \mathbf{u}_D(\boldsymbol{\theta}_m, \mathbf{p}_m) \right\|_D \quad , \quad \text{s.t.} \quad \mathbf{p}_m \in Ad(\mathbf{p}_m) \tag{7.69}$$

$$AE_m = \left\| \mathbf{g}_\mathbf{E}(\boldsymbol{\theta}_A, \tilde{\mathbf{p}}_A) - \mathbf{g}_\mathbf{E}(\boldsymbol{\theta}_m, \tilde{\mathbf{p}}_m) \right\|_E \tag{7.70}$$

where m is the parameter dimension; $\|\cdot\|_E$ and $\|\cdot\|_D$ are L_2 norms defined in the objective space E and observation space D, respectively; $Ad(\mathbf{p}_m)$ is the admissible set of $\mathbf{p}_m$, and $(\boldsymbol{\theta}_m, \tilde{\mathbf{p}}_m)$ is the solution of Equation (5.69).

Step 6. If $AE_m > \varepsilon$ and $RE_m > 2\eta$, increase the parameter dimension m by 1 and repeat the above procedure to find $\boldsymbol{\theta}_{m+1}$, where ε is the specified accuracy requirement and η is the upper bound of the norm of observation error measured in the observation space.

Step 7. When m increases, the value of AE_m will decrease and the value of RE_m will decrease.

Thus, finally we must arrive at one of the following three situations: (i) $AE_m < \varepsilon$ but $RE_m \geq 2\eta$; (ii) $AE_m < \varepsilon$ for all $RE_m < 2\eta$; or (iii) $AE_m \geq \varepsilon$ but $RE_m < 2\eta$. If cases (i) and (ii) occur, we can conclude that the design is robust and sufficient. Otherwise, when case (iii) occurs, the design is insufficient.

As mentioned before, a robust design is a conservative design. It provides the maximum information for identifying the most difficult parameter in the admissible parameter region. Based on the WCP, Sun and Yeh (2007b) proposed a heuristic design procedure for finding a robust but sub-optimal experimental design.

7.8 Summary and Conclusions

i. We can classify parameter structure identification into the extended inverse problem (EIP) and the generalized inverse problem (GIP). The EIP considers the identification of parameter structure (parameter dimension, parameter pattern, and parameter values) based on maximum likelihood estimation. The GIP broadens the EIP by incorporating model prediction or management along with the maximum likelihood estimation. Model discrimination and model selection criteria, including the parsimony principle, parameter uncertainty, structure error, prediction error and statistical information criteria, are introduced to determine the best parameter structure.

ii. Parameterization is a necessary step to represent a distributed parameter (e.g., the hydraulic conductivity discussed in this chapter) for groundwater modeling and define parameter structure for parameter structure identification. We introduced a generalized parameterization (GP) method to integrate the zonation method and the interpolation method to increase parameterization flexibility. This approach is important because a parameter estimate is not necessarily limited to a piece-wise constant or smooth distribution. Moreover, the GP method extends the traditional geostatistical framework to characterize spatially correlated parameters for stochastic inverse modeling.

iii. Interpolation point selection plays an important role in parameter structure identification. We introduced the indicator generalized parameterization (IGP) to determine the selection of sample data points for interpolation. The IGP focuses on an unsampled location's relevance to its neighboring sampled locations. This is more general than the traditional approach of focusing on the influential range of sample data on its neighboring locations where parameter value needs to be determined.

iv. In contrast to model selection and model discrimination, we introduced Bayesian model averaging (BMA) to address the non-uniqueness of parameterization methods. Using a single parameterization method for parameter structure identification is likely to underestimate estimation uncertainty. Based on the law of total probability, the BMA weighs candidate models by the evidence of data. Then, the spatial statistics of an estimated parameter using multiple parameterization methods can be obtained by the BMA expectation and covariance.

v. Experimental design complements the inverse theory. If the existing data are found to be insufficient for parameter structure identification, experimental design can be used to determine sampling strategies with the objective of reducing model uncertainty. We introduced several optimality criteria for optimizing experimental design for collecting informative data.

Acknowledgements

The first author was supported in part by the Department of the Interior, U.S. Geological Survey under Grant No. 05HQGR0142 and 06HQGR0088. The views and conclusions contained in the document are those of the authors and should not be interpreted as necessarily representing the official policies, either expressed or implied, of the U.S. Government.

7.9 References

Akaike, H. (1973). "Information theory and an extension of the maximum likelihood principle." *Proceeding of the second international symposium on information theory*, B. N. Petrov and F. Csaki, eds., Akademiai Kiado, Budapest, 267-281.

Akaike, H. (1974). "A new look at statistical model identification." *IEEE Transactions on Automatic Control*, AC-19(6), 716-723.

Akaike, H. (1981). "Likelihood of a model and information criteria." *Journal of Econometrics*, 16(1), 3-14.

Aschenbrenner, F., and Ostin, A. (1995). "Automatic parameter-estimation applied on a groundwater model: The problem of structure identification." *Environmental Geology*, 25(3), 205-210.

Atkinson, A. C. (1981). "Likelihood ratios, posterior odds and information criteria." *Journal of Econometrics*, 16(1), 15-20.

Berger, J. O. (1985). *Statistical decision theory and bayesian analysis*, Springer-Verlag, New York.

Bernardo, J. M., and Smith, A. F. M. (1994). *Bayesian theory*, Wiley and Sons, New York.

Boots, B. N. (1986). *Voronoi (thiessen) polygons*, Geo Books, Norwich, UK.

Bowyer, A. (1981). "Computing Dirichlet tessellations." *Computer Journal*, 24(2), 162-166.

Bozdogan, H. (2000). "Akaike's information criterion and recent developments in information complexity." *Journal of Mathematical Psychology*, 44(1), 62-91.

Brown, P. J., Vannucci, M., and Fearn, T. (2002). "Bayes model averaging with selection of regressors." *Journal of the Royal Statistical Society Series B-Statistical Methodology*, 64, 519-536.

Burnham, K. P., and Anderson, D. R. (2002). *Model selection and multimodel inference - A practical information-theorectic approach*, Springer-Verlag, New York.

Burnham, K. P., and Anderson, D. R. (2004). "Multimodel inference: Understanding AIC and BIC in model selection." *Sociological Methods and Research*, 33(2), 261-304.

Carrera, J., and Neuman, S. P. (1986a). "Estimation of aquifer parameters under transient and steady-state conditions 1. Maximum-likelihood method incorporating prior information." *Water Resources Research*, 22(2), 199-210.

Carrera, J., and Neuman, S. P. (1986b). "Estimation of aquifer parameters under transient and steady-state conditions 2. Uniqueness, stability, and solution algorithms." *Water Resources Research*, 22(2) 211-227.

Carrera, J., and Neuman, S. P. (1986c). "Estimation of aquifer parameters under transient and steady-state conditions 3. Application to synthetic and field data." *Water Resources Research*, 22(2), 228-242.

Cassiani, G., and Medina, M. A. (1997). "Incorporating auxiliary geophysical data into ground-water estimation." *Ground Water*, 35(1), 79-91.

Certes, C., and de Marsily, G. (1991). "Application of the pilot point method to the identification of aquifer transmissivities." *Advances in Water Resources*, 14(5), 284-300.

Chang, L.-F., Sun, N.-Z., and Yeh, W.-G. (2005). "Optimal observation network design for parameter structure identification in groundwater modeling." *Water Resources Research*, 41(3), W03002, doi:10.1029/2004WR003514.

Chow, G. C. (1981). "A comparison of the information and posterior probability criteria for model selection." *Journal of Econometrics*, 16(1), 21-33.

Claeskens, G., and Hjort, N. L. (2003). "The focused information criterion." *Journal of the American Statistical Association*, 98(464), 900-916.

Cleveland, T. G., and Yeh, W. W.-G. (1990). "Sampling network design for transport parameter-identification." *Journal of Water Resources Planning and Management*, 116(6), 764-783.

Clifton, P. M., and Neuman, S. P. (1982). "Effects of kriging and inverse modeling on conditional simulation of the Avra Valley aquifer in southern Arizona." *Water Resources Research*, 18(4), 1215-1234.

Cooley, R. L., Konikow, L. F., and Naff, R. L. (1986). "Nonlinear-regression groundwater flow modeling of a deep regional aquifer system." *Water Resources Research*, 22(13), 1759-1778.

Cressie, N., and Wikle, C. K. (1998). "The variance-based cross-variogram: You can add apples and oranges." *Mathematical Geology*, 30(7), 789-799.

de Marsily, G. (1986). *Quantitative hydrogeology : Groundwater hydrology for engineers*, Academic Press, Orlando, Fla.

Delhomme, J. P. (1979). "Spatial variability and uncertainty in groundwater flow parameters: A geostatistical approach." *Water Resources Research*, 15(2), 269-280.

Dirichlet, G. L. (1850). "Uber die reduction der positeven quadratischen formen mit drei unbestimmten ganzen zahlen." *Journal of fur die Reine und Angewandte Mathematik*, 40, 209-234.

Distefano, N., and Rath, A. (1975). "Identification approach to subsurface hydrological systems." *Water Resources Research*, 11(6), 1005-1012.

Doherty, J. (2003). "Ground water model calibration using pilot points and regularization." *Ground Water*, 41(2), 170-177.

Draper, D. (1995). "Assessment and propagation of model uncertainty." *Journal of the Royal Statistical Society Series B-Methodological*, 57(1), 45-97.

Emsellem, Y., and de Marsily, G. (1971). "An automatic solution for the inverse problem." *Water Resources Research*, 7(5), 1264-1283.

Eppstein, M. J., and Dougherty, D. E. (1996). "Simultaneous estimation of transmissivity values and zonation." *Water Resources Research*, 32(11), 3321-3336.

Fedorov, V. V. (1972). *Theory of optimal experiments*, translated and edited by W. J. Studeen and E. M. Klimko, Academic, San Diego, Calif.

Foster, D. P., and George, E. I. (1994). "The risk inflation criterion for multiple-regression." *Annals of Statistics*, 22(4), 1947-1975.

Gaganis, P., and Smith, L. (2001). "A Bayesian approach to the quantification of the effect of model error on the predictions of groundwater models." *Water Resources Research*, 37(9), 2309-2322.

Gelfand, A. E., and Dey, D. K. (1994). "Bayesian model choice: Asymptotics and exact calculations." *Journal of the Royal Statistical Society Series B-Methodological*, 56(3), 501-514.

Gelman, A., Meng, X. L., and Stern, H. (1996). "Posterior predictive assessment of model fitness via realized discrepancies." *Statistica Sinica*, 6(4), 733-760.

George, E. I., and McCulloch, R. E. (1993). "Variable selection via Gibbs sampling." *Journal of the American Statistical Association*, 88(423), 881-889.

Gomez-Hernanez, J. J., Sahuquillo, A., and Capilla, J. E. (1997). "Stochastic simulation of transmissivity fields conditional to both transmissivity and piezometric data 1. Theory." *Journal of Hydrology*, 203(1-4), 162-174.

Gotway, C. A., Ferguson, R. B., Hergert, G. W., and Peterson, T. A. (1996). "Comparison of kriging and inverse-distance methods for mapping soil parameters." *Soil Science Society of America Journal*, 60(4), 1237-1247.

Green, P. J., and Sibson, R. (1978). "Computing Dirichlet tessellations in plane." *Computer Journal*, 21(2), 168-173.

Hannan, E. J., and Quinn, B. G. (1979). "Determination of the order of an autoregression." *Journal of the Royal Statistical Society Series B-Methodological*, 41(2), 190-195.

Hansen, B. E. (2007). "Least squares model averaging." *Econometrica*, 75(4), 1175-1189.

Hill, M. C., Colley, R. L., and Pollock, D. W. (1998). "A controlled experiment in ground water flow model calibration." *Ground Water*, 36(3), 520-535.

Hoeting, J. A., Madigan, D., Raftery, A. E., and Volinsky, C. T. (1999). "Bayesian model averaging: A tutorial." *Statistical Science*, 14(4), 382-401.

Hsu, N.-S., and Yeh, W. W.-G. (1989). "Optimum experimental design for parameter identification in groundwater hydrology." *Water Resources Research*, 25(5), 1025-1040.

Hurvich, C. M., and Tsai, C. L. (1989). "Regression and time-series model selection in small samples." *Biometrika*, 76(2), 297-307.

Hyndman, D. W., and Gorelick, S. M. (1996). "Estimating lithologic and transport properties in three dimensions using seismic and tracer data: The Kesterson aquifer." *Water Resources Research*, 32(9), 2659-2670.

Hyun, Y., and Lee, K. K. (1998). "Model identification criteria for inverse estimation of hydraulic parameters." *Ground Water*, 36(2), 230-239.

Isaaks, E. H., and Srivastava, E. H. (1989). *An introduction to applied geostatistics*, Oxford University Press.

Jeffreys, H. (1961). *Theory of probability*, Clarendon Press, Oxford.

Jiao, J. J., and Lerner, D. N. (1996). "Using sensitivity analysis to assist parameter zonation in ground water flow model." *Water Resources Bulletin*, 32(1), 75-87.

Karpouzos, D. K., Delay, F., Katsifarakis, K. L., and de Marsily, G. (2001). "A multipopulation genetic algorithm to solve the inverse problem in hydrogeology." *Water Resources Research*, 37(9), 2291-2302.

Kashyap, R. L. (1982). "Optimal choice of AR and MA parts in autoregressive moving average models." *IEEE Transactions on Pattern Analysis and Machine Intelligence*, 4(2), 99-104.

Kass, R. E., and Raftery, A. E. (1995). "Bayes factors." *Journal of the American Statistical Association*, 90(430), 773-795.

Kass, R. E., and Wasserman, L. (1995). "A reference Bayesian test for nested hypotheses and its relationship to the Schwarz criterion." *Journal of the American Statistical Association*, 90(431), 928-934.

Kitanidis, P. K. (1996). "On the geostatistical approach to the inverse problem." *Advances in Water Resources*, 19(6), 333-342.

Kitanidis, P. K. (1997). *Introduction to geostatistics applications in hydrogeology*, Cambridge University Press.

Kitanidis, P. K., and Lane, R. W. (1985). "Maximum-likelihood parameter-estimation of hydrologic spatial processes by the Gauss-Newton method." *Journal of Hydrology*, 79(1-2), 53-71.

Kitanidis, P. K., and Vomvoris, E. G. (1983). "A geostatistical approach to the inverse problem in groundwater modeling (steady-state) and one-dimensional simulations." *Water Resources Research*, 19(3), 677-690.

Knopman, D. S., and Voss, C. I. (1988). "Discrimination among one-dimensional models of solute transport in porous media: Implications for sampling design." *Water Resources Research*, 24(11), 1859-1876.

Knopman, D. S., and Voss, C. I. (1989). "Multiobjective sampling design for parameter estimation and model discrimination in groundwater solute transport." *Water Resources Research*, 25(10), 2245-2258.

Knopman, D. S., Voss, C. I., and Garabedian, S. P. (1991). "Sampling design for groundwater solute transport: Tests of methods and analysis of Cape Cod tracer test data." *Water Resources Research*, 27(5), 925-949.

Kohn, R., Smith, M., and Chan, D. (2001). "Nonparametric regression using linear combinations of basis functions." *Statistics and Computing*, 11(4), 313-322.

Koltermann, C. E., and Gorelick, S. M. (1996). "Heterogeneity in sedimentary deposits: A review of structure-imitating, process-imitating, and descriptive approaches." *Water Resources Research*, 32(9), 2617-2658.

Kuo, L., and Mallick, B. (1997). "Bayesian semiparametric inference for the accelerated failure-time model." *Canadian Journal of Statistics-Revue Canadienne de Statistique*, 25(4), 457-472.

LaVenue, A. M., RamaRao, B. S., de Marsily, G., and Marietta, M. G. (1995). "Pilot point methodology for automated calibration of an ensemble of conditionally simulated transmissivity fields 2. Application." *Water Resources Research*, 31(3), 495-516.

Leamer, E. E. (1978). *Specification searches: Ad hoc inference with non-experimental data*, Wiley, New York.

Luis, S. J. and McLaughlin, D. (1992). "A stochastic approach to model validation." *Advances in Water Resources*, 15(1), 15-32.

Madigan, D., and Raftery, A. E. (1994). "Model selection and accounting for model uncertainty in graphical models using Occam's window." *Journal of the American Statistical Association*, 89(428), 1535-1546.

Madigan, D., and York, J. (1995). "Bayesian graphical models for discrete data." *International Statistical Review*, 63(2), 215-232.

Mallows, C. L. (1973). "Some comments on Cp." *Technometrics*, 15(4), 661-675.

Mayer, A. S., and Huang, C. L. (1999). "Development and application of a coupled-process parameter inversion model based on the maximum likelihood estimation method." *Advances in Water Resources*, 22(8), 841-853.

McLaughlin, D., and Townley, L. R. (1996). "A reassessment of the groundwater inverse problem. *Water Resources Research*, 32(5), 1131-1161.

Neuman, S. P. (2003). "Maximum likelihood Bayesian averaging of uncertain model predictions." *Stochastic Environmental Research and Risk Assessment*, 17(5), 291-305.

Neuman, S. P., and Yakowitz, S. (1979). "Statistical approach to the inverse problem of aquifer hydrology 1. Theory." *Water Resources Research*, 15(4), 845-860.

Nishikawa, T., and Yeh, W. W. G. (1989). "Optimal pumping test design for the parameter-identification of groundwater systems." *Water Resources Research*, 25(7), 1737-1747.

Okabe, A., Boots, B., Sugihara, K., and Chiu, S. N. (2000). *Spatial tessellations: Concepts and applications of Voronoi diagrams*, John Wiley & Sons, Chichester, UK.

Olea, R. A. (1999). *Geostatistics for engineers and earth scientists*, Kluwer Academic Publishers, Boston.

Poeter, E. P., and Anderson, D. (2005). "Multimodel ranking and inference in ground water modeling." *Ground Water*, 43(4), 597-605.

Poeter, E. P., and McKenna, S. A. (1995). "Reducing uncertainty associated with groundwater-flow and transport predictions." *Ground Water*, 33(6), 899-904.

Posada, D., and Buckley, T. R. (2004). "Model selection and model averaging in phylogenetics: Advantages of Akaike information criterion and Bayesian approaches over likelihood ratio tests." *Systematic Biology*, 53(5), 793-808.

Raftery, A. E. (1995). "Bayesian model selection in social research." *Sociological Methodology*, 25, 111-163.

Raftery, A. E. (1996). "Approximate Bayes factors and accounting for model uncertainty in generalised linear models." *Biometrika*, 83(2), 251-266.

Raftery, A. E., Madigan, D., and Hoeting, J. A. (1997). "Bayesian model averaging for linear regression models." *Journal of the American Statistical Association*, 92(437), 179-191.

RamaRao, B. S., LaVenue, A. M., de Marsily, G., and Marietta, M. G. (1995). "Pilot point methodology for automated calibration of an ensemble of conditionally simulated transmissivity fields 1. Theory and computational experiments." *Water Resources Research*, 31(3), 475-493.

Rubin, Y., and Hubbard, S. S. (2005). *Hydrogeophysics*, Springer, Dordrecht, New York.

Sambridge, M., Braun, J., and McQueen, H. (1995). "Geophysical parametrization and interpolation of irregular data using natural neighbors." *Geophysical Journal International*, 122(3), 837-857.

Schwarz, G. (1978). "Estimating dimension of a model." *Annals of Statistics*, 6(2), 461-464.

Seghouane, A. K., and Bekara, M. (2004). "A small sample model selection criterion based on Kullback's symmetric divergence." *IEEE Transactions on Signal Processing*, 52(12), 3314-3323.

Shah, P. C., Gavalas, G. R., and Seinfeld, J. H. (1978). "Error analysis in history matching: Optimum level of parameterization." *Society of Petroleum Engineers Journal*, 18(3), 219-228.

Sibson, R. (1981). "Chapter 2: A brief description of natural neighbour interpolation." *Interpreting multivariate data*, V. Barnett, ed., Wiley.

Steinberg, D. M., and Hunter, W. G. (1984). "Experimental design: Review and comment." *Technometrics*, 26(2), 71-97.

Stone, M. (1974). "Cross-validatory choice and assessment of statistical predictions." *Journal of the Royal Statistical Society Series B-Methodological*, 36(2), 111-147.

Suchard, M. A., Weiss, R. E., and Sinsheimer, J. S. (2003). "Testing a molecular clock without an outgroup: Derivations of induced priors on branch-length restrictions in a Bayesian framework." *Systematic Biology*, 52(1), 48-54.

Sugiura, N. (1978). "Further analysis of data by Akaike's information criterion and finite corrections." *Communications in Statistics Part A-Theory and Methods*, 7(1), 13-26.

Sun, N.-Z. (1994). *Inverse problems in groundwater modeling*, Kluwer Academic, Dordrecht, Boston.

Sun, N.-Z., Jeng, M.-C., and Yeh, W. W.-G. (1995). "A proposed geological parameterization method for parameter-identification in 3-dimensional groundwater modeling." *Water Resources Research*, 31(1), 89-102.

Sun, N.-Z., and Sun, A. Y. (2002). "Chapter 9: Parameter identification of environmental systems." *Environmental fluid mechanics: Theories and applications*, H. Shen et al., eds., ASCE, Reston, Virginia, 297-337.

Sun, N.-Z., Yang, S.-L., and Yeh, W. W.-G. (1998). "A proposed stepwise regression method for model structure identification." *Water Resources Research*, 34(10), 2561-2572.

Sun, N.-Z., and Yeh, W. W.-G. (1985). "Identification of parameter structure in groundwater inverse problem." *Water Resources Research*, 21(6), 869-883.

Sun, N.-Z., and Yeh, W. W.-G. (2007a). "Development of objective-oriented groundwater models 1. Robust parameter identification." *Water Resources Research*, 43(2), W02420, doi:10.1029/2006WR004887.

Sun, N.-Z. and Yeh, W. W.-G. (2007b). Development of objective-oriented groundwater models 2. Robust experimental design. *Water Resources Research* 43(2), W02421, doi:10.1029/2006WR004888.

Tan, C. C., Tung, C. P., and Tsai, F. T.-C. (2008). "Applying zonation methods and tabu search to improve the ground-water modeling." *Journal of the American Water Resources Association*, 44(1), 107-120.

Thiessen, A. H. (1911). "Precipitation averages for large areas." *Monthly Weather Review*, 39(7), 1082-1084.

Tierney, L., and Kadane, J. B. (1986). "Accurate approximations for posterior moments and marginal densities." *Journal of the American Statistical Association*, 81(393), 82-86.

Tsai, F. T.-C. (2006). "Enhancing random heterogeneity representation by mixing the kriging method with the zonation structure." *Water Resources Research*, 42(8), W08428, doi:10.1029/2005WR004111.

Tsai, F. T.-C. (2009). "Indicator generalized parameterization for interpolation point selection in groundwater inverse modeling." *Journal of Hydrologic Engineering*, 14(3), 233-242.

Tsai, F. T.-C. and Li, X. (2008a). "Chapter 18: Conditional estimation of distributed hydraulic conductivity in groundwater inverse modeling: Indicator-generalized parameterization and natural neighbors, *Practical hydroinformatics:*

Computational intelligence and technological developments in water applications, R. J. Abrahart et al., eds., Water Science and Technology Library, 68, Springer, 245-257.

Tsai, F. T.-C., and Li, X. (2008b). "Inverse groundwater modeling for hydraulic conductivity estimation using Bayesian model averaging and variance window." *Water Resources Research*, 44, W09434, doi:10.1029/2007WR006576.

Tsai, F. T.-C., and Li, X. (2008c). "Multiple parameterization for hydraulic conductivity identification." *Ground Water*, 46(6), 851-864.

Tsai, F. T.-C., Sun, N.-Z., and Yeh, W. W.-G. (2003a). "A combinatorial optimization scheme for parameter structure identification in ground water modeling." *Ground Water*, 41(2), 156-169.

Tsai, F. T.-C., Sun, N.-Z., and Yeh, W. W.-G. (2003b). "Global-local optimization for parameter structure identification in three-dimensional groundwater modeling." *Water Resources Research* ,39(2), 1043, doi:10.1029/2001WR001135.

Tsai, F. T.-C., Sun, N.-Z., and Yeh, W. W.-G. (2005). "Geophysical parameterization and parameter structure identification using natural neighbors in groundwater inverse problems." *Journal of Hydrology*, 308, 269-283.

Tsai, F. T.-C., and Yeh, W. W.-G. (2004). "Characterization and identification of aquifer heterogeneity with generalized parameterization and Bayesian estimation." *Water Resources Research*, 40, W10102, doi:10.1029/2003WR002893.

Usunoff, E., Carrera, J., and Mousavi, S. F. (1992). "An approach to the design of experiments for discriminating among alternative conceptual models." *Advances in Water Resources*, 15(3), 199-214.

Voronoi, G. (1908). "Nouvelles applications des parametres continus a la theroie des formes quadratiques, deuxieme memoire, recherches sur les parallelloedres primitifs." *Journal fur die Reine und Angewandte Mathematik*, 134, 198-287.

Wagner, B. J., and Gorelick, S. M. (1987). "Optimal groundwater quality management under parameter uncertainty." *Water Resources Research*, 23(7), 1162-1174.

Wasserman, L. (2000). "Bayesian model selection and model averaging." *Journal of Mathematical Psychology*, 44(1), 92-107.

Watson, D. F. (1981). "Computing the n-dimensional Delaunay tessellation with application to Voronoi polytopes." *Computer Journal*, 24(2), 167-172.

Ye, M., Neuman, S. P., and Meyer, P. D. (2004). "Maximum likelihood Bayesian averaging of spatial variability models in unsaturated fractured tuff." *Water Resources Research*, 40(5), W05113, doi:10.1029/2003WR002557.

Ye, M., Neuman, S. P., Meyer, P. D., and Pohlmann, K. (2005). "Sensitivity analysis and assessment of prior model probabilities in MLBMA with application to unsaturated fractured tuff." *Water Resources Research*, 41(12), W12429, doi: 10.1029/2005WR004260.

Yeh, W. W.-G. (1986). "Review of parameter identification procedures in groundwater hydrology: The inverse problem." *Water Resources Research*, 22(2), 95-108.

Yeh, W. W.-G. (1992). "Systems analysis in ground-water planning and management." *Journal of Water Resources Planning and Management*, 118(3), 224-237.

Yeh, W. W.-G., and Yoon, Y. S. (1976). "A system identification procedure for the identification of inhomogeneous aquifer parameters." *Advances in groundwater hydrology*, Z. A. Saleen, ed., American Water Resources Association, Middleburg, Va., 77-83.

Yeh, W. W.-G., and Yoon, Y. S. (1981). "Aquifer parameter identification with optimum dimension in parameterization." *Water Resources Research*, 17(3), 664-672.

Yeh, W. W.-G., Yoon, Y. S., and Lee, L. S. (1983). "Aquifer parameter identification with kriging and optimum parameterization." *Water Resources Research*, 19(1), 225-233.

Zheng, C., and Wang, P. (1996). "Parameter structure identification using tabu search and simulated annealing." *Advances in Water Resources*, 19(4), 215-224.

CHAPTER 8

DEVELOPMENT OF GROUNDWATER RESOURCES

Zhuping Sheng[1], **Jiang Li**[2], **Phillip J. King**[3] and **William Miller**[4]
[1]Texas AgriLife Research Center at El Paso, Texas A&M University System,
[2]Department of Civil Engineering, Morgan State University
[3]Department of Civil Engineering, New Mexico State University
[4]William J. Miller Engineers, Inc., New Mexico

8.1 Introduction

Current urban population growth rates present a challenge to municipalities across the U.S. and elsewhere and demand comprehensive strategies for groundwater development. The planning and implementation of any groundwater development project require a detailed assessment to ensure that the objectives of project are achieved within all economic, legal, and environmental constraints. Conducting such an assessment requires a fundamental understanding of aquifer yield concepts. Decisions on well field design, development of recharge facilities, groundwater withdrawals and conjunctive use of surface and groundwater all depend on aquifer yield. Issues such as land subsidence, brackish and seawater intrusion, groundwater level decline and associated impacts on surface water systems have to be properly assessed, and mitigated accordingly.

This chapter explores the key concepts that are needed for developing groundwater resources. Concepts such as well yield, perennial yield, and mining yield are covered in this chapter. Effects of groundwater development are described, including water level decline, depletion of surface water, brackish and seawater intrusion, and land subsidence. Regional scale development of groundwater, conjunctive use of surface and groundwater and artificial recharge, coastal aquifer development, and brackish groundwater development are discussed as well.

8.2 Aquifer Yield and Groundwater Availability

Even though water is a renewable resource, the rates of aquifer recharge versus the rates of groundwater withdrawal are uneven. Considering that aquifers are recharged mainly by precipitation, the amount of water considered as renewable is limited. It is important for water resources management projects to find equilibrium between aquifer recharge and groundwater withdrawal. Aquifer yield concepts are foundational to devising strategies to maintain such equilibrium.

Groundwater availability is simply the amount of groundwater that is available for use from an aquifer (Mace et al. 2001). Although easy to define in words, groundwater availability is much more difficult to quantify in practice. There are no simple formulas or equations to define groundwater availability. An estimate of groundwater availability requires not only scientific characterization of the aquifer, but it also depends on the applicable water-use policy. Hydrological science helps understand dynamics of water in an aquifer (recharge, storage, flow, and discharge) and provides the tools such as numerical groundwater flow models for calculation of the aquifer yield (Bredehoeft 2002). At the same time, policy provides the guidance and defines the bounds that science can use to calculate groundwater availability. Policy involves management goals, environmental issues, and rules and regulations. Although science is required to quantify the aquifer yield, policy is the most important factor that influences the groundwater availability. Depending on policy, groundwater availability can range from a very large number to zero for the same aquifer (Mace et al. 2001).

8.2.1 Well Yield

Well yield is the volume of water per unit of time discharged from a well, either by pumping or free flow (Driscoll 1995). It is measured commonly as a pumping rate in cubic meters per day (m^3/d) or gallons per minute (gpm). When a water well is pumped, the quantity of water discharged initially is derived from casing storage and then immediately from the aquifer storage around the well. As pumping continues, more water must be derived from the storage at greater distances from the well bore. This creates a cone of depression that expands as water in storage moves towards the well. In an unconfined aquifer, water is released from the pores of the geologic formation, which results in dewatering of the aquifer and a reduction in the saturated thickness. In a confined formation, water is released due to depressurizing of the porous media and the expansion of water as a result of the reduction in water pressure induced by pumping. In a confined aquifer, the thickness of the aquifer is generally not reduced during pumping unless groundwater mining changes its confined condition to unconfined.

Equilibrium occurs when the cone of depression enlarges to the point at which one or more of the following happen (Driscoll 1995): (1) it intercepts enough of the natural discharge from the aquifer to equal the pumping rate; (2) it intercepts a body of surface water from which enough water will enter the aquifer to equal the pumping rate; (3) enough vertical recharge from precipitation occurs within the radius of influence to equal the pumping rate; or (4) sufficient leakage occurs through the overlying or underlying formations to equal the pumping rate. Depending on the hydrogeologic setting, the time to reach equilibrium can span from hours to decades.

As defined by steady-state equations derived from the Darcy's law (Slichter 1899; Turneaure and Russell 1901; Thiem 1906), the well yield for equilibrium conditions is a function of hydraulic head difference between the static water level in the formation and pumping water level in a well, diameter of the well bore, radius of influence, and hydraulic conductivity of geological formation.

The Dupuit-Forchheimer well discharge formula for an unconfined nonleaky aquifer (Bear 1979) is,

$$Q = \pi K\left(H^2 - h^2\right)/\log\left(R/r\right) \tag{8.1}$$

while that for a confined nonleaky aquifer (Thiem 1906; Bear 1979) takes the form of

$$Q = 2\pi Kb\left(H - h\right)/\log\left(R/r\right) \tag{8.2}$$

where: Q is the well yield or pumping rate; K is the hydraulic conductivity of the water bearing formation; H is the static head measured from bottom of the aquifer; h is the pumping head (or depth of water) in a fully penetrated well while pumping; R is the radius of the depression (or influence); r is the radius of the well; and b is the thickness of the confined aquifer.

Properly conducted pumping tests can reveal important facts about the groundwater reservoir that cannot be reliably determined by any other means. A water well may be pump-tested for information concerning the performance and efficiency of the well being pumped, or for basic data from which principal factors of aquifer performance can be calculated.

Equations (8.1) and (8.2) are useful for studying the relationship of various parameters to each other and to well yield. For example, if all other parameters are equal, well yield is directly proportional to hydraulic conductivity of the water bearing formation. A formation with twice the hydraulic conductivity of another should provide twice the yield.

For a well in an unconfined aquifer, the part of the formation within the cone of depression is actually dewatered during pumping. This affects the ratio of drawdown to yield as reflected in Equation (8.1). When the drawdown is doubled, the well yield is less than doubled because the saturated thickness is reduced. Figure 8.1 shows the relationship between drawdown and yield for an unconfined aquifer. The maximum drawdown is defined as the lowering of the water level to the bottom of the well. The maximum yield is the quantity of water that a well will produce at the maximum or 100% drawdown. For example, suppose that a well 35 m (114.8 ft) deep has a static water level of 5 m (16.4 ft) below land surface and the saturated thickness of the formation is 30 m (98.4 ft). During the test, the water was pumped at 1,000 m^3/d (0.264 mgd) and the pumping water level stabilized at 13.8 m (45.3 ft) below land surface or 8.8 m of drawdown. In this case, 100% drawdown is 30 m (98.4 ft). The 8.8 m (28.9 ft) drawdown is thus 29% of the total possible drawdown. The curve in Figure 8.1 shows that at 29% drawdown, the yield is 50% of the obtainable maximum yield. Therefore, the maximum yield is 2,000 m^3/d (0.528 mgd). A drawdown of 20 m (65.6 ft) is 67% of the total possible drawdown. The curve shows that at this drawdown, the well will produce at approximately 90% of the maximum yield. To obtain the remaining 10% of the maximum yield, another 1/3 of the total possible drawdown is required. Obviously, the extra pumping costs would be out of proportion to the increase in yield. Therefore, it is uneconomical to operate a well with a drawdown greater than 67% of the maximum drawdown.

For a well in a confined aquifer, the well yield is directly proportional to drawdown, $(H-h)$, as long as the drawdown does not exceed the distance from the static potentiometric surface to the top of the aquifer. In other words, as long as the aquifer is not dewatered, or the saturated thickness of the aquifer does not change with the increased drawdown, theoretically, this means that if drawdown is doubled, the yield is doubled. Therefore, the yield of a well in the confined condition reaches its maximum once the drawdown reaches $(H-b)$. Once the drawdown exceeds $(H-b)$, part of the aquifer will be in an unconfined condition.

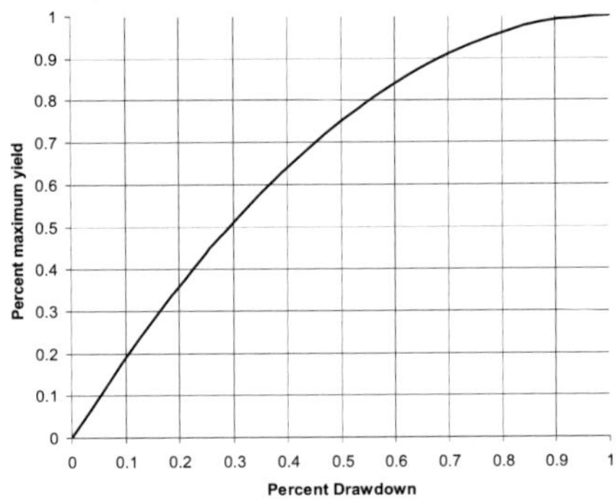

Figure 8.1: Comparison of yield with drawdown in an ideal unconfined aquifer that is fully penetrated and open to the well (Reprinted by permission of Johnson Screens/a Weatherford Company; Driscoll, 1995).

Primary factors that control the initial maximum well yield are reflected in the equilibrium equations, including hydrological properties of the aquifer (thickness, static water levels, hydraulic conductivity) and well design (location, depth, diameter, interference of neighboring wells). In addition, the performance of a well often declines during the life of the well. The performance decline may be due to one or more of the following factors (Driscoll 1995):

i. Lowering of the static water level due to lack of recharge during drought or groundwater mining;
ii. Inefficient pumping caused by worn, corroded, or plugged pumps;
iii. Blockage of the well screen and gravel pack by mud, sand, and silt;
iv. Deposits of encrustation, corrosion products, and microorganism growths at the well screen, at the aquifer face, and/or in the pump; and
v. Failure of the well casing and/or screen resulting in sand pumpage.

8.2.2 Aquifer Yield and Groundwater Availability

Several terms, such as safe yield, perennial yield, sustained yield, mining yield and others have been used to define groundwater availability. Some of them can be easily defined and understood, while others are not. They were discussed in details in following section.

8.2.2.1 *Perennial yield*

The perennial yield of a groundwater reservoir is defined as "the practicable rate of withdrawing water from it perennially for human use" (ASCE 1987). The word "practicable" is the key to the definition: adverse or deleterious side effects of groundwater development, such as seawater intrusion, land surface subsidence, saltwater coning, and others are included. Perennial yield is often defined as the maximum quantity of water that can be continuously withdrawn from a groundwater basin without adverse effects. Perennial yield is determined for a specified set of operating conditions. Any change in conditions, such as changes in economics, land use, or importation of new water supplies, would require calculation of a new yield.

The concept of perennial yield of a groundwater reservoir stems from the principle of the renewability of the resources as shown in the hydrologic cycle, and is based on an analogy to surface reservoir operation. The perennial yield from a designated reservoir would be achieved if the artificial discharge by wells were patterned so as to reduce discharge from the reservoir and induce recharge in an equal amount, and if storage were utilized only to provide regulation of the fluctuating inflow to meet the demand on the wells. In practice, three complicating factors must be taken into account.

First, if the aquifer is close to the land surface, either initially or during the operating period, some recharge to the water-bearing formation may be rejected. That is, as water cannot be added to storage when a reservoir is full, the excess water must be evaporated or wasted over the spillway. Development of a previously unused or underused aquifer will usually result in increased recharge from surface streams and precipitation, or a reduction in groundwater discharge to a surface stream. Thus, use of an aquifer replenished by meteoric water may diminish surface runoff so the yield from an aquifer reduces the water available from surface sources. Jenkins (1968) has discussed the theory of stream depletion by the operation of wells and given detail of sample calculations. Conover (1954, 1961a, 1961b) has portrayed the effects of groundwater use on surface water and the benefits of coordinated surface water-groundwater use.

Second, an aquifer contains a supply of water that can be mined, regardless of whether the aquifer is replenished. Thus the perennially-available water supply may be substantially augmented for a very long time indeed – 50 years of mining is not an unusual circumstance.

Third, the patterns of distribution and volume of pumpage from wells in a developed groundwater reservoir commonly differ from the ideal pattern for obtaining the maximum perennial replenishment. The maximum yield then may depend on the practicability of locating or relocating wells rather than on any characteristic of the natural resource. Pumping from a groundwater reservoir in quantities greater than the perennial yield could then result in a reduction the supply available to other water users.

The perennial yield of an aquifer, in some instances, can be substantially augmented by engineering controls. For example, more water can be made available through artificial recharge by spreading or injecting wells, or by lowering groundwater levels to reduce evapotranspiration, to capture rejected recharge, or to capture surface water from streams. The concept of the perennial yield of an area requires the application of water management, including the services of specialists in hydrology, geology, and engineering for its evaluation. The level of water production quantified by the phrase "perennial yield" is fixed at any point in times only in the sense that no more money will be available for engineering construction, or that legally no more water may be obtained from any source. Should these constraints be changed, for example, by importation of water and utilization of groundwater storage, the perennial yield could be increased. Under such circumstances, the concept of hydrologic quantity for the perennial yield is not fixed, but is an ever-changing quantity depending on the hydrologic constraints of the aquifer and on particular configuration of wells. It involves consequences of pumping, such as economics, environmental effects, change in quality, and deformation of aquifers, which are deemed acceptable or safe. In contrast, overdraft involves any factor that may be termed unsafe; it involves more than just continuing decline in storage, that is, groundwater mining.

The term "safe yield" is often used by hydrologists instead of the preferred perennial yield. Even though different definitions have been proposed for this term, they are all based on more or less the same approach. When originally introduced, the safe yield was defined as the maximum annual withdrawal from an aquifer that still maintains the sustainability of the aquifer (Lee 1915; Meinzer 1920, 1923; Williams and Lohman 1949; Alley et al. 1999; Alley and Leake 2004). In other words, water resources of an aquifer could be maintained indefinitely if pumping did not exceed the safe yield.

Based on concerns of numerous hydrogeologists (e.g., Conkling 1946; Banks 1953), the definition of safe yield was later amended to address the undesired results, including impairment of water quality such as saltwater and/or brackish water intrusion, infringement of rights of other users in the same (or adjacent) basin (i.e., legal consideration) including the effect on base flow in streams, increased energy costs by lowered water levels (i.e., economic consideration), and environmental damages such as land subsidence and drying up of springs (Bear 1979; Fetter 1994). In other words, an aquifer pumped at or less than the safe yield would be able to be maintained indefinitely with the minimum impairment on water-quality and other environmental damages.

Because of its ambiguity, safe yield is not a popular concept in groundwater resource management adopted by scientists and engineers (e.g., Thomas 1951, 1952a, 1955; Muckel 1955; Kazmann 1956; Anderson and Berkebile 1977; Sophocleous 1997;

Alley et al. 1999; Alley and Leake 2004). This is because policy is required to define how much land subsidence, base flow decline, water quality degradation, or spring flow decline is acceptable. Quantifying safe yield requires inputs from economists, engineers, engineering geologists, plant and wildlife ecologists, and lawyers (Fetter 1994). Safe yield is also difficult to determine because it has no unique or constant value depending on the spacing and location of wells and their influence on the aquifer and environment (Domenico and Schwartz 1990; Sophocleous 1997, 2000).

Once we understand the dynamic nature of a groundwater system, the large number of factors involved, the need to manage the system to achieve various goals, etc., it becomes obvious that the essentially hydrologic static concept of a safe yield is insufficient. The important entity in determining how a groundwater system reaches a new equilibrium is capture (Bredehoeft 2002). Capture is independent of the recharge; it depends on the dynamic response of the aquifer system to development. The concept of optimal yield, or operational yield (Renshaw 1963; Burt 1964, 1967a, 1967b; Bear and Levin 1967; Domenico et al. 1968; Bredehoeft and Young 1970; Domenico 1972; Freeze 1971), incorporates in it both the hydrological features of the physical groundwater system and those related to water-resource management policy. Optimal yield is the determination of a yield using optimization theory considering socio-economic issues (Freeze and Cherry 1979). The optimal yield of an aquifer is determined by selecting the optimal management approach that best meets socio-economic goals. An optimal yield can lead to depletion or complete conservation of an aquifer, but is usually somewhere in-between (Freeze and Cherry 1979). An important distinction between optimal yield and safe yield is that an optimal yield can be safe yield or it can result in depletion of an aquifer. Optimization theory is a mathematical technique for determining the most advantageous approach out of a set of possible alternatives.

The term "sustained yield" is most useful when applied to well field in situations that Thomas (1957) has termed "water-course aquifers." Such an aquifer generally underlies the floodplain of a major river system and is in hydraulic contact with the water in the river. The sustained yield is the minimum rate of pumpage sustainable under all conditions of river discharge (and temperature) by a specified well field that taps the alluvial aquifers. Sustained yield is not a fixed quantity. It can be changed by water management efforts even if the results of constructing and operating injection wells, spreading grounds, recharge pits to improve recharge conditions are excluded. Additional wells can be constructed and operated and evapotranspiration reduced, and the yield of the system (assuming the permeability of the area of river bed is not restricting the natural recharge) can be increased, albeit in progressively smaller, more expensive steps (Conover 1954).

Another concept is that of deferred perennial yield. In managing a groundwater basin or district, the quantity of water available for all uses might be limited by legal restrictions or the price of water. It is possible, however, that a significant quantity of water could be obtained from aquifer storage. The withdrawal of this water might be cheaper than importing an equal quantity from outside the basin. Lowering the potentiometric surface often produces no undesirable side effects. Under such circumstances, extractions at a rate greater than the ultimate steady-state supply of groundwater are desirable. Thus initially, for a period of years, the pumpage may be far greater than the anticipated steady-state supply of water. The economic use of water that was withdrawn from a shortage frequently results in a strengthened economy, which is capable of importing water from a greater distance and at a higher cost than was originally anticipated. Also, the strengthened economy may sustain conversion of uses requiring large amounts of water to uses with low water requirements, such as light manufacturing or high technology activities. Such an operation, in basin management, is termed "planned overdraft".

Maximum perennial yield means the maximum quantity of water perennially available if all available methods and sources are utilized for recharging an aquifer and the extractions are located so that the maximum reduction in the aquifer discharge is achieved. In the final analysis this depends on the quantity of water economically (or economically, legally and politically) available for capture by the user, which is not fixed. The problem then becomes one of the groundwater basin management. To achieve the maximum perennial yield, it becomes necessary to "unitize" the aquifer for the most efficient and economical production of water. Instead of individually operated well fields, each well field owner receives a share in a common pool. The manager specifies pattern of well development, negotiates for imported water on behalf of the users, and operates the entire system in a coordinated fashion. The only limit to the operation is the ability to import and distribute more water. Admittedly, such a unit operation is complex and highly technical, and it becomes increasingly complicated as the quality of water becomes more critical and disposal or reuse of water receives more attention.

8.2.2.2 *Mining yield*

Water is being mined from an aquifer as long as the water table or potentiometric surface is consistently lowered in elevation on a year-by-year basis as a result of a withdrawal of groundwater in excess of the induced recharge plus the reduced groundwater discharge. This assumes that proper allowances are made for climatic variations.

In many areas – the High Plains of Texas, West Texas, and New Mexico, for instance – the replenishment of groundwater is not only negligible but additional recharge cannot be induced by lowering the water table. Further, the natural discharge is so diffuse, small, and distant from the centers of pumpage that, in practical terms, the discharge will be little affected by the pumpage. Thus the water is essentially being mined. Water in storage was estimated by the Texas Water Development Board (1968) as about 345 billion m^3 (280 million ac-ft), of which possibly 222 billion m^3 (180 million ac-ft) to 247 billion m^3 (200 million ac-ft) was considered recoverable. This quantity is sufficient to supply water for a 40-to-50-year period at the rate of mining in 1969, with no adverse side effects such as land surface subsidence or saltwater encroachment. In the 2002 Texas Water Plan (TWDB 2002), the current groundwater availability from the Ogallala as assessed by the Planning Groups was determined to be about 6.4 million ac-ft/yr under drought condition. Thus, the principal water management issue in this instance involves keeping track of the groundwater pumpage and the quantity remaining in the aquifer, and leading an educational campaign to reduce use and waste of water and a political effort to import a new, replacement water supply. It might also appear that some more economic uses of water, such as for industry rather than agriculture, and some arbitrary reduction in the rate of pumpage would be of value, or that reduction in the rate of pumpage would be of value, or that the establishment of well-spacing regulations would be useful.

Aquifers are rarely deposited on a smooth slopping surface. The base of a sand aquifer usually resembles normal erosional topography with hills, ridges, valleys, plains, and drainage channels, all of which were buried by the deposition of aquifer materials. The upper surface of a sand aquifer also commonly shows erosional features. The aquifer itself thus varies in thickness and depth, but usually in a reasonable, mappable fashion. A water-mining operation will, consequently, be variable in rate and duration. Those property owners whose wells tap the lowest elevation of the aquifer – the deepest portions of the buried valleys – will normally enjoy wells possessing the largest yields and longest lives. Those whose properties overlie the hills and plateaus of the aquifer floor will find that their wells lose capacity more quickly.

8.2.3 Evaluation of Yield or Groundwater Availability from a Groundwater Basin

Water is universally classified as a renewable resource. Renewal of water supplies is inherent in the hydrologic cycle, and it is obvious that precipitation replenishes soil moisture and streamflow. In concept groundwater is also recognized as a generally renewable resource because virtually all usable supplies are of atmospheric origin. Detailed studies of many groundwater areas have shown conclusively that groundwater supplies are recharged by precipitation. There are no known examples of fresh groundwater resource areas that are isolated from replenishment.

Nevertheless, there are limits to the rate at which groundwater can be replenished. The average annual precipitation on the land area of the U.S. is about 6,000 km^3 (4,864 million ac-ft) (ASCE 1987; NCDC, 2010), of which a portion evaporates, a portion is used in transpiration, and a portion runs off to the oceans. Only a small fraction of the 6,000 km^3 (4,864 million ac-ft) is available for groundwater recharge. Total freshwater and saline-water withdrawals for the year of 2000 were estimated to be 1.55 billion m^3/d (408,000 mgd) or 564 billion m^3/yr (457 million ac-ft/yr) http://pubs.usgs.gov/circ/2004/circ1268/htdocs/table01.html (Hutson et al. 2004). Given that there is about 250,000 km^3 (202,677 million ac-ft) of fresh groundwater within 1,000 m (3,281 ft) of the land's surface in the U.S. (ASCE 1987), it is apparent that the quantity of groundwater that can be considered to be available as a renewable resource is limited by groundwater recharge from precipitation.

A perennial supply of water can be assured by not withdrawing more water than it is available over the long term. This holds true for either groundwater or surface water projects. However, determining the perennial yield from groundwater sources is more difficult than determining yield from surface water projects because it is difficult to differentiate replenishable groundwater from accumulated stored groundwater.

8.2.3.1 *Hydrologic balance*

A hydrologic balance is simply a statement of the conservation of mass applied to a groundwater basin. All water entering a groundwater basin during any given period of time must either go into storage within its boundaries, be consumed, be exported, or flow out, either on the surface or underground, during that period. The hydrologic balance may be expressed as follows (ASCE 1987, Alley et al. 1999):

Total water inflow - Total water outflow = Change in water storage

The terms of this balance equation generally consist of the following components:

Inflow

i. Surface inflow into basin
ii. Subsurface (groundwater) inflow
iii. Precipitation
iv. Imported water

Outflow

i. Surface outflow
ii. Subsurface outflow
iii. Consumptive uses including human consumption, evaporation, and
iv. Exported water

Storage

i. Decreased/Increased surface storage (streams, rivers, reservoirs, lakes, snowpack)
ii. Decreased/Increased soil moisture storage (the unsaturated soil zone)
iii. Decreased/Increased groundwater storage (aquifers)

The hydrologic balance for a groundwater basin that does not import or export water would include all inflow and outflow terms except term (iv). In a short-term budget – a month, a season, a year, or even several years – substantial differences may occur between natural inflows and outflows, terms (i) through (iii). The difference is accounted for in the balance by changes in storage, terms (v) through (vii). Budgets for successively short periods of time would show an unsteady state resulting from short-term climatic changes. The longer the time period chosen for the budget, the more nearly the budget will indicate the steady-state conditions of the average climate. Over a long enough period of time changes in storage, terms (v) through (vii), become insignificant, with terms (i) through (iii) expressing the overall natural balance between inflow and outflow.

If a groundwater basin is in hydrologic balance, how is it possible to increase the amount of water taken from the basin to supply human demands without adversely affecting the basin's long-term hydrologic balance? If groundwater is to be developed in an area to supply increased human demands, the consumptive use in the area, outflow term (iii), will be increased. Such an increase can be expected to upset the hydrologic balance of the basin unless the increase is compensated for elsewhere. Either the inflow to the area must be increased or other outflows must be decreased, or there must be a combination of the two. Changes in storage, terms (v) through (vii), cannot be considered in the long-term balance. Surface, sub-surface, and precipitation inflows, terms (i) through (iii), cannot normally be increased, although cloud seeding or climate change has increased precipitation in some areas. Water can be imported, inflow term (iv), but would be more likely to be utilized directly to meet increased human need rather than to sustain groundwater levels.

The long-term hydrologic balance of a basin supplying increased human demands can more easily be preserved by reducing surface and subsurface outflows and non-beneficial consumptive uses in the area, outflow terms (i) through (iii). As groundwater is pumped to supply increased human demands, the groundwater table is usually lowered. Lowering the groundwater table lowers the lake and swamp levels and reduces their areas, thereby reducing surface and subsurface outflows from the area, reducing the availability of water to non-beneficial consumptive uses, outflow terms (i) through (iii). As pointed out by Bredehoeft (2002) the size of a sustainable groundwater development usually depends on how much of the discharge from the system can be captured by development. Reductions in surface and subsurface outflows from an area may adversely affect adjacent areas and may have to be compensated for in the form of increased exports from the area, outflow term (iv). In this situation the only source of water to supply increased human demands comes from reduction in non-beneficial consumptive uses.

A hydrologic budget defining water supply in terms of annual or seasonal precipitation is needed in order to evaluate existing and potential development of dependable water supplies in a given area. Precipitation records are more readily available for longer periods of time than rare groundwater records. If the relationship between groundwater conditions and precipitation can be established, the long-term ability of a groundwater source to supply human demands can be better evaluated by means of the precipitation records. Unfortunately, such a relationship is difficult to establish. Basic data, such as precipitation, pumping, and groundwater levels, may not be concurrently available for an extended period of time. Available basic data may not be representative of the overall area. Groundwater-level data, for example, may be affected by nearby pumped wells and not representative of the groundwater basin. Some basic data items may be very accurately measured, while other items may be poorly measured (e.g., evapotranspiration) or may not exist (e.g., recharge).

When relatively small inflow or outflow terms have to be derived from differences in large terms, gross errors may occur due to the inherent uncertainty in estimating hydrologic data. Such errors can lead to poor relationships between precipitation and available water supply.

Natural drainage basins are usually selected as areas in which the hydrologic balance is developed relating precipitation to yield. The natural basin limits frequently represent physical boundaries of both surface and groundwater flow direction. Where the groundwater flow direction differs from that of surface water, it may be more useful to use the groundwater basin limits rather than the surface water basin limits, depending on the individual characteristics of the area being studied. Once an area has been selected, it is necessary to assemble all available hydrologic records to evaluate the inflow and outflow terms and obtain a useful hydrologic balance. The result of accumulating all available information is frequently referred to as a hydrologic inventory. An inventory covering only items required for a groundwater budget is referred to as a groundwater inventory.

An independent evaluation of each of the inflow and outflow terms of the hydrologic balance is desirable in a comprehensive hydrologic investigation. If the inflows and outflows so determined are not in balance when compared (within the limits of accuracy of the data involved), adjustments must be made. The value or values inherently subject to the largest probable error are usually adjusted to affect a balance. If the lack of balance exceeds reasonable limits of error in the basic data, further investigation and special studies may be required. It is frequently impossible to evaluate independently all of the inflow an outflow items of the hydrologic inventory. As a result, engineering judgment is required in preparing a hydrologic balance and evaluating the dependable water supply from a specific groundwater area.

Time lag must be recognized and adjusted in using basic hydrologic data. If, for example, water levels in a number of wells are used in determining changes in groundwater storage, the levels taken during periods of minimum pumping and minimum or stable natural recharge are usually the most reliable, because these can be expected to be periods of least change in groundwater storage. While precipitation may fluctuate rapidly with time, the effect of an unusual period of precipitation may not be reflected in a change in the groundwater level for months or even years.

In most cases the dependable water supply of a specific groundwater area is needed for long-time average climatic conditions as well as for severe drought and excessive precipitation periods. It is normally impossible to make a direct estimate of long-tem conditions because of the relatively short periods for which basic data are available. The climatic and other conditions prevailing during the period for which basic data are available must be compared with long-term climatic and other conditions that are expected to prevail, and study period results must be adjusted accordingly.

8.2.3.2 ***Estimates of perennial yield***

Perennial yield of an aquifer is withdrawal of water at a new equilibrium established by pumping. Perennial yield exceeds recharge or discharge of an aquifer under natural equilibrium by the amount of induced recharge captured from evapotranspiration, precipitation, and from surface water. Groundwater studies provide estimates of the perennial yield of a groundwater area. The perennial yield is used to plan the development of a used groundwater reservoir.

Perennial yield may be dictated by considerations other than the physical and climatic conditions of the area. Pre-existing development and established water withdrawal patterns may determine perennial yield. The cost of well construction and pumping may limit the degree of water level lowering that can be expected. Pumping may have to be limited to preclude damage to overlying developed areas from possible land surface subsidence. Excessive lowering of the water level may cause saltwater or other unsuitable water intrusion. Also, legal restrictions may exist that prohibit interference with established development or water rights.

Perennial yield may change with time and with changing economic or social conditions. A change of use from irrigation to domestic, municipal, or industrial use changes the value of the water and may increase economical pumping lifts. Greater pumping lifts may increase perennial yield by increasing storage capacity, increasing recharge and reducing in the value of irrigated crops, with the same results.

It is preferable to reserve the concept of perennial yield for use with an entire groundwater reservoir or a natural subdivision thereof. Perennial yields have been estimated for single wells or for relatively small well fields in a large groundwater reservoir. The accuracy of such estimates is obviously limited by the assumptions that must be made concerning the effects of development of the reservoir that may be made by others. If the total yield of the reservoir is estimated, the share of the total that should reasonably be produced by a single well, or a small number of wells, could be more reasonable estimated more reasonably.

Perennial yield is difficult to estimate even for a fully developed groundwater reservoir because of inadequate data. It is almost always necessary to estimate some of the required data by indirect methods. As a rule, the older, more developed groundwater areas will have more and better hydrologic and geologic data available than the undeveloped areas. Therefore, the approach used to determine perennial yield is dictated by the data available. Several examples are provided below to illustrate different approaches to determining perennial yield.

i. Calculation by hydrologic budget (ASCE 1987)**:** Calculation of perennial yield from a groundwater reservoir is similar to determining the yield of a surface reservoir. The inflow into a reservoir minus the outflow from a reservoir equals the change in storage in the reservoir. Estimated inflows into a groundwater reservoir for a series of discrete time periods are compared with a series of trial yield patterns of withdrawal from the reservoir to obtain fluctuations in the groundwater reservoir levels. Surface and subsurface outflows have to be estimated for each reservoir level. The maximum trial yield that can be withdrawn without producing undesirable results is the perennial yield.

Subsurface inflow and outflow from a groundwater reservoir can seldom be estimated accurately. It is usually sufficient to use an average subsurface inflow or outflow for the average groundwater reservoir level for a given range of water-level fluctuation.

If it is known from previous experience that the groundwater reservoir has sufficient capacity within a desirable water level operating range to allow the established long-term mean withdrawal to be continued without undesirable consequences, then it is necessary to calculate only the long-term mean withdrawal rate, and no reservoir operation studies are needed. Here the perennial yield is the surplus (increase) of water in storage plus net recharge. Pumpage may be used in lieu of net recharge.

ii. Calculation from limited data (ASCE 1987)**:** Sometimes it is necessary to make reconnaissance estimates of groundwater availability in sparsely populated regions where little or no groundwater data are available. There may be no exploratory drilling information, no operating wells, no detailed geologic studies, and even the existence of a groundwater aquifer may be in doubt. Reconnaissance perennial yield estimates can be derived only indirectly from the other hydrologic cycle constituents (precipitation, runoff), and an indirect estimate of natural groundwater discharge based on observation of springs, seepage areas, marshes, ponds, and vegetation sustained by groundwater. If such reconnaissance estimates are favorable for an area where water use is growing, additional investigations of the extent and characteristics of a potential aquifer should be undertaken.

In more heavily populated areas where limited groundwater development has taken place, available groundwater records may be limited to water levels in a few wells. The records may or may not indicate the depth and extent of an aquifer and specific yields are usually unavailable. Such records can only be used to derive mean or weighted mean water levels during an inventory period. Arithmetic mean water levels would be appropriate for evenly spaced wells and a weighted mean for unevenly distributed wells.

Average water-level changes in an area may be correlated with discharge from the groundwater reservoir or the difference between discharge and recharge. In the simplest case, where it is found that there is no net change in average water levels from the beginning to the end of an inventory period, the average annual net draft during the period is the perennial yield — provided that the average water supply to the groundwater reservoir during the inventory period is equal to the long-term water supply under conditions of future development, including stream-flow modifications; provided that, if a pressure aquifer exists, land surface subsidence during the inventory and the reduction in aquitard water storage are determined; and provided further that there has been no rejection of potential recharge during the inventory period because groundwater levels were too high.

In an area of steadily falling groundwater levels, all points may fall below zero change in water level. The yield that would produce no significant change in water level would have to be inferred by extrapolation of the line fitting the plotted points. Care should be exercised in applying such extrapolation, because the apparent relationship may not be linear over an extended water-level range.

Perennial yield was determined for an area near Cedar City Valley in southwestern Utah (Thomas 1952b) by comparing average annual inflow with average annual draft. The inflow, or recharge, to the groundwater reservoir was found to come chiefly from Coal Creek, and draft consisted primarily of water pumped by irrigation wells. It was found that the groundwater levels remained unchanged if the annual runoff in Coal Creek was 10,000 ac-ft (12.3 million m^3) greater than the irrigation pumping draft, for the period 1934 through 1951. The 10,000 ac-ft (12.3 million m^3) appears to be a nearly constant natural groundwater loss. The perennial groundwater yield, therefore, fluctuates with the runoff in Coal Creek.

iii. Calculation from pressure trough (ASCE 1987)**:** A direct method of determining perennial yield of a confined aquifer was developed in an investigation of saltwater intrusion into the Salinas River Valley in California (CDPW 1935). This method is based on Darcy's law but does not require a separate determination of the aquifer's hydraulic conductivity and cross-sectional area.

Seawater intrusion was confirmed by discovery of a trough in the potentiometric surface in the 180-ft (54.9-m) thick, partially confined aquifer and by well water quality analysis. When such a trough exists, all water pumped from the aquifer on the inland side of the bottom of the trough is being supplied by groundwater flowing toward the ocean. Water pumped between the bottom of the trough and the ocean shoreline is being supplied by groundwater moving inland from the ocean. As the 180-ft (54.9-m) thick aquifer is not sealed off from the ocean, seawater intrusion may approach, but cannot extend beyond, the farthest inland position of the bottom of the trough, except for the toe of the seawater wedge (ASCE 1987).

When pumping exceeds the perennial yield of the aquifer, a pressure surface trough begins to form at a short distance inland from the shoreline. The trough deepens and the bottom of the trough moves inland with increased summer season pumping. When pumping decreases, the trough becomes shallower and the bottom of the trough moves toward the shoreline. The perennial yield of the aquifer is that yield that does not cause the formation of a trough and that does not cause excessive loss of freshwater into the ocean. It corresponds to the hydraulic gradient that reaches the shoreline just above the ocean levels and that has no inland pressure trough.

The total pumping at the time the pressure trough was detected was about 245 cfs. Total pumping from the aquifer Q reduced by seaward pumping between the bottom of the trough and the shore Q_o equals the pumping from the inland freshwater source Q_i.

$$Q_i = Q - Q_o \tag{8.3}$$

Pressure trough elevations were taken and its average location determined to obtain distance L and head differential Δh from the forebay region of the aquifer to the bottom of the trough. The quantity Q_i can then be estimated from

$$Q_i = kA\frac{\Delta h}{L} \tag{8.4}$$

in which k is the average hydraulic conductivity, A is the average net cross-section area of the aquifer, Δh is the average difference in elevation between the forebay and trough levels, and L is the average distance from the forebay to the bottom of the trough.

The average kA was computed to be 311,000 for three trough locations using Equation (8.4). The perennial yield Q of the freshwater aquifer is computed as 230 cfs (6.5 m^3/sec) by using this value of kA, the distance L (135,000 ft) from the forebay to the shore, and Δh (100 ft) corresponding to the gradient for sustained freshwater yield. The maximum draft on the aquifer was 330 cfs (9.34 m^3/sec), a rate 100 cfs (2.83 m^3/sec) more than the perennial yield. The annual overdraft rate was estimated to be 20,000 ac-ft (24.7 million m^3) for the prevailing pumping periods.

The foregoing calculated overdraft is greater than the seaward overdraft, the draft between the bottom of the trough and the shore, Q_0, which was 55 cfs (1.56 m^3/sec). The calculated perennial yield is based on reestablishing a freshwater hydraulic gradient that would reduce the freshwater flow from the forebay while at the same time preventing seawater encroachment.

The perennial yield calculated by this method is qualified by the existing pattern of pumping draft. In this example there was evidence that the yield could be increased by changing the pattern of withdrawal.

iv. Calculation by hydraulic formula (ASCE 1987)**:** Perennial yield may be determined by Darcy's law where the hydraulic conductivity, hydraulic gradient, and cross-sectional area of the water-bearing formation are known. This method was used to determine that the perennial yield of a well field near Hamilton, Ohio and adjacent to the Miami River could be substantially increased by pumping more water and lowering the groundwater level, thereby inducing more recharge from the Ohio River.

Three interconnected aquifers were separately analyzed in a quantitative study of the Roswell, New Mexico, groundwater reservoir (Hantush 1955). These aquifers include a limestone artesian aquifer overlain by a leaky formation, an unconfined limestone aquifer to the west of the artesian aquifer, and an unconfined alluvial aquifer overlying the artesian aquifer. The unconfined limestone aquifer constitutes the 7,000 mi^2 (18,130 km^2) recharge area for the artesian aquifer. The overlying alluvial aquifer is recharged by upward leakage from the artesian aquifer as well as by infiltration from the Pecos River, irrigation, and precipitation. The study showed the perennial yield from the Roswell groundwater reservoir to be limited by the distribution of wells in both the alluvial and artesian aquifers, the need to allow the alluvial aquifer to continue to discharge water to the Pecos River to satisfy existing surface water rights, and the need for salinity control near the eastern edge of the artesian aquifer

8.2.3.3 *Estimates of mining yield*

Mining of a groundwater reservoir is said to occur when water is removed from a reservoir without the expectation of its being replenished. The total water available by mining is the amount of water in the reservoir. The amount mined, however, may be limited to that which can be economically extracted from the reservoir, or the amount that can be legally extracted. Annual mining yield is the annual draft on the reservoir that is expected to exhaust the reservoir in a specified number of years.

Mining occurs whenever the long-term average draft on a groundwater reservoir exceeds the perennial yield. Limited groundwater mining also occurs during the initial development of a groundwater reservoir. The groundwater level is initially lowered a limited amount to reduce or cut off flows from points of natural discharge in order to direct more water to beneficial use. Such initial groundwater level lowering may dry up springs, ponds, and flowing wells and may, thereby, significantly modify the local environment. Initial mining may be limited by legal restrictions.

Groundwater mining can also occur intermittently in a reservoir whose withdrawals are regulated to the perennial yield, which is conservatively determined to minimize shortages during dry periods. Higher than normal withdrawals may be allowed for such a reservoir during infrequent and protracted wet periods.

The Ogallala Aquifer is a good example of a mined groundwater reservoir (Banks et al. 1954; McGuire et al. 2003) The aquifer area underlies most of Nebraska, western Kansas, eastern Colorado, western Oklahoma, eastern New Mexico, and northern Texas. The region underlain wholly or in part by the Ogallala Aquifer totals about 570,000 km^2 (220,000 mi^2). In the area that overlies the High Plains aquifer, farmers began extensive use of groundwater for irrigation in the 1940's. The estimated irrigated acreage increased rapidly from 1940 to 1980 and did not change greatly from 1980 to 1997. About 850,000 ha (2.1 million acres) were irrigated in 1949, 5.5 million ha (13.7 million acres) in 1980, and 5.5 million ha (13.9 million acres) in 1997. These irrigated acreages were estimated from farmer surveys or from satellite images (Heimes and Luckey 1982; McGuire et al. 2003; Thelin and Heimes 1987;

U.S. Department of Agriculture 1999). Annual pumpage from the High Plains aquifer for irrigation increased from 4.9 to 23.6 billion m^3 (4 to 19 million ac-ft) from 1949 to 1974; annual pumpage did not change greatly from 1974 to 1995 (Heimes and Luckey 1982; McGuire et al. 2003). Irrigation efficiency improvements reduced the unit application of irrigation water by about 30% from 2 to 1.4 ac-ft per acre during the same period. Annual natural aquifer recharge is estimated to vary from 10 to 24 mm (less than 0.5 to 1 in.) over much of the area to about 150 mm (6 in.) in some sandy surface areas in Nebraska. The volume of water in storage in the High Plains aquifer in 2000 was about 3,700 billion m^3 (3 billion ac-ft). This storage amount is expected to fall if present trends continue. In 2000, about 11% of the High Plains aquifer area had more than a 25% decrease in predevelopment saturated thickness, and 4% of the aquifer area had more than a 50% decrease. Kansas and Texas had the largest areas with more than a 25% decrease in predevelopment saturated thickness (McGuire et al. 2003).

8.2.3.4 *Accuracy of estimates*

When estimating the yield of a groundwater basin using the principles of groundwater hydrology, the uncertainty surrounding the basic data must be considered as is discussed by Baker (1951):

> "Groundwater hydrology is not an exact science, not so much because of a present lack of scientific or technical knowledge concerning the principles in regard to the behavior of groundwater as because of lack, in any estimates made, of basic data of a quality and a quantity which will allow results to be achieved of a precision attained in other engineering or scientific computations. If groundwater had a value — per unit of quantity — approaching that of petroleum, it would be entirely feasible to make such field investigations and study as would be necessary to allow the determination of safe yield or other phenomena in a fairly precise manner. Under present conditions, however, the groundwater hydrologist usually must accept the basic data which he finds available — much of it of poor or uncertain quality — and must endeavor to supplement this with such additional data as it is feasibly and financially possible for him to secure. To such data as he ultimately secures, he must apply sound and mature judgment; and through utilization of all of these, arrive at his preliminary estimate. To this estimate must be applied a factor of safety which should vary inversely with the quality and quantity of basic data that has been used. With time, more data of better quality usually becomes available, and subsequent estimates of greater accuracy can be made."

It should be recognized that any estimate of groundwater yield will probably be superseded in the future as better data and better methods of analysis become available. Groundwater yield estimates based on numerous long-term records and sound estimates of future development can be expected to be more dependable and less subject to future change than estimates based on less representative data and less accurate estimates of future development. Human activities, which create the need for water in the first place, modify, and will probably continue to modify, the hydrologic balance of an area. Changes that may affect the hydrologic balance include the extent and timing of the groundwater draft, the transfer of water from agricultural applications to urban supplies, increases in impermeable areas, modifications of surface runoff patterns, modifications in subsurface inflow and outflow, the import or export of water, and alteration of groundwater gradients. These and other possible changes in hydrologic conditions have to be evaluated when planning methods of groundwater development and in estimating groundwater yield. Agencies managing groundwater basins must continuously study the hydrologic and physical conditions that affect the performance of the basin. A number of these conditions are considered in following sections of this book.

8.2.3.5 *Legal, environmental and economic constraints*

Three questions must be answered before any water project is undertaken: "Should we do it?" "Can we do it?" and "Can we afford it?" (ASCE 1987).

The first question refers to the social desirability of the project. What improvement or degradation in the environment will result? Will it satisfy water needs in an effective and economic manner? What are the human and ecological benefits and detriments? Among the benefits might be a more secure source of water, an improvement in water quality or quantity, new areas for recreation, improved fire protection, and a possible reduction in population density. Subsidence from substantial lowering of groundwater levels would be a detriment.

The second question refers to the physical, or engineering, feasibility of the project. Can the work be done by use of proven technology? Is water available at the present time and in the future for the project? Is the hydrogeology favorable? Is the water quality satisfactory for the end use of the water, or must it be upgraded prior to use? Are there any legal or environmental restrictions that would affect the engineering justification of any plan? Will the project produce beneficial or deleterious side effects, and if so, what are they?

The third question refers to the economic and financial feasibility. Can we afford such project? Is the project financially feasible? What are values of groundwater? A number of economic and institutionally-related factors affect groundwater use. Groundwater is used most efficiently when it is extracted at rates that maximize net benefits. Costs include the cost of extracting and delivering the groundwater and opportunity, or user, cost. The benefits are determined by the uses to which the water is put (NRC 1997).

i. Legal considerations (ASCE 1987)**:** Apart from the inventory of individual water rights and the evaluation of their individual and aggregate constraints upon basin management, there are certain broad conceptual areas of a legal nature that must be understood by the basin manager if legal pitfalls between planning and implementation are to be avoided.

Many of the legal problems which inhibit groundwater planning are attributable in part to lawyers or judges who have oversimplified engineering concepts. "Safe yield" constitutes a prime example of this type of problem. As previously noted, engineers are continually refining and seeking to understand the meaning and implications of safe yield. The search has been long and at times fruitless, and there have been suggestions that the whole concept should be abandoned. Yet lawyers, legislators, and other laymen to the field of hydrology have ignored the complexities of the engineering concept and adopted "safe yield" as though it were an end in itself, simply and precisely derived. Stripped of its real physical and theoretical complexities, the natural safe yield is viewed as "the long-term average annual supply" to a hydrologic unit.

As the reasoning goes, extraction of water in excess of natural safe yield would destroy the groundwater resources if carried on for a long enough time. Courts soon recognized that "pumping in excess of safe yield," i.e., "overdraft," is harmful, detrimental, and threatens irreparable injury. The hypothesis having been thus established, there have been numerous adjudication proceedings whose object has been to seek injunctive relief to equate extractions to the natural safe yield. Such a pattern of strict safe yield management, if adopted at the wrong time in the rainfall cycle, for instance, may well force the loss of runoff from the area in subsequent heavy rainfall years because groundwater storage is full. In addition, substantial quantities of water in storage may never be put to beneficial use if strict safe yield operation is followed. Also, the benefits of artificial recharge, salvage of nonbeneficial use of water, and recycling of water may never be realized if extractions sometimes fail to override the compelling force of the "safe yield" cliché.

The coordinated utilization of the aquifer for storage of local and imported water supplies presents some interesting theoretical problems. Because the aquifer is often used for both storage and transmission, as well as filtration and mixing of water, problems are always present of quality, quantity, and location of recharge of imported water as compared to extractions. The problem becomes significant when the agency recharging the groundwater with imported water is seeking to extract equivalent quantities of water at some other point in space and time.

Water rights historically have related to the use of the natural supply of both surface water and groundwater. What of the rights to utilize the unused groundwater storage capacity in a basin or aquifer? It has been argued for some western states that such reservoir capacity belongs to the state and should be subject to appropriation in the same manner as unappropriated water. In contrast, both legal theory and precedent exist for the argument that the overlying owners, in the aggregate, own the storage capacity under their surface lands, much as the owners of land within a surface reservoir own the super adjacent air space wherein the waters will be impounded.

Most groundwater is not in a static state and should be described as transient, i.e., water that is moving downgradient in the aquifer. The right to use, especially recharged water, may be temporary. Management must account for the storage space required for recharge of groundwater from all sources, especially if the reservoir is shared with other water users. If it is assumed that a groundwater management plans results in a conscious alteration in groundwater levels, it may be anticipated that some damages will result to individual parties. Perhaps the simplest example would be that of homeowners, or even gravel pit operators, who are flooded out as water levels rise. At the other extreme is the water user with shallow wells on the fringe of a groundwater basin. If the basin management plan involves drawing water levels down to use water in storage, it is probable that water will be drained out from under the shallow fringe wells and a replacement water supply will have to be furnished to these injured parties.

ii. Organizational considerations (ASCE 1987)**:** For the water management process to succeed, certain powers must be entrusted to the agency responsible for management. The objectives of the program, once firmly determined, should be protected by a grant of adequate authority to attain and sustain the program. The vagaries of rights to groundwater are such that the legislative authority must be broad enough to be effective; however, care must be taken both to prevent overlap of responsibility or authority with existing agencies, and to assure that conflicts of responsibility and authority are avoided. Because the planning and implementation of groundwater management programs assume some preliminary determination as to political and social definition, an organizational decision is best made at an early stage. Political forces will make a logical or scientific solution impossible if an appeal to negative emotionalism succeeds after a plan has been formulated and submitted for adoption.

To implement and maintain a comprehensive management program, the manager's authority must include the power to purchase water from either local sources or sources outside the agency's boundaries. The statute should enable the agency to manage the water through distribution, recharge, storage, recapture, reclamation, purification, treatment, and other controls that might be foreseen, to provide for the beneficial use of water for persons or property within the agency's jurisdiction.

Authority to protect the basin or basins within the management of the governing agency from diminution of water quantity and water quality, either from within the agency's boundary or from external sources, must be included. These powers should also permit construction and operation of saline water intrusion barriers, regulation of disposal of liquid or solid wastes that might tend to degrade groundwater quality, and restriction of groundwater extraction practices that might adversely affect the quality of groundwater. If water-quality control organizations are established prior to the development of the groundwater management unit, the division of responsibilities for quality control must be agreed upon before an adequate management program can be achieved. The groundwater basin manager should have the power to restrict pumping or to require that extractions be made, in the event that rising groundwater levels cause waterlogging and damage.

In summary, an efficient groundwater management program implies that the governing agency will have the legal authority to:

i. Purchase water supplies.
ii. Spread water for recharge.
iii. Acquire-lands and improvements by eminent domain.
iv. Protect the basin with regard to water levels and water quality.
v. Influence pumping practices.
vi. Obtain revenue.

iii. Environmental constraints: Besides hydrological balance, economic and legal constrains, environmental concerns become more and more important in assessment of groundwater availability. Historically, water availability could be viewed simply as an issue of quantity, and water management could focus largely on controlling or alleviating impacts of droughts or floods. With escalating population growth and increasing demands for multiple water users, it is now clear that is quality is equally critical to the long-term sustainability of the human communities and ecosystems (Alley et al. 1999).

Water quality deterioration caused by brackish water intrusion and seawater intrusion could reduce freshwater availability or even destroy freshwater resources. For example, as a result of long-term groundwater mining, brackish water moves into the freshwater zone in the Hueco Bolson aquifer in the west Texas. Some wells have had to be abandoned because of poor quality, while others have required the installation of reverse-osmosis (RO) units to make the water suitable for potable use.

Groundwater pumping can reduce instream flow by increasing infiltration from the streambed or reducing inflow from groundwater discharge. Reduced instream flow or lowered groundwater baseflow into surface water stream, and impaired water quality can also impact aquatic biota, including endangered species. Examples include the silvery minnow in the Rio Grande, Texas Blind Salamander (*Typhlomolge rathbuni*), Fountain Darter (*Etheostoma fonticola*), San Marcos Gambusia (*Gambusia georgei*), Texas Wild Rice (*Zizania texana*), Comal Springs Riffle Beetle (*Heterelmis comalensis*), Comal Springs Dryopid Beetle (*Stygoparnus comalensis*), and Peck's Cave Amphipod (*Stygobromus pecki*) in the Edwards aquifer system, Texas.

A variety of factors have the potential to contribute to the degradation in quality of groundwater supplies, such as point and non-point pollution sources, shallow aquifer depths, and unprotected groundwater supplies. Furthermore, it is recognized that groundwater and surface water are interrelated in that groundwater discharging to surface water can affect surface water quality. However, with the possible exception of temperature impacts, groundwater quality is likely not a major contributor to surface water quality in comparison with the many other factors affecting surface water.

Land use is a major factor to consider while developing strategies to protect groundwater quality. The characteristics of the sources play an important role also in the vulnerability or risk of source contamination. Assessment is needed to both prioritize areas for protection and to select the most appropriate actions or strategies to address those risks.

iv. ***Economic constraints:*** Groundwater can be considered a natural asset. The value of such an asset resides in its ability to create flows of service over time. There are two broad categories of resources services provided by groundwater: extractive and in situ (NRC 1997). The extractive services include agricultural application, municipal supply, and industrial uses; while in situ services include buffer value, waste assimilation, subsidence avoidance, saltwater intrusion avoidance and ecological service.

To simplify the economic analysis, the economic life of the project is assumed as a fixed, determinable number of years. Forty years generally is considered a reasonable period for a municipal or state water-management project that involves major construction. In industry, depending on the rate of technological change, shorter periods are used. For industrial water service, an economic life of as little as 15 or 20 years may be postulated. Note that a distinction is made between the physical and economic life of a project.

In order to make all costs comparable, the operating costs incurred each year must be evaluated in terms of their present worth, and the entire stream of costs can be added to the original, immediate capital investment to determine the over-all cost of the project. Such costs may also include damages that might be expected if the project is built, properly evaluated as to their equivalent present worth. The distinction between the "nominal" interest rate and the "effective" interest rate should be recognized and taken into account in the calculations of both costs and benefits.

A groundwater management program is usually justified by its ability to prevent damage that inevitably will occur under existing practice, for instance, saltwater encroachment, subsidence, water depletion, or increased cost of water production. It may also be justified by potential improvement in the utilization of the aquifer as a storage reservoir or local water distributary, and by the invulnerability of the aquifer, as compared to a surface reservoir of equal cost, with respect to pollution, radioactive contamination, enemy action, or civil disorder. Although all of these benefits are tangible, the engineer must work closely with economists to estimate the benefits which are achieved.

In as much as all of the benefits resulting from a project do not occur at the same time and the magnitude of the benefits may change with the passage of time, it is both desirable and necessary to determine the present net worth of the benefit stream. The computations involved are based on year-by-year estimates of the stream of benefits, as is done in computing costs (Shoemaker 1963; Essley 1965; NRC 1997).

A crucial decision in such evaluation is selecting the discount rate (interest rate) to be used in the computation. Haveman (1969) has outlined a computational procedure that might be used to determine the appropriate discount rate for a publicly financed project. The premise on which the need for the determination is based is that the social return on public investments must at least equal that derived from the spending displaced in the private sector. This implies that the use of a social discount rate for public projects that is less than the discount rate used in the private sector would entail a significant diversion of resources from high productivity activities to low.

It should be pointed out that the economic comparison of alternative management plans is greatly simplified if the goal of all plans is to satisfy future water needs. Under such conditions, benefits will be common to all plans and need not be computed. The plan with the least present worth would, therefore, be the most economical plan and would: produce the greatest net benefits (gross benefits less costs). Several methods for economic evaluation are summarized by the NRC (1997).

In the final comparison of costs and benefits, the project manager, and advisory staff, should attempt to evaluate (ASCE 1987):

i. The long-term water requirements to satisfy water demands at several levels of water cost, and to prevent or alleviate such spillover effects as saltwater encroachment into aquifers, land-surface subsidence, and increase in TDS of groundwater
ii. A drought contingency water-supply plan, along with its savings over meeting full water needs in all years.
iii. The differences between cost to the nation and cost to the local agency (owner).
iv. The feasible alternatives both structural and nonstructural (for example, prohibition or reduction of groundwater extractions, or increase in pumpage charge) to accomplish the water-management objectives.
v. The qualitative descriptions of the diffuse benefits and costs accruing to each proposed procedure. (Note: The difficulty involved in this evaluation is recognized - it falls in the category of, "It can't be done, but you've got to do it anyway.")

The project manager must recognize that groundwater resources, like surface water resources, even with the best possible management, are not unlimited. It follows, therefore, that sooner or later structural measures will become uneconomic and, later, physically inadequate. Thus, any final project plan should be flexible within the project area so that the relatively limited perennial supply of groundwater may be used to maximum economic advantage. The economic utilization of any natural resource, including water, cannot be established once and for all, for it changes from time to time as technology and the values of commodities change. Thus, within the geographical limits of the groundwater basin, the sale and purchase of water rights should be made as easy as, or not more difficult than, the sale and purchase of land, equipment, or services.

When changes in the patterns of water withdrawal do occur, the manager's purpose must be to save the agency from additional capital and operating expense, and to impose such charges on those making substantive changes in patterns of pumping as might be needed to achieve this result.

8.2.3.6 *Financial feasibility*

Financial feasibility should not be confused with economic justification. A project may be economically attractive and financially infeasible. Such a situation will result when the benefits are diffuse and the cost can be neither allocated nor paid. Financial feasibility can be defined as the willingness and ability of the beneficiaries to pay for the services and products of the project. Moreover, a project may be economically attractive, yet the borrowing power of the sponsoring organization may be so limited by statute that the project cannot be financed.

The management of groundwater resources usually can be financed by taxation of one sort or another, by pumpage charges, or by some combination of these. Sometimes, in addition, payments may be made by some governmental unit whose operations would benefit by aquifer management. For example, if the effluent from a municipal treatment plant could be used for aquifer replenishment, and if such use would delete the cost of a long outfall sewer from the sewage treatment project, then the sanitary district might find it desirable to contribute financially to the operation of the groundwater replenishment district.

The financial feasibility of a project will vary with the interest rate, the value of the water, and imputed value of damage prevented. The financial feasibility will be greatest when the project area is populous and industrialized, and least in agricultural areas.

There is an argument that, in as much as an aquifer management project enhances the value of all real estate within the project area, any such program should utilize an *ad valorem* tax as part of its financial structure. If it were possible to assign the increase in property value attributable solely to the aquifer management program, this suggestion might have some merit. A property owner who does not operate a well and who lives on the outskirts of the geographic area being managed, however, may experience a rise in land value (if it rises at all) that is only directly and remotely attributable to the project. The effect of an *ad valorem* tax would be to tax that property's total value for the monetary; benefit of the well owner situated in the heart of the aquifer management area. A pump age charge, on the other hand, will be paid in proportion to the rate and quantity of water use, whether directly by the well owner or indirectly by the consumer when the well owner increases the price of water to include the tax. The financial feasibility of a tax-supported project, moreover, is always greater than that of an identical project supported by pumpage charges that are used to finance a revenue bond issue.

8.3 Effects of Groundwater Development

The development of groundwater resources can result in adverse impacts. These impacts can take the form of a lowering of groundwater levels, which results in increased energy costs to lift the water and costs to deepen wells and reset or modify pumps. Surface water resources also can be adversely impacted due to interactions between surface water and groundwater. For example, surface water may be depleted as a consequence of groundwater development for systems sustained by surface water. In coastal areas, groundwater utilization can result in saltwater intrusion into coastal aquifers, impacting potable water-supply wells. Land subsidence can also occur as a result of groundwater development, causing settlement of structures and increasing flood risk. Each of these effects is discussed in more detail below.

8.3.1 Long Term Effects on Regional Groundwater Levels (Water Level Decline)

As long as water is mined (a withdrawal of groundwater in excess of the induced recharge plus the reduced groundwater discharge) from an aquifer, the water levels or potentiometric surface will be consistently lowered in elevation on a year by year basis. Water-level declines increase pumping lift, decrease well yields, and limit development of the groundwater resource. To illustrate these impacts, a discussion of the High Plains aquifer is provided below.

The High Plains aquifer is a 174,000-square-mile (450,658 km^3) area underlying parts of eight States from South Dakota to Texas. Irrigation water pumped from the aquifer has made this region of the country an important agricultural area. The intense use of groundwater has caused major water-level declines and reduced the saturated thickness of the aquifer (the groundwater remaining in storage) in some areas to a level at which it is no longer economical to use the aquifer as a water supply. The changes are particularly evident in the central and southern High Plains (Figure 8.2), where more than 50% of the predevelopment saturated thickness has been dewatered in some areas. The net amount of water removed from storage in the aquifer is estimated to have been 220 million ac-ft (270 km^3) through 1999. This is a very large volume of freshwater—equal to more than half the volume of water in Lake Erie. If the total volume depleted from this single, multi-state aquifer was spread over the surface of the oceans, it would raise sea level about 0.75 mm (0.03 in).

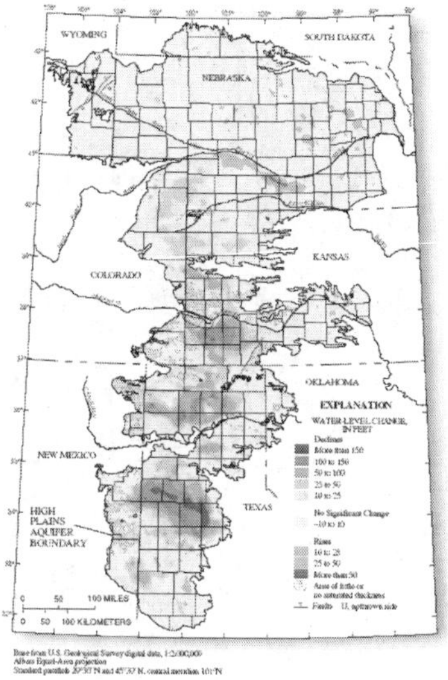

Figure 8.2: Changes in groundwater levels in the High Plains aquifer from before groundwater development to 2001 (McGuire 2003).

In response to declines in water levels and groundwater storage (Figure 8.3), a monitoring program was begun across the High Plains in 1988 to assess annual groundwater-level changes in the aquifer. Water-level measurements have been made each year in more than 7,000 wells. This substantial effort requires collaboration among numerous Federal, State, and local water-resource agencies. Water levels continue to decline in many areas of the aquifer, but the monitoring program indicates overall reductions in the rate of decline during the past two decades in some areas. This change is attributed to decreases in irrigated acreage, reduced water needs because of improved irrigation and cultivation practices, and above-normal precipitation and recharge during this period.

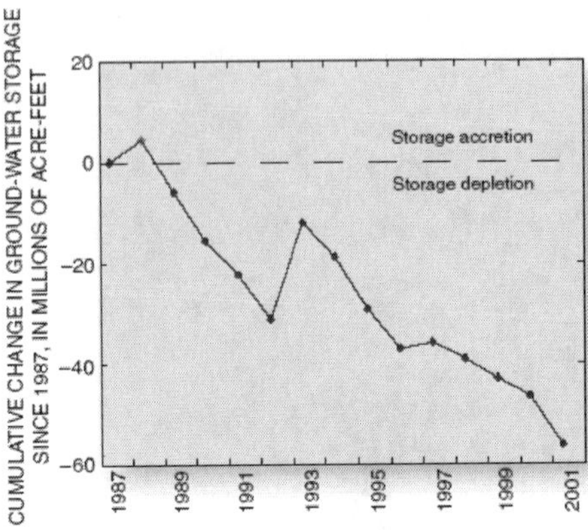

Figure 8.3: Cumulative changes in groundwater storage in the High Plains aquifer system since 1987 (McGuire 2003).

8.3.2 Depletion of Surface Water

Surface and groundwater systems are in continuous dynamic interaction. Groundwater can be a source for surface water by supplying base flow and maintaining wetlands in times of low precipitation. In other situations, surface water can recharge groundwater via seepage through the stream bed. On average about 40% of the river flow nationwide depends on groundwater (Conservation Technology Information Center 2010). Over-pumping of aquifers can lead to lowered stream and lake levels, and to desiccation of wetlands.

Streams either gain water from inflow of groundwater (i.e., a gaining stream, Figure 8.4a) or lose water by outflow to groundwater (i.e., a losing stream, Figure 8.4b). Many streams do both, gaining in some reaches and losing in other reaches. Furthermore, the flow directions between groundwater and surface water can change seasonally as the altitude of the groundwater table changes with respect to the stream-surface altitude or can change over shorter timeframes when increases in stream stage during high flow periods cause recharge to the streambank. Under natural conditions, groundwater makes some contribution to streamflow in most physiographic and climatic settings. Thus, even in settings where streams are primarily losing water to groundwater, certain reaches may receive groundwater inflow during some seasons. Losing streams can be connected to the groundwater system by a continuous saturated zone (Figure 8.4b) or can be disconnected from the groundwater system by an unsaturated zone (Figure 8.4c).

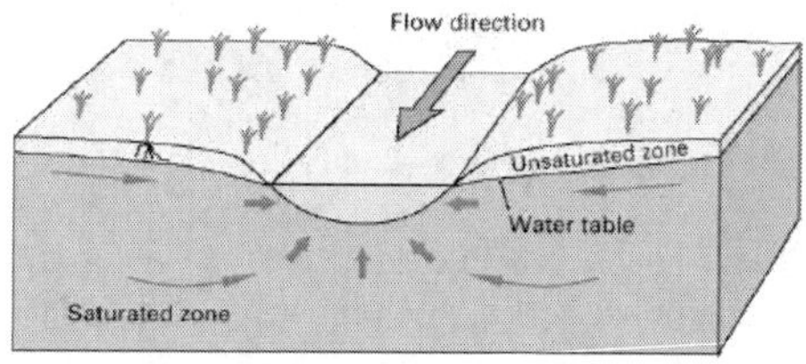

(a) Gaining stream

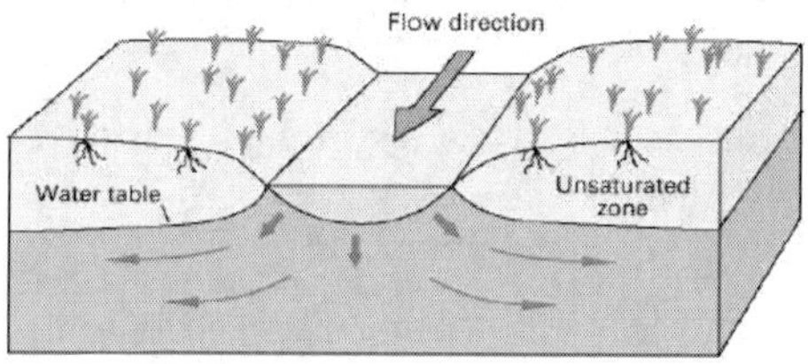

(b) Losing stream

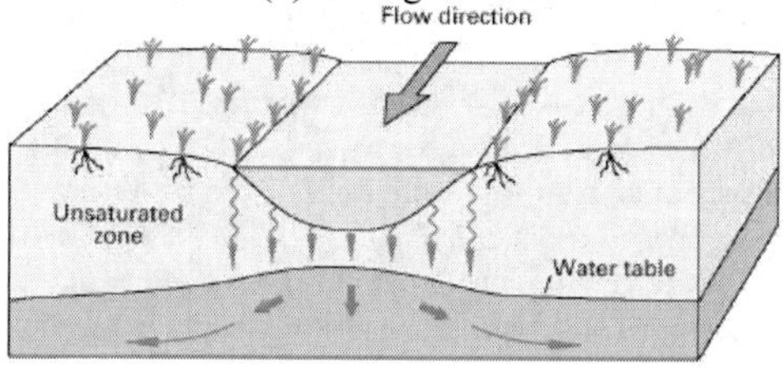

(c) Losing stream that is disconnected from the groundwater surface

Figure 8.4: Interaction of streams and groundwater (from Alley et al. 1999).

For a gaining stream, the groundwater contribution to streamflow may be reduced by groundwater pumping. When the aquifer water level is lowered by pumping, the hydraulic head difference between stream and the aquifer will be reduced, thus resulting in less groundwater discharge to the stream or even turning the gaining stream into a losing stream when pumping stresses become large enough to reverse the flow direction between the stream and aquifer.

In a losing stream, stream seepage (streamflow losses to the aquifer) may be affected by groundwater pumping and natural variations in aquifer water level. When the aquifer water level is near land surface, seepage from the river is partially controlled by the height of the aquifer water level (see Figure 8.4b). Activities or events that result in a lowering of the groundwater surface, such as groundwater pumping, induce more seepage from the stream. Conversely, events that cause the aquifer water level to rise (recharge events) will result in a decrease in stream seepage. If aquifer water levels rise above the level of the stream, what was previously a losing stream reach will become a reach that is gaining water from the aquifer.

Another hydrologic condition exists that is very important in understanding surface water and groundwater interaction. A surface water body is perched above an aquifer when aquifer water levels are well below the bed of the river, stream, or lake (Figure 8.4c). Under these conditions, water will seep from the surface water body to the groundwater, but the surface water body will not be affected by aquifer water levels and consequently streamflow does not change in response to groundwater pumping. Nearby groundwater pumping will cause a lowering of the water table, but will not affect surface water supplies.

The Big Lost River is an example of a losing stream supplying an aquifer (Johnson et al. 1998). The river flows out of a mountain valley on the northwest margin of the Snake River Plain and entirely disappears through seepage into the permeable lava of the plain. The underlying Snake River Plain aquifer flows to the southwest, ultimately discharging in the form of springs along the wall of the Snake River canyon.

The Snake River provides an excellent example of a gaining stream. As the Snake River flows across southern Idaho much of the flow is diverted for irrigation. At Shoshone Falls, about 30 miles (48.28 km) downstream of Milner Dam, the river may nearly dry up due to irrigation diversions. In the next 40 miles (64.37 km) downstream, the river is again reborn in the impressive Thousand Springs area, where springs collectively discharge more than 5,000 cfs (141.58 m^3/sec). Niagara Springs is an example of the many scenic springs in the Thousand Springs area. These river gains provide the majority of the downstream flow during summer.

Discharges from springs are often relatively constant, but may fluctuate with the season and from year to year, depending upon natural weather patterns and man-induced effects of pumping and irrigation. For example, discharge of Blue Lakes Spring along the Snake River near Twin Falls shows both seasonal and long-term variation (Figure 8.5). Much of the short and long-term variation in the flow of Blue Lakes Spring is due to distribution and application of water from the Snake River for irrigation and groundwater pumping.

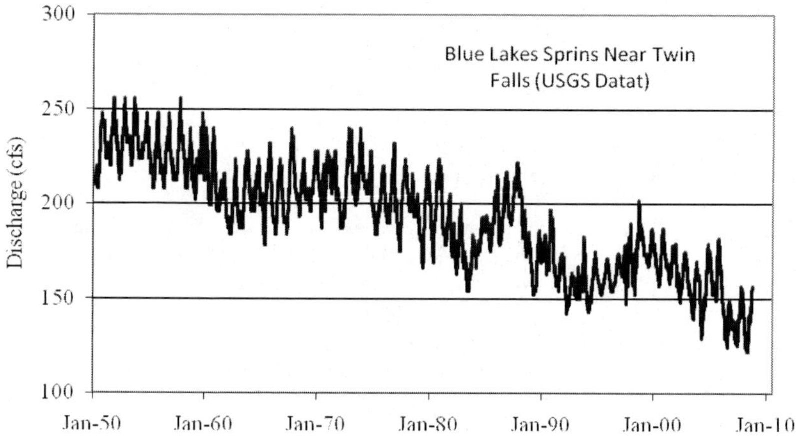

Figure 8.5: Discharge at Blue Lakes Springs near Twin Falls (Johnson et al. 1998; with permission, updated from USGS NWIS, 13091000 Blue Lakes Spring NR Twin Falls, ID, 09S 17E 28DBA1S).

8.3.3 Saltwater Intrusion

In coastal areas, the water-bearing deposits of many groundwater basins outcrop offshore. Under natural conditions it is presumed that there is subsurface outflow from such deposits into the ocean. In some areas, these deposits extend considerable distances under the ocean and store large volumes of freshwater. As these basins are developed and groundwater levels are drawn below sea level, highly saline waters will begin to encroach into the freshwater-bearing deposits. If groundwater basin development is on a sustained yield basis and encroachment is controlled so that seawater does not reach active wells, this effect can be desirable. In this type of operation, seawater that enters the aquifers during dry periods can be driven out when water levels are increased during ensuing wet periods. However, for effective development of this storage, wells should not be so located as to cause extreme hydraulic gradients or cones of depression adjacent to the coast. Development of such cones of depression will have the effect of increasing the rate of seawater intrusion at localized points, with the result that the valuable freshwater storage capacity may be impaired.

In some areas, geologic formations contain entrapped brines, remaining from some past geologic age when seawater covered the area involved, perhaps resulting from evaporation of water and deposition of salts. Subsequently, freshwaters have moved through and, to a great extent, flushed out many of the brines originally contained in the deposits. However, because of unusual conditions, certain of these brines were trapped and have remained in place over the ensuing ages. In some of these areas, the influence of groundwater extractions can lower water levels or otherwise alter the regimen with the result that these deep-seated brines can escape, commingle with the existing freshwater supplies, and cause serious water-quality problems. Once these brines have commingled and caused degradation, an extended period will be required for the removal of the degraded water. To prevent this occurrence, extensive groundwater development, particularly in the form of heavy pumping drafts, should be avoided in the vicinity of such brines.

Small water level declines in coastal aquifers can result in a significant amount of saltwater intrusion. Under hydrostatic conditions, the weight of a unit column of freshwater extending from the water table to the interface is balanced by a unit column of saltwater extending from sea level to the same depth as the point on the interface (Freeze and Cherry 1979; Driscoll 1995), which is depicted in Figure 8.6. For hydrostatic conditions in a homogeneous, unconfined coastal aquifer, a 1 m reduction in the water table elevation will cause the saltwater interface to rise by 40 m.

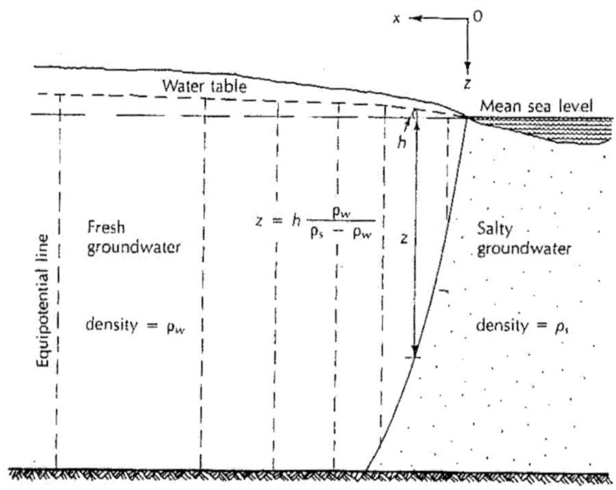

Figure 8.6: Saltwater interaction in an unconfined coastal aquifer according to the Ghyben-Herzberg relation (Freeze and Cherry 1979, reprinted by permission of Prentice Hall).

8.3.4 Land Subsidence and Earth Fissures

Land subsidence is a gradual settling or sudden sinking of the Earth's surface owing to subsurface movement of earth materials (Galloway et al. 2000). More than 80% of the identified subsidence in the U.S. is a consequence of our exploitation of groundwater, and the increasing development of land and water resources threatens to exacerbate existing land subsidence problems and initiate new ones. In many areas of the arid Southwest, and in more humid areas underlain by soluble rocks such as limestone, gypsum, or salt, land subsidence is an often-overlooked environmental consequence of our land- and water- use practices.

Subsidence of the land surface has been observed at a number of locations throughout the world (Poland 1984). It is related, in large part, to several human activities, including:

i. Extraction of groundwater.
ii. Withdrawal of oil, gas, and associated brackish and saline waters.
iii. Application of water to unconsolidated moisture-deficient deposits.
iv. Dewatering of peat soils and deeper peat deposits.
v. Loads impose by engineering structures causing foundation settlement.

Land subsidence can also result from natural causes such as dissolution and weathering of rock, tectonic activity, and volcanic activity; however, a discussion of these natural processes is beyond the scope of this book.

Agencies and individuals involved in planning the management of groundwater reservoirs, or in the day-to-day mechanics of operating a groundwater reservoir, should familiarize themselves with the types of subsidence, and with the cause, areal extent, magnitude, and chronology of damage associated with each type of subsidence. Subsidence may be the critical factor in managing some groundwater reservoirs. For such reservoirs, any subsidence monitoring program must be designed so that cause and effect will be associated; there is no use from a subsidence standpoint in monitoring the water levels in a deep aquifer if the subsidence is due to the application of water on unconsolidated moisture-deficient deposits.

Land-surface subsidence due to intensive and prolonged pumping of groundwater (a phenomenon sometimes described as "deep subsidence") is occurring in several areas of the world — China, Japan, Mexico, England, and the U.S. (Poland 1984). It may also be occurring in other areas where groundwater is being pumped. Its identification requires an adequate vertical control monitoring program, or visible damage to engineering structures.

Land-surface subsidence due to continued extraction of groundwater and the associated decline of piezometric levels results in a gradual, regional settlement of the land surface. Decrease in pressure head in compressible confined aquifer systems results in increased effective stress (grain-to-grain load) on the confined sediments. The sediments compact in response to the added load, and the land surface subsides. The magnitude of subsidence depends on the magnitude of change in head and on the compaction characteristics and thickness of the sediments. The greater the number of clayey beds in the aquifer system, the greater may be compaction. Continuous measurement of compaction of materials in deep holes indicates rapid response to head change at most places in the subsiding areas. Subsidence can be slowed down or stopped by raising the level of the potentiometric surface or the pressure head sufficiently.

Slow regional subsidence poses serious problems in the operations of many types of engineering structures, particularly those involved in the storage, transport, and pumping of water. For example, tilting of the land surface can appreciably reduce the flow of water in low-gradient gravity canals. This has occurred in the U.S. Bureau of Reclamation's Delta-Mendota canal along the west side of the San Joaquin Valley in California. In addition, smaller structures such as drains and sewers can be affected, and even the channel capacity of streams may be altered. Tilting can also affect the operation of pumping plants as such plants may be highly sensitive to minute tilting of the land surface. In a critical area such as a coastal bay, where bordering lands subside, levees may have to be built. This has been done in the southern San Francisco Bay area in California to prevent flooding of adjacent agricultural, urban, and .industrial areas by saline bay waters (Fowler 1981). In addition, when the consolidating sediments are deep, casings of the water-supply wells are compressed and frequently ruptured, requiring expensive maintenance and replacement. As the sediments continue to compact in a groundwater reservoir, reduction in groundwater storage capacity and even in the permeability of the sediments may be occurring, though the usual case is for the fine-grained materials (primarily clays) to consolidate rather than the more important granular materials of the aquifers. The legal aspects of land-surface subsidence caused by groundwater withdrawal are additional concerns facing water-resource managers (Kopper and Finlayson 1981).

The magnitude of subsidence associated with water-level decline appears to be related in large part to geologic factors such as: (i) differences in mineral composition; (ii) particle size; (iii) sorting; (iv) degree of consolidation; (v) degree of cementation; and (vi) degree of confinement of the deposits in the groundwater reservoir. Thus, the ratio of subsidence to head decline will vary between groundwater reservoirs, and even within a single groundwater reservoir. For example, measured ratios of subsidence to head decline vary from 0.008 to 0.1. Annual measured rates of land-surface subsidence in groundwater reservoirs range from a fractional value to about 0.5 m (1.5 ft). To reduce the long-term water level decline, the technology of aquifer storage and recovery (ASR) or artificial recharge (ASCE 2001) becomes popular and has been widely applied to management and development of freshwater resources in the U.S. Land subsidence due to ASR applications has been investigated and discussed by Li and Helm (2000, 2001a, 2001b) and Li (2003).

The maximum measured subsidence throughout the world ranges from less than 0.3 m (1 ft) to 8.8 m (29 ft). Poland and Davis (1969), Poland (1969, 1981), Hwang and Wu (1969), and Leake (1997) report the following subsidence values: London, England, 21 cm (0.7 ft); Savannah, Georgia, 22.7 cm (0.75 ft) ; Denver, Colorado, 38 cm (1.25 ft); Las Vegas, Nevada, 2.44 m (8 ft); Houston-Galveston, Texas, 2.74 m (9 ft); Taipei, Taiwan, 1.3 m (4.26 ft); Eloy-Picacho, Arizona, 4.57 m (15 ft); Arvin-Maricopa, California, 2.4 m (8.0 ft); Osaka, Japan, 2.7 m (9.0 ft); Tulare-Wasco, California, 3.7 m (12.0 ft); Tokyo, Japan, 3.7 m (12.0 ft); Santa Clara, California, 4.0 m (13.0 ft); Los Banos-Kettleman City, California, 8.8 m (29 ft); and Mexico City, Mexico, 7.9 m (26.0 ft).

Summaries of the case histories of subsidence in many areas of the world are presented in articles by Marsden and Davis (1967), Poland (1969; 1981), Poland and Davis (1969), and Galloway et al. (2000). Detailed evaluations of specific problem areas in California have been compledted for the following areas: (1) Los Banos-Kettleman City area, Western Merced and Fresno Counties, San Joaquin Valley (Poland and Davis 1956; Poland 1969); (2) Tulare-Wasco area, Tulare and Kern Counties, and San Joaquin Valley (Lofgren 1969a, 1969b; Poland and Davis 1956, 1969; Poland 1969); (3) Arvin-Maricopa area, Kern County, and San Joaquin Valley (Lofgren 1963, 1969a, 1969b; Lofgren and Klansing 1969; Poland 1969; Poland and Davis 1969); and (4) San Jose area, Alameda and Santa Clara Counties, and Santa Clara Valley (Poland and Green 1962; Poland 1969, 1981; Poland and Davis 1969).

Similar detailed subsidence studies in the U.S. have been made of conditions in the Houston-Galveston area (Gabrysch 1967; Neighbors 1981; Coplin and Galloway 1999; Michel 2006); in the New Orleans area (Kazmann and Heath 1968); in the Baton Rouge area (Kazmann and Heath 1968; Louisiana Water Resources Research Institute 1968); and in the Savannah, Georgia area (Davis et al. 1963). General discussions of subsidence in Denver area, Colorado, Eloy-Picacho area, Arizona, and Las Vegas area, Nevada are presented by Poland and Davis (1969) and Poland (1981). Poland and Davis (1969) and Poland (1981) also discussed subsidence in Mexico City, London, and Tokyo, Osaka, and Nagoya areas in Japan.

At the same time, subsidence-related ground failures can occur as a result of groundwater withdrawal under appropriate geological conditions (Holzer 1984; Helm 1994a, 1994b; Sheng et al. 2003; Li 2007a, 2007b, 2007c). Ground failures can be classified into earth fissures and surface faults (Holzer 1984). An earth fissure results from a tensile failure of a geological mass (i.e., the opposing sides have a major displacement component perpendicular to the plane of failure). A surface fault is a shear failure of a geological mass (i.e., the opposing sides have a major displacement component parallel to the plane of failure). Earth fissures normally start as small (millimeter-scale) tension cracks in the buried sediment, and then grow due to mechanical piping and additional pumping stress, eventually breaking through the sedimentary cover to reach the ground surface (Sheng and Helm 1998). Ground failures and land subsidence are closely related to aquifer movement. Ground failure is not necessarily the direct result of land subsidence, or at least land subsidence is not an essential cause of ground failure. They co-exist and occur contemporaneously because they have the same cause, namely aquifer movement (Sheng and Helm 1998; Sheng et al. 2003).

Earth fissures that have ruptured the land surface and widespread land subsidence in deep alluvial basins of southern Arizona are related to groundwater overdrafts. Since 1900 groundwater has been pumped for irrigation, mining, and municipal use, and in some areas more than 500 times the amount of water that naturally replenishes the aquifer systems has been withdrawn (Schumann and Cripe 1986). The resulting groundwater-level declines—more than 183 m (600 ft) in some places—have led to increased pumping costs, degraded the quality of groundwater in many locations, and led to the extensive and uneven permanent compaction of compressible fine-grained silt- and clay-rich aquitards. A total area of more than 7,770 km^2 (3,000 mi^2) has been affected by subsidence, including the expanding metropolitan areas of Phoenix and Tucson and some important agricultural regions nearby.

Earth fissures have been observed in the Las Vegas Valley since 1925 (Bell and Price 1993). They were first recognized as effects of groundwater withdrawal in the late 1950's. Several studies followed to document and map fissures (Conwell 1965; Patt and Maxey 1978; Holzer and Pampeyan 1979; Bell 1981; Bell and Price 1993; Helm 1994a, 1994b; Sheng and Helm 1998; Sheng et al. 2003; Li 2007a, 2007b, 2007c). Bell and Price (1993) examined the spatial relation between earth fissures and faults and found that about 50 % of all fissures occur within 152 m (500 ft) of the nearest mapped fault and 90 % occur within 610 m (2,000 ft). Incidence of fissuring was found to be increasing with time, with fissures continuing to form in most of the previously recognized fissuring zones.

Several long and continuous fissures (individually over 100 to 400 m [330-1,300 ft] long) were observed on different unconsolidated surficial deposits chiefly dominated by silt, sand, and pebble gravel within Escalante Valley during the flood of January 8-12, 2005 (Lund et al. 2005). After the flood these fissures were subsequently enlarged by infiltrating floodwater and surface and subsurface groundwater flow (piping), forming linearly-aligned centimeter- to decimeter-scale depressions and piping holes up to 3 m (10 ft) wide by 2 m (6.6 ft) deep. Lund et al. (2005) concluded that the formation of the earth fissures in the Escalante Valley was a consequence of groundwater pumping that has resulted in a permanent decline in groundwater levels. Dewatering of the upper portion of the basin-fill aquifer allowed the aquifer to compact, resulting in subsidence of the ground surface over the western margin of the basin beneath the Escalante Valley and lateral changes in lithology that created horizontal tension and ground surface cracks.

Ground failures related to groundwater withdrawal have been observed in Fremont Valley, San Jacinto Valley, and other areas of California (Fett et al. 1967; Holzer 1984; Shlemon and Davis 1992; Shlemon and Hakakian 1992). It is worth mentioning that the 1987 Temecula fissures were probably caused by groundwater withdrawal and associated increases in tensile stresses along nearby faults (Shlemon and Davis 1992; Shlemon and Hakakian 1992); whereas the Murrieta fissures, no more than few kilometers from those in Temecula, were likely related to an increase in groundwater levels due to irrigation. These results indicate that both a fall and a rise in groundwater levels can accompany the aquifer movement, which in turn causes earth fissuring.

When a water planner formulates alternative management plans involving water from several sources, potential damages resulting from subsidence of the land surface as a consequence of their implementation must be evaluated. Operating a groundwater reservoir on the basis of deferred perennial yield is the plan most likely to cause subsidence, because it results in the greatest lowering of groundwater levels. However, net benefits from such a plan, even when considering subsidence damage, may be greater than from other plans because groundwater is usually cheaper than water from other sources, and subsidence could be expected to be slow and regional in nature. Results of many-land-surface subsidence studies can be summarized as follows:

i. The cause of land subsidence is over-withdrawal of groundwater and associated long-term decline in potentiometric surface, resulting in increased effective stress (grain-to-grain load) and consolidation of clays.
ii. Land-surface subsidence can be in excess of 10 m (30 ft), with a maximum annual decline about of 0.5 m (1.5 ft).
iii. Land subsidence can be mitigated by reducing the net groundwater extraction by either reducing pumping or by increasing artificial recharge.

8.4 Regional Scale Development of Groundwater

As surface water networks and groundwater reservoirs are increasingly being considered by water resources managers and planners as integrated systems that affect diverse geopolitical entities, it is imperative that water resources managers devise strategies to manage water resources from a regional scale in spite of the cultural and legal challenges that the undertaking of such an approach to water resources entails. After identifying the problems that threaten the groundwater supply of an area, and deciding that they should be solved, the next step for the water managers is to develop the objectives for an action program to avoid, solve, or mitigate the difficulties. Advice from many specialists — engineering (including hydrologic), legal, fiscal, geologic, environmental, and economic — is required in this difficult task. Questions that may arise during the deliberations are (ASCE 1987).

i. Should the problems be approached on the basis of political or hydrologic boundaries?
ii. How much of the cost of water service should be paid by the user through water bills, and how much through ad valorem or pumping taxes?
iii. Are the occurrence and use of ground and surface waters sufficiently unrelated so they can be managed separately, or are they sufficiently related to require coordinated management?

iv. When should legal, environmental, organizational, and political influences be considered before, during or after formulation of alternative management plans?

Non-economic forces will inevitably influence the implementation of the plan and may even override economic factors. Many planners elect to disregard non-economic aspects until the alternatives are compared on the basis of costs and benefits. In this manner, the decision makers can evaluate the extra cost of revisions required by legal, institutional, social, and political forces.

Formulation of management objectives is often more complex than the technical task of developing alternative plans to meet them. Among the major obstacles will be the misconceptions mentioned previously. There will be many who oppose implementation of a management plan because it breaks with traditional practices. Others will object to any plan because it connotes "control" and an infringement on what they consider to be their prerogatives or rights. Management, if it truly performs its function, will consider the effect of various objectives on voter and water-user reaction, economic growth, capital investment, water rights, and reservation of native resources.

8.4.1 Elements of management plans

Once management objectives are adopted, activities and costs of a planning investigation must be developed in sufficient detail to obtain authorization and funds for the study. These require identification of the elements of the management plan or the variables to be considered. Pumping and artificial recharge locations and schedules may be important aspects of any plan, as could be the manner and extent of the use of surface water, both local and imported. The possibilities of alternative sources of water, such as water conservation and waste water reuse, should be identified. Determination of the boundary conditions that are not only geologic and hydrologic but also political is needed. The legal structure of the management agency, including its authority and responsibilities, must be appraised in light of the management objectives and existing local, state, and federal law. The financing of any management plan must be analyzed. Some type of monitoring program will be essential after implementation of the selected plan, to confirm hydrologic assumptions or judgments, and to measure the performance of the operational program and its ability to satisfy the management objectives.

All the information related to the proposed management plan should be obtained, even if it is available only in qualitative terms. When the information is presented, executive and legislative public bodies and water agency boards of directors should be able to visualize clearly the difficulties to be overcome, and the objectives, general work activities, costs, and schedule of the recommended water resources management study.

8.4.2 Political Boundary vs. Aquifer Boundary

For effective groundwater management, the unit controlled should be a complete hydrologic unit, including the places of natural and artificial recharge, the places of natural or artificial discharge, and all the intervening media through which water may move. In concept this would mean control of a total drainage basin, including streams and lakes as well as groundwater. For large river basins such control may not be practical, but it could be approached for portions of those basins. As an example, in Florida, where the Floridian Aquifer underlies the whole state, the problem of management of surface water and groundwater statewide was resolved by creating five water management districts that cover the state and whose boundaries are based upon groundwater and surface water basins. This concept is explicit in the enabling legislation for the groundwater conservation districts in Texas, in which it is provided that the area of a district must be coterminous with the area of an underground reservoir or subdivision thereof. In the long-continued regulation of the Roswell groundwater reservoir in New Mexico, the original area of control in 1927 included only the region in which artesian wells had been drilled. The controlled area has been progressively increased until, by about 1940, it embraced the entire groundwater reservoir, including all areas of recharge and natural discharge.

If the term "groundwater reservoir" is generally accepted as denoting complete groundwater units (without limitation as to size), several other terms are available for denoting any subdivisions that may be necessary because of compartmentalization created by relatively impermeable rocks. Among these, the term "aquifer" is already used rather widely to denote subdivisions discriminated on the basis of depth [e.g., the 1,400 ft (426.72 m) aquifer in the vicinity of Memphis, the Dakota sandstone in North and South Dakota]. Similarly, the term "groundwater basin" has been given areal or geographic connotations in numerous instances (e.g., the Raymond Basin or the Chino Basin in southern California).

Closed basins in the southwestern U.S. are generally rimmed by mountain ranges and have centripetal drainage, forming a lake in the lowest part of the basin whenever runoff from precipitation is sufficient. Many of these basins, however, are in arid regions where there is no streamflow except after occasional storms, and where water principally moves from the bordering mountains toward the lowlands as groundwater. In such basins, the mountains may constitute groundwater divides as well as topographic divides, and the low part of the basin constitutes a sink from which groundwater is discharged by evapotranspiration. Thus, a groundwater basin will exist which is coextensive with a drainage basin. This may, however, be the exception rather than the rule. Many groundwater basins do not conform with surface drainage basins, either in area or in pattern of flow.

If groundwater basins are to be subdivided, their subbasin boundaries would logically be the relatively impermeable barriers that make such subdivision necessary. For management purposes, a groundwater basin may have to be subdivided using barriers to flow and political boundaries even though groundwater flows through such political divides. Such approaches are not desirable but may be required for practical reasons.

Interstate Management Approaches to Development of Groundwater - Ogallala Aquifer Case Study: The High Plains aquifer underlies parts of eight states—Colorado, Kansas, Nebraska, New Mexico, Oklahoma, South Dakota, Texas, and Wyoming. In 2000, the High Plains aquifer contained about 3,676 km^3 (2,980 million ac-ft) of water in storage. The volume of water in storage by state ranged from about 49 km^3 (40 million ac-ft) in New Mexico to 2,467 km^3 (2,000 million ac-ft) in Nebraska. The volume of water in storage in the High Plains aquifer in 2000 was about 243 km^3 (197 million ac-ft), or about 6%, less than the total water in storage prior to development of the aquifer for irrigation. About 95% of storage loss occurred over a 69,000 km^2 (17-million-acre) area with greater than 7.6 m (25 ft) of water-level declines. In this 69,000 km^2 (17-million-acre area), the water lost from storage represents about 34% of the predevelopment storage. The change in aquifer storage ranges from an overall increase of 4.9 km^3 (4 million ac-ft) in Nebraska to an overall decrease of 153 km^3 (124 million ac-ft) in Texas (McGuire et al. 2003).

The fate of the High Plains has been a policy issue since depletion of the Ogallala became apparent in the 1970s. Policy makers have wrestled with how—and whether—to conserve the groundwater resource. This debate has recently returned to the fore at both state and federal levels. The renewed interest within the High Plains states partly reflects the recent drought and exhaustion of the aquifer's usable economic life in many areas. The debate at the federal level has changed over time. Debates in the 1970s focused on conserving the Ogallala for national and international food security. During the 2002 Farm Bill debate, the focus shifted to the regional impacts of federal policies through their effects on water use. Because water is so fundamental to the region's livelihood, policies affecting water use ultimately change the scope and distribution of economic activity as well as the use of land and other natural resources. Federal policies affecting water use in the High Plains include commodity price programs, the Conservation Reserve Program, and cost-share programs for investments in new technologies. Central to policy debates is why water conservation should (or should not) be a policy priority (Peterson et al. 2003).

Some of the states—Colorado, parts of Kansas, New Mexico, Oklahoma, and Texas—formally recognize that, in some areas, water is being withdrawn from the aquifer at rates greater than the aquifer is being replenished. Therefore, the states implemented groundwater-management plans to prevent aquifer depletion, to manage aquifer development, and to ensure the availability of aquifer resources for at least a specified number of years. In Kansas, Nebraska, and Wyoming, some groundwater-management plans were also designed to limit water-level declines to try to maintain an acceptable amount of groundwater discharge to surface water.

All states share a groundwater management philosophy in that the water should be used for beneficial purposes, which generally implies that the water is used for a reasonable purpose and that the water is not wasted. Water conservation is required and implemented at different degrees from one state to another. The differences among these approaches to groundwater management include the legal doctrines, which form the basis for groundwater allocations.

On Federal and Indian lands, water rights are based on the reserved-rights doctrine. Under the reserved-rights doctrine, the Federal agencies and Indian Tribes are entitled to sufficient water to fulfill the purpose for which the lands were set aside. There are about 1.9 million acres of Indian land, mainly in South Dakota, and 2.5 million acres of Federal land that overlie the High Plains aquifer (Getches 1997; McGuire et al. 2003).

The U.S. Supreme Court has held that groundwater is an article of interstate commerce and, therefore, is subject to constitutional restrictions on state regulation of such commerce (U.S. Supreme Court 1982). These constitutional restrictions allow states to limit interstate transfers of groundwater if the limitations are based on legitimate needs of the state's citizens and not on arbitrary criteria.

Four legal doctrines form the basis for States' rules on allocations of groundwater in the High Plains aquifer. These legal doctrines are rule of capture, reasonable use, correlative rights, and prior appropriation (Getches 1997; Ashley and Smith 1999; McGuire et al. 2003).

i. According to the rule of capture or absolute ownership doctrine, landowners can withdraw groundwater for use without regard to the effect of the withdrawals on wells owned by adjacent landowners.

Texas is the only state that uses the rule of capture for groundwater allocation. In Texas, no one has title to groundwater until the water is withdrawn from the aquifer. The landowners whose land overlies the aquifer own the right to withdraw available water from the aquifer, subject to limitations. In addition, landowners can sell the withdrawn water for use at other locations. Limitations to the rule of capture in the High Plains aquifer in Texas are that (1) the person withdrawing the water cannot dewater a neighbor's well for malicious reasons or willfully waste the water, and (2) in areas with Groundwater Conservation Districts, the district can enact some limitations on the groundwater withdrawal rate, the total amount withdrawn, and the well spacing (State of Texas Statutes, Chapters 35 and 36). Fourteen Groundwater Conservation Districts regulate to a varying extent more than 81% of the area that overlies the High Plains aquifer in Texas (TWDB 2010).

ii. According to the reasonable use doctrine, landowners are entitled to withdraw a reasonable amount of the groundwater that underlies their land, as long as the landowners use the water for beneficial purposes.

In Oklahoma, groundwater is owned by the landowners whose property overlies the aquifer. Use of groundwater is governed by the reasonable-use doctrine and is subject to regulation by the Oklahoma Water Resources Board (OWRB). The High Plains aquifer in Oklahoma was subdivided into two administrative areas by the OWRB. The maximum annual yield in the Panhandle Region, which includes most of Beaver, Cimarron, and Texas Counties, is 2.0 ac-ft per acre. The maximum annual yield in the Northwest Region, which includes parts of Ellis, Harper, Dewey, and Woodward Counties, has been recently reduced from 2.0 ac-ft per acre to 1.4 ac-ft per acre for new permits; however, existing permit holders are allowed to continue to withdraw 2.0 ac-ft per acre (OWRB 2002).

iii. Under the correlative rights doctrine, landowners are entitled to withdraw a reasonable amount of the groundwater that underlies their land, as long as the landowners use the water for beneficial purposes. The amount of water that can be withdrawn by a landowner is limited by the needs of others whose land overlies the aquifer and who have an equal right to a reasonable amount of the water.

In Nebraska, groundwater is a public resource, and the use of groundwater is governed by a modified correlative rights doctrine and by numerous legislative enactments. In those areas where no local or state regulations supersede, the Nebraska correlative rights doctrine provides for proportionate sharing whenever the natural supply of groundwater becomes insufficient for all overlying landowners' needs. Twenty-three Natural Resource Districts (NRDs) created by the legislature in the early 1970's were assigned multiple responsibilities relative to the management of Nebraska's soil and water resources. Each NRD prepared a groundwater-management plan that was approved by the Nebraska Department of Natural Resources. Each groundwater-management plan identified groundwater depletion and water-quality issues in the district and specified the regulatory or other procedures that the NRD would enact to address these issues. As of April 2003, groundwater-management areas are located in 18 of the 21 NRDs that overlie the High Plains aquifer in Nebraska; a goal of eight of these groundwater-management areas was to address groundwater supply problems. Approaches to addressing these issues included moratoriums on installation of new wells and restrictions on the amount of groundwater that can be withdrawn (McGuire et al. 2003).

iv. According to the prior appropriation doctrine, a landowner must apply for a permit from the state to use groundwater. If the state approves the permit application, the state grants the landowner a right to a specified amount of water. Landowners whose water right predates the state's permit system can receive a grandfathered right. If there is not sufficient water to meet the needs of all appropriators, water rights are enforced by seniority. A simplified way to explain this system is "first in time, first in right."

Five other states including Colorado, Kansas, New Mexico, South Dakota and Wyoming use the appropriation doctrine to varying degrees. Groundwater use is governed jointly by state water laws and groundwater management districts (GMDs) controlled by local irrigators. Controls differ by state and district.

The water in the High Plains aquifer in Colorado is considered designated groundwater. Designated groundwater is defined by the Colorado Legislature as "groundwater which in its natural course would not be available to and required for the fulfillment of decreed surface rights, or groundwater in areas not adjacent to a continually flowing stream wherein groundwater withdrawals have constituted the principal water usage for at least fifteen years." (Colorado, Statutes, Title 37, Article 90). In 1967, depletion criteria for the Northern High Plains Designated Basin were established such that no new large-capacity well permits could be issued if the proposed withdrawal, after 25 years, was projected to deplete the aquifer by more than 40% within a 3-mile radius of the well. In 1990, depletion criteria were amended to 40% in 100 years. Depletion criteria were not established for the Southern High Plains Designated Basin because of the complex geology of the area.

Five GMDs in Kansas created by eligible voters and the State Engineer overlie most of the High Plains aquifer in state. GMD boards can propose regulations, which must not conflict with state law, and develop plans for the management of local groundwater supplies for nondomestic use within the district. The Chief Engineer and GMD boards manage water resources by using either the concept of "safe yield" or "allowable depletion," depending on conditions in the area. The "safe yield" concept considers existing appropriations within a specified radius of the proposed well and limits total appropriations to a percentage of the estimated recharge to the aquifer in that radius. The "allowable depletion" concept considers existing appropriations within a specified radius of the proposed well and limits total appropriation to a level that will deplete the aquifer by a specified amount in a specified timeframe in that radius (Sophocleous 1998; Kansas Department of Agriculture 2009).

In New Mexico six groundwater basins were designated that encompass most of the High Plains aquifer in state (New Mexico Office of the State Engineer 2005). Within a designated groundwater basin, a water user must obtain permit for a new water right or for changes to an existing water right from the State Engineer. The State Engineer generally approves permits for livestock, lawn and garden irrigation, and other domestic purposes. For other types of water uses, the State Engineer approves permits for new or revised water rights if (1) no objections are filed, (2) unappropriated water exists in the basin, (3) no infringement on the water rights of prior appropriators occurs, and (4) it is not detrimental to the public welfare or the water conservation goals of the state. The State Engineer determines whether unappropriated water exists in the basins by monitoring water-level declines within a 9 to 25 mi^2 area, depending on aquifer properties, around the site of a proposed new appropriation. If the annual water-level decline is exceeding 2.5 ft, the State Engineer will not approve the permit because the rate of decline is considered excessive (Templer 1992; Ashley and Smith 1999).

Management of water quantity in South Dakota is accomplished through the water-rights permit program, established in 1955, and State Water Plan, established in 1972, and is administered by the South Dakota Department of Environment and Natural Resources. However, most of the land overlying the High Plains aquifer in South Dakota lies within the boundary of either the Pine Ridge or Rosebud Indian Reservations, and the Tribes assert the right to control the development and use of groundwater resources within reservation boundaries (South Dakota, Codified Laws, Title 46; South Dakota Administrative Rules, Article 74:02).

Wyoming water law, with some exceptions, is based on the doctrine of prior appropriation. The exceptions are that (1) municipalities can acquire water rights through the power of eminent domain if proper procedures are followed and the water-right holder is compensated, and (2) domestic and livestock use customarily have precedence over other water users. The State Engineer administers state laws and regulations related to water-right permits. The State Engineer generally issues water rights to anyone who plans to make beneficial use of the water. Recognized beneficial uses include, but are not limited to, irrigation, municipal, industrial, power generation, recreational, stock, domestic, pollution control, and instream flows. Water-rights holders for groundwater are not limited to a specified amount of water (Wyoming State Engineer's Office 2005).

8.4.3 Issues in the Management of Transboundary Aquifers

Transboundary groundwater refers to a continuous groundwater reservoir (generally an aquifer) that underlies or whose water flows beneath two or more political jurisdictions and can be exploited by each jurisdiction (Campana 2005). Even though the jurisdictions are not necessarily different nations, here it refers to internationally shared aquifers. To promote studies and joint international cooperation in developing policy for the management and sound utilization of transboundary aquifers, the International Association of Hydrogeologists (IAH) established a Commission on Transboundary Aquifer Resources Management (TRAM). The International Shared Aquifer Resources Management (ISARM) program was endorsed by the Intergovernmental Council of the UNESCO International Hydrological Program (IHP), at its Fourteenth Session held in June 2000 with support of UNESCO and IAH in co-operation with the United Nations Food and Agricultural Organization (FAO) and the United Nations Economic Commission for Europe (UNECE). One of the drivers of the ISARM program is to support cooperation among countries to develop their scientific knowledge and to eliminate potential for conflict, particularly where conceptual differences might create tensions. It aims to train, educate, inform and provide inputs for policies and decision making, based on good technical and scientific understanding (Puri and Aureli 2005).

The ISARM Program has identified five key focus areas that require attention for sound development of transboundary aquifers (Puri 2001). These include scientific-hydrogeological, legal, the socio-economic, the institutional and the environmental aspects. Some transboundary aquifers contain huge freshwater resources, enough to provide safe and high quality drinking water for the needs of all humanity for many tens of years (Puri 2001). Many of the largest internationally shared aquifers have not yet been intensively used, while many of the smaller internationally shared aquifers are being used intensively. However, the effects of use are currently contained within national boundaries. Concern is increasing that conflicts will eventually arise over the use of shared aquifers as the competition for water increases.

In many places the rush to gain access to water from a shared resource may result in its inefficient use. Additionally, the consequences of lowered water tables on ecosystem function and on the long-term sustainability of the resource may be ignored. The ISARM program recognizes that sound utilization and management of internationally shared aquifers requires a multidisciplinary approach, which includes not only a hydrogeologic understanding, but also an understanding of the legal, socioeconomic, institutional, and environmental issues of all the countries sharing the groundwater resource. Since its inception, several regional initiatives hasve been undertaken, including the ISARM-Americas Program (coordinated by the Organization of American States); the ISARM-Europe Programme (coordinated by the UNECE), and ISARM-Balkans Program (Coordinated by UNESCO-INWEB). Similar ISARM initiatives for Africa, Latin America, and the Arab states have been proposed (Puri and Aureli 2005). The experience gained from ISARM activities in several regions confirms that there would be considerable value in developing a multidisciplinary toolkit, or a guidance package, that might consolidate the best practice, guidance, and information on transboundary resources management. A flowchart representing the approach to development of the toolkit us shown in Figure 8.7 (Puri and El Naser 2003; Puri and Aureli 2005).

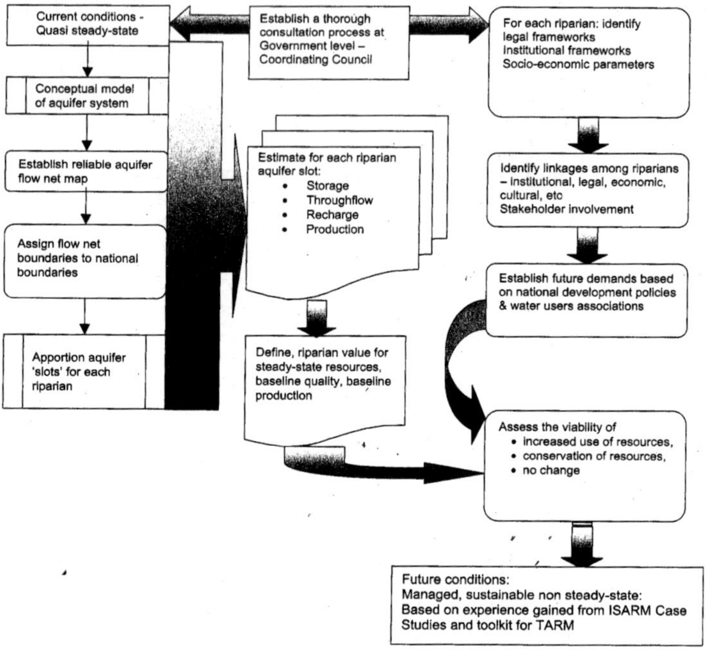

Figure 8.7: Suggested development strategy for ISARM case studies – toward a toolkit (Puri and El Naser 2003).

Geopolitical Issues of Groundwater Management across Transboundary Aquifers – Middle East Case Study: In addition to the current political issues in the Middle East, water scarcity remains as an underlying barrier to the economic development of the region. Since the 1970s the region as a whole has been in net deficit, with nation-states consuming more litres per year than the indigenous sources can supply. These sources include limited precipitation, groundwater and, most importantly, the river systems. Governments are, on the whole, currently able to meet basic household needs. However, this constitutes only 10% of the total amount of water required by the economies of the region. Since the 1970s governments have struggled to provide the remaining 90% necessary to sustain agricultural self-sufficiency (Jobson 2003).

Water scarcity in the Middle East is nothing new. History tells an ongoing story of water management complexity, including the development of ingenious storage, transfer and irrigation systems. However, recent population growth, the development of industry and new consumer demands have put pressure both on finite supplies themselves and on those tasked with managing the supplies within the states of the region. Tensions between countries have also intensified. Most states rely on transboundary river systems and underground aquifers that extend across political boundaries (Jobson 2003). Such water resources are often a contentious issue requiring negotiation between states. As their needs grow, the development of water resources by one state has increasingly impacted on those of others. Indeed, hydropolitics is today one of the most charged and complex geopolitical issues facing the region. What is notable, given the intensity of demand and exceptional implications of water scarcity, and despite recurring alarm calls and references to impending "water wars", is that the region as a whole has avoided outright armed conflict over this issue. Exceptions include hostilities between Syria and Israel in the 1960s over water extraction from the upper Jordan tributaries. It is hard to overstate the significance of water within the region. Projections of future use and supply are worrying. Since the 1970s states have been largely unable to meet their strategic water needs for industrial use and food production.

A range of strategic responses to this regional water crisis has been attempted, including (Jobson 2003):

i. Intrastate management policies, such as improving water storage facilities and distribution networks, and the allocation of water where it can be used most efficiently in economic terms;
ii. Improvements in the productivity of irrigation systems, the recycling of wastewater and the use of desalinated water, which all suggest that nations/states are looking to preserve the old sources of water and to find new ones;
iii. Bilateral precedents for the management of transboundary river systems within the region and possible scope for multilateral agreements not only manage such basin systems but also to facilitate the long-distance transfer of water between basins.

In addition to the above responses, the concept of "virtual water" has been introduced. Since the 1970s the governments of the water-scarce states of the Middle East have gradually realized that the water supplies of the region are finite and inadequate. Unable to harness the massive quantities of water needed to sustain self-sufficiency in agriculture and food production, most states began importing the commodities that were the most water intensive to produce, such as grain. In so doing, they were essentially importing "virtual water" – the water embedded in water-intensive commodities – and thereby freeing up their own scarcer water resources for domestic and industrial use, both of which have a higher economic return than agriculture. The supply of virtual water has ensured regional water and food security in the Middle East since the 1970s. This in turn has allowed the lack of self-sufficiency in water to be de-emphasized in national and regional politics. In 2000 the Middle East as a whole imported about 25% of its requirements as virtual water; this is forecast to rise to 50% by 2025. It is clear, therefore, that virtual water has, and will continue to have a huge role in remedying the Middle East's severe water shortage.

8.5 Conjunctive Use of Surface Water and Groundwater

Often in the past, individuals and agencies responsible for developing water supplies have ignored groundwater in their planning. This approach to water development ignores the requirement that successful management of the world's water resources must consider the integrated use of both surface water and groundwater. Sustainable development of the world's water resources cannot be achieved merely by building more dams, reservoirs, and canals, or merely by pumping more groundwater. Successful management demands full consideration of all implications of aquifer development, water quality, hydrologic and geochemical aspects, and, where appropriate, the coordinated operation of surface water and groundwater supplies and storage capacities.

Regions across the southwestern U.S. face the challenge of dwindling surface water and groundwater supplies. Under such a scenario, conjunctive use of surface water and groundwater has become imperative. In some areas, farmers are pumping groundwater to make up for the shortage of their allotted surface water volume. In order to develop effective systems that use groundwater and surface water conjunctively, surface water conveyance channels and the underlying groundwater reservoirs must be considered as interconnected hydrologic units.

8.5.1 Conjunctive Use of Surface Water and Groundwater as a Strategy for Basin Water Resources Management

Conjunctive use of surface water and groundwater requires combining both sources of water to optimize the water demand/supply balance, while minimizing the undesirable physical, environmental and economical impacts. Usually conjunctive use of surface water and groundwater is considered within a river basin management program, i.e., both the river and the aquifer belong to the same basin. These river basins serve as a spatial basis for the formulation of water management strategies integrating all cross-sectorial issues such as water resources conservation, environment, water resources allocation, water demand management, etc. The conjunctive use of surface water and groundwater is one of the strategies of water supply management that should be considered to optimize the development, management and conservation of water resources within a basin, and artificial recharge of aquifers is certainly one of the tools to be used for that purpose.

The use of the river basin as the spatial unit for analyzing the interactions and interrelations between the various components of the system, and for defining the water management policy, is well justified, and is increasingly becoming common practice:

i. In China, river basin plans have legal status and development projects are required to be consistent with the provisions of the plans.

ii. In Indonesia, the government recently adopted new water management policies in order to prepare spatial management plans and to link water and land use through river basin plans, to center water management at the river basin level, and to centralize water management responsibilities through a more effective participation and collaboration of beneficiaries.
iii. In Italy, a 1989 law introduced the river basin as a management unit to regulate the programs of the various sectorial and regional institutions.
iv. In the United Kingdom, Spain, France, and in most European countries, water resources management is now essentially centered on river basins.
v. In the southwestern region of the U.S., comprehensive management alternatives are being developed to use conjunctively surface water and groundwater.

The adoption of an integrated river basin management approach for establishing policies and strategies of water resources development, management and conservation facilitates an integrated management approach of surface water and groundwater one system and avoids a water resources development approach focused only on surface water. This approach also facilitates the management of the resource itself, allowing a better understanding of the hydrological issues involved.

Assuming that conjunctive use is accepted policy, several hydrologic aspects need to be carefully studied before selecting the different options and elaborating a program of conjunctive use of surface water and groundwater:

i. Underground storage availability;
ii. Production capacity of the aquifer(s) in term of potential discharge;
iii. Natural recharge of the aquifer(s);
iv. Induced natural recharge of the aquifer(s); and
v. Potential for artificial recharge of the aquifer(s).
vi. Comparative economic and environmental benefits derived from the various possible options.

In order to use the underground reservoir to store a significant volume of water - possibly of the same order of magnitude as the annual runoff with the intent to use it at a later stage, it is necessary to ascertain the potential storage capacity of the groundwater reservoir as well as its suitability for being recharged by the surface water and for easily returning the stored water when needed. The groundwater reservoir should present sufficient free space between the ground surface and the water table to accommodate and retain the water to be recharged, for the period during which water is not needed. This condition requires accurate hydrogeological investigations including geological mapping, geophysics and reconnaissance drilling, in order to determine the configuration and the storage capacity of the underground reservoir.

Should a significant natural recharge of the aquifer occur from the surface runoff and the deep percolation of rainfall, and should the average annual amount of recharge be of the same order of magnitude as the water demand, there would not be the need for any additional human intervention. On the contrary, any tentative modification of the natural course of surface water may significantly alter the groundwater renewable resources. There are many examples of double counting of water resources as if surface water and groundwater were independent. Because of this wrong approach, dams have been built with the intent to store the surface water and to create an additional resource. The resulting situation has often been catastrophic with rapid depletion of aquifers not longer being recharged, destruction of the ecosystem based on the groundwater and extreme difficulty to go back to initial conditions.

Induced natural recharge occurs when intensive exploitation of groundwater close to a river results in an important depression of the groundwater level and subsequent water inflow from the river. This phenomenon is well known in temperate climates where rivers flow all year long; but it may also occur in semi-arid climates where a depression of the piezometric level of an aquifer underlying an ephemeral river facilitates aquifer recharge during floods.

8.5.2 Design of Conjunctive Uses of Surface Water and Groundwater

Successful conjunctive use of surface and subsurface water resources combines the groundwater and surface water facilities to the extent possible, allowing their coordinated use. The coordinated operation of a groundwater basin involves the use of a basin for transmitting and storing varying portions of the area's water supply by coordinating the aquifer functions with surface facilities, such as reservoirs and pipelines, to meet the water requirements of an area.

Before establishing the coordinated operation of the groundwater basin, an analogy between physical characteristics of the groundwater basin and surface water system should be described. The rate of deep percolation and subsurface flow into the groundwater reservoir is, roughly analogous to the rate of inflow into a surface reservoir. The storage capacity of the groundwater basin is analogous to the storage capacity of the surface reservoir. The transmission characteristics of the aquifers in the groundwater basin may be compared to the characteristics of the distribution system in the area of use. Finally, the potentiometric surface and water table in the groundwater basin are analogous to the hydraulic grade line in the surface distribution system. By use of equations that numerically describe the flow characteristics of the groundwater basin and surface distribution network, it is possible to calculate and integrate capabilities of these water delivery media to meet the physical objectives.

Facilities must be provided so that the largest flow rate requirement is satisfied at all locations and at required pressures. In this way, all lesser flow rate requirements will be satisfied. In brief, the coordinated operation of a groundwater basin must include determining the changing requirements for surface water and groundwater supplies and facilities to meet the largest flow rate requirement of each year. In instances where the water use of an area is met by both imported and locally pumped groundwater, changes in the amount of use of water from one source affects the amounts required from the other. At the same time, capacities of facilities for delivering imported water through the surface system also affect the capacities of booster pumps and storage facilities of the local distribution system required for regulating the water supplies to meet the fluctuating water use. The cost of operation depends on the number and type of facilities provided, the manner of operating these facilities; and the amounts of water used from both imported and groundwater sources.

In developing a water-supply system for municipal and industrial purposes, it is considered good practice to utilize more than one source of supply. In rapidly developing areas where there are space limits and use of water is high, operation of a groundwater system in coordination with a surface water supply is mandatory. The proper balance between use of a groundwater system and supplementary surface features for storage and transmission for any given region will be unique to each area and a function of the amounts of water and space available. A coordinated system of supply using groundwater and surface water is desirable, because it provides the means of capitalizing on the inherent advantages of each, which are hydrologically complementary (Ryder 1978).

Generally, for practical and economic reasons, groundwater supplies are used in the early stages of the development of an area, for either agriculture or some limited urban use. As the transition from rural to urban use and industrialization occurs, and the requirements for water supply increase and become more complex, a supplemental imported supply usually is needed. By establishing traditions of coordinated use of all available supplies as they occur or as they become available, the manager will greatly enhance the opportunity for optimization of the total water system.

The manager must evaluate the effect that variations in operational plans will have on existing and future import and distribution facilities; how implementation of the plan will affect the staging, or itself be influenced by the timing, of future import facilities; and the degree of influence each additional water-supply source, such as reclaimed wastewater and saline water conversion, will have on the overall plan.

Sufficient emergency storage should be provided within the groundwater basin so that any prolonged failure of importation facilities will not seriously jeopardize the local water supply of the area. Prior to developing any plan, it is advantageous to project conservative estimates of future water-supply conditions, and not to depend on nature to provide mean or above-mean water supply at all times. Plans of operation should not cause undue hardship to individuals or inhabitants of local areas within the water-supply service district without providing some economic relief.

In addition to utilizing the underground reservoir in conjunction with an imported water supply, the basin can be coordinated with other sources of surface water to control both quantity and quality. Development of surface reservoirs can be undertaken to store normal and flood flows for later artificial recharge of the groundwater basin Reservoirs located in the mountainous area upstream of the basin will require downstream spreading facilities. In California, for example, Santa Felicia Dam in Ventura County and Twitchell Dam in San Luis Obispo County were constructed to store water for later recharge of aquifers. Reservoirs constructed overlying groundwater basins and built without linings or cutoffs will also provide additional direct recharge to the groundwater basin. Small dams have been constructed in southern Texas on water courses to enhance the recharge of the Edwards limestone. Wadi Hanifah and Wadi Nimar dams in Saudi Arabia provide similar recharge capability.

The aquifer system can also be used as a treatment facility. Imported and local surface water usually must be treated prior to delivery to the customer. In lieu of passage through a treatment plant and direct delivery, these waters can be spread and later repumped, thus utilizing the groundwater basin as a treatment and transmission facility. This is not to say that wastewater containing organic or toxic chemicals can be adequately treated in this manner. The groundwater supply can be contaminated by infiltration of surface waters containing hazardous materials.

Numerous facilities have been constructed for flood control with the additional benefits of conservation of flood flows through artificial recharge. Projects constructed by the U.S. Army Corps of Engineers and local flood-control districts fall into this category. There are also many methods of improving the environment through development of a water-resources management system. Among these is the development of recreational lakes associated with artificial recharge projects. These replenishment facilities can provide swimming, boating, and fishing, and, with proper beautification by planting trees and appropriate landscaping, become attractive regional parks. Replenishment water, in certain areas, can be released in intermittent streambeds to rejuvenate ecosystems by providing running water year round, and thereby enhance a naturally depressed environment.

Conjunctive use of the groundwater system reduces the need for surface storage tanks and surface distribution systems, which can be unsightly. Unlined channels should be encouraged wherever possible instead of completely lined channels. They not only provide increased natural infiltration, but also offer a more natural setting. Landscaping, hiking, riding trails, and other recreational facilities can be provided adjacent to storm and recharge channels for public use. To reduce noise pollution, submersible pumps and underground facilities can be substituted for the use of surface pumping facilities.

8.5.3 Managed Aquifer Recharge

The optimum use of available surface water and groundwater resources is an essential element of any comprehensive plan for water-supply development. In many instances, underground storage capacity can be used for cyclic storage of conserved surface water supplies, both local and imported. Managed Aquifer Recharge (MAR), denoting the planned introduction of surface water and/or recycled water into permeable underground formations through aquifer storage and recovery (ASR) wells and infiltration basins, may be accomplished independently of, or in conjunction with, naturally occurring infiltration and percolation to permeable aquifers (CSIRO 2005; EPA 2005). MAR offers a potential solution to many of the complex problems associated with restoring and operating groundwater reservoirs, and of preventing or minimizing overdraft and its attendant adverse consequences.

8.5.3.1 *Artificial recharge, aquifer storage and recovery system, and managed aquifer recharge*

Even though artificial recharge has been studied and widely used for water resources management around the world for almost half a century, the ASR concept was introduced about three decades ago (ASCE 1987; Johnson and Pyne 1994; ASCE 2001; Bouwer 2002; Dillon 2002; Sheng 2005; NRC 2008). Artificial recharge is the deliberate process of replenishment of the water in groundwater storage through works provided primarily for that purpose. It should be noted that intent is an essential element in this definition. Many projects designed for disposal of return irrigation waters, sewage, cooling water, or other wastes will also augment groundwater supplies. However, groundwater replenishment is generally incidental to the primary function of these works, and they are not considered true artificial recharge projects.

Many ASR systems have been implemented across the U.S. (Pyne 1995; ASCE 2001). The primary purposes of most MAR activities are to: (1) combat adverse conditions resulting form overdevelopment of groundwater; (2) increase conservation of local and imported waters; (3) increase the quantity, or yield, of groundwater available for use; and (4) allow the groundwater basin storage to be used for peak use periods. Specific objectives include the following (ASCE 1987; Pyne 1995; NRC 2008):

i. Augmenting natural recharge to the groundwater reservoir to compensate for human activities that tend to reduce natural recharge. These activities include: (a) lining of stream channels for flood protection or erosion control; b) collecting sewage and industrial wastes of suitable water quality and discharging them to the saline waters or out of the groundwater basin; (c) sealing of natural recharge areas with impervious sidewalks, streets, airports, parking lots, and buildings; and (d) diverting surface water supplies that might otherwise infiltrate naturally in stream channels.
ii. Coordination of surface water and groundwater reservoir operation.
iii. Reducing groundwater overdraft.
iv. Mitigating problems of adverse salt balance, controlling seawater intrusion and local bodies of poor-quality water, and eliminating problems of groundwater pollution.
v. Maintaining or raising water levels to avoid deepening wells, increased pumping costs, and obsolescence of equipment.
vi. Reducing or stopping critical land surface subsidence problems in areas where draft results in compaction of sediments.

vii. Storing off-season water supplies in groundwater basins for use during peak water demand periods.

Artificial replenishment of groundwater is accomplished primarily through works designed to maintain high infiltration capacities, increase the wetted area, and lengthen the period of infiltration beyond that which exists under natural conditions. Although no two projects are identical, most utilize variations or combinations of basins, modified streambeds, pits, ditches and furrows, flooding, and injection wells. Upstream surface reservoirs are often provided to regulate flood flows, temporarily store floodwaters, limit flows to the absorptive capacity of the spreading works, and remove a large portion of the detritus.

Another method of artificial recharge utilizes the operational flexibilities of water-supply systems that have surface water supplies available. This is the "in lieu" method. The "in lieu" method leaves water underground which would have been extracted by supplying surface water directly to the agency or individuals who would have extracted the groundwater. Not taking water out of the ground accomplishes in a very efficient and effective manner the same thing as putting it underground.

For storage purposes, underground reservoirs have many advantages over surface reservoirs (ASCE 1987):

i. They cost nothing to construct. However, there is a cost for installing wells and pumping the water out for use, and a cost for maintenance of wells and pumps.
ii. They do not silt up, but surface siltation of replenishment areas may significantly curtail infiltration rates.
iii. They have no evaporation losses, except in areas of shallow water table where vegetative use of water may be high and saline residues may accumulate.
iv. They have generally relatively uniform temperature and mineral quality; however, there may be large variations from place to place within the reservoir.
v. They are free of turbidity, except in some limestone areas and volcanic areas with high secondary permeability, and where large underground conduits occur.
vi. They do not preempt the land surface that is useful for other purposes.
vii. They are not subject to eutrophication.
viii. They have indefinite life if properly managed. However, indiscriminate overdevelopment or poor practices can dewater aquifers and cause compaction of sediments, resulting in land surface subsidence, and degradation of water quality.

An aquifer is generally suitable for recharging provided that it has the following characteristics:

i. the surface material is highly permeable so as to allow water to percolate easily;
ii. the unsaturated zone has a high vertical permeability, and vertical flow of water should not be restrained by less permeable clayey layers;
iii. the depth to water level is not less than 5 to 10 m; and

iv. the aquifer transmissivity is high enough to allow water to move rapidly from the mound created under the recharge basin but not too high (as in karstic channels) so that water cannot be recovered. An adequate transmissivity for recharge is also a good indicator of the aquifer capacity to have high well yields that allow efficient recovery of the water stored.

The ASR concept was first introduced as one type of groundwater recharge by Pyne (1995), in which ASR is defined as the storage of water in a suitable aquifer through a well during times when water is available, and recovery of the water from the same well during times when it is needed. Sheng (2005) expanded the definition such that an ASR system is defined as a framework in which treated or untreated surface water as well as reclaimed wastewater is stored in a suitable aquifer through spreading basins, infiltration galleries and recharge wells, and part or all of the stored water is recovered through recharge wells themselves or neighboring production wells, or by increasing the base flow in the stream as needed. Besides dual-functional wells, the recovery component of the ASR system also uses neighboring production wells as recovery wells, and adjacent streams as receivers of increased discharge from surrounding aquifers. Such configuration integrates production of native groundwater with recovery of stored water as well as marginal quality water that otherwise may not be used directly. Using this approach, all subsystems interact with each other and depend on each other. For example, stored source water should be compatible with native groundwater so that resulting water does not deteriorate the underground storage space, and screens of production and recharge wells do not become clogged by adverse geochemical reactions such as precipitation of calcium carbonate (calcite), and iron and manganese oxide hydrates (Pyne 1995; ASCE 2001; Bouwer 2002). The stored water may also cause swelling or dispersion of clay particles in the aquifer, which can reduce the storage space and the hydraulic conductivity of the aquifer (Pyne 1995; ASCE 2001; NRC 2008). The influence of injection on existing production wells cannot be ignored or overlooked because it becomes one of the important components in the design, construction and operation of an ASR system.

MAR has the potential to be a key component in many integrated water management systems; providing storage, and in some cases, water treatment. Treated wastewater and stormwater are increasingly being used as sources for MAR (CSIRO 2005; EPA 2005). Demand for water also needs to be managed to balance the recharge to the aquifer, be it natural, managed or incidental. Demand management can take many forms including more efficient use, limiting the depth of wells, education, moves to a lower water use economy (i.e., away from irrigated agriculture) as well as fiscal controls.

8.5.3.2 *Design and operation of managed recharge projects*

An MAR system is a water management tool that helps economically sustain recoverable water supply by redistributing available water over a period of time and hydrologically suitable space. Such system consists of four primary elements: source water to be stored, storage space (aquifer), recharge facilities accompanied by necessary delivery pipelines and/or channels, and recovery facilities with an adequate distribution network (Sheng 2005). The uses of the recovered water define purposes of an underground storage of recoverable system or MAR system (NRC 2008). The MAR system integrates these elements and also includes different recharge methods, namely spreading basins and infiltration galleries, irrigation networks, seepage from canals and laterals, infiltration from agricultural irrigation, and run-off collection ponds, injection wells and ASR wells (Bouwer 1991, 2002; Johnson and Pyne 1994; Barnett et al. 2000; Dillon 2002; Sheng 2005).

MAR projects should be designed and constructed to provide the desired utility at a reasonable cost with due allowance for environmental concerns. The project concept may range from a complex plan carefully prepared and executed to an informal project developed in the field. In general, the more complex project will be most suitable for the capture, diversion, and spreading or injecting of a more or less continuous supply of water in highly urbanized areas, while simplified projects are appropriate for a predominantly agricultural or undeveloped area. In most cases, one, or a combination, of the basic methods of artificial recharge can be adapted to meet the needs of a project. Operations are generally classed as basin, modified stream bed, pit and shaft, ditch and furrow, flooding, and injection well recharge projects or methods.

Generally, facilities are required to: (i) divert water from a source of supply; (ii) upgrade the quality of water; (iii) convey water to the recharging area; (iv) contain and control the flow of water within the spreading or recharge area; (v) provide efficient operation and maintenance of the project; and (vi) convey excess water from spreading area back to the stream. In addition, it is desirable to install measuring devices to determine the quantity of water recharged and the overall efficiency of the recharge operation. It is desirable also to monitor the impact of the project recharge upon the basin water table and groundwater quality.

The detailed design of project facilities is based upon hydraulic and hydrologic principles, with due regard to structural limitations of available construction materials. In all instances, advantage should be taken of available natural conditions, including native material, vegetation, soil conditions, and topography at the project site. When construction is on soils of fine texture, minimum disturbance of the existing soil profile should be a goal of earth-moving operations, unless the fine materials can be removed completely.

8.5.3.3 *Source water consideration*

The method and, in part, the degree of utilization of the groundwater reservoir depend not only on the source of replenishment but also on the characteristics of this source. The source may be rainfall, surface runoff, return flows from applied water (irrigation and waste disposal), imported water spread and stored for regulatory or replenishment purposes, or reclaimed wastewater. Characteristically, the supply may range from a relatively uniform flow obtained from controlled releases from a reservoir project to uncontrolled discharge that may occur for only short periods at erratic and possibly protracted intervals.

Under conditions for which water supply for replenishment is relatively uniform in time, only short-term underground storage is normally required. Selection of the location of extraction facilities normally relates to the place of use, with due regard to the subsurface aquifer conditions and the areas where replenishment can and will most readily occur. However, the amounts of water extracted should be balanced to prevent overdevelopment of any part of the basin with consequent adverse effects.

In areas where the surface water supply is extremely variable, extensive development of groundwater is advantageous and extraction facilities should be placed to permit maximum utilization of available underground storage capacity. The placement of facilities to accomplish this must consider the location and regime of replenishment operations in the area. The most permeable areas, where the greatest quantities of extraction and recharge can occur, should be given preference within the limits permitted by other factors. In this manner, a maximum amount of water can be stored during the short period of time when water is available.

The NRC (1998), in a report, entitled "Issues in Potable Water Reuse," concluded that reclaimed wastewater can be used to supplement drinking water sources, but only as a last resort and after a thorough health and safety evaluation. Municipalities first must fully assess health impacts from likely contaminants and develop comprehensive systems for monitoring, testing, and treatment. Other water sources and conservation measures also should be tried to the extent practical, before turning to reclaimed wastewater. When considering reclaimed wastewater for public water supplies, the NRC (1998) concluded that the public distinguishes between direct and indirect use. Adding highly treated wastewater directly into a water supply without storing it first in a reservoir is usually not a viable option. Indirect use, however, was determined to be generally viable. Indirect use augments the drinking water supply by adding reclaimed treated water first to a lake, reservoir, or underground aquifer. The mixture of natural and reclaimed water is then subjected to normal water treatment before it is distributed as drinking water for the community. Since the 1960s, Los Angeles County in California has operated an indirect-use system in which wastewater, mixed with stormwater and surface water, supplies about 16% of total flow into groundwater basins. This mixture then is used as a source for drinking water supplies. The NRC (1998) describes other reclaimed-water projects currently operating in the U.S., including those supplying northern Virginia, Orange County (Los Angeles County Flood Control District 1951), California, and Phoenix, as well as feasibility studies conducted by the cities of San Diego and Tampa. Limited data from projects and studies nationwide show that highly-treated reclaimed wastewater produced drinking water of excellent quality, and that no obvious health effects have been found in animal tests or in communities where reclaimed water has been used.

8.5.3.4 *Storage characterization*

Storage is one of the key components in feasibility analysis and operation of an MAR system. To select storage locations, following factors should be evaluated.

i. Surface hydrologic considerations (ASCE 1987): The quantity, quality, and character of the supply to be infiltrated should be carefully evaluated. In addition to the selection of a site and method of artificial recharge, the amount of water available for recharge has direct bearing on the need for, and size of a project.

The water supply available may vary widely within a short period of time or may remain almost constant for a long period. This can usually be ascertained from a review of recorded streamflow for the selected supply source. Generally, highly variable streamflows are problematic for project operation, as large volumes of water containing high concentrations of silt and debris must be handled within a relatively short period of time. Operational criteria may also require bypassing flows of unacceptable quality if infiltration rates can be maintained only by other means.

Any of the spreading methods can be used for replenishment when the source is a fluctuating flow. Excessive fluctuations can best be handled by increases in the storage capacity of the project. The basin and pit techniques have the advantage of providing temporary storage capacity, dependent, of course, on the area covered and depth of water. At the other extreme, injection wells have very little storage capacity and require a rather uniform supply of water of high chemical, biological, and physical quality.

Water quality, primarily with respect to silt and clay content, has an important bearing on site selection and recharge method. This is particularly true with regard to injection well methods, where almost all silt and clay must be removed for successful operation. Quality of water used in connection with deep pits or shafts, with no facilities for desilting, should be nearly as good. Relatively good quality water is desired or basin projects. On the other hand, properly designed ditch and furrow projects have been used advantageously with waters of high silt and clay content.

ii. Geologic and subsurface hydrologic considerations (ASCE 1987): Where surface recharge methods are to be used, the selection of a proposed artificial recharge site or the evaluation of the effectiveness of an existing site should be based on a detailed knowledge of the geologic conditions at the site. Physical properties of surface and subsurface deposits determine, to a large extent, the sustained infiltration capacities that can be maintained, and the volumes of water that can be recharged to the groundwater body. In addition, location of geologic structures and subsurface deposits that might form a complete or partial barrier to groundwater movement, and the position and hydraulic gradient of the existing water table or potentiometric surface, affect the efficiency and suitability of an artificial recharge project.

Infiltrating water first must pass through the surface deposits prior to reaching the water table. Therefore, the properties of the various soil horizons are critical in the establishment of initial and sustained infiltration capacities. The more important chemical, physical, and biological properties of the soil include: (i) texture; (ii) permeability; (iii) clay-pan and iron or lime-hardpan development in the subsoil horizon; (iv) depth of soil profile; (v) organic matter; (vi) degree of compaction due to use of construction equipment; and (vii) the chemistry of the interaction between soil and recharge water. In general, the coarser the texture of the surface deposits, the higher are the initial and sustained infiltration capacities. In many cases, development of clay-pan and iron-hardpan layers in the subsoil horizon is the major limitation to maintenance of high sustained capacities. In the western U.S., recharge activities are conducted primarily on unconsolidated alluvial fan, flood or coastal plain, and inland alluvial valley deposits.

Aside from the influence of the soil horizons, the determination of sustained infiltration capacities, maximum storage capacity, and effective transmission rates of groundwater from beneath the recharge site to other parts of the groundwater basin, is based on various physical and structural features of the underlying subsurface deposits. The important features of the subsurface deposits that must be evaluated include:

i. The physical character and permeability of subsurface deposits above the water table. This determines the allowable sustained infiltration capacity.
ii. The permeability, specific yield, thickness of the deposits, and position and allowable fluctuation of the water table. These factors establish the subsurface water storage capacity in the vicinity of the project.
iii. The transmissivity of the aquifer and the hydraulic gradients of the water table. These factors affect the rate of movement of groundwater from spreading sites to areas of groundwater draft.
iv. The underground structural and lithologic barriers existing in the area. Such barriers will form an impediment to both vertical and lateral movement of groundwater.
v. The chemistry of the native groundwater in relationship to that of the recharge water must be considered, to minimize problems associated with possible chemical reactions or ion exchange phenomena.

iii. Infiltration capacity (ASCE 1987): One of the major problems associated with the evaluation of an artificial recharge site is the determination of infiltration capacity. Such capacities are critical, for they dictate the method and size of the recharge site, and the operation and maintenance techniques to be utilized. Many factors affect infiltration capacities at artificial recharge projects. Natural factors include the composition of surface soils, and the geologic and subsurface hydrologic conditions discussed previously. The quality of the water and the procedures used in the construction, operation, and maintenance of projects also affect infiltration capacities, although these factors can ordinarily be managed to maintain favorable capacities.

Infiltrometers and small experimental plots have been used successfully to determine the infiltration characteristics of the surface soils at a site. For a given quality of water and type of material, infiltration capacities are affected by the length of recharge period, size of test plot, condition of soil, and length of drying period for the spreading area. Care should be taken in extrapolation of preliminary test measurements to estimate of potential project performance. In general, it has been found that the initial infiltration capacity will decrease rapidly, followed by an increase to nearly the initial value, after which a gradual continuous reduction will occur until basins are taken out of service for drying and maintenance. The familiar S-shaped curve of infiltration capacities is explained as follows (Muckel 1959):

i. The initial decrease in permeability or infiltration capacities is believed to be caused by dispersion and swelling of the soil particles. This is much more pronounced in some soils than others.
ii. The increase in permeability following the initial decrease accompanies the elimination of entrapped air from the soil. This air is slowly dissolved in the water passing through the soil.
iii. The gradual decrease in permeability that follows is due primarily to biological activity in the soil.

Considerable data on infiltration capacities in various types of artificial recharge projects throughout the world are available. In most instances, however, correlation of capacities is difficult to establish owing to sparse information regarding specific factors, such as geologic and hydrologic conditions, type and size of project, method of treating recharge waters, and related items. Since infiltration capacities vary with time, the long-time infiltration capacity has been selected to exemplify the effects of natural ground slope, surface soils, and physiographic positions. The long-time infiltration capacity is defined as that capacity which will exist after spreading water for a period of from two to four weeks, depending upon the character of the surface and subsurface conditions.

The soil type may be used as a guide in obtaining an estimate of the infiltration capacities to be expected for a project site. In general, it has been found that the coarser the texture of the soil, the higher the initial and sustained infiltration capacity. Soil types may be identified from available soil maps, or by field reconnaissance supplemented by hand auger borings to determine the soil profile. Considerable data are available regarding long-time infiltration capacities for soils found on alluvial fans, in recent stream channels on alluvial fans, and on floodplains. In general, for a given natural ground slope, the long-time infiltration capacity is relatively higher for projects located in recent stream channels on alluvial fans than for projects in other portions of alluvial fans. The long-time infiltration capacities of projects located on floodplains are relatively high, which may be due to a high standard of maintenance for the projects studied.

After the applied recharge water has passed through the soil zone, geologic and subsurface hydrologic conditions control the sustained infiltration capacity, because layers of less permeability will become the controlling hydraulic resistance in the profile and override the initial surface restriction state. Subsurface conditions should be thoroughly investigated to determine the total usable storage capacity and the rate of movement of water from the spreading grounds to the areas of groundwater draft.

8.5.3.5 *Recharge facilities*

During the initial phase of a groundwater resources management study, the investigators should inventory the existing groundwater extraction and distribution systems and the existing artificial recharge projects, if any. These data, together with information on the capability of modifying the existing system, are essential to the planner in subsequent phases of the study when alternative management plans are devised. The inventory will answer the question: what existing facilities are available for withdrawing water from and adding water to the saturated zone?

The relationship between existing recharge facilities and the water supply system as regards capacity and location is fundamental in the preparation of plans for the utilization of groundwater resources in meeting future water demands. A detailed inventory of existing facilities is needed and should include the evaluation of such characteristics as: (1) short- and long-term recharge rates; (2) capacity of delivery systems; (3) limitations on capacity resulting from seasonal use for recharge of local water; and (4) the ability of the underground formations to accept and transmit recharged water to the areas of pumpage. A thorough evaluation of existing capabilities will provide a base for a wide range of alternative management plans and will help to establish requirements for new to improved facilities necessary to satisfy management objectives.

Artificial recharge of aquifers can be achieved using three different methods, namely surface spreading, watershed management (water harvesting) and recharge wells (ASCE 1987). Each of these methods is described in more detail below.

1. Surface spreading (ASCE 1987): The surface spreading method consists of increasing the surface area of infiltration by releasing water from the source to the surface of a basin, pond, pit or channel. This is certainly the most efficient and most cost-effective method for aquifer recharge. However, only phreatic (unconfined) aquifers can be recharged by the spreading method, which also requires large surface areas to accommodate the recharge scheme, allowing water to evaporate if percolation in the ground is slow. Surface spreading usually needs two structures: the diversion structure and the infiltration scheme.

Diversion structures are the same as those used for spate irrigation. The traditional methods, based on centuries of experience, are well adapted to the conditions of arid land wadis. They consist of the construction of earthen bunds (ogmas) and deflectors across the wadi to divert the flow into the fields. But large spates usually destroy the ogmas and reduce irrigation of the fields. Furthermore, the very high sediment content of spate water tends to fill the diversion canals, which have to be cleaned regularly. So, although the ogmas are relatively inexpensive to rebuild, the overall cost of seasonal maintenance and repair of the scheme is high.

Permeable soils and particularly coarse alluvial deposits have vertical hydraulic conductivity ranging from 1 m/d (3.28 ft/d) (fine loamy sands) to 10 m/d (32. ft/d) (sand and gravel mixes). However, because of clogging, the actual infiltration rate of recharge basins generally varies from 0.3 to 3 m/d (0.98 to 9.8 ft/d). The quantity of water infiltrated obviously varies with the duration of the source of water which may be limited to a few days per year if no surface water storage facility exists. In any case, the total duration of the infiltration process will seldom exceed 100 days per year because, if a surface reservoir controls the floods, infiltrating water in the ground would be competitive only if it can be achieved in less than 100 to 150 days. The volume that one can reasonably expect to be infiltrated in basins will therefore range approximately from 100,000 to 3,000,000 m^3/yr/ha.

The infiltration scheme may consist of basins, channels or pits depending on the local topography and on land use. The most common system consists of a number of basins each one having an area ranging from 0.1 to 10 ha (1,196 to 119,600 yd^2) according to space availability. Each basin must have its own water supply and drainage so that each basin can be flooded, dried and cleaned according to its best schedule. Basins should never be in series, because in such a system, they cannot be dried and cleaned individually. Often the first basins are used as pre-sedimentation facilities.

In the vicinity of urban agglomerations, pits may have been dug in ancient paths of wadis in order to extract construction material (gravel, sand). The depth of these pits may range from 2 to 3 m (from 6.56 to 9.84 ft) up to 30 to 40 m (98.43 to 131.23 ft). Pits may also be excavated for the specific purpose of artificial recharge. Aquifer recharge simply consists of diverting water from the main channel to the pit. Even with a deep pit, it may be advisable to have a smaller settling pit between the main channel and the larger recharge pit. Both recharge and settling pits should be fenced and have a suitable inlet so that the inflowing water does not erode the walls of the pits.

ii. Watershed management (ASCE 1987): Watershed management offers an effective method to intercept dispersed runoff. Many techniques of water conservation have been developed along hill slopes with the intention of preventing soil erosion and reducing surface runoff, then increasing the infiltration in the ground, thus recharging the aquifers. Traditional terraced agriculture is certainly one of the most common water harvesting methods in arid areas and particularly in the Near East. Where the terraces are well maintained, they effectively control runoff and improve aquifer recharge but, once allowed to fall into disuse, they progressively lead to gully erosion, collapse of the retaining walls, destruction of the whole system and severe modification of the hydrological regime. Therefore, whatever the economic virtues of such terraces, it should be recognized that their abandonment on a large scale can upset the hydrological conditions within a basin for a considerable period of time.

Because of the siltation problems in the surface reservoirs resulting from soil erosion in the upper catchment, large program of soil and water conservation as well as forestation have been undertaken in several semi-arid countries. Although the primary objective of the watershed management is to limit the soil erosion and therefore to reduce sediment accumulation in the surface reservoirs downstream, the effect of these practices may become significant on the aquifer recharge when large areas are included in the programs. However there are few examples of quantitative analysis of the modification of the water cycle in a catchment where soil and water conservation has been practiced.

iii. Recharge wells or ASR wells: Artificial recharge by injection consists of using a conduit access, such as a tubewell, shaft or connector well, to convey the water to the aquifer. It is the only method for artificial recharge of confined aquifers or deep-seated aquifers with low permeable overburden. The recharge is instantaneous and there are no transit or evaporation losses. The method is very effective in the case of highly fractured hard rocks and karstic limestone aquifers.

Recharge wells or "injection" wells are similar in construction to pumped wells, using screened sections. The great difficulty in using recharge wells is their susceptibility to rapid clogging. While a basin may clog within years and in any case may easily be reconditioned, a recharge well may clog in a few days or weeks and is always difficult to maintain in good operating condition. There are several possible causes of clogging (Johnson 1981; ASCE 1987).

i. First there is suspended matter present in the water; it reduces the pore space in the gravel pack and in the formation at the interface with the gravel pack. This causes clogging to be more severe in aquifers with finer grain size.
ii. High organic matter content may result in bacteriological growth. This is why clogging phenomena vary during the year as temperature of injected water changes.
iii. Clogging may also occur due to gas or air bubbles in the water, especially in shallow wells with low water pressure. It is essential to prevent underpressure in recharge pipes, valves and connections. This problem can be overcome by the use of small diameter recharge pipes in order to ensure the pipe-full conditions
iv. Mixing of chemically dissimilar waters causes these waters to react and to form precipitates. Another form of clogging is caused by the swelling of clay particles that may be present in the target aquifer.

The most economical way to recharge by injection consists of using dual purpose wells (injection and pumping, i.e. ASR wells) so that cleaning of the aquifer may be achieved during the pumping period (Pyne 1995). However, pretreatment to eliminate the suspended matter is always necessary.

The spacing of injection wells depends upon the range of influence of a well, which in turn depends upon the amount of water to be recharged through it, and the acceptance rate of the aquifer. The acceptance rate is a function of the aquifer hydraulic conductivity, hydraulic gradient, the length of perforated casing or screen penetrating the aquifer, and the open area of casing perforations or screen.

In general, it has been found that gravel-packed wells operate more efficiently and require less maintenance in alluvial aquifers than do wells without gravel-pack. At Manhattan Beach, California, a 600 mm (24 in.) diameter gravel-packed well with 200 mm (8 in.) casing was found most desirable for recharging purposes. On Long Island, New York, where cooling water is returned to the groundwater basin, a minimum casing size of 200 mm (8 in.) and a minimum gravel packing thickness of 50 mm (2 in.) was used successfully. In addition, where water is being injected under pressure, it has been found that a concrete seal should be provided on the outside of the casing at a point where it passes through the relatively impermeable confining bed, to prevent the upward movement of water outside the casing (ASCE 1987).

The casing perforations or screens should be carefully designed to minimize the movement of formation materials into the well when injection pressure is relieved. It has been found that locating the screen or casing perforations below the normal water table lessens the incidence of chemical incrustation. When recharge wells are installed near a pumping well, the screen or perforations of the recharge wells should be at, an elevation somewhat different than that of the pumping well in order to increase the length of the percolation path of the recharged water.

A well header assembly must be provided to bring the water to the recharge well and to regulate the flow of water into the well. In general, the water to be recharged should be supplied at a relatively constant pressure, and should not be allowed to fall freely into the well, as the resulting aeration adversely affects acceptance rates. The design of the assembly will vary depending upon the purpose for which the project will be used. Figure 8.8 (a) shows a plan for recharge through active irrigation wells, used in the High Plains area in Texas to recharge stored natural runoff artificially. Figure 8.8 (b) shows a typical recharge assembly used for injecting water under positive pressure to prevent the encroachment of seawater into freshwater-bearing deposits at Manhattan Beach, California.

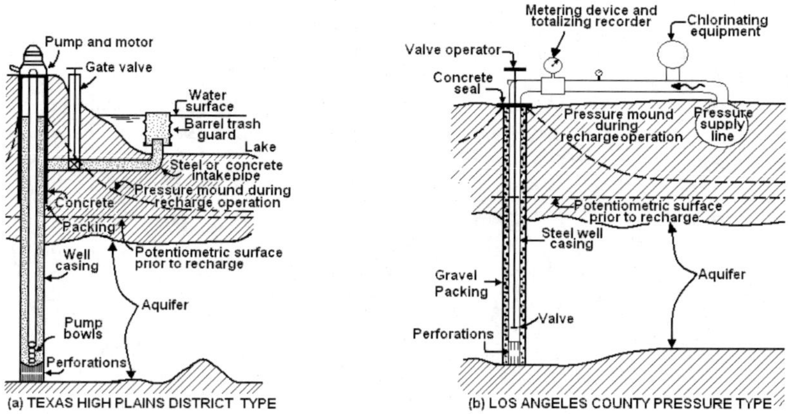

Figure 8.8: Typical recharge wells (from ASCE 1987).

It should be emphasized that long-term use of injection wells in alluvial aquifers requires treatment of the injected water. Sediment must be completely removed. The clear water should be treated with chlorine, calcium hypochlorite, or copper sulfate to prevent the growth of bacterial slime and algae. The water must also be free of dissolved gases that may be released into the formation and cause "bubble" binding (reduced permeability). In addition, care should be taken to assure that in formations containing appreciable proportions of base exchangeable clay, water containing a high percentage of sodium will not be used, as this will cause deflocculation of aquifer sediments and a rapid decrease in transmissivity. These treatment requirements may not be necessary in highly permeable limestone and volcanic aquifers. Gravity injection of stormwater and treated wastewaters into the limestone Floridan aquifer in Florida has been effective for many years with no evidence of significant plugging. However, risk of introducing contaminants such as microbiological constituents, heavy metals, organics, solvents, and synthetic compounds in the injected water should be assessed (Bloetscher et al. 2005).

8.5.3.6 *Recovery facilities*

Ideally, the amount and areal pattern of artificial extractions of groundwater should be arranged so as to develop the full sustained yield of the basin and take advantage of the available storage capacity. In order to accomplish this, wells should be spaced and operated in such a manner that, in any given area, the amount of water withdrawn is commensurate with replenishment, with due regard to the avoidance of undesirable effects such as saline water intrusion and excessive pumping lifts. The design of a plan to accomplish the foregoing objective requires knowledge of the physical characteristics of the particular basin and their effect upon extraction of water. These characteristics include the geological aspects of aquifer and aquitard structure, transmissivity, and storage capacity; their vertical and horizontal variation; and the hydrologic aspects of source, occurrence, movement, and use of the groundwater supply. However, although they are enumerated separately, it must be realized that each of the foregoing considerations is closely related to and dependent upon changes or alterations in the other items listed. A model of the groundwater basin can be a useful tool in testing the effects of well spacings and locations on water levels and basin yield.

8.5.3.7 *Operations and maintenance*

The details of administration and organization involved in the operation and maintenance of artificial recharge projects vary with the purpose of the project, the method of recharge employed, the character of the water, and the location. However, a number of basic operating problems or conditions can be anticipated, such as silt, weeds, rodents, mosquitoes, maintenance of infiltration capacities, and maintenance of structures. Some of these problems, their manifestations, and corrective actions that may be taken have been discussed (ASCE 1987; NRC 2008). A set of operating instructions should be prepared by the engineering section of the recharge project operating agency. Specific operating instructions for maintenance personnel are extremely helpful, particularly when the project is large and complex, and when spreading is periodic, as in the conservation of storm runoff. Such instructions should include, among other matters, the purpose of the project, a clear definition of the duties and responsibilities of the operating personnel, instructions relating to the regulation of inflow and outflow rates, enumeration of the items to be measured and details of operation of measuring devices, and instructions on the use of special equipment. In addition, the instructions must include guidance relating to actions to be initiated in case of emergencies or failures.

In an area where spreading is contemplated on a continuous basis, design and construction practices are often well defined. Personnel to operate the grounds are usually assigned on a more or less permanent basis, and operation is usually relatively routine. Under these conditions, maintenance of a project is closely associated with operation, and schedules must include an occasional shutdown period of short duration to permit proper maintenance of equipment and structures and to sustain infiltration rates. For this type of operation, it is generally desirable to provide some type of treatment of water prior to spreading. The number and duration of periods of shutdown should be scheduled to obtain the maximum infiltration capacities for the spreading area.

In contrast, a spreading area utilizing uncontrolled stormwater usually encounters more operational problems than projects using controlled flows. The personnel who operate the grounds are not normally required to attend the grounds until runoff occurs. Usually, maintenance of spreading grounds used to infiltrate uncontrolled stormwater need not be closely timed to operations. Normally, these grounds are out of operation during much of the year, and maintenance necessary to the project can be accomplished during these periods.

Silting and the resultant reduction in infiltration capacity in artificial recharge projects is a common operation and maintenance problem. The silt content of source water should be determined and monitored throughout the recharge operation. It has been found that the maximum turbidity of the recharging water should not exceed 25 Jackson Turbidity Units. If it is higher, desilting facilities should be employed or low infiltration rates can be expected. The allowable silt content in the water diverted for spreading may vary with the method of spreading, vegetative cover, surface soil, and relative cost of removing silt from the spreading grounds.

In diverting water to basin spreading grounds for the first time, or after an extended nonoperating period, care must be taken to fill the basins slowly to prevent breakdown of uncompacted dikes and other earth-fill structures. In addition, a constant patrol must be maintained to observe and correct sloughing, leaking, erosion, or other signs of structural failure. Where basins are formed by levees the potential for failure started by animal burrows is great. Burrowing animals must be controlled at these facilities.

8.5.4. Case Study - ASR Applied to Everglade Restoration, Everglade, Florida

In April 1999, the Comprehensive Everglades Restoration Plan (CERP) proposed large-scale development of ASR facilities as the preferred method of providing the additional freshwater storage required for overall restoration success (USACE 1999). The proposed CERP ASR system includes a total of 333 ASR wells and related surface facilities at the general locations shown on Figure 8.9. The proposed ASR wells have a target capacity of 5 mgd (18,927 m^3/d) with water treatment facilities included in the conceptual CERP ASR components. Total cost of the proposed CERP ASR System is approximately $1,700,000,000, about one-fifth of the total estimated cost of CERP. The Jacksonville District of the U.S. Army Corps of Engineers (USACE) and the South Florida Water Management District (SFWMD) are 50/50 cost sharing partners for design studies required prior to implementation of any large scale CERP ASR facilities.

In cooperation with a multi-agency project delivery team, the USACE and SFWMD have been tasked with evaluating the feasibility of the proposed CERP ASR projects individually and through the development of an ASR regional study (USACE and SFWMD 2003). One component of this study addresses the effect of injecting 1.67 billion gal/day (6.32 million m^3/d) to the Upper Floridan Aquifer System and overlying Hawthorn Group sediments, on the piezometric pressure and the associated hydraulic fracturing potential.

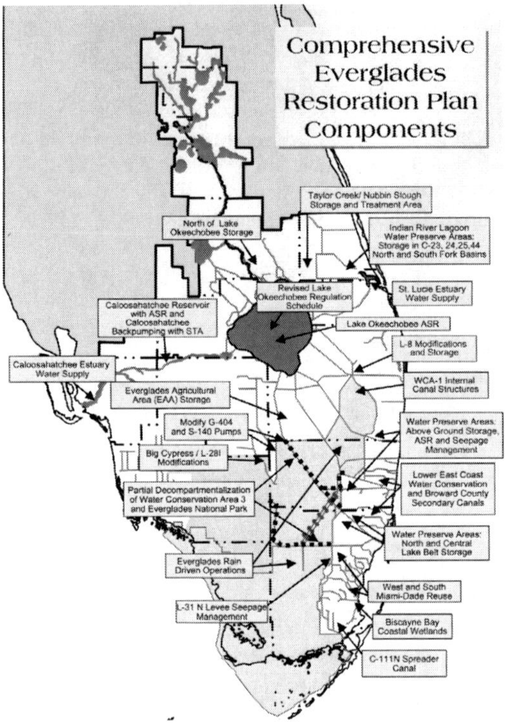

Figure 8.9: Comprehensive Everglades Restoration Plan components (USACE 1999).

The magnitude of the increase or decrease in piezometric pressure within the Upper Florida Aquifer System during recharge and recovery cycles is highly dependent upon numerous factors such as aquifer transmissivity, well spacing, and aquifer porosity. During the recharge, increases in static piezometric (hydraulic) head of 100 to 200 ft (30.48 to 60.96 m) near the pumping wells are certainly possible based upon both analytical and numerical models. Conversely, during the recovery, decreases in static head of similar magnitudes are possible. These anticipated pressure changes may present planning and engineering constraints that limit ASR development. Upper Floridan Aquifer System heads substantially higher than the current regional flow system could lead to changes in flow direction or velocity. For ASR design purposes, the pressure changes may constrain wellhead design or pump selection. In addition, continually injecting at high pressures could result in unsustainably large costs for electricity.

A secondary effect of large-scale ASR operations is the potential of hydraulic fracturing of the limestone rock of the Upper Floridan Aquifer System or overlying

Hawthorn Group sediments. In addition, during recovery operations, settlement or subsidence of the overlying Hawthorn Group clays is a possibility that requires further examination. Conversely, during ASR recharge cycles, expansion or "lengthening" of the Hawthorn Group clay is a remote possibility.

8.6 Coastal Aquifer Development

Wherever a freshwater aquifer is in hydraulic connection with the ocean or other large body of saline water, the possibility exists that such water may invade the freshwater body (CDWR 1958; Henry 1959; Cooper et al. 1964; Lee and Cheng 1974; Newport 1975). Seawater, because of its greater density, will tend to move inland under the freshwater body and a freshwater-saltwater interface will be created. This interface will occur at a depth below sea level generally about forty times the elevation of the groundwater table, based on seawater specific gravity of 1.025 (Ghyben-Herzberg principle). In coastal areas at which an unconfined groundwater aquifer, with a depth extending well below sea level, is in contact with the ocean, the fresh groundwater will tend to form a lens-shaped body floating on the seawater (Ghyben-Herzberg lens). This phenomenon is found in the Hawaiian Islands, in the coastal sand dunes of the Netherlands and Belgium, on Long Island, and in many other coastal areas.

The successful development of groundwater in a Ghyben-Herzberg lens system hinges largely on meticulous control of pumping to avoid creation of a condition of persisting deep drawdown of the water table (Dagan and Bear 1968; Schmorak and Mercado 1969; Shamir and Dagan 1971; Sahni 1972; Chandler and McWhorter 1975; Das Gupta 1983). As has been previously indicated, every 30.5 cm (1 ft) of draw down will induce a rise of 13 m (40 ft) in the underlying saltwater level, although the response may be greatly delayed. The encroachment of seawater can be largely avoided by utilization of judiciously located shallow wells, infiltration galleries, or tunnels, where drawdowns are controlled.

A modified Ghyben-Herzberg relationship may exist where an artesian aquifer is in hydraulic connection with the ocean (Rumer and Shiau 1968; Collins and Gelhar 1971). Under natural conditions, the groundwater may be under considerable hydrostatic pressure generated from a remote inland recharge area, with potentiometric levels well above sea level. The freshwater head will usually be adequate to limit the intrusion of seawater to a relatively short distance inland from the point of contact. When the groundwater basin is developed and freshwater levels drop, there will be a tendency for seawater to move landward, displacing the freshwater in the aquifer. A landward sloping freshwater-saltwater interface will be formed, with a depth governed by the Ghyben-Herzberg relationship. As water levels continue to decline the prism of seawater, which has been designated the seawater wedge, will continue to advance until conditions of equilibrium are reestablished (Baumann 1957; CDWR 1960a, 1960b; Bear and Dagan 1962a, 1962b; CDWR 1963; Bear and Dagan 1964; Water Resources Engineers 1969; Pinder and Cooper 1970;

Mercer et al. 1980). This type of seawater intrusion has occurred in the West Coast Basin in southern California (CDPW 1951).

In coastal aquifers, another major concern is land subsidence. In most cases, the aquifer material is unconsolidated. As groundwater levels are lowered, the effective stress can easily reaches or exceed the preconsolidation stresses of clay layers, which will result in additional compaction of unconsolidated geological formation. Such compaction is usually permanent or nonrecoverable. Refer to Section 8.3.4 for details.

8.6.1 Measures for Control Seawater Intrusion and Land Subsidence

Vulnerability to seawater intrusion does not necessarily preclude the utilization and operation of an underground reservoir. However, it does introduce a factor that must be taken into consideration in formulation and development. Methods of controlling seawater intrusion that may be available to the water manager are described below.

i. Reduction of groundwater extractions (ASCE 1987): One method of preventing seawater intrusion is to reduce pumpage to a planned level. Of course, this implies that water needs can be reduced or that an alternative supply of water can be arranged. When pumping is reduced, the gradient from the recharge source flattens, the landward gradient associated with intrusion is replaced with a slight seaward gradient, and seaward flow of freshwater resumes, as shown in Figure 8.10. If sufficient information is available on the water levels, geology, and hydrology, pumping can be controlled so that a maximum amount can be pumped while the intrusion of seawater is halted. Note that if the pumped water is for an overlying use, such as supplying a small farm, the reduction in pumping that occurs as the water becomes saline acts automatically to apply this system of control.

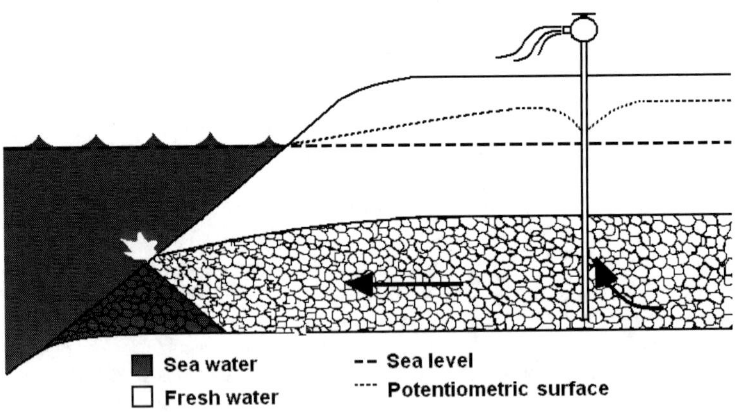

Figure 8.10: Confined aquifer with controlled pumping (modified from ASCE 1987).

ii. Artificial recharge by spreading (ASCE 1987): Another method of control is to increase recharge. An additional source of water must be available for this purpose and the physical conditions must be present so that recharge can be increased. Figure 8.11 shows the results of this method. The gradient from the recharge area steepens to carry the additional flow. Again, the landward gradient between the pumping wells and the sea changes to a seaward gradient, and a seaward flow of freshwater resumes. Pumping continues at the previous level. If sufficient information is available, the recharge and pumping can be controlled to minimize the waste of freshwater to the ocean. An advantage of this system is that increasing recharge by spreading is relatively inexpensive. A possible limitation is that the aquifer may not have adequate capacity to carry the required additional flow.

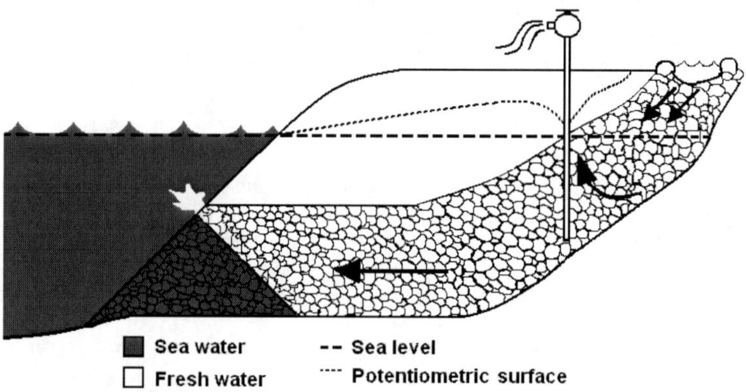

Figure 8.11: Confined aquifer with surface recharge (modified from ASCE 1987).

iii. Physical barrier (ASCE 1987): Another method of control is a physical barrier, illustrated in Figure 8.12. An impermeable membrane of some type is placed between the seawater source and the pumping wells. The method of construction for such a membrane might be an excavated trench backfilled with bentonite clay, or a closely spaced line of wells through which impermeable grout is injected. It is likely that such a barrier could be used only in relatively shallow formations. It would be important to monitor the effectiveness of the barrier to determine the magnitude of leakage of seawater. If the physical barrier is effective, this system has the advantage of allowing significant drawdowns of water levels, which might allow either the utilization of large amounts of stored freshwater or the steepening of the gradient from the recharge area, with a consequent increase in supply. The method has the disadvantage of preventing subsurface outflow from the basin which is sometimes needed to maintain a favorable salt balance. In some situations, it may be possible to use the former outflow for some purpose, such as some industries, and then discharge the waste to the ocean or salt sink to maintain a favorable salt balance in the aquifer.

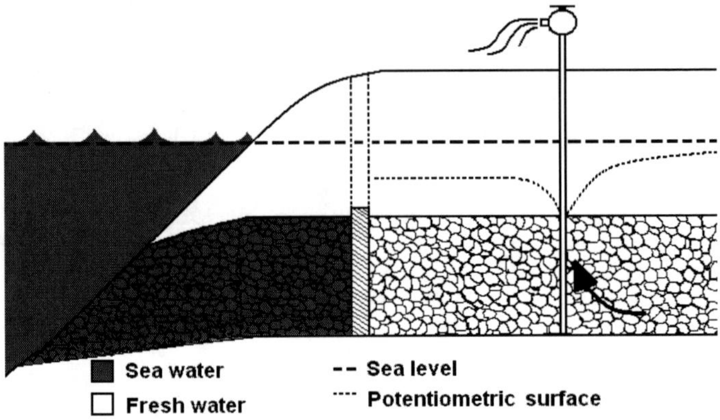

Figure 8.12: Confined aquifer with physical barrier (from ASCE 1987).

iv. Pumping trough (ASCE 1987): The pumping trough is another method of control (Figure 8.13). It consists of a line of control pumping wells located between the water-supply wells and the seawater source. The control wells must be pumped at rates that will intercept all the seawater moving landward toward the supply wells. After equilibrium is reached, the groundwater levels in the trough will generally be lower than at any other point in the basin. For this system eventually to reach an equilibrium condition, pumping must be reduced by at least an amount slightly larger than the rate at which seawater was initially intruding. The important items in this method are monitoring the groundwater levels at the pumping trough and determining the amount of water to be pumped at the control wells. The control-pumping requirement will vary widely with the actual conditions, but will be substantially larger than the amount of water that was originally intruding from the sea. The more closely the control wells are located to the sea, the greater will be the pumping requirement. For example, if the control wells are located halfway between the sea and the original trough caused by the supply wells, then the control-pumping requirement will be more than twice the amount of the original intruding seawater. If equilibrium conditions are established at lower levels to increase the gradient, and the supply of freshwater, the control-pumping requirement will be increased similarly. An experimental extraction-type barrier was tested in the field in Ventura County (CDWR 1970).

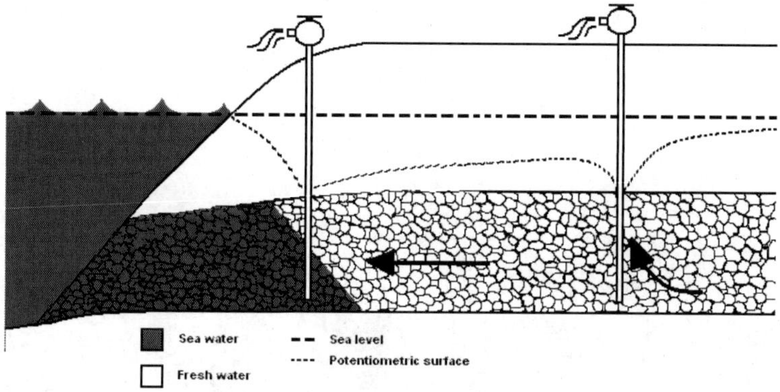

Figure 8.13: Confined aquifer with pumping-trough type barrier (from ASCE 1987).

v. Hydraulic ridge (ASCE 1987): The pressure barrier method of controlling intrusion depends on the resistance to movement that water encounters as it moves underground. As additional water is introduced, water levels rise to provide the necessary gradient to carry it away. In the case of confined aquifers, additional water can be introduced through injection wells. Figure 8.14 shows the pressure barrier method being used for control. Freshwater is injected through a line of injection wells which separates the seawater source from the pumping wells. Because of the difference in density between freshwater and seawater, a column of freshwater 13 m (41 ft.) high is needed to counterbalance a column of 12 m (40 ft) of seawater under static conditions. Therefore, to control the intrusion of seawater, it is necessary to determine the elevation of the base of the formation which contains the freshwater supplies of interest. The pressure barrier is then formed by causing a pressure rise along the line of injection walls to an elevation of about 0.76 m (2.5 ft) above sea level for every 30.5 m (100 ft) below sea level that the base of the aquifer lies. Where a pressure barrier is being formed by injection wells, the critical control location is the point approximately halfway between the injection wells, where the pressure is lowest. After the pressure barrier comes to equilibrium, the amount of freshwater that will flow inland from the barrier must be approximately the same as the amount of seawater previously intruding, assuming the same inland pumping. However, to maintain the dynamic balance between seawater and freshwater, a small freshwater flow toward the ocean is necessary. The amount of this flow will vary according to conditions, but should be on the order of 10% of the amount flowing inland. The number of injection wells required depends on the rate at which a well will accept water, per foot of water level rise. This can be roughly estimated as the specific capacity of a pumping well of similar size in the particular formation. In shallow unconfined aquifer areas where surface-water supplies are plentiful, such as in Florida, a hydraulic ridge effect can be produced in the aquifer by use of salinity control structures in the drainage canals near the coast. Imported surface water from the canal upstream of the control structures moves into the aquifer and keeps the seawater from intruding into the aquifer. The system of salinity controls in canals in southeastern Florida has been in use since the 1930s and is very effective (Klein 1965; Sherwood and Grantham 1966). A hydraulic pressure ridge has been in operation since 1960 in Los Angeles County, California (Bruington and Seares 1965).

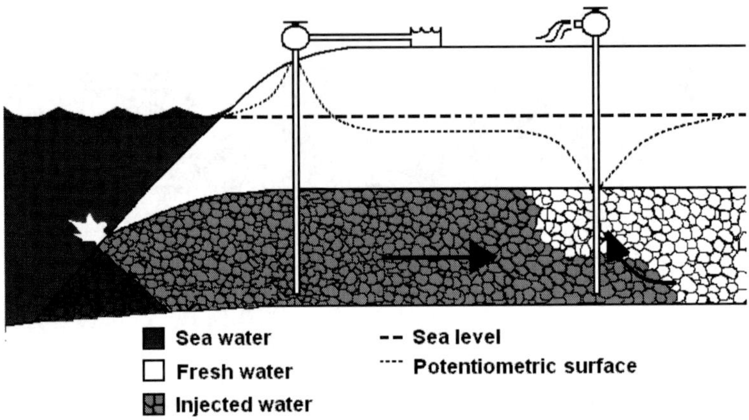

Figure 8.14: Confined aquifer with injection barrier (from ASCE 1987).

vi. Combination barrier (ASCE 1987): This type of barrier is a combination of the pumping trough and the hydraulic ridge. The pumping trough is operated on the seaward side of the injection wells, and the hydraulic ridge on the landward side of the pumping trough. The combination barrier requires about one-third as much extraction to achieve the desired effect as a pumping trough, and requires injection of a smaller total quantity of freshwater to achieve the effects of a hydraulic ridge. A combined barrier composed of the two systems operating in conjunction minimizes the effects of land surface subsidence and water logging. In addition, the combination system avoids many other side effects, and allows for greater flexibility of operation than any of the single operating systems previously described. Such a system was installed in Orange County, California (CDWR 1966; Montgomery 1967).

Each of the aforementioned methods for control of seawater intrusion has certain characteristics that affect groundwater basin operation. Maintenance of water levels above sea level would reduce pumping costs. However, they would severely restrict the utilization of groundwater storage in coastal aquifers, many of which are hundreds of feet thick and extend far below sea level. This could tend to limit natural groundwater recharge, both from subsurface inflow and percolation of surface runoff, as available storage might not be adequate for full conservation of local water resources. The pumping trough overcomes these disadvantages to some extent, as water levels on the landward side can be maintained below sea level. However, the depth of the trough limits the elevation to which the water table can be lowered, and thus the usable groundwater storage. The groundwater ridge and subsurface barrier both would permit full utilization of underground storage, inasmuch as land groundwater levels could be maintained as low as desired. The greatest latitude in groundwater basin operation is provided by either the groundwater ridge or subsurface barrier; the pumping trough permits limited degree of latitude, but severe restrictions are imposed on utilization of groundwater basin by maintenance of water levels above sea level.

Many other factors are involved in evaluating the feasibility of protecting a groundwater basin from seawater intrusion. These include the capital and annual costs of physical works required (such as wells, supply and distribution lines, subsurface barriers, and so on); the availability and cost of supplemental water supplies; water rights and other legal considerations; drainage and salt balance requirements; potentialities for conservation of flood runoff and wastewaters; and the value of the storage capacity involved. All of these factors must be taken into consideration in determining the most advantageous method of seawater intrusion control.

Similar methods such as MAR and reduction of groundwater pumpage can be used to control land subsidence. It should be noticed that land subsidence can be controlled, however, the lowered land surface cannot be restored even the water levels are restored to historic level because the permanent compaction of clay layer. There is a time lag in controlling land subsidence, which should also be considered in designing and implementing controlling measures.

8.6.2 Case Study – Prevention of Seawater Intrusion in Orange County, California

Seawater intrusion has occurred in the West Coast Basin in southern California. The Silverado aquifer, a principal source of groundwater in the area, is an artesian aquifer that receives freshwater from an inland recharge area. It extends seaward for some distance beneath the ocean and outcrops on the floor of Santa Monica Bay. This basin has been a prolific source of water supply for many years, despite long-standing intrusion of seawater.

Water levels have long been depressed below sea level and intrusion was noted as early as 1912 in the West Coast Basin. A large part of the extracted groundwater consisted of freshwaters stored in submarine extensions of artesian aquifers. Under natural conditions, these aquifers were recharged from inland recharge areas. With water levels below sea level, the direction of flow in these submarine aquifers was reversed and the previously stored groundwaters began moving landward in quantities equal to the excess of withdrawals over inland replenishment (overdraft).

The landward flow was about 43 million m^3 (35,000 ac-ft) per year for the period 1932-1950. Groundwater levels in the basin dropped to 30 m (90 ft) or more below sea level and seawater intrusion advanced inland more than 4.8 km (3 miles). In order to preserve the groundwater reservoir, several protective measures were initiated. Groundwater pumpage was restricted by voluntary agreement and, to a large extent, imported water supplies were substituted. Additionally, plans were formulated to augment the freshwater recharge to the basin and a hydraulic barrier to stop encroachment of seawater was constructed.

Located in an arid region where water resources are precious, the Orange County Water District (OCWD) currently purchases imported water from the Colorado River and the State Water Project. The high costs associated with imported water, combined with an increasing population that is expected to reach 2.8 million by 2020, drove the OCWD to consider a more reliable, cost effective, micro-filtration water treatment system. The micro-filtration system will allow the district to effectively manage its water resources with little or no increase in cost. In addition, it can employ a new water source that is completely independent of the imported water. High-quality, dependable water will be processed with half the energy that it takes to import the water.

In 1997, the water and sanitation districts of Orange County formed the Groundwater Replenishment (GWR) system, a joint project specifically created to purify highly treated secondary wastewater currently released into the ocean. The GWR system will process the wastewater through an intricate membrane treatment plant to be constructed on the Fountain Valley water campus, occupied by both Orange County's water district and sewer district. The treatment plant will include a micro-filtration unit, reverse osmosis system, and ultraviolet disinfection equipment, as well as supplementary pumping, power and chemical facilities. Following treatment, the water will be injected into the local groundwater basin to prevent seawater intrusion. The remaining water not directly injected into the basin will be introduced into a groundwater aquifer through the district's Santa Ana River percolation basins, located in the cities of Anaheim and Orange.

8.7 Development of Brackish Groundwater

Desalination is any technique that can be used to reduce salts from brackish groundwater, surface water, or seawater. According to the International Desalination Association, there are currently 13,600 desalination plants worldwide, producing 6.8 billion gallons of water per day (22 million m^3/d). The U.S. is ranked as having the second largest total desalination capacity of any country in the world. This is due to the numerous inland desalination plants that are used to treat brackish surface water and groundwater. In the continental U.S., two seawater desalination plants were installed in Santa Barbara, California and in Key West, Florida, but neither one is operating on a continuous basis. In Texas there are more than 100 desalination units producing about 40 mgd (151,000 m^3/d). All desalination plants in Texas currently use either brackish surface water or groundwater as their raw water source. Municipal desalination accounts for 23 mgd (87,000 m^3/d) while industrial desalination is approximately 17 mgd (64,352 m^3/d). Prominent municipal desalination sites in

Texas using surface water as their raw water source include Sherman (Lake Texoma), Robinson (Brazos River), and Lake Granbury, while Fort Stockton and Kennedy use brackish groundwater.

Regardless of the technique employed, desalination offers many benefits and advantages over other conventional forms of water resource development. The most important advantage is that desalination provides a relatively drought-proof water resource. There is no need to build expensive dams or reservoirs or deal with issues such as land submergence and flooding. In addition, issues such as water rights and concentrate disposal may not be as complex as with surface water desalination plants.

About a dozen new desalination projects may be constructed in Florida, California, and Texas in the next few years. In Texas there are currently several private and public partnerships in desalination projects. The El Paso Water Utilities (EPWU) and Fort Bliss in west Texas are collaborating to build the country's largest inland desalination plant for municipal use. The 27.5 mgd (104,000 m^3/d) Eastside Brackish Groundwater Desalination Plant will desalinate brackish water from the Hueco Bolson aquifer and will provide about one-fourth of the city's water requirement (Burkstaller et al. 2001; USACE 2004). The City of Corpus Christi is considering a 5 mgd (18,927 m^3/d) plant on Mustang Island using brackish groundwater, and may later build a 25-30 mgd (95,000-113,000 m^3/d) plant in conjunction with the Barney Davis Power Station. The Regional Water Authority in Cameron County has completed a 7.5 mgd (28,000 m^3/d) desalination plant using brackish groundwater. In the 2002 Texas State Water Plan (TWDB 2002), desalination is recommended as a water management strategy to produce additional water supplies in four regions. In the Far West Texas Region (E) and Coastal Bend Region (N), desalination of brackish groundwater is used as a strategy. With exciting advances in technology that make desalinated water more price competitive, it is possible that within this decade the late President Kennedy's dream may become a reality.

A very important consideration when devising brackish groundwater development strategies is to frame them in the context of maximizing aquifer yield by exploiting brackish water with freshwater conjunctively. The environmental issue arisen from brackish water development is the unavoidable concentrate produced as a byproduct. Alternatives to dispose of brine, which can be varied, have to be sensitive to the environment. Therefore, to fully utilize brackish water within affordable costs and minimum environmental impacts, water resources engineers need to have a better understanding of brackish water distribution, water quality variation, relationship between freshwater and brackish water, desalination technologies, and possible alternatives for management of the concentrate.

8.7.1 Definition of Brackish Water

Brackish water is water that is saltier than freshwater, but not as salty as seawater. It may result from mixing of seawater with freshwater, as in estuaries, or it may occur naturally, as in brackish fossil aquifers. Technically, brackish water contains between

500 and 30,000 milligrams of salt per liter. Thus, the term brackish covers a range of salinity regimes and is not considered a precisely defined condition. It is characteristic of many brackish surface waters that their salinity can vary considerably over space and/or time.

Saline water is formally defined as water containing greater than 1,000 mg/L total dissolved solids (TDS) (U.S. Public Health Service 1962). Saline waters are further classified by TDS content as follows (USBR 2003):

Mildly brackish, 1,000 - 5,000 mg/L;
Moderately brackish, 5,000 - 15,000 mg/L;
Heavily brackish, 15,000 - 35,000 mg/L; and
Seawater and brine, greater than 35,000 mg/L

Historically, the Texas Water Development Board (TWDB) has defined aquifer water quality in terms of TDS concentrations and has classified water into four broad categories (LBG-Guyton Associates 2003):

Fresh, less than 1,000 mg/L;
Slightly-saline, 1,000 - 3,000 mg/L;
Moderately-saline, 3,000 - 10,000 mg/L; and
Very-saline, 10,000 - 35,000 mg/L.

In a recent assessment of brackish water conducted by the TWDB, brackish water was defined as water with a TDS ranging from 1,000 mg/L to 10,000 mg/L (LBG-Guyton Associates 2003). In summary, brackish water can be defined as water that contains 1,000 to 35,000 mg/L of salt. It is saltier than freshwater, but less than seawater.

8.7.2 Desalination as a Management Strategy to Maximize Aquifer Yield

In the southwestern U.S. the availability of fresh groundwater is limited by precipitation, while brackish water is plentiful. Through desalination, brackish water can effectively increase the aquifer yield. To the maximum extent feasible, existing production wells should be used to extract brackish water. Appropriate desalination technology should be selected based on the water chemistry and efficiency of treatment techniques. Permeated water can be blended with untreated brackish water to maximize production of potable water (Burkstaller et al. 2001; USACE 2004). Under certain conditions, pumping of brackish water wells can develop a pumping trough to protect freshwater from further intrusion of brackish water (Sheng and Devere 2005). Additional legal and regulatory frameworks should also be developed if current fresh groundwater laws or regulations do not address issues related to brackish water development.

Reverse osmosis is a method for removing salts from water by applying pressure to the brine, forcing the water through a semi-permeable membrane which leaves the salts behind and collects the purified water on the other side of the membrane. Reverse osmosis is capable of treating feed waters of up to 45,000 mg/L TDS. Most reverse osmosis applications involve brackish feed waters ranging from about 1,000 to 10,000 mg/L. Reverse osmosis is also less efficient than lime softening; so more raw water is needed to produce finished water.

Electrodialysis is a method for removing salts from water using electric currents and membranes to separate the ions from the water. Electrodialysis reversal is a similar process but provides for the reversing of the electrical current which causes a reversing in the direction of ion movement. Electrodialysis and electrodialysis reversal are useful in desalting brackish water with total dissolved solids feedwater concentration of up to 10,000 mg/L. However, electrodialysis/electrodialysis reversal is generally not considered to be an efficient and cost-effective organic removal process.

8.7.3 Concentrate Management

Concentrate disposal is an important factor in the economics of desalination in inland areas. Every desalination process yields a concentrate as a byproduct, which must be managed in an environmentally appropriate manner. The method used to dispose of concentrate is a major decision in designing and planning the overall desalination strategy. The ability to estimate the quantity and quality of the concentrate stream allows proper selection of the disposal process and subsequent regulatory permitting. In coastal areas, the concentrate stream is usually returned to the ocean where it is diluted with seawater. In landlocked areas such as New Mexico and west Texas, concentrate management options include evaporation ponds, deep subsurface injection, or byproduct recovery. These methods can significantly increase the cost and environmental impact of inland desalination operations.

HDR and others (2000) identify potential approaches for brine disposal and the typical requirements for obtaining regulatory approval for brine disposal. These approaches include surface water discharge, pre-discharge mixing, discharge to municipal wastewater systems, deep well injection, and land application. Table 8.1 summarizes the potential advantages and constraints for different types of brine disposal. The major cost considerations for each of these brine disposal methods are also discussed by HDR and others (2000). However, it is difficult to estimate generic disposal cost relationships because the options vary significantly between projects. Prior to project implementation, a thorough review of pertinent regulations regarding brine disposal and associated water quality issues should be completed to ensure that proposed brine disposal methods and cost estimates are appropriate for planning purposes.

Table 8.1: Concentrate Disposal Options Summary (after HDR and others, 2000).

Disposal Option	Advantages	Constraints
	Brackish Desalination	
1. Direct surface water discharge	• Low cost upfront	• Requires available receiving water body • Future regulations may restrict • Monitoring program
2. Pre-discharge mixing	• Low to medium cost upfront	• Requires adequate mixing source • Monitoring program
3. Municipal wastewater system	• Low cost (if co-located) • Additional source for reclaimed water	• Higher wastewater treatment costs • Impacts to treatment process
4. Deep well injection	• Can handle large volume • May be available to inland plants	• Difficult permitting, high upfront cost
5. Land Application	• Best suited for small facilities	• Difficult to site

The U.S. Bureau of Reclamation documented membrane concentrate disposal practices and the regulations that impact disposal systems and techniques (Mickley 2001). This report was based on the findings from a detailed survey of 149 membrane plants that included 84% of the utility desalting plants (reverse osmosis, electrodialysis reversal, and nanofiltration) built in the U.S. between 1993 and 1999. The survey also included 44% of the utility low-pressure membrane (microfiltration and ultrafiltration) plants built during the same period. The report describes cost considerations for concentrate disposal to deep well injection, evaporation ponds, spray irrigation, and zero liquid discharge.

8.7.4 Case Study - Desalination of brackish groundwater in El Paso, Texas

In west Texas, a joint project of the EPWU and Fort Bliss, the Kay Bailey Hutchison Desalination Plant is the largest inland desalination plant for municipal use in the U.S. (EPWU 2008). The desalination facilities produces 27.5 mgd (104,099 m^3/d) of freshwater by desalinating brackish groundwater, making it a critical component of the region's water portfolio and providing about one-fourth of the city's current water supply (USACE 2004; EPWU 2008). The plant was put into operation started in 2007. The EPWU completed a desalination facility and production wells analysis, implemented a monitoring well program, conducted a pilot plant verification study, performed site investigations, and provided a basis of design report. The facilities, which include groundwater collection, reverse osmosis treatment, bypass, utilities and interconnecting piping, will reliably produce 27.5 mgd (104,099 m^3/d) of potable water and will have an expansion capacity to 40 mgd (151,417 m^3/d).

El Paso is located in the arid Chihuahua desert with an average annual precipitation less than 9 inches. El Paso's water sources include groundwater from bolsons (aquifers) and surface water from the Rio Grande. Water from the Rio Grande is released from the Elephant Butte Reservoir and Caballo Reservoir in New Mexico and is only available during the spring, summer and early fall months and is further limited in years of drought. The Hueco Bolson, providing majority of water supplies for El Paso historically, is also the source of water for Ciudad Juárez in México and other communities in the area. Historically, pumping from the bolson has exceeded the natural recharge rate, resulting decline of groundwater levels and deterioration of water quality due to brackish water intrusion (Heywood and Yager 2003; Sheng and Devere 2005). EPWU (2008) recognized the need to diversify its resources and to reduce reliance on groundwater from the bolsons and has made significant strides in recent years toward that objective. EPWU's conservation initiatives have been very successful, setting benchmarks for cities across the western U.S. However, knowing that additional freshwater sources would be needed, EPWU began exploring the idea of desalinating the brackish water in the bolsons in the early 1990s. The amount of brackish water in the Hueco Bolson exceeds the amount of potable water by approximately 600%. The brackish water contains more salt than is allowed in drinking water, but significantly less than seawater.

When new technology reduced the cost of the reverse osmosis process, EPWU began to plan the construction of a desalination plant. Because Fort Bliss was considering a similar facility, a public-public partnership was formed. The two entities recognized the benefits of a partnership in the complex process of building what will become the world's largest inland desalination plant with the capacity to meet the needs of both Fort Bliss and EPWU. This is the largest public-public project of its kind in the country involving the Department of Defense and a local community (USACE 2004; EPWU 2008).

In 1997, the EPWU and the Juárez water utility, the Junta Municipal de Aqua y Saneamiento, along with other agencies on both sides of the border, commissioned the U.S. Geological Survey to conduct a detailed analysis of the amount of freshwater remaining in the Hueco Bolson, the amount of brackish water available, plus a determination of the flow patterns (Heywood and Yager 2003). The data from the model was used to determine where to locate the desalination plant and source wells (Hutchison 2004; EPWU 2008). Additional analysis was carried out to identify existing wells that might be used to supply the desalination facility. Test wells were drilled and monitored to further characterize a section of the aquifer selected to provide a stable and consistent supply of brackish water as well as blend water. Additional hydrogeological studies were conducted to determine the flow of the brackish water in the Hueco Bolson. A reverse osmosis pilot plant was constructed to test the chemicals, filters, and membranes used in the reverse osmosis process and determine which work best with local water.

An inland desalination plant presents significantly more challenges than desalination plants near oceans because of limited supply of consistent water and challenges in disposal of the concentrate. Therefore, the most complex studies were directed toward the problem of concentrate disposal. Considerable testing, studies and pilot projects were performed to determine the most economically and environmentally sound means of disposing of the concentrate (EPWU 2008). A comprehensive initial study examined six alternatives for disposal (Burkstaller et al. 2001). Two methods were determined to be the most feasible: evaporation and deep-well injection. The EPWU then tested evaporation methods, including conventional evaporation ponds, evaporation misting equipment and evaporation ponds with concentrators. Deep-well injection was eventually selected as the preferred method of disposal, and the concentrate is placed in porous, underground rock through wells. The sites would confine the concentrate to prevent migration to freshwater, provide storage volume sufficient for 50 years of operation and meet all the requirements of the Texas Commission on Environmental Quality (USACE 2004). Deep-well injection entailed extensive study of local geological and hydrological conditions as well as the examination of existing data, including seismic analysis and water samples. A geophysical study was conducted to create a geologic model and four wells were drilled to test for geological formation (EPWU 2008; Hutchison 2008).

The desalination plant uses reverse osmosis to obtain potable water from brackish water drawn from the Hueco Bolson. Raw water from new and existing wells is pumped to the plant and filtered before being sent to reverse osmosis membranes. Through a pressurized process, raw water will pass through fine membranes that separate salts and other contaminants from the water. Approximately 83% of the water is recovered while the remainder is output as a concentrate (EPWU 2008). At the last stage of the reverse osmosis process, the permeate, or desalted water, is piped to a storage tank and the concentrate is routed to a disposal facility. The permeate is blended with water from newly constructed wells. Following pH adjustment and disinfection, the finished water is sent to the distribution system.

8.8 References

Anderson, M.P., and Berkebile, C.A. (1997). "Discussion of: Hydrogeology of the South Fork Long Island, New York." by C. W. Fetter, *Geol. Soc. of Amer. Bull.* 88, 895-896.

Alley, W.M., Reilly, T.E., and Franke, O.L. (1999). *Sustainability of groundwater*, U.S. Geological Survey Circular 1186, 79 p.

Alley, W.M., and Leake, S.A. (2004). "The journey from safe yield to sustainability." *Ground Water*, 42(1), 12-16.

American Society of Civil Engineers (ASCE) (1987). *Ground water management,* 3rd edition, ASCE Manual and Reports on Engineering Practice No. 40, New York, 281 p.

ASCE (2001). *Standard guidelines for artificial recharge of ground water*, EWRI/ASCE 34-01, ASCE, Reston, Va., 106 p.

Ashley, J.S., and Smith, Z.A. (1999). *Groundwater management in the West*, University of Nebraska Press, Lincoln, Nebr., 310 p.

Baker, D.M. (1951). "Safe yield of groundwater reservoirs." *Assemblee Generale de Bruxelles, Assoc. Inti. D'Hydrologie Scientifique,* 2, 160-164.

Banks, H.O. (1953). "Utilization of underground storage reservoirs." *Transactions, ASCE*, 118, 220.

Banks, H.O., et al. (1954). "Artificial recharge in California." *ASCE Hydraulics Division Conference*, Austin, Tex.

Barnett, S. R., Howles, S., Martin, R. R., Gerges, N. Z. (2000). "Aquifer storage and recharge: Innovation in water resources management." *Australian Journal of Earth Sciences*, 47(1), 13–19.

Baumann, P. (1957). "Basin recharge." *Transactions, ASCE*, 122, 458-502.

Bear, J. and Dagan, G. (1962a). *The transition zone between fresh and salt waters in coastal aquifers,* Progress Report No. 1, Technical Research and Development Foundation, Haifa, Israel.

Bear, J. and Dagan, G. (1962b). *The transition zone between fresh and salt waters in coastal aquifers*, Progress Report No. 2, Technical Research and Development Foundation, Haifa, Israel.

Bear, J. and Dagan, G. (1964). "Moving interface in coastal aquifers." *Journal of the Hydraulics Division, ASCE*, 90(HY4), 193-216.

Bear, J. and Levin, O. (1967). "The optimal yield of an aquifer." *Intern. Association of Sci. Hydrol. Bulletin,* 72, 401-412.

Bear, J. (1979). *Hydraulics of groundwater*, McGraw-Hill, New York, 569 p.

Bell, J. W. (1981). *Subsidence in Las Vegas Valley*, Nevada Bureau of Mines and Geology Bulletin 95, 83 p.

Bell, J.W. and Price, J.G. (1993). *Subsidence in Las Vegas Valley, 1980–91*, Final Project Report, Nevada Bureau of Mines and Geology Open-File Report 93-4.

Bloetscher, F., Muniz, A. and Witt, G.M. (2005). *Groundwater injection: Modeling, risks, regulations*, McGraw-Hill, New York, 337 p.

Bouwer, H. (1991). "Groundwater recharge with sewage effluent." *Water Science and Technology,* 23, 2099-2108.

Bouwer, H. (2002). "Artificial recharge of groundwater: Hydrogeology and engineering." *Hydrogeology Journal,* 10, 121-142.

Bredehoeft, J.D. (2002). "The water budget myth revisited: Why hydrogeologists model." *Ground Water*, 40(4), 340-345.

Bredehoeft, J.D. and Young, R.A. (1970). "The temporal allocation of groundwater-A simulation approach." *Water Resources Research*, 6(1), 3-21.

Bruington, A.E. and Seares, F.D. (1965). "Operating a sea water barrier project." *Journal of the Irrigation and Drainage Division, ASCE*, 91(IR1), 117-140.

Burkstaller J., Sheng, Z., Fierro, E., and Morgan, S. (2001). "Elements of a brackish water resource master plan for El Paso, Texas." *Proc., New Horizons In Drinking Water, AWWA 2001 Annual Conference, Washington, D.C.*, AWWA, Denver, Colo., 11 p.

Burt, O.R. (1964). "Optimal resource use over time with an application to groundwater." *Management Sci.*, 11, 80-93.

Burt, O.R. (1967a). "Temporal allocation of groundwater." *Water Resources Research*, 3(1), 45-66.

Burt, O.R. (1967b). "Groundwater management under quadratic criterion functions." *Water Resources Research*, 3(3), 673-682.

California Department of Public Works (CDPW) (1935). *Salinas Basin investigation*, Division of Water Resources, Bulletin No. 52.

CDPW (1951). *Proposed investigational work for control and prevention of sea-water intrusion into groundwater basins*, Division of Water Resources.

California Department of Water Resources (CDWR) (1958). *Sea-water intrusion, California*, Bulletin. No. 63.

CDWR (1960a). *Laboratory and model studies of sea-water intrusion, Appendix C*, Bulletin No. 63.

CDWR (1960b). *Intrusion of salt water into the groundwater basins of southern Alameda County*, Bulletin 81.CDWR (1963). *Alameda County investigation*, Bulletin No. 13.

CDWR (1966). *Santa Ana Gap salinity barrier, Orange County*, Bulletin 147-1.

CDWR (1970). *Oxnard Basin experimental extraction type barrier*, Bulletin 147-6.

Campana, M.E. (2005). "Forward: Transboundary ground water." *Ground Water*, 43(5), 646-650.

Chandler, R.L. and McWhorter, D.B. (1975). "Upconing of the salt-water-fresh-water interface beneath a pumping well." *Ground Water,* 13(4), 354-359.

Collins, M.A. and Gelhar, L.W. (1971). "Seawater intrusion in layered aquifers." *Water Resources Research,* 7(4), 971-979.

Commonwealth Scientific and Industrial Research Organisation (CSIRO) (2005). *Managed aquifer recharge: Frequently asked questions*, <http://www.csiro.au/resources/pf49.html>, (Oct. 2005).

Conkling, H. (1946). "Utilization of groundwater storage in stream system development." *Transactions, ASCE*, Ill, 275.

Conover, C.S. (1954). *Ground-water conditions in the Rincon and Mesilla valleys and adjacent areas in New Mexico*, U.S. Geological Survey Water Supply Paper 1230.

Conover, C.S. (1961a). "Groundwater in the arid and semi-arid zone: Its source, development, and management." *Ground-water in arid zones*, Publication 57, International Association of Hydrologic Sciences, 513-529.

Conover, C.S. (1961b). *Groundwater resources development and Man*, U.S. Geological Survey Circular 442.

Conservation Technology Information Center (2010). *Groundwater & surface water: Understanding the interaction*, <http://www.conservationinformation.org/?action=learningcenter_kyw_groundsurfacewater> (Apr. 2010).

Conwell, F.R. (1965). *Engineering geology and foundation condition survey of Las Vegas, Nevada*, John A. Blume and Associates unpublished report to the U.S. Department of Energy, 49 p.

Cooper, H.H., et al. (1964). *Sea water in coastal aquifers*, U.S. Geological Survey Water Supply Paper 1613-C, 84 p.

Coplin, L.A. and Galloway, D. (1999). "Houston-Galveston, Texas: Managing coastal subsidence." *Land Subsidence in the United States*, Galloway, Jones, and Ingerbritsen, eds., U.S. Geological Survey Circular 1182, 35-48.

Dagan, G. and Bear, J. (1968). "Solving the problem of local interface up-coning in a coastal aquifer by the method of small perturbations." *Journal* of *Hydraulic Research,* 6(1), 15-44.

Das Gupta, A. (1983). "Steady interface upconing beneath a coastal infiltration gallery," *Ground Water,* 21(4), 465-474.

Davis, G.H., Small, J.B. and Counts, H.B. (1963). "Land subsidence related to decline of artesian pressure in the Ocala Limestone at Savannah, Georgia." *Engineering Geology Case Histories,* No.4, Geological Society of America, 1-8.

Dillon, P.J. (2002). "Management of aquifer recharge for sustainability." *Proc., 4th Intl. Symposium on Artificial Recharge of Groundwater, ISAR4, Adelaide*, A. A. Balkema, 567 p.

Domenico, P.A. (1972). *Concepts and models in groundwater hydrology,* McGraw-Hill, New York.

Domenico, P.A., Anderson, D.V. and Case, C.M. (1968). "Optimal ground-water mining." *Water Resources Research*, 4(2), 247-255.

Domenico, P.A. and Schwartz, F.W. (1990). *Physical and chemical hydrology*, John Wiley & Sons, New York, 824 p.

Driscoll, F.G. (1995). *Groundwater and wells*, 2nd edition, USF Johnson Screens, 1089 p.

El Paso Water Utilities (2008). *El Paso Water Utilities – Public Service Board, Desalination plant*, <http://www.epwu.org/water/desal_info.html> (Aug. 28, 2008).

Environmental Protection Authority (EPA) (2005). *s16(e) managed aquifer recharge*, <http://www.epa.wa.gov.au/template.asp?ID=45&area=Reviews&Cat= s16(e) +Managed +Aquifer+Recharge> (Oct. 2005).

Essley, P.L. Jr. (1965). "The difference between nominal and effective interest tables and nominal and effective rates of return." *Journal of Petroleum Technology,* 17(8), 911-918.

Fett, J.D., Hamilton, D.H., and Fleming, F.A. (1967). "Continuing surface displacements along the Casa Loma and San Jacinto faults in San Jacinto Valley, Riverside County, California." *AEG Bulletin*, IV, 1-22.

Fetter, C.W. Jr. (1994). *Applied hydrogeology*, 4th edition, Prentice-Hall, Upper Saddle River, N.J., 691 p.

Fowler, L.C. (1981). "Economic consequences of land surface subsidence." *Journal of the Irrigation and Drainage Division,* ASCE, 107(IR2).

Freeze, R.A. (1971). "Three-dimensional, transient saturated-unsaturated flow in a groundwater basin." *Water Resources Research*, 7, 347-366.

Freeze, R.A., and Cherry, J.A. (1979). *Groundwater*, Prentice-Hall, Englewood Cliffs, N.J., 604 p.

Gabrysch, R.K. (1967). *Development of groundwater in the Houston District, Texas, 1961-65*, Report 63, Texas Water Development Board.

Galloway, D.L., Jones, D.R. and Ingebritsen, S.E. (2000). *Measuring land subsidence from space*, U.S. Geological Survey Fact Sheet 051-00, 4 p.

Getches, D. H. (1997). *Water law in a nutshell,* West Publishing Co., St. Paul, Minn., 455 p.

Hantush, M.S. (1955). *Preliminary quantitative study of the Roswell groundwater reservoir, New Mexico*, New Mexico Inst. of Mining and Tech., Socorro, N.M.

Haveman, R.H. (1969). "The opportunity cost of displaced private spending and the social discount rate." *Water Resources Research,* 5(5), 947-957.

HDR Engineering, Inc., Water Resources Associates, Malcolm Pirnie, Inc.. and PB Water (2000). *Desalination for Texas water supply. Part A: Membrane technologies and costs. Part B: Economic importance of siting factors for seawater desalination*, 188 p.

Heimes, F.J., and Luckey, R.R. (1982). *Method for estimating irrigation requirements from groundwater in the High Plains in parts of Colorado, Kansas, Nebraska, New Mexico, Oklahoma, South Dakota, Texas, and Wyoming*, U.S. Geological Survey Water-Resources Investigations Report 82–40, 64 p.

Helm, D.C. (1994a). "Horizontal aquifer movement of a Theis-Thiem confined system." *Water Resources Research*, 30(4), 953-964.

Helm, D.C. (1994b). "Hydraulic forces that play a role in generating fissures at depth." *AEG Bulletin*, 31(3), 293-304.

Henry, G.R. (1959). "Salt intrusion into fresh water aquifers." *Journal of Geophysical Research,* 64(2), 1911-1919.

Heywood, C.E., and Yager, R.M. (2003). *Simulated ground-water flow in the Hueco Bolson, an alluvial-basin aquifer system near El Paso*, Texas, U. S. Geological Survey Water- Resources Investigations Report 02-4108, 73 p.

Holzer, T.L. (1984). "Ground failure induced by groundwater withdrawal from unconsolidated sediments." *Reviews in engineering geology*, 6, T. L. Holzer, ed., Geol. Soci. of Am., 67-105.

Hutchison, W. (2004). *Hueco Bolson groundwater conditions and management in the El Paso area*, El Paso Water Utilities Hydrogeology Report 04-01.

Hutchison, W.R. (2008). Deep-well injection of desalination concentrate in El Paso, Texas, *Southwest Hydrology*: Vol.7, No. 2, pp.28-30

Holzer, T.L. and Pampeyan, E.H. (1979). "Earth fissures and localized differential subsidence." *Water Resources Research*, 17(1), 223-227.

Hutson, S.S., Barber, N.L., Kenny, J.F., Linsey, K.S., Lumia, D.S., and Maupin, M.A. (2004). *Estimated use of water in the United States in 2000*, U.S. Geological Survey Circular 1268, 46 p.

Hwang, J.M. and Wu, C.M. (1969). "Land subsidence problems in Taipei Basin." *Land Subsidence,* International Association of Scientific Hydrology, Vol. I, Publication No. 88.

Jenkins, C.T. (1968). "Techniques for computing rate and volume of stream depletion by wells." *Ground Water,* 6(2), 37-46.

Jobson, S. (2003). *Water of strife: The geopolitics of water in the Euphrates-Tigris and Jordan River basins*, Royal Institute of International Affairs, Briefing Paper No. 4, 12 p.

Johnson, A.I. (1981). "Some factors contributing to decreased well efficiency during fluid injection." *Proc., Second Symposium on Water for Subsurface Injection,* ASTM-STP 735, American Society for Testing and Materials, Baltimore, Md., 89-101.

Johnson, G., Cosgrove, D. and Lovell, M. (1998). "What is meant by surface and ground water interaction?" *Surface water and ground water interaction*, <http://www.if.uidaho.edu/~johnson/ifiwrri/sr3/swgw.html#topone> (Oct. 2005).

Johnson, A.I. and Pyne, R.D. (1994). *Artificial Recharge of Groundwater II, Proceedings of the Second International Symposium on Artificial Recharge of Ground Water, Walt Disney World Swan, Florida, July 17-22*, ASCE, Reston, Va., 938 p.

Kansas Department of Agriculture (2009). *Kansas Water Appropriation Act.* <http://www.ksda.gov/includes/statute_regulations/appropriation/KSWaterAppropriationAct82a_701.pdf> (Apr. 10, 2010).

Kazmann, R.G. (1956). "Safe yield in ground-water development, reality or illusion?" *Journal of the Irrigation and Drainage Division,* ASCE, 82(IR3).

Kazmann, R.G., and Heath, M.M. (1968). "Land subsidence related to groundwater off take in the New Orleans area." *Transactions, Gulf Coast Assoc. of Geo. Societies*, 18, 108-113.

Klein, H., (1965). "Salt Intrusion Can Be Controlled," *Leaflet* 7, Florida Board of Conservation, Division of Geology.

Kopper, W. and Finlayson, D. (1981). "Legal aspects of subsidence due to well pumping." *Journal of the Irrigation and Drainage Division, ASCE*, 107(IR2).

LBG-Guyton Associates (2003). *Brackish groundwater manual for Texas regional water planning groups*, prepared for Texas Water Development Board, Austin, Tex., 188 p.

Leake, S.A. (1997). "Land subsidence from ground-water pumping." *Impact of climate change on the southwestern United States*, U.S. Geological Survey.

Leake, S.A. (2004). *Land subsidence from ground-water pumping*, <http://geochange.er.usgs.gov/sw/changes/anthropogenic/subside/> (Apr. 7, 2010).

Lee, C.H. (1915). "The determination of safe yield of underground reservoirs of the closed basin type." *Transactions of American Society of Civil Engineers*, 78, 148-151.

Lee, C. and Cheng, R.T. (1974). "On seawater encroachment in coastal aquifers." *Water Resources Research,* 10(5), 1039-1043.

Li, J. (2003). "A nonlinear elastic solution of 1-d subsidence due to aquifer storage and recovery applications." *Hydrogeology Journal,* 11, 646-658.

Li, J. (2007a). "Transient radial movement of a confined leaky aquifer due to variable well flow rate." *Journal of Hydrology*, 333(2-4), 542-553.

Li, J. (2007b). "Analysis of radial movement of an unconfined leaky aquifer due to pumping and injection." *Hydrogeology Journal,* 15(6), 1063-1076.

Li, J. (2007c). "Analysis of radial movement of a confined aquifer due to pumping and injection." *Hydrogeology Journal,* 15(3), 442-458.

Li, J. and Helm, D.C. (2000). "A nonlinear viscous model for aquifer compression associated with ASR applications." *Proc. 6th Int. Symp. on Land Subsidence*, L. Carbognin et al., eds., Ravenna, Italy, 2, 319-330.

Li, J. and Helm, D.C. (2001a). "A nonlinear elasto-viscous model for subsidence due to artificial recharge-discharge." *Proc. 6th Symp. on Geological Engineering and Geotechnical Engineering*, Las Vegas, Nev., 437- 446.

Li, J. and Helm, D.C. (2001b). "Using an analytical solution to estimate the subsidence risk caused by ASR applications." *Journal of Environmental & Engineering Geosciences,* 7(1), 67-79.

Lofgren, B.E. (1963). *Land subsidence in the Arvin-Maricopa area, San Joaquin Valley, California*, U.S. Geological Survey Professional Paper 475-8, Article 47, 171-175.

Lofgren, B.E. (1969a). "Field measurement of aquifer-system compaction, San Joaquin Valley, California, USA." *Land Subsidence,* Vol. 2, International Association of Scientific Hydrology, Publication 88, 272-284.

Lofgren, B.E. (1969b). "Land subsidence due to the application of water." *Reviews in Engineering Geology* 2, Geological Society of America, 271-303.

Lofgren, B.E. and Klansing, R.L. (1969). *Land subsidence due to groundwater withdrawal, Tulare-Wasco area, California*, U.S. Geological Survey Professional Paper 473B.

Los Angeles County Flood Control District (1951). *Report on tests for creation of fresh water barriers to prevent salinity intrusion performed in West Coastal Basin, Los Angeles County, California.*

Louisiana Water Resources Research Institute (1968). *Saltwater encroachment into aquifers*, Bulletin 3, Louisiana State University, Baton Rouge, La.

Lund, W. R., DuRoss, C.B., Kirby, S.M., McDonald, G.N., Hunt, G., and Vice, G.S. (2005). *The origin and extent of earth fissures in Escalante Valley, Southern Escalante Desert, Iron County, Utah*, Special Study 115, Utah Geological Survey, Utah Department of Natural Resources, 30 p.

Mace, R.E., Mullican, W.F., III, and Way, T. (S.-C.). (2001). "Estimating groundwater availability in Texas." *Proc., 1st annual Texas Rural Water Association and Texas Water Conservation Association Water Law Seminar. Water Allocation in Texas: The Legal Issues, Austin, Texas, January 25-26, 2001*, Section 1, 16 p.

Marsden, S. and Davis, N. (1967). "Geological subsidence." *Scientific Am.,* 216(6), 93-100.

McGuire, V.L. (2003). *Water-level changes in the High Plains aquifer, Predevelopment to 2001, 1999 to 2000, and 2000 to 2001*, U.S. Geological Survey Fact Sheet FS–078–03, 4 p.

McGuire, V.L., Johnson, M.R., Schieffer, R.L., Stanton, J.S., Sebree, S.K. and Verstraeten, I.M. (2003). *Water in storage and approaches to ground-water management, High Plains aquifer, 2000*, U.S. Geological Survey Circular 1243, Reston, Va.

Meinzer, O.E. (1920). "Quantitative methods of estimating ground-water supplies." *Bulletin of the Geological Society of America*, 31(2), p. 329-338.

Meinzer, O.E. (1923). *Outline of groundwater in hydrology with definitions*, U.S. Geological Survey.

Mercer, J.W., Larson, S.P., and Faust, C.R. (1980). "Simulation of saltwater interface motion." *Ground Water,* 18(4), 374-385.

Michel, T.A. (2006). "100 years of groundwater use and subsidence in the Upper Texas Gulf Coast." *Aquifers of the Gulf Coast of Texas*, Texas Water Development Report 365, 139-148.

Mickley, M.C. (2001). *Membrane Concentrate Disposal: Practices and Regulation.* U.S. Bureau of Reclamation, Desalination and Water Purification Research and Development Program Report No. 69. <http://www.usbr.gov/pmts/water/media/pdfs/report069.pdf> (Apr. 2010).

Montgomery, J.M. and Toups, J.M. (1967). *Pilot wastewater reclamation and injection study for Orange County Water District.*

Muckel, D.C. (1955). "Pumping groundwater so as to avoid overdraft." *Water-Yearbook of the Department of Agriculture*, 294-295.

Muckel, D.C. (1959). *Replenishment of groundwater supplies by artificial means*, Technical Bulletin 1195, U.S. Soil and Water Conservation Service.

National Climatic Data Center (NCDC) (2010). "Annual precipitation Contiguous United States", *Climate at A Glance,* <http://climvis.ncdc.noaa.gov/cgi-bin/cag3/hr-display3.pl> (Apr. 2010).

National Research Council (NRC) (1997). *Valuing groundwater*, National Academy Press, Washington, D.C.

NRC (1998). *Issues in potable water reuse: The viability of augmenting drinking water supplies with recycled water.* National Academy Press, Washington, D.C.

NRC (2008). *Prospects for managed underground storage of recoverable water*, Committee on Sustainable Underground Storage of Recoverable Water, Water Science and Technology Board, National Academy Press, Washington, D.C., 337 p.

Neighbors, R. (1981). "Subsidence in Harris and Galveston Counties, Texas." *Journal of the Irrigation and Drainage Division, ASCE*, 107(IR2).

New Mexico Office of the State Engineer (2005). *Underground water basins in New Mexico*, <http://www.ose.state.nm.us/PDF/Maps/underground_water.pdf> (Oct. 2005).

Newport, R.D. (1975). *Salt water intrusion in the United States*, R. S. Kerr Environmental Research Laboratory, Ada, Okla.

Oklahoma Water Resources Board (OWRB) (2002). *2001 Annual Report*, Oklahoma University Printing Services, Norman, Okla., 18 p.

Patt, R.O. and Maxey, G.B. (1978). *Mapping of earth fissures in Las Vegas Valley, Nevada*, Publication No. 41501, Water Resources Center, Desert Research Institute, University of Nevada-Reno.

Peterson, J.M., Marsh, T.L. and Williams, J.R. (2003). "Conserving the Ogallala aquifer: Efficiency, equity, and moral motives." *Choices*, 1, 15-18.

Pinder, G.F. and Cooper, H.H. (1970). "A numerical technique for calculating the transient position of the saltwater front." *Water Resources Research,* 6(3), 875-882.

Poland, J.F. (1969). "Status of present knowledge and need for additional research on compaction of aquifer systems." *Land Subsidence,* Vol. 1, International Association of Scientific Hydrology, Publication 88, 11-21

Poland, J.F. (1981). "Subsidence in United States due to groundwater withdrawal." *Journal of the Irrigation and Drainage Division, ASCE*, 107(IRl), 115-135.

Poland, J.F. (1984). *Guidebook to studies of land subsidence due to groundwater withdrawal*, UNESCO.

Poland, J.F. and Davis, G.H. (1956). “Subsidence of the land surface in the Tulare-Wasco (Delano) and Los Banos-Kettleman City area, San Joaquin Valley, California.” *Transactions, American Geo physical Union*, 37(3), 287-296.

Poland, J.F. and Davis, G.H. (1969). “Land subsidence due to withdrawal of fluids.” *Reviews in Engineering Geology*, 2, Geological Society of America.

Poland, J.F. and Green, J.H. (1962). *Subsidence in the Santa Clara valley, California, a progress report*, U.S. Geological Survey Water Supply Paper 1619-C.

Puri, S. (ed.) (2001). *Internationally shared (transboundary) aquifer resources management – A framework document*, IHP-VI, IHP Non Serial Publications in Hydrology, UNESCO, Paris, 71 p.

Puri S. and Aureli, A. (2005). “Transboundary aquifers: A global program to assess, evaluate, and develop policy.” *Ground Water*, 43(5), 661-668.

Puri, S. and El Naser, H. (2003). “Intensive use of groundwater in transboundary aquifers.” *Intensive Use of Groundwater*, R. Llamas and E. Custodio, eds., Balkema Publishers, Lisse, The Netherlands, 425-438.

Pyne, R.D.G. (1995). *Groundwater recharge and wells,* Lewis Publishers, Boca Raton, Fla.

Renshaw, E. F. (1963). “The management of groundwater reservoirs.” *J. Farm Econ.*, 45, 285-295.

Rumer, R.R. Jr., and Shiau, J.C. (1968). “Salt water interface in a layered coastal aquifer.” *Water Resources Research,* 4(5), 1235-1247,

Ryder, P.O. (1978). *Model evaluation of the Cypress Creek wellfield in West-Central Florida*, U.S. Geological Survey Water Resources Investigations 78-79.

Sahni, B.M. (1972). *Salt water coning beneath fresh-water wells*, Water Management Tech. Report No. 18, Colorado State University, Fort Collins, Colo.

Schmorak, S. and Mercado, A. (1969). “Upconing of fresh water-sea water interface below pumping wells: Field study.” *Water Resources Research,* 5(6), 1290-1311.

Shamir, V. and Dagan, G. (1971). “Motion of the sea water interface in coastal aquifers: A numerical solution.” *Water Resources Research,* 7(3), 644-657.

Sheng, Z. (2005). “An aquifer storage and recovery system with reclaimed wastewater to preserve native groundwater resources in El Paso, Texas.” *Journal of Environmental Management*. 75(4), 367-377.

Sheng, Z. and Devere, J. (2005). “Systematic management for a stressed transboundary aquifer: The Hueco Bolson in the Paso del Norte region.” *Journal of Hydrogeology*, 13(5-6), 813-825.

Sheng, Z. and Helm, D.C. (1998). “Multiple steps of earth fissuring caused by groundwater withdrawal.” *Case Studies of Land Subsidence and Current Research, Proc., Joseph F. Poland Symp., Sacramento, Calif., October, 1995*, J. W. Borchers, ed., Star Publishing, Belmont, Calif., 149-154.

Sheng, Z., Helm, D.C., and Li., J. (2003). “Mechanisms of earth fissuring caused by groundwater withdrawal.” *Journal of Environmental & Eng. Geoscience*, IX(4), 313-324.

Sherwood, C.B. and Grantham, R.G. (1966). "Water Control vs. Sea Water Intrusion, Broward County, Florida," *Leaflet* 5, Florida Board of Conservation, Division of Geology..

Shlemon, R.J. and Davis, P. (1992). "Ground fissures in the Temecula area, Riverside County, California." *Engineering Geology Practice in Southern California*, B. W. Pipkin and R. J. Proctor, eds., AEG Special Publication 4, 275-288.

Shlemon, R.J. and Hakakian, M. (1992). "Fissures produced both by groundwater rise and groundwater fall: A geological paradox in the Temecula-Murrieta area, southwestern Riverside County, California." *Proc., 35th Annual Meeting of the Association of Engineering Geologists*, M. L. Stout, ed., 143-150.

Shoemaker, R.P. (1963). "A graphical short-cut for rate of return determination." *World Oil,* July 1963, 72-84; August 1963, 69-73; September 1963, 64-68.

Schumann, H.H. and Cripe, L.S. (1986). "Land subsidence and earth fissures caused by groundwater depletion in southern Arizona, U.S.A.." *Proc., International Symposium on Land Subsidence, 3rd, Venice, 1984*, A. I. Johnson, L. Carbognin, and L. Ubertini, eds., International Association of Scientific Hydrology Publication 151, 841–851.

Slichter, C.S. (1899). *Theoretical investigations of the motion of groundwaters.* 19th Annual Report, Part II, U.S. Geological Survey.

Sophocleous, M. (1997). "Managing water resources systems: Why safe yield is not sustainable." *Ground Water*, 35(4), 561.

Sophocleous, M.A. (1998). "Water resources of Kansas: A comprehensive outline." *Perspectives on Sustainable Development of Water Resources in Kansas*, M.A. Sophocleous, ed., Kansas Geological Survey Bulletin 239, 239 p.

Sophocleous, M.A. (2000). "From safe yield to sustainable development of water resources: The Kansas experience." *Journal of Hydrology*, 235, 27-43.

Templer, O.W. (1992). "The legal context of groundwater use." *Groundwater exploitation in the High Plains,* D. E. Kromm and S. E. White, eds., University of Kansas Press, Lawrence, Kans., 240 p.

Texas Water Development Board (TWDB) (1968). *The Texas water plan*, Austin, Texas.

TWDB (2002). *Water for Texas - 2002, Austin, Texas,* http://www.twdb.state.tx.us/publications/reports/State_Water_Plan/2002/FinalWaterPlan2002.asp (Apri. 8, 2010).

TWDB (2010). *Groundwater conservation districts*, <http://www.twdb.state.tx.us/mapping/maps/pdf/gcd_only_24x24.pdf> (Apr. 8, 2010).

Thelin, G.P. and Heimes, F.J. (1987). *Mapping irrigated cropland from Landsat data for determination of water use from the High Plains aquifer in parts of Colorado, Kansas, Nebraska, New Mexico, Oklahoma, South Dakota, Texas, and Wyoming*, U.S. Geological Survey Professional Paper 1400–C, 38 p.

Thiem, G. (1906). *Hydrologische methodern*, Leipzig, Geghardt, 56 p.

Thomas, H.E. (1951). "Fluctuations of groundwater levels." *International Union of Geophysics and Geodesy*, Brussels.

Thomas, H.E. (1952a). "Ground water regions of the United States: Their storage facilities." U.S. 83rd Congress, Interior and Insular Affairs Committee, U.S. House of Representatives, 76 p.

Thomas, H.E. (1952b). "Cedar City valley." *Status of development of selected groundwater basins in Utah*, Engineering Tech. Publication 7, Utah State University, 22-34.

Thomas, H.E. (1955). *Water rights in areas of groundwater mining*, U.S. Geological Survey Circular 347.

Thomas, H.E. (1957). *The conservation of groundwater,* McGraw-Hill, New York.

Turneaure, F. E., and Russell, H. L. (1901). *Public water supplies*, John Wiley & Sons, New York, 269 p.

United States Army Corps of Engineers (USACE) (1999). *C&SF restudy final integrated feasibility report and programmatic environmental impact statement (PEIS), Jacksonville, Florida.*

USACE (2004). *Proposed leasing of lands at Fort Bliss, Texas for the proposed siting, construction, and operation by the city of El Paso of a brackish water desalination plant and support facilities: Draft environmental impact statement*, prepared for U.S. Army Air Defense Artillery Center and Fort Bliss.

USACE and South Florida Water Management District (2003). *Central and southern Florida project, comprehensive Everglades restoration plan: Project management plan, wastewater reuse technology pilot project, final integrated project, implementation report and environmental impact statement.* USACE, Jacksonville, Fla.

U.S. Bureau of Reclamation (USBR). 2003. *Desalting handbook for planners*, 3rd Edition, USBR Washington, D.C.

U.S. Department of Agriculture (1999). *Census of agriculture, v. 1, geographic area series*, (CD-ROM AC97–CD–VOL1–1B), National Agricultural Statistics Service.

U.S. Public Health Service (1962). *Drinking water standards*, U.S. Public Health Service Pub. 956, 61 p.

U.S. Supreme Court (1982). "Sporhase v. Nebraska ex rel. Douglas." *United States Reports*, v. 458, Government Printing Office, Washington, D.C., 941–965.

Water Resources Engineers, Inc. (1969). *An investigation of salt balance in the upper Santa Ana River basin: A report* to *the state of California Water Resources Control Board.*

Williams, C.C. and Lohman, S.W. (1949). *Geology and ground-water resources of a part of south central Kansas*, Geological Survey of Kansas, Bulletin 212, 79 p.

Wyoming State Engineer's Office (2005). "Regulating and administering Wyoming's water resources." *About the state engineer's office*, <http://seo.state.wy.us/about.aspx> (Oct. 2005).

CHAPTER 9

SUBSURFACE AND SURFACE WATER FLOW INTERACTIONS

Mohamed M. Hantush[1], **Latif Kalin**[2], and **Rao Govindaraju**[3]
[1]National Risk Management Research Laboratory, USEPA, Cincinnati, Ohio, USA
[2]Auburn University, Auburn, Alabama, USA
[3]Purdue University, Lafayette, Indiana, USA

9.1 Introduction

Groundwater and surface water are major system components in the hydrologic cycle that interact continuously in a wide variety of landscapes and climate conditions. Due to these interactions, overdraft of water from one system has the potential to deplete water from the other, and contamination of one system can lead to degradation of water quality of the other. Following storm events, saturated riparian banks release stored water gradually to sustain streamflow during prolonged no-rain and/or drought periods. The water so stored supplies necessary moisture to sustain riparian vegetation and aquatic ecosystems both inland and along riparian stream corridors and coastlines. Shallow restrictive geologic strata cause some of the infiltrating water to move parallel to the soil surface as interflow often aided by preferential flow paths to contribute to the stream as subsurface storm flow. In some hillslopes, a part of the interflow water may reappear on the surface as return flow. During and after intense rainfall events, the water table near a stream rises to ground surface and contributes saturation overland (or saturation excess) flow to surface runoff from variable source areas (Freeze 1974). Thus, surface and subsurface waters interact in complex ways to generate streamflow. Lakes and wetlands can recharge the water table and ultimately contribute to baseflow, or serve as the locus points and/or areas of groundwater discharge. Groundwater and surface water interactions in lakes and wetlands are complex phenomena; these systems sometimes act as flow-through systems where groundwater flows into the system from one side and discharges downgradient from the other side (Anderson and Munter 1981). Without adequate management, excessive pumping of groundwater from an aquifer can result in the depletion of streams and wetlands and impairment of ecosystems. These various modes of interactions exhibit complex spatiotemporal behaviors that are scale dependent. An understanding of the basic principles and physical laws governing exchanges between groundwater and surface water, therefore, is needed, including tools for modeling the interactions at multiple scales, not only for the sound management of these resources, but also for the maintenance of healthy ecological systems.

In this chapter we present basic concepts and principles underlying the phenomena of groundwater and surface water interactions. Fundamental equations and analytical and numerical solutions describing stream-aquifer interactions are presented in hillslope and riparian aquifer environments. Analytical and numerical techniques for solving flow routing problems in channels with contiguous alluvial aquifers are covered. We also discuss classical and recent developments in modeling stream depletion due to groundwater pumping. Several sections are devoted to concepts and models for surface runoff and subsurface water interactions and variably saturated flow. Physically-based baseflow models derived from the well-known Boussinesq equations and their relationship to the familiar exponential and hyperbolic empirical recession curves are presented, including implications on up-scaling aquifer parameters to catchment scales. Concepts behind different modes of coupling subsurface and surface waters, with emphasis on the widely used MODFLOW package and variations thereof are discussed. A brief discussion of the analytical element method and watershed-scale modeling of groundwater and surface water interactions is presented. The chapter ends with a case-study on watershed-scale modeling of ground and surface water interaction. The topic of saltwater intrusion into coastal aquifers, another form of interaction between surface water and groundwater, is addressed in Chapter 10.

9.2 Subsurface Flow in Near Stream Environments

Most of the streams in the world exist in alluvial environments and are fed by water table aquifers comprised of granular material such as sand and gravel. These streams typically exchange flow with the adjacent unconfined aquifers and are surrounded by riparian vegetative and wooded buffer strips that extract and transpire water from the underlying shallow water table or directly from the stream. Although not as common, some streams cut through impervious layers and partially penetrate and come into contact with confined aquifers. Such streams receive water under artesian conditions. During dry periods, or in arid zone areas, streamflow is primarily derived from groundwater discharge and sometimes supplemented by regulated reservoir releases. Understanding groundwater-surface water interactions is not only important for water supply management, but also for the maintenance of healthy aquatic ecological systems and riparian vegetation (Sophocleous 2002). Impacts of these interactions are also relevant to problems involving the legal rights of communities whose main water supply depends on the flow of the surface streams.

Streams whose water levels are lower than water table elevations (or piezometric levels in the case of confined aquifers) in the adjacent aquifer are called *effluent streams*: they receive groundwater from the aquifer (Figure 9.1a). On the other hand, if the water level in the adjacent aquifer is below the water level in a stream, the water flows from the stream to the aquifer (Figures 9.1b and 9.1c). Such a stream is called *influent stream*. The same stream cross section may be receiving water from one side and losing water from the other side, hence an influent and effluent stream or *flow-through stream* (Figure 9.1d). Furthermore, the same stream may be effluent in one reach and influent in another reach. Hydrogeologic conditions such as the regional hydraulic head gradient and slope of underlying bedrock can produce flow-through-type streams.

A significant portion of the water delivered to the surface during precipitation and snowmelt events infiltrates into the soil and, after filling the soil moisture deficit, percolates downward below the root zone to replenish the water table in unconfined aquifers. Groundwater mounds in response to increased recharge and the water table rises above water levels in adjacent surface water bodies, such as streams, rivers, and lakes. As a general example, in near-stream environments the resulting hydraulic head gradient produced by water table mounding directs groundwater flow toward the streams. If the precipitation or snowmelt event continues and whenever soil and hydrogeological conditions permit, the water table will continue to rise and the hydraulic head gradient will further increase resulting in greater groundwater discharge to the stream. This process continues until precipitation or snowmelt ceases and water table accretion starts to diminish. During dry periods, the water table declines and the hydraulic gradient decreases as groundwater moves from higher elevations towards the stream. Consequently, less water will be discharged to the stream as drainage continues and the water table accretion rate decreases (Singh 1992). This process is called *baseflow recession*, and the rate at which streamflow is sustained in the absence of precipitation and in the absence of artificial storage water works is referred to as *baseflow* (Brutsaert 2005). The phenomenon described above typically occurs in lower order streams with shallow and rapidly draining channels. However, in higher order channels that receive greater channel inflow during storm or snowmelt events, an additional process comes into play: temporary storage of flood waters in bank sediments and the gradual release thereafter - a phenomenon known as "bank storage" - influences considerably the dynamics of groundwater flow at the stream-aquifer interface and supplements pre-event baseflow and newly recharged groundwater as components of total groundwater discharge to streams. Basic fundamental equations and key results describing the above surface-subsurface drainage phenomena are presented in the following sections.

9.2.1 Basic Groundwater Flow Equation

Groundwater flow in a heterogeneous, anisotropic unconfined aquifer, where the principal axes are in the x, y, and z directions, can be described by (Bear 1979)

$$\nabla \cdot \mathbf{K} \nabla \varphi + W = S^* \frac{\partial \varphi}{\partial t}; \quad \nabla = \frac{\partial}{\partial x}\mathbf{i} + \frac{\partial}{\partial y}\mathbf{j} + \frac{\partial}{\partial z}\mathbf{k}; \tag{9.1}$$

$$\mathbf{K} = \begin{bmatrix} K_x & 0 & 0 \\ 0 & K_y & 0 \\ 0 & 0 & K_z \end{bmatrix} \tag{9.2}$$

where $\varphi = z + p/\gamma$ is hydraulic head (piezometric head in confined aquifers) [L]; p is pore-water pressure [$ML^{-1}T^{-2}$]; γ is specific weight of water [$ML^{-3}T^{-2}$]; $\mathbf{K}$ is the anisotropic hydraulic conductivity tensor; W is net volumetric source/sink term [$L^3L^{-3}T^{-1}$] representing water added to or removed from the medium (i.e., positive for sources and negative for sinks); S^* is the aquifer *specific storativity* referred to also as *elastic storativity* [L^{-1}]; K_x, K_y, and K_z are, respectively, the hydraulic conductivities along the x, y, and z directions [LT^{-1}]; ∇ is the gradient operator; **i**, **j**, and **k** are, respectively, the x, y, and z unit vectors; $\nabla \cdot$ **a** is the divergence of vector **a**; x, y, and z are Cartesian coordinates with z axis typically taken along the vertical [L]; and t is time [T]. The specific discharge vector is given by Darcy's law,

$$\mathbf{q} = -\mathbf{K}\nabla\varphi\ ; \quad \mathbf{q} = q_x i + q_y j + q_z k \tag{9.3}$$

where q_x, q_y, and q_z are the specific discharge components in the x, y, and z directions, respectively [LT^{-1}]. Note that the off-diagonal elements of the second rank conductivity tensor **K** are zeros because Equation (9.2) is expressed in a principal axes system.

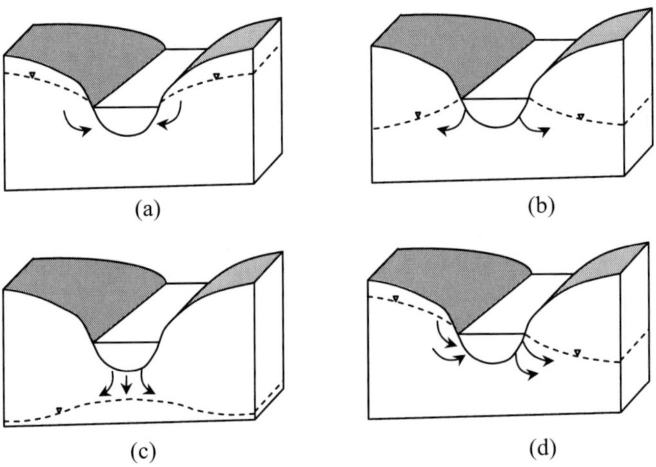

Figure 9.1: Stream-aquifer interrelationships: (a) effluent stream, receiving water from aquifer; (b) influent stream, stream connected to water table and losing water to the aquifer; (c) influent stream, stream losing water to a deep, disconnected water table; and (d) flow-through stream, receiving water from one side and losing water from the opposite side.

While the term W in Equation (9.1) aims generically to account for all sources and sinks in the flow domain, point and simple line sources and sinks (e.g., wells, springs, straight thin stream channels, tile drains) in W can be explicitly accounted for using the well-known Dirac delta, δ, or the Heaviside step function, $H(x)$. Irregular line and area sources/sinks (e.g., drainage galleries, lakes, leakage losses/recharge, etc.), however, are typically treated as boundary conditions in detailed three-dimensional analyses. Expressions for line, areal, and volume source are provided by Gunduz and Aral (2005a) using the properties of the Dirac delta function and parametric representation of a curve in a three-dimensional space. The expressions allow for explicit accounting of discontinuous forcing terms in the governing partial differential equation and facilitate easy use of their integrals in the numerical solution of the groundwater flow equation, such as the Galerkin finite-element method.

Equation (9.1) applies to both confined and unconfined aquifers except that in unconfined aquifers the flow domain is bounded from above by the water table, which is a *free surface* where the pressure is atmospheric, conveniently taken as $p = 0$ and forms a moving boundary separating the saturated flow domain from the upper unsaturated zone. The free surface boundary condition with water table accretion is given by the following nonlinear, unsteady equation (Hantush 1964; Bear 1979),

$$n_e \frac{\partial \varphi}{\partial t} = \left[K_x \left(\frac{\partial \varphi}{\partial x} \right)^2 + K_y \left(\frac{\partial \varphi}{\partial y} \right)^2 + K_z \left(\frac{\partial \varphi}{\partial z} \right)^2 - (K_z + N) \frac{\partial \varphi}{\partial z} + N \right]; \quad at\ z = \varphi(x, y, z, t) \qquad (9.4)$$

where n_e is the effective porosity and sometimes called the specific yield [L^3L^{-3}]; and N is water table replenishment (also, recharge/accretion) rate per unit area [LT^{-1}]. The quantity n_e should be distinguished from the porosity of the unconfined aquifer since drainage is never complete when the water table is falling. A certain amount of water in the pores is retained against gravity by capillarity and other forces (Brutsaert 2005). At the free surface, $\varphi(x,y,z,t) = z$ because $p = 0$. Equation (9.4) decouples the saturated zone from the unsaturated soil and neglects the capillary fringe above the free surface. The nonlinear and unsteady free surface equation underscores the difficult task of solving Equation (9.1) subject to boundary condition (9.4) - the position of the water table at time t must be known a priori for Equation (9.1) to be solved for the head at any point in the flow domain, but the latter must be known beforehand for the position of the water table to be determined at time t.

In confined aquifers, free surface conditions do not exist and the flow is instead constrained from above by an impermeable boundary or a semi-pervious layer separating the main formation from an overlying water table aquifer and/or a surface water body (streams and lakes). The semi-pervious layer could be a relatively thick (e.g., silt and clay) layer called an *aquitard*, or made of fine sediments typical to fluvial depositional environments such as river beds and lake bottoms. Through these layers leakage occurs into or out of the aquifer. An aquifer overlain by an aquitard is referred to as a leaky confined aquifer, with leakage rate and direction controlled by the magnitude of hydraulic head change across the semi-pervious layer and its geometric and hydraulic characteristics. Of course, confined and unconfined aquifers may also rest on top of semi-pervious layers through which the aquifers hydraulically communicate and exchange groundwater with deeper formations (also referred to as source beds).

Equation (9.1) can be solved for $\varphi(x,y,z,t)$ over the flow domain D whenever its initial values $\varphi(x,y,z,0)$ are known at all points inside D at $t = 0$ and boundary conditions on $\varphi(x,y,z,t)$ and/or its gradient are imposed over the entire domain boundaries, Γ_D.

Equation (9.4) is one such boundary condition, but applies to the free surface in unconfined aquifers. Groundwater flows parallel to bedrocks and hydraulic divides, with negligible flow components normal to the boundary (Figure 9.2):

$$\mathbf{q}\cdot\mathbf{n}=0 \quad ; \quad -\mathbf{K}\nabla\varphi\cdot\mathbf{n}=0; \ (x,y,z)\in\Gamma_1 \tag{9.5a}$$

where $\mathbf{n}$ is the unit-outward vector normal to the boundary surface $\Gamma_1 \subset$ D; and $\mathbf{q}$ is the specific discharge vector evaluated on all points of the boundary surface, Γ_1. The left-hand-side of Equation (9.5a) is the specific discharge normal to the surface. A prescribed flux (Neumann condition) may also be imposed on some boundaries,

$$\mathbf{q}\cdot\mathbf{n}=q_0 \ ; \quad (x,y,z)\in\Gamma_1 \tag{9.5b}$$

where q_0 is the specified flux at Γ_1. However, at the interface between porous media and open water bodies, q_0 is usually unknown and iterative coupling numerical schemes are used to compute this term internally (Huang and Yeh 2009).

The piezometric head φ may also be specified on portions of D; in this case a prescribed head (Dirichlet) boundary condition may be imposed

$$\varphi(x,y,z,t)=\varphi_0 \ ; \quad (x,y,z)\in\Gamma_2 \tag{9.6}$$

where φ_0 is a specified head on all points of the boundary surface $\Gamma_2 \subset$ D; it need not be a constant and could be an arbitrarily specified functional relationship, e.g., linear head variation along Γ_2.

In river beds clogged with silt and clayey sediments, continuity of flux is often imposed across a thin, semi-pervious layer; this is also known as Cauchy or third-type boundary condition. While the use of pre-specified head boundary condition is not uncommon to flow problems involving a surface water body in contact with porous media (e.g., Theis 1941; Glover and Balmer 1954; and Hantush 1964, 1967), field evidence shows that in depositional environments a relatively thin layer of significantly lower hydraulic conductivity separates surface water bodies from the porous media, and that streambed conductivity can be as much as one to three orders of magnitude lower than aquifer conductivity (Fox et al. 2002, citing Larkin and Sharp 1992). Solutions to governing groundwater flow equations that are based on prescribed head at the stream-aquifer interface and which fail to account for the semi-pervious layer separating the two media could lead to serious errors in the estimation of stream-aquifer flow interactions (Sophocleous et al. 1995).

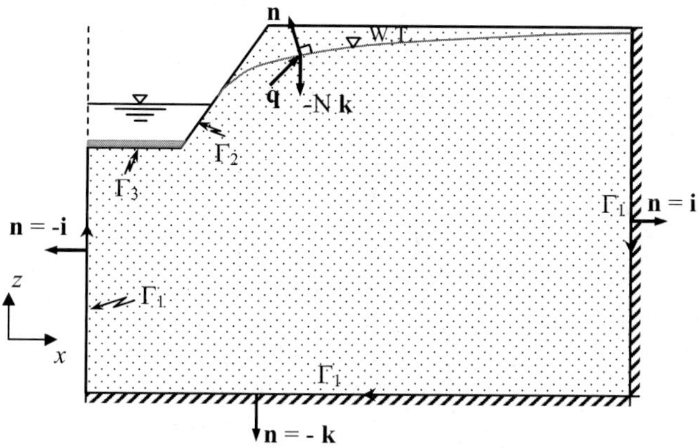

Figure 9.2: Cross-sectional stream-aquifer system with free-surface, no-flow and prescribed head, and semi-pervious boundaries.

If we neglect storage in the semi-pervious thin layer and let φ denote the piezometric head in the aquifer side of the layer and φ_0 denote the head elevation in the surface-water body side, then flow from the aquifer is,

$$\mathbf{q}\cdot\mathbf{n}=\frac{K'}{b'}(\varphi-\varphi_0);\quad (x,y,z)\in\Gamma_3 \tag{9.7}$$

where K' and b' are, respectively, the hydraulic conductivity [LT^{-1}] and thickness [L] of the thin semi-pervious layer; and $\Gamma_3 \subset$ D denotes all points of the boundary surface occupied by the semi-pervious layer. K'/b' is the streambed conductance [T^{-1}], and its reciprocal, b'/K', is the resistance of the semi-pervious layer [T]. Note that $\mathbf{q}\cdot\mathbf{n}=q_n$ is the outward-normal component of the specific discharge vector, hence, points in the direction from the aquifer to the surface water body. This equation is widely used in both analytical and numerical models to calculate the flux of water in stream-aquifer systems, which is proportional to the head gradient across the semi-pervious layer and its conductance. The rate of leakage therefore depends on the depth of water in the stream (or surface water body) and on the permeability and thickness of the riverbed. It is also widely used to describe leakage flux in three-dimensional flow in leaky confined/unconfined aquifers (Hantush 1960, 1967).

The mixed boundary condition (also called Cauchy or third-type boundary condition) has greater appeal in applications involving hydraulically-connected, surface water-groundwater systems. The MODFLOW RIVER and STREAM packages (Harbaugh and McDonald 1996), and many other analytical models, utilize Equation (9.7) to model stream-aquifer interactions. Furthermore, it has also been used in flow situations where the aquifer head drops below the bottom of the streambed and both the stream and the aquifer become hydraulically disconnected (Figure 9.1d). In this case, both MODFLOW packages assume that seepage becomes exclusively dependent on the water level in the stream and the streambed thickness and conductivity. In essence, pressure head is taken as $p = 0$ (i.e., atmospheric) at the bottom of streambed sediments, and recharge to the water table therefore becomes constant. In reality, however, the flow underneath the stream is more complex than that; when the water table drops below the streambed, an unsaturated flow regime develops between the two boundaries (e.g., Bouwer 1969; Rovey 1975; Osman and Bruen 2002; Fox and Durnford 2003). The use of streambed conductance to simulate river-aquifer interaction in regional groundwater models and the development of unsaturated flow regime below streams has received significant attention recently in the literature (e.g., Osman and Bruen 2002; Fox and Durnford 2003; Rushton 2007; Fox and Gordji 2007), and will be discussed further in a later section.

9.2.2 Hydraulic Approach

In relatively thin water table aquifers, i.e., aquifers whose thicknesses are much smaller than their areal extents, groundwater flow is essentially horizontal (i.e., in the x and y directions) and equipotential surfaces are assumed to be nearly vertical. The assumption of essentially horizontal flow was first introduced by Dupuit (1863), by recognizing that in most aquifers, the slope of the water table is very small and vertical fluxes are negligible. The Dupuit assumptions allow the integration of Equation (9.1) along the z axis, after making use of Equation (9.4), to yield vertically-averaged, two-dimensional partial differential equation, with recharge/discharge expressed as distributed and/or point sources/sinks (Bear 1979):

$$S_y \frac{\partial h}{\partial t} = \frac{\partial}{\partial x}\left[K_x (h-\eta)\frac{\partial h}{\partial x}\right] + \frac{\partial}{\partial y}\left[K_y (h-\eta)\frac{\partial h}{\partial y}\right] + N(x,y,t) - P(x,y,t) \qquad (9.8)$$

where $S_y = n_e$ is the specific yield of the aquifer [L^3L^{-3}]; h is the water table elevation [L]; η is the impervious/semi-pervious bottom elevation; $N(x,y,t)$ is the replenishment rate per unit area [LT^{-1}] from such sources as water table accretion, leakage from underlying artesian aquifer, streambed infiltration, etc.; and $P(x,y,t)$ is the sum of potential point/distributed losses [LT^{-1}] such as well pumping discharge rates, losses to streams, leakage to underlying aquifers, tile drain losses, etc. The averaging of the three-dimensional flow equation along the thickness of the aquifer is called the *hydraulic approach* (Bear 1979; Brutsaert 2005); it has also been referred to as the Dupuit-Forchheimer theory (e.g., Brutsaert 2005). Again, Equation (9.8) neglects capillary effects and seepage faces that usually develop at the stream-aquifer interface. Furthermore, elastic storativity, S^*, is often neglected since storativity due to drainage from the pore space is much larger than that resulting from the elasticity of the water and the solid matrix, $S_y >> S^* h$. A more general form of Equation (9.8) where leakage occurs across the aquifer bed is derived by Hantush (1964, p. 30).

For a horizontal impervious base, η = 0, and dropping N and P for the moment, Equation (9.8) reduces to

$$S_y \frac{\partial h}{\partial t} = K_x \frac{\partial}{\partial x}\left[h\frac{\partial h}{\partial x}\right] \tag{9.9}$$

for a homogeneous one-dimensional case. This equation is also known as the Boussinesq equation. Equation (9.8) can also be applied to flow in confined aquifers by replacing S_y with aquifer storativity $S_y = S^*B$ and setting $h - \eta = B(x,y)$, and noting that $T_{xx} = K_xB$ and $T_{yy} = K_yB$ are aquifer transmissivities [L^2T^{-1}] in the x and y directions, respectively. The main mechanism responsible for release of water in unconfined aquifers is drainage from pore space induced by lowering the water table, whereas in confined formations, it is the outcome of water and matrix compressibility.

Equations (9.8) or (9.9) show that the hydraulic approach still yields nonlinear flow in unconfined aquifers due to the moving upper flow boundary condition (free surface), but with the advantage that the nonlinear, free surface Equation (9.4) is now built into Equation (9.8) and no longer appears as a nonlinear boundary condition. Despite the simplification introduced using the hydraulic approach, the nonlinearity of the resulting Boussinesq equation continues to pose a challenge, especially for coupled and dynamic stream-aquifer interactions. However, with the advent of digital computing and the rapidly developing computing speed and efficiency, development of stable and efficient numerical methods (e.g., Huyakorn and Pinder 1983; Bear and Verruijt 1987), and availability of groundwater computer packages such as MODFLOW (Harbaugh and McDonald 1996), increasingly efficient and accurate solutions for three-dimensional flow and two-dimensional Boussinesq equation are feasible even for complex subsurface-surface water interaction problems. Linearized forms of Boussinesq Equation (9.9) constitute the basis for numerous analytical and numerical treatments of stream-aquifer interactions (e.g., Werner and Noren 1951; Haushild and Kruse 1962; Cooper and Rorabaugh 1963; Singh 1969; Hall and Moench 1972; Hunt 1990; Govindaraju and Koelliker 1994; Hantush et al. 2002; Hantush 2005). Numerical solutions (e.g., Yeh 1970; Pinder and Sauer 1971; Zitta and Wiggert 1971; Mariño 1975; Gureghian 1978) and exact/approximate solutions (Polubarinova-Kochina 1962, p.507, cited by Brutsaret 2005; Parlange et al. 1985; Parlange et al. 1998; Tolikas et al. 1990; Hogarth et al. 1997; Serrano and Workman 1998) have been developed for the nonlinear Boussinesq equation. Bruggeman (1999) provides a compendium of analytical solutions to steady-state and transient stream-aquifer flow under confined, phreatic, and leaky flow conditions, and for various recharge and geometric configurations.

9.2.3 Subsurface Drainage and Baseflow

Drainage from a Horizontal Aquifer: Assuming groundwater flow is predominantly normal to the channel length, groundwater flow in an aquifer hydraulically connected to a stream can be described by Equation (9.9). The outflow rate Q_0 from the aquifer into the stream (negative for inflow rate) is given by Darcy's equation

$$Q_0 = K_x h \frac{\partial h}{\partial x}\bigg|_{x=0} \qquad (9.10)$$

The negative sign has been omitted because Q_0 is positive in the direction of decreasing x (see Figure 9.3).

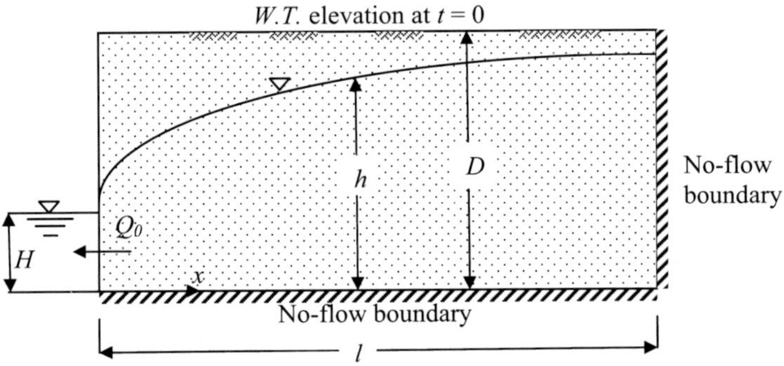

Figure 9.3: Diagrammatic representation of a horizontal aquifer hydraulically connected to a stream. Water table is initially at ground surface elevation.

By means of Boltzmann's transform, Brutsaert (2005, p. 394) presents a solution to the Boussinesq Equation (9.9) and provides a convenient formula for the outflow rate into an empty stream (H = 0) from an initially fully saturated aquifer of infinite breadth ($l \to \infty$) and resting on a horizontal impervious bed

$$Q_0 = 0.33206\sqrt{\frac{K_x S_y D^3}{t}} \qquad (9.11)$$

where D is the initial, uniform saturated thickness of the aquifer; and n_e as defined above is the drainable porosity ($n_e = S_y$). Brutsaert (2005) stresses that the physical significance of the drainable porosity remains ambiguous and that the parameter was introduced to compensate for the neglect of the partially saturated flow and capillary effects above the water table in the free-surface approximation in Equation (9.4). Equation (9.11) is also applicable for sufficiently short time where water table drawdown at x = 0 is not felt further away from the channel and where the no-flow boundary at $x = l$ has yet to influence drainage.

The long-time behavior of drainage from an initially fully saturated aquifer was obtained by Boussinesq (1904), by solving the nonlinear Boussinesq equation over a finite aquifer width (Brutsaert 2005)

$$Q_0 = \frac{0.862 K_x D^2}{l}\left(1 + \frac{1.115 K_x D}{S_y l^2} t\right)^{-2} \tag{9.12}$$

This equation is of a hyperbolic form that is widely used in baseflow recession analysis for periods without significant precipitation during which withdrawal of water is mainly from groundwater storage (e.g., Singh 1992). Equation (9.12) is based on a negligible flow depth in the connecting stream.

A linear approximation of Equation (9.9) is possible provided that water table fluctuations are small compared to the average saturated thickness of the aquifer, $\bar{h}$,

$$\frac{\partial h}{\partial t} = \frac{K_x \bar{h}}{S_y} \frac{\partial^2 h}{\partial x^2} \tag{9.13}$$

The solution of this equation for an initially fully saturated, semi-infinite aquifer width and shallow channel flow can be obtained using the method of separation of variables (Brutsaert 2005, pp. 400-404). For $t > 0.2\, n_e\, l^2/K_x \bar{h}$, Q_0 is approximated by retaining the leading term of the infinite series summation

$$Q_0 = 2\frac{K_x \bar{h} D}{l} exp\left(-\frac{\pi^2 K_x \bar{h}}{4 S_y l^2} t\right) \tag{9.14}$$

This formula corresponds to the most widely-used, empirically-based exponential baseflow recession, $Q = Q_0 e^{-ct}$ (e.g., Maillet 1905; Barnes 1939; Singh 1992; Tallaksen 1995, Dingman 2002; Kresic 2007), except here, Equation (9.14) is expressed in terms of parameters that have physical meaning and relate to initial water table depth and aquifer geometric and hydraulic characteristics. Scaling up aquifer hydraulic parameters to the catchment scale by means of Equations (9.12) and (9.14) will be discussed in a later section.

Of fundamental significance is the problem of groundwater flow under steady-state conditions in unconfined aquifer situated between two streams at different elevations (Figure 9.4). This problem is also of practical significance when computing groundwater discharge to drainage ditches fully penetrating the soil and reaching the bottom impervious layer (Ritzema 2006, citing Hooghoudt 1940 and Donnan 1946). Figure 9.4 shows a typical cross section of such drainage systems. For partially penetrating streams or drains, flow lines are essentially horizontal further away but will converge to radial flow in the vicinity of the stream/drain, and flow is thus no longer horizontal. For a sufficiently wide aquifer or distance between the drains, flow is essentially horizontal and Dupuit-Forchheimer theory can be applied to obtain solutions for the Boussinesq Equation (9.9) or its linearized form Equation (9.13). In Figure 9.4 we extend the problem to the case of a leaky bed where flow in the aquifer is either supplemented by a source bed or subject to leakage losses to an underlying formation. The steady-state solution (Dingman 2002; Ritzema 2006) is based on an impervious bed and similar water levels in the drain.

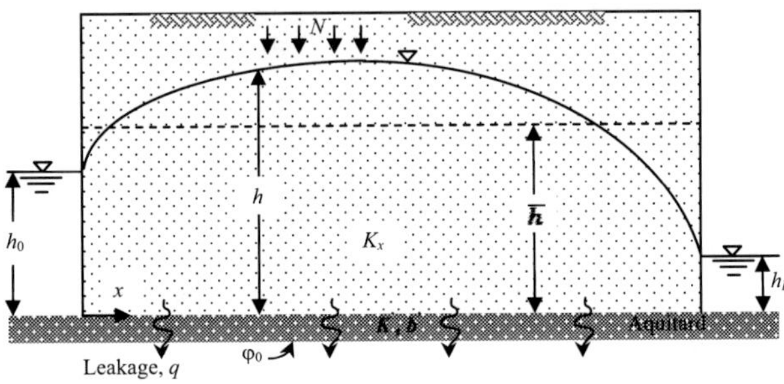

Figure 9.4: Schematic representation of a water table aquifer situated between two streams and underlain by a semi-pervious layer (aquitard).

The equation of steady-state flow toward open drains in an unconfined aquifer underlain by a semi-permeable layer can be obtained from Equation (9.9) by dropping the partial derivative with time and adding a vertical leakage term of the form given by Equation (9.7), resulting in

$$K_x \frac{\partial}{\partial x}\left[h\frac{\partial h}{\partial x}\right] - \frac{K'}{b'}\left(h-\varphi_0\right) + N = 0 \qquad (9.15a)$$

with

$$h(0) = h_0, \quad h(L) = h_L \qquad (9.15b)$$

An exact analytical solution of this boundary value problem was obtained by Emich (1962) using elliptic integration. A simpler, but approximate, nonlinear solution can be obtained by assuming small variation of h relative to some average aquifer thickness $\bar{h}$ and a constant aquifer transmissivity $T = K_x\bar{h}$. A linear form of Equation (9.15a) is

$$\frac{d^2H}{dx^2} - \frac{2}{\lambda^2(1+\beta)}H + \frac{2N}{K_x} = 0 \qquad (9.16a)$$

in which

$$H = \left(h^2 - \varphi_0^{\,2}\right), \quad \lambda = \sqrt{(b^{'} / K^{'})T}, \quad \beta = \varphi_0 / \bar{h} \tag{9.16b}$$

where φ_0 is piezometric head in the underlying formation [L]; and λ is called the leakage factor [L] that determines the areal distribution of the leakage. Storage in the semi-permeable layer is neglected and leakage flux is proportional to the hydraulic head difference across the layer.

The solution of Equation (9.16a) with Equation (9.15b) is

$$h^2(x) = \left(h_L^2 - \varphi_0^2 - \frac{2N}{K_x a}\right)\frac{sinh\left(\sqrt{a}\,x\right)}{sinh\left(\sqrt{a}\,L\right)} + \left(h_0^2 - \varphi_0^2 - \frac{2N}{K_x a}\right)\frac{sinh\left(\sqrt{a}\,(L-x)\right)}{sinh\left(\sqrt{a}\,L\right)} + \frac{2N}{K_x a} \tag{9.17a}$$

in which

$$a = \frac{2}{\lambda^2 (1+\beta)} \tag{9.17b}$$

This result shows that the importance of leakage increases as λ decreases in magnitude.

Discharge per unit stream/drain length at $x = 0$ is obtained by substituting the solution (9.17a) into Equation (9.10) and evaluating the derivative

$$Q(0) = -\frac{K_x}{2}\sqrt{a}\left[h_L^2 - \varphi_0^2 - \frac{2N}{K_x a}\right]csch\left(\sqrt{a}\,L\right) + \frac{K_x}{2}\sqrt{a}\left[h_0^2 - \varphi_0^2 - \frac{2N}{K_x a}\right]coth\left(\sqrt{a}\,L\right) \tag{9.18}$$

At $x = L$, we have (flow to the stream)

$$Q(L) = -\frac{K_x}{2}\sqrt{a}\left[h_L^2 - \varphi_0^2 - \frac{2N}{K_x a}\right]coth\left(\sqrt{a}\,L\right) + \frac{K_x}{2}\sqrt{a}\left[h_0^2 - \varphi_0^2 - \frac{2N}{K_x a}\right]csch\left(\sqrt{a}\,L\right) \tag{9.19}$$

Note the symmetry when $h_L = h_0$.

Equations (9.18) and (9.19) show that aquifer discharge to the drainage outlets depends on such climatic and hydrogeologic factors as average net water table accretion rate, drain spacing, heads in drains, hydraulic and geometric properties of both aquifer and semi-pervious bottom layer, and hydraulic head in the aquifer below the semi-permeable layer. On this account, quick observation reveals that $Q(0)/K_xL$ and $Q(L)/K_xL$ scales with the following dimensionless variables: L/λ, N/K_x, $\varphi_0/\bar{h}$, $\left(h_0^2 - \varphi_0^2 / L\right)$ and $(h_L^2 - \varphi_0^2)/L^2$; they summarize the interaction among basin and hydrogeologic characteristics and climatic conditions on baseflow.

It can be seen that both groundwater discharge Equations (9.18) and (9.19) are useful in estimating baseflow at the catchment scale; it is not uncommon that streams cut through different elevations in the same catchment. Equation (9.18) or (9.19) describes steady-state baseflow discharge per unit length of stream from one side of the basin.

For the specific case of negligible leakage, $\lambda \to \infty$ and $h_L = h_0$, either of the equations greatly simplifies to the rather intuitive result $Q(0) = Q(L) = NL/2$. Note that unlike the case of leakage, discharge to the stream in this case is independent of the aquifer hydraulic conductivity. The configuration of the water table and drainage rate to streams for the case of no leakage is a fundamental one and can be shown equal to that derived from Equation (9.17a), by taking the limit $\lambda \to \infty$ and applying *l'Hospital's* rule, albeit, repeatedly.

Finally, in cases where streams partially penetrate aquifers and open drains do not reach the impervious layer, or even for pipe drains, flow lines contract as they converge toward the drain and become longer than for essentially horizontal flow. Consequently, extra head loss is required to maintain the same flow rate into the stream or drain, which results in a higher water table than that predicted on the basis of Equation (9.17a). Hooghoudt (1940) suggested adding an imaginary impervious layer on top of the bottom layer and replacing the pipe drains, or partially penetrating ditches, by imaginary ditches with their bottoms on the imaginary impervious layer. The effect of adding an impervious hypothetical layer is to force an artificially higher water table and maintain transmissivity required to conduct the same amount of water flow as in the actual situation. Equation (9.17a) thus can still be used, but, of course, with the thickness of the artificial impervious layer treated as an empirical parameter.

Drainage from an Aquifer with Sloping Bed: This problem is relevant to catchment-scale aquifers underlain by inclined bedrocks and to a smaller-scale, hillslope riparian aquifer resting on an inclined impervious bed.

Two frameworks have been utilized in solving problems involving subsurface drainage from aquifers with sloping beds. Figure 9.5 illustrates the two approaches. The first, originally proposed by Childs (1971), is widely used to describe hillslope drainage wherein subsurface flow is essentially parallel to the sloping bed and equipotential surfaces are normal to the bed (Wooding and Chapman 1966; Brutsaert 1994; Koussis et al. 1998; Akylas and Koussis 2007). The alternative approach makes direct use of Equation (9.8) by substituting $\eta = (\tan\theta)\, x$ (Polubarinova-Kochina 1962; Hantush 1964; Bear 1972; Hantush and Mariño 2001; Upadhyaya and Chauhan 2001); it assumes essentially horizontal flow, along the x axis, with $q_z \approx 0$, and vertical equipotential surfaces, along the z axis.

For the case where x and z axis, respectively, are aligned parallel and normal to the sloping bed, the equation for subsurface drainage is

$$S_y \frac{\partial \eta}{\partial t} = K_x cos\theta \frac{\partial}{\partial x}\left(\eta \frac{\partial \eta}{\partial x}\right) + K_x sin\theta \frac{\partial \eta}{\partial x} + K_y \frac{\partial}{\partial y}\left(\eta \frac{\partial \eta}{\partial y}\right) + N \tag{9.20}$$

where, here, η is water table elevation normal to the bed; and θ is bed slope angle (Figure 9.5a). Bed slope angle here is limited to plain rotation around the y axis (i.e., normal to the figure). For predominantly cross-sectional flow, or in case of hillslope width much smaller than the channel length, variation with respect to y, represented in the third term on the right-hand side of Equation (9.20), is usually neglected, and the equation reduces to the well-known Boussinesq equation.

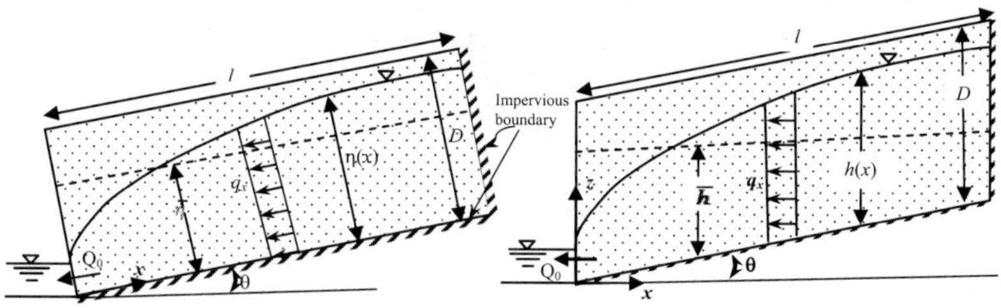

Figure 9.5: Schematic diagram of a hillslope aquifer hydraulically connected to a stream; the aquifer is resting on a slightly inclined, impervious bed: (a) flow is parallel to the bed (Childs 1971); and (b) flow is essentially horizontal.

Alternatively, flow in aquifer resting over an inclined bed, with x along the horizontal axis and z along the vertical (i.e., positive or negative in the direction of gravity, see Figure 9.5b), can be described by Equation (9.8) and making use of $\eta = (\tan\theta)\,x$

$$S_y \frac{\partial h}{\partial t} = K_x \frac{\partial}{\partial x}\left(h \frac{\partial h}{\partial x}\right) + K_x \beta \frac{\partial h}{\partial x} + K_y \frac{\partial}{\partial y}\left(h \frac{\partial h}{\partial y}\right) + N \tag{9.21}$$

where $\boldsymbol{h}$ here is water table depth measured along the vertical; and $\beta = \tan\theta$ is aquifer bed slope.

Equations (9.20) and (9.21) should yield comparable results for relatively small angles of inclination, θ, where $\beta = \tan\theta << 1$, $\sin\theta \approx \tan\theta$ and $\cos\theta \approx 1$. The first term on the right-hand-side of either of these equations is diffusive in nature, whereas the second term is advective, hence, they are advection-diffusion equations. For the specific case $N = 0$ and neglecting the diffusive terms along the x and y axes, a kinematic wave approximation to Equation (9.20) yields the rather simple solution for θ sufficiently larger than zero: $q \approx \eta\ K\ \sin\theta$ (Akylas and Koussis 2007, citing Boussinesq 1877).

For the second approach, the outflow rate Q_0 from the aquifer into the stream (negative for inflow rate) is given by Equation (9.10); for the case where flow depth is measured normal to the inclining bed, the outflow rate is given by

$$Q_0 = K_x \eta \left(cos\theta \frac{\partial \eta}{\partial x} + sin\theta \right) \Bigg|_{x=0} \tag{9.22}$$

Similarly, the negative signs have been omitted in both Darcy's equations because Q_0 is positive in the direction of decreasing x (Figure 9.5).

The linearized boundary-value problem describing drainage from a fully saturated hillslope aquifer block to a shallow stream is

$$\frac{\partial \eta}{\partial t} = \frac{K_x \bar{\eta} cos\theta}{S_y} \frac{\partial^2 \eta}{\partial x^2} + \frac{K_x sin\theta}{S_y} \frac{\partial \eta}{\partial x} \tag{9.23}$$

and

$$\left. \begin{array}{lr} \eta = 0\,, & x = 0\,, \quad t \geq 0 \\ \bar{\eta} \cos\theta \dfrac{\partial \eta}{\partial x} + \eta \sin\theta = 0\,, & t \geq 0 \\ \eta = D\,, & 0 \leq x \leq l\,, t = 0 \end{array} \right\} \tag{9.24}$$

where $\bar{\eta}$ is some average saturated depth normal to the inclined bed [L]; D is thickness of hillslope aquifer (normal to sloping bed) [L]; and x here is distance along the slope whose length is l. An impervious boundary is assumed at $x = l$. Expressed in terms of dimensionless variables, the solution of this boundary-value problem yields this dimensionless outflow rate formula (Brutsaert 1994)

$$q_+ = 2\sum_{n=1}^{\infty} \frac{z_n^2 \left[\left(2e^{H_i/2} cos\,(z_n) \right) - 1 \right] e^{-\left(z_n^2 + H_i^2/4\right)t_+}}{\left(z_n^2 + \dfrac{H_i^2}{4} + \dfrac{H_i}{2} \right)} \tag{9.25}$$

where $q_+ = lQ_0/(K_x \bar{\eta} D \cos\theta)$; $H_i = l \tan\theta / \bar{\eta}$ is the so-called hillslope number; $t_+ = [(K_x \bar{\eta} \cos\theta)/(S_y l^2)]t$; and z_n is the nth root of tan(z) = -2 z/H_i. This equation applies to a horizontal aquifer bed by noting that in this case: $\theta = 0$, $H_i = 0$, and $z_n = (2n - 1)\pi/2$.

Brutsaert (1994) expresses Equation (9.25) in terms of dimensionless variables and shows that aquifer discharge is related to two fundamental coefficients, the first z_n reflects the diffusive character of the flow, and the second is the hillslope number H_i which reflects the relative importance of the slope (i.e., the effect of gravity) versus diffusive effects. H_i thus reflects the kinematic effect of the flow. Figure 9.6 plots q_+ versus t_+ in a semi-logarithmic scale and shows exponential decay of Q_0 for large values of t, but at a rate magnified by the kinematic effect, as indicated by the exponent term in Equation (9.25). As H_i becomes larger, the flow in the aquifer becomes less diffusive and more kinematic (i.e., increasingly driven by gravity on account of the slope). Initially, diffusive effects dominate causing the spread of the water table near the outlet, but for longer times, diffusive effects diminish and kinematic effects take over, causing the appearance of a "hump" in the hydrograph during the transition between the two regimes (Brutsaert 2005, pp. 411-412). Small H_i values cause sustained aquifer discharge, whereas large H_i value causes rapid hillslope drainage and relatively quick baseflow depletion.

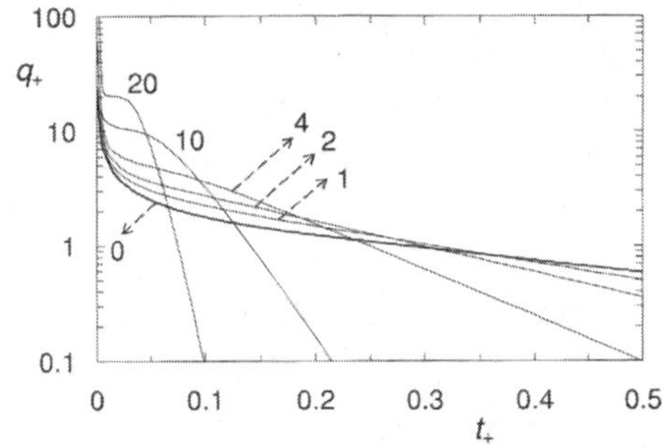

Figure 9.6: Scaled discharge hydrograph from an initially saturated aquifer computed by Equation (9.25), for the values of the hillslope flow number H_i = 0, 1, 2, 4, 10 and 20; H_i = 0 represents the horizontal case. From Brutsaert (2005), with permission of Cambridge University Press.

The short-time behavior is (Brutsaert 1994)

$$Q_0 = -\left(K_x \bar{\eta} S_y cos\theta / \pi\right)^{1/2} D t^{-1/2} \tag{9.26}$$

This formula can be obtained by solving Equation (9.23) in a semi-infinite aquifer width, $l \to \infty$, without the second (advective) term on the right. Clearly, Equation (9.26) shows that drainage from an initially saturated hillslope aquifer baseflow is diffusion-dominated, but only for a relatively short period in time.

A few exact analytical solutions have been presented for nonlinear unconfined flow over an inclined bed (e.g., Bear 1972). Of fundamental and practical value are the nonlinear analytical solutions and approximation thereof derived by Hantush and Mariño (2001) for steady-state flow in unconfined aquifer with a sloping bed (Figure 9.7). The aquifer is hydraulically interacting with a stream through a thin resistive layer, and receives net water table accretion (positive for recharge and negative for net evapotranspiration) and a prescribed flux at the right boundary. This model applies to a hillslope riparian aquifer receiving net flow from upper catchment drainage and subject to recharge/extraction within the riparian buffer strip. The solution is equally applicable to small catchments, where recharge is catchment-averaged and a no-flow boundary condition exists at the hydraulic divide. An approximation to the otherwise highly nonlinear exact solution is provided in much simpler closed forms for small aquifer bed slopes.

Figure 9.7 is a schematic presentation of a water table aquifer with a sloping bed. The angle of inclination, θ, is positive rotating clock-wise from the horizontal axis (i.e., rising bottom along the x axis), and negative rotating counter-clock-wise (i.e., downward bottom slope in the x direction). The aquifer receives net recharge rate of magnitude N per unit width per unit channel length and a volumetric flow rate per unit channel length of magnitude Q_0 at the right boundary. Assuming flow is essentially horizontal with negligible vertical velocities, the equation describing steady-state unconfined flow is (Hantush and Mariño 2001)

$$K\frac{\partial}{\partial x}\left(\zeta\frac{\partial\zeta}{\partial x}\right)+K\beta\frac{\partial\zeta}{\partial x}+N=0 \ , \tag{9.27}$$

and

$$\left.\begin{aligned} -K\zeta\frac{dh}{dx}&=Q_0 \ , && x=0 \\ -K\zeta\frac{dh}{dx}&=wK'\frac{h-H}{b'} \ , && x=l \end{aligned}\right\} \tag{9.28}$$

where $h(x) = \zeta(x) + \beta x$ is the hydraulic head relative to a datum (Figure 9.7); ζ is the vertical water table elevation above the sloping bed; $\beta = \tan\theta$ is the bed slope (positive for upslope and negative for downward bottom slope, see Figure 9.7); K is the effective hydraulic conductivity of the aquifer; K' and b' are, respectively, the hydraulic conductivity and thickness of the semi-permeable thin layer; H is channel stage measured relative to the datum; l is aquifer width; and w here is half the wetted perimeter of the channel. For a fully penetrating stream, half the wetted perimeter is equal to $\zeta(l)$. However, in either case, and because partially penetrating streams are more common than fully penetrating streams, w should be treated as an empirical coefficient to account for additional resistance to flow caused by converging flow lines and the extra distance water has to travel toward partially penetrating streams. In fact, if the streambed conductance (K'/b') is unknown, the product wK'/b' on the right of Equation (9.28) may be viewed as an effective streambed conductance requiring calibration. Note that, an alternative form to the boundary condition at $x = l$ is $-K\zeta dh/dx = Q_0 + Nl$, which follows from the continuity of volumetric flow rate.

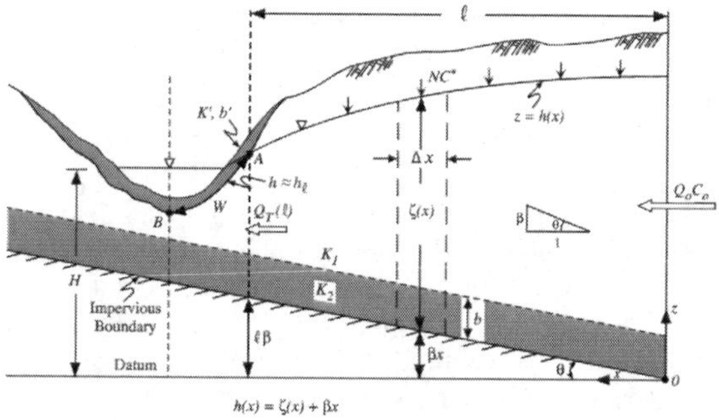

Figure 9.7: Schematic illustration of aquifer-stream system. The aquifer is underlain by an inclined bed and subject to recharge from above and the right boundary. From Hantush and Mariño (2001), used with permission of the American Geophysical Union.

The solution of the above boundary-value problem under the condition $Nx + Q_0 \geq 0$ is (Hantush and Mariño 2001)

$$\left|\frac{\zeta^2 + (\beta/N)(Nx+Q_0)\zeta + (1/KN)(Nx+Q_0)^2}{\zeta_l^2 + (\beta/N)(Nl+Q_0)\zeta_l + (1/KN)(Nl+Q_0)^2}\right| = \exp\left\{-\frac{\beta}{\gamma N}\left[g\left(\frac{\zeta_l/(Nl+Q_0)+\beta/(2N)}{\gamma}\right) - g\left(\frac{\zeta(x)/(Nx+Q_0)+\beta/(2N)}{\gamma}\right)\right]\right\} \tag{9.29}$$

where g is the inverse of familiar trigonometric and hyperbolic functions

$$g(y) = \begin{cases} \tan^{-1} y, & \gamma^2 > 0 \\ \coth^{-1} y, & \gamma^2 < 0 \ \ and \ \ |y| > 1 \\ \tanh^{-1} y, & \gamma^2 < 0 \ \ and \ \ |y| < 1 \end{cases} \tag{9.30}$$

and

$$\gamma^2 = 1/(KN) - \beta^2/(4N^2), \quad \zeta_l = \zeta(l) \tag{9.31a}$$

$$\zeta_l = \frac{Nl + Q_0}{w\left(K'/b'\right)} + H - \beta l \tag{9.31b}$$

where tan, coth and tanh, respectively, are the tangent, hyperbolic cotangent, hyperbolic tangent function; and ζ_l is ζ at $x = l$. Note that for a horizontal aquifer bed, $\beta = 0$, consequently, the term in Equation (9.29) is equal to 1. For the specific case of a horizontal aquifer bed, $\beta = 0$, and no recharge, $N = 0$, Equation reduces to the simpler result (Bruggeman 1999, p. 23)

$$\zeta^2(x) = \zeta_l^2 - (2Q_0 / K)(x - l) \tag{9.32}$$

Equation (9.29) can serve as a benchmark for testing the adequacy of numerical schemes and linearized analytical solutions of the Boussinesq equation. Although highly nonlinear, it can be solved using standard iterative procedures, such as the Newton-Raphson method or by bracketing-bisection root finding technique. A useful approximate, but closed-form, solution to Equation (9.29) was also obtained for a sufficiently small β using a straightforward expansion of $\eta(x)$ in the form of a power series in β. See Hantush and Mariño (2001) for more details and the solution.

The solution given by Equation (9.29) reveals a distinct behavior for drainage below a riparian hillslope during wet and dry periods. During a wet period the unsaturated zone is filled with moisture and thus continuously supplies the water table with moisture at a rate $N > 0$; in this case, $g(y) = \tan^{-1}(y)$. During prolonged no-rain, extended dry periods, the riparian hillslope switches from a supply mode to that with net extraction by evapotranspiration, and the dynamics of drainage over the sloping bed differs from that during the wet period: in this case $N < 0$ and $g(y) = \coth^{-1}(y)$ or $\tanh^{-1}(y)$, depending on whether $|y|$ is greater than or less than one, respectively, provided that $Nx + Q_0 \geq 0$. The dependence of subsurface flow on the function $g(y)$ is entirely attributed to the presence of the sloping bed and the consequent effect of gravity on the flow regime. Further, this effect interacts with N and Q_0 and, in addition to β, depends on both K and l. The specific case of $Nx + Q_0 \leq 0$ is more complex and is not treated here. For example the special case of $Q_0 = 0$ or Q_0 too small relative to evapotranspiration demand (i.e., $Q_0 + NL << 0$ and $N < 0$) could lead to a partially dry riparian aquifer.

Figure 9.8 fits the exact solution given by Equation (9.29) to observed heads in two riparian watersheds in Kent County, Maryland (Hantush and Mariño 2001): Morgan Creek, underlain by a positively inclined bed (β = 0.0052 > 0), and Chesterville Branch with β = -0.0052. There, Q_0 = 0, since the hydraulic divide, separating the two catchments, is situated at x = 0. The area is underlain by a relatively shallow unconfined aquifer, which is predominantly sand and gravel of fluvial origin. The aquifer is separated from a lower confined aquifer by a relatively thick layer that dips at an angle $\theta = 0.3^\circ$. While field evidence indicated that the layer is semi-pervious, it is assumed to be impervious for both simplicity and use of Equation (9.29). The bottom layer is shallow in the riparian zone of Morgan Creek, but relatively deep below the Chesterville Branch. Using a uniform hydraulic conductivity of magnitude 16.76 m/d inferred from an independent study, the analytical solution was calibrated for the recharge and streambed conductance at both sites, by matching simulated head values with observed counterparts.

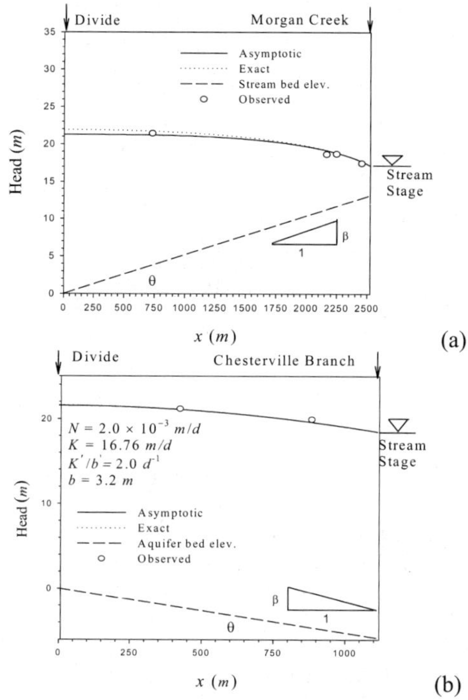

Figure 9.8: Observed versus computed heads: (a) Morgan Creek; and (b) Chesterville Branch. The asymptotic solution almost perfectly matches the exact solution. From Hantush and Mariño, (2001), used with permission of the American Geophysical Union.

The calibrated recharge rate, *N*, values were such that their spatial average was about 16 in./yr and within the range of recharge rate values reported for the northern Atlantic coastal plain aquifer, which range between 10 and 25 in./yr (Meisler 1986). The calibrated streambed conductance, K'/b', values at both sites, however, were 2.0 d^{-1}, an order of magnitude greater than the 0.2 d^{-1} value obtained by Drummond (1998) by calibrating a regional groundwater flow model for Kent County. Such discrepancies are not unexpected and may be attributed to uncertainty in the data and model and various simplifying assumptions. Nonetheless, differences between the exact solution and the approximate closed-form expansion were very minimal.

Spring Flow: A spring is a point or a relatively small area through which groundwater is discharged from an aquifer to the ground surface. Springs emanate from shallow flow systems and from deeper formations through fissure or fault zones reaching the ground surface (Bear 1979). Springs occur in localized depressions when a high water table intersects the ground surface, or when the impervious layer underlying a perched water table intersects the ground surface. Fractured carbonate or karst environments produce springs when they are exposed to the surface. Confined aquifers exposed by localized deep incisions or subcropping in riverine and coastal areas can also be drained from a spring.

In general, spring flow exhibits highly variable discharge in response to recharge events and often displays seasonal variations, especially in karst and fractured environments (e.g., Manga 1996; Bahr and Mishra 1997). However, steady-flow spring flows are not uncommon in areas where discharge is supplied by means of faults from deep groundwater reservoirs (Swanson and Bahr 2004, citing Bryan 1919), or from shallow sandstone aquifers receiving steady diffuse recharge (Swanson and Bahr 2004).

A spring flows if the water table elevation in the vicinity is higher than the elevation of the discharging point, and dries up when the water levels fall below its elevation. During wet periods, discharge is sustained by a combination of releases from aquifer storage and groundwater recharge. In dry periods, spring discharge is derived from water stored in the aquifer with no inflow and decreases in time as available groundwater storage depletes. This process is also known as flow recession; it constitutes the falling limb of a hydrograph for times greater than the point of inflection on the receding limb.

Figure 9.9 is a schematic diagram of a spring system. The simplest spring discharge model is that where the discharge, Q [L^3T^{-1}], is proportional to the average head in the aquifer (measured relative to the spring water level), h [L], $Q = \alpha h$. For an area A [L^2] of the aquifer discharging at a rate Q and receiving recharge at a rate R per unit area [LT^{-1}], mass balance requires that

$$RAdt - Qdt = dS\,; \quad Q = \alpha h \tag{9.33}$$

and

$$dS = S_y A dh \tag{9.34}$$

where S is aquifer storage [L^3]; and α is a constant [L^2T^{-1}]. Equation (9.33) can be integrated after substituting for dS and Q, with $Q = Q_0 = \alpha h_0$ at $t = t_0$, to obtain

$$Q(t) = Q_0 exp\left[-\frac{\alpha}{S_y A}(t - t_0)\right] + \frac{\alpha}{S_y}\int_{t_0}^{t} e^{-(\alpha/(S_y A))(t-\tau)} R(\tau)\,d\tau \tag{9.35}$$

For a constant $R(\tau) = R$, the integral on the right-hand side can be evaluated rather easily. The recharge contribution - the second term on the right hand side - may be dropped if recharge during a storm event causes a rapid rise in the water table throughout the spring catchment area and where spring discharge rate becomes dominated by releases from the contributing aquifer pore space. Indeed, this is the case during dry periods and prolonged interstorm periods where spring discharge is derived from aquifer storage releases only ($R \approx 0$). Under these conditions, Q follows a simple exponential recession model.

Although treated above as an empirical parameter, for an aquifer length l, an estimate for α is $2TA/l^2$, where $T = Kh$ is average aquifer transmissivity. The extension of Equation (9.35) to one of the most widely used baseflow recession models will be covered later.

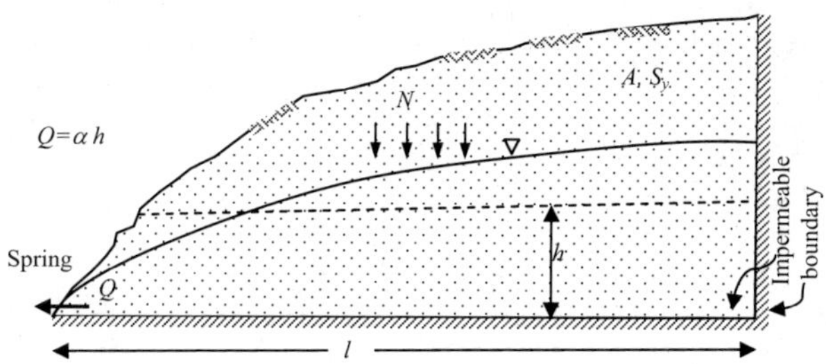

Figure 9.9: Schematic spring-aquifer system. A is the spring catchment area.

Aquifer discharge into springs can also be described on the basis of Darcy's law and continuity of groundwater flow in aquifers (e.g., Swanson and Bahr 2004). Hunt (2004) derives an analytical solution describing the depletion by pumping from well of a circular spring draining a water table aquitard. The pumped well penetrates a confined leaky formation overlain by the water table aquitard.

9.2.4 Bank Storage

In higher order channels such as alluvial rivers, the cumulative flow into a channel segment draining upstream catchments and stream reaches can be large enough to cause a rapid rise of the river stage over the water table level in the adjacent floodplain, even though the water table might be receiving supplemental recharge during the precipitation event. The hydraulic gradient in this case is toward the river banks causing water in the river to infiltrate into the surrounding alluvial aquifer from both sides of the channel to recharge, and be stored temporarily in, the alluvial (floodplain) aquifer. When precipitation or other input has ceased and all surface runoff has departed from the drainage basin, the river stage recedes to its initial level and the hydraulic head gradient is reversed when the stage falls below the water table in the adjacent alluvial sediments, causing gradual release of the stored water, entirely or portions thereof, back into the channel. Portions of the water so stored might be lost to the atmosphere by riparian plant transpiration, evaporation, or by extraction due to pumping from nearby wells.

The phenomena of water storage in alluvial banks during high streamflow events and the slow release from storage and drainage back into the channel is known as "bank storage"; it is a natural flood-control process whereby alluvial banks absorb some of the flood waters and store the water temporarily, thus, reducing the magnitude of peak flow and delaying peak arrival time of the flood wave that would otherwise have occurred in response to a precipitation event. The lateral exchange of the water between the river and bank and the extent of flood wave modification depends on channel geometry, thickness and hydraulic properties of bed and bank sediments, aquifer extent and hydraulic properties, water table elevation near the bank, and magnitude and duration of the flood event. Groundwater releases from bank storage can sustain streamflow for a relatively long period following the decline of the river stage. While various authors have referred to baseflow as low flow or drought flow, a broader definition should also include periods following runoff events during which bank storage contributes a significant portion of groundwater discharge to the stream. Maximum baseflow therefore occurs immediately after a storm event because precipitation has just supplied recharge to the water table, which increases groundwater discharge that is further supplemented by releases from bank storage.

Analysis of bank storage problem may be classified into three distinct approaches: (1) solution of unconfined flow equation for water table fluctuations in response to sudden or continuous perturbation of stream stage (Werner and Noren 1951; Cooper and Rorabaugh 1963, Hantush 1967; Yeh 1970; Hall and Moench 1972; Mariño 1975; Dever and Cleary 1979; Gill 1985; Govindaraju and Koelliker 1994; Whiting and Pomeranets 1997; Workman et al. 1997; Zlotnik and Huang 1999; Moench and Barlow 2000); (2) coupled channel flow dynamics and aquifer flow where one-dimensional Saint-Venant equations of continuity and momentum, describing flood wave movement down the channel, and aquifer flow equation are solved together (Pinder and Sauer 1971; Zitta and Wiggert 1971; Hunt 1990; Perkins and Koussis 1996; Gunduz and Aral 2005b); and (3) coupled linear/nonlinear Muskingum routing equations and aquifer flow (Morel-Seytoux 1975; Illangasekare and Morel-Seytoux 1982; Hantush et al. 2002; Birkhead and James 2002; Hantush 2005).

Predetermined Stream-Stage Fluctuations: Cooper and Rorabaugh (1963) solved Equation (9.13) subject to these initial and boundary conditions: $h(x,0) = 0$, $0 \le x \le l$, $\partial h(l,t)/\partial x = 0$, and $h(0,t) = Ne^{\delta t}(1-\cos\omega t)$ when $0 \le t \le T_0$, and 0, when $t > T_0$; h_0 is the maximum rise in stage [L]; t is the time since the beginning of the flood wave [T], T_0 is the duration of the wave [T]; $\omega = 2\pi / T_0$ is flood wave frequency [T^{-1}]; $N = e^{\delta t} / (1-\cos t_c)$ is a constant that causes the flood curve to peak at height h_0; $\delta = \omega\cot(\omega t_c / 2)$; and t_c is time to peak [T]. This equation approximates the shapes of many flood stage hydrographs. Figure 9.10 depicts a stream-aquifer system and a typical stage hydrograph. The solution is obtained by means of Laplace transformation. The semi-infinite aquifer case corresponds to the case where the impermeable valley wall has no effect.

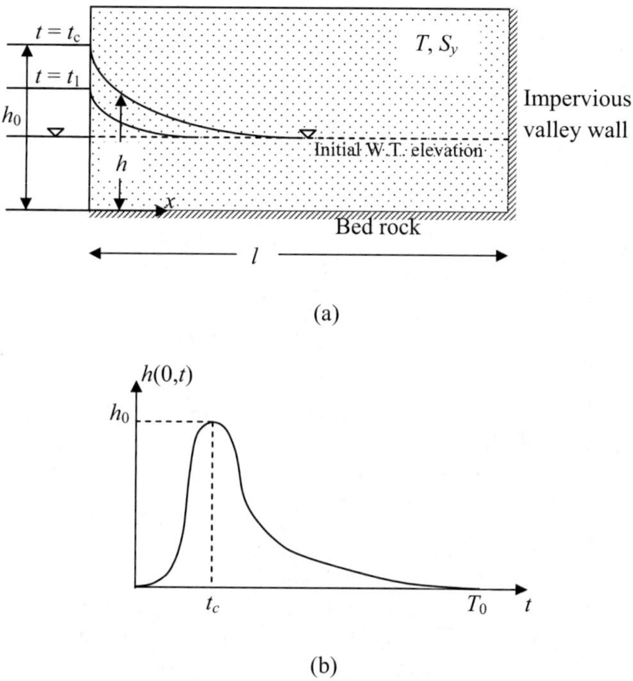

Figure 9.10: Bank storage problem: (a) stream-aquifer system; and (b) stream-stage hydrograph.

The flow per unit length into the aquifer is computed using Darcy's law:

$$Q = -T\frac{\partial h(0,t)}{\partial x} \tag{9.36}$$

The integration of Q [$L^3T^{-1}L^{-1}$] in time is net volumetric storage per unit length of bank, V [L^3L^{-1}]

$$V = \int_0^t Q(\tau)\,d\tau \tag{9.37}$$

For large time, $t >> T_0$, finite aquifer breadth, and symmetric flood hydrograph, $\delta = 0$, Q approaches

$$Q = -\frac{2h_0\sqrt{(\beta\omega TS_y)(e^{2\pi\beta}-1)}}{\pi(1+\beta^2)}e^{-\beta\omega t}, \quad t >> T_0 \tag{9.38}$$

where $\beta = \pi^2 T / \left(4S_y l^2 \omega\right)$. Note the correspondence between this equation and Equation (9.14); both have the exponential form $Q = Q_0\, e^{-ct}$ (e.g., Barnes 1939; Tallaksen 1995) often used to describe baseflow recession. On the other hand, for a semi-infinite aquifer breadth and $\delta = 0$, this approximate formula was obtained:

$$Q = -\frac{h_0\sqrt{\omega T S_y}}{2}\left\{\frac{1}{\sqrt{\pi\omega t - 2\pi^2}} - \frac{1}{\sqrt{\pi\omega t}}\right\}, \quad t >> T_0 \tag{9.39}$$

which does not have the exponential form. Solutions for asymmetric flood hydrograph ($\delta \neq 0$) are also provided (Cooper and Rorabaugh 1963).

Numerical computations for different values of the dimensionless variable β, with $\beta =$ 0 corresponding to the semi-infinite aquifers, showed that bank storage in finite aquifers declines very rapidly when β is large and declines slowly when β is small. In a semi-infinite aquifer bank storage persists for a sufficiently long time after the flood wave has subsided. Numerical results showed that bank storage is as large as 14% of the maximum at t = 10 T_0 and 4% of the maximum at t = 100 T_0. The effect of asymmetry in the stage oscillation was also investigated, with $\delta = 0$ corresponding to the case of a symmetric flood stage. It is shown that asymmetry causes the rate of flow out of the stream to be large during the early part of the flood wave owing to the more rapid rise in stage; but, because it also shortens the duration of this flow, it lessens the maximum bank storage.

Hantush (1967) extended the bank storage problem to thick, leaky confined or unconfined aquifers that are cut through by pervious stream channels (Figure 9.11). In Figure 9.11a, the leaky confined aquifer is overlain by a semi-pervious layer on the top of which the head is maintained uniform. In Figure 9.11b, the leaky water table aquifer overlies a semi-pervious layer at the bottom of which the head is maintained constant. In both cases the flow is two-dimensional in the cross-section and the systems are semi-infinite in areal extent. Interestingly, assuming the rise of the water table is small compared to the initial depth of saturation, both systems can be described by a similar boundary-value problem, but with different storage coefficients and orientation of the vertical, z, axis

$$\frac{1}{\nu}\frac{\partial\varphi}{\partial t} = \frac{\partial^2\varphi}{\partial x^2} + \frac{\partial^2\varphi}{\partial z^2} \tag{9.40}$$

$$\varphi(x,z,0) = 0 \tag{9.41a}$$

$$\varphi(0,z,t) = \delta \tag{9.41b}$$

$$\varphi(\infty,z,t) = 0 \tag{9.41c}$$

$$K\frac{\partial\varphi(x,0,t)}{\partial z} = \frac{K'}{b'}\left[\varphi(x,0,t) - \varphi_0\right] \tag{9.41d}$$

$$K \frac{\partial \varphi(x,b,t)}{\partial z} = 0 \tag{9.41c}$$

in which

$$\left.\begin{aligned} \nu &= K / S^* , && \text{for the confined system} \\ \nu &= K(2\varphi_0 + \delta) / 2S_y , && \text{for the unconfined system} \end{aligned}\right\} \tag{9.42}$$

where $\varphi(x,z,t)$ is the hydraulic head measured relative to the base of the aquifer [L]; φ_0 is the initial head in the system [L]; K is the hydraulic conductivity of the main aquifer [LT^{-1}]; S^* is the specific storativity of the confined aquifer [L^{-1}]; S_y is the specific yield of the unconfined aquifer; K' [LT^{-1}] and b' [L] are, respectively, the hydraulic conductivity and thickness of the aquitard; b is thickness of the confined aquifer [L], and for the unconfined aquifer it is equal to the initial water table elevation, $b = \varphi_0$; and δ is sudden rise in stream stage [L].

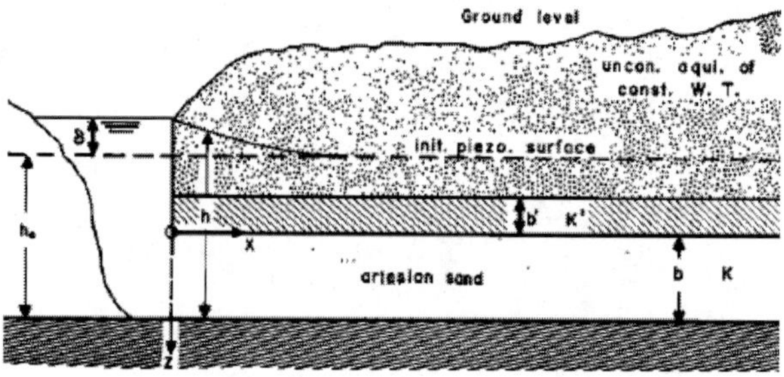

(a)

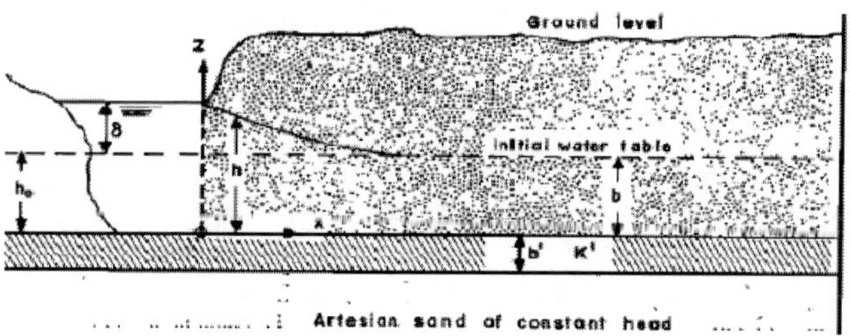

(b)

Figure 9.11: Schematic representation of stream-aquifer systems: (a) leaky confined aquifer; and (b) leaky water table aquifer [h is φ in Equation (9.40)]. From Hantush (1967), used with permission of the American Geophysical Union.

Hantush (1967) hypothesized that water added into (or withdrawn from) storage due to increasing (or lowering) the water table takes place within the main aquifer and no-flow boundary is artificially introduced at $z = b$ to maintain water balance. In other words, he introduces an effective specific storativity equal to the drainable porosity or specific yield, and replaces the free-surface boundary Equation (9.4) with a hypothetical impervious boundary at $z = \varphi_0$. Thus, by replacing S^* with S_y, transmissivity, T, with $K(2\varphi_0 + \delta)/2$, b with φ_0, z with $(b - z)$, the solution obtained for the leaky confined aquifer is equally applicable to the leaky unconfined aquifer. In this case, flow per unit channel length toward the aquifer, Q [L^2T^{-1}], is given by integrating Darcy's equation evaluated at $x = 0$ along the saturated thickness

$$Q = -K \int_0^b \frac{\partial \varphi(0,z,t)}{\partial x} dz \tag{9.43}$$

The solution of the boundary-value problem defined by Equations (9.40) and (9.41) is obtained by Laplace transformation with respect to t and the Fourier sine transformation with respect to x. From the solution, expressions for Q and bank storage volume per unit channel length, V [L^3L^{-1}], were obtained by carrying the integrals in Equations (9.43) and (9.37), respectively. Approximations were also obtained for $b/B \leq 0.1$, where $B = \sqrt{Kbb'/K'}$ is the leakage factor.

For the more general case of flow through a semi-permeable streambank, the boundary-value problem is similar to that described by Equations (9.40) and (9.41), except that boundary condition (9.41b), instead, takes this form

$$-K \frac{\partial \varphi(0,z,t)}{\partial x} = \frac{K_a}{b_a} \left[\delta + \varphi_0 - \varphi(0,z,t) \right] \tag{9.44}$$

where K_a [LT^{-1}] and b_a [L] are the hydraulic conductivity and thickness of the semi-permeable layer, respectively.

Dever and Cleary (1979) solve the boundary-value problem given in Equations (9.40)-(9.41), with (9.44) replacing (9.41b), by applying Fourier transform with respect to x and the Finite Fourier transform with respect to z and solving the resulting ordinary differential equation in time, followed by respective Fourier inverse transforms. As described above, by use of superposition principle in conjunction with solutions describing the response of a stream-aquifer system to a step rise in stream level, H, the response of the system to an arbitrary time variable stage hydrograph can be obtained. They compare the case of a streambank as permeable as the aquifer, i.e., $R = b_a K / K_a = 0$, where R is the retardation factor, to the case of a 10 cm thick semi-pervious layer with hydraulic conductivity of 10 cm/d (R = 15 m) and show 60% greater bank storage in the permeable streambank case. Therefore, neglecting the semi-permeable property of a bank in a mathematical model could grossly overestimate the amount of flood wave modification by bank storage.

This result underscores the importance of accounting for semi-pervious layers in a mathematical model based on which flood control strategies are evaluated.

The solutions for φ, Q, and V can be extended to arbitrary stage fluctuations, $H(t)$, by noting the linearity of the system given in Equations (9.40)-(9.41) and recognizing that φ/H, Q/H, and V/H represent the corresponding *unit step response* functions, g, i.e., solutions for a sudden unit step increase in the river stage, and whose derivative with respect to time is the instantaneous unit *impulse response* function $u(t) = dg/dt$, or equivalently, $g(t) = \int_0^t u(\tau)d\tau$. The instantaneous unit impulse response function, u, can be convoluted with the stage equation $H(t)$ to yield the corresponding solutions (Carslaw and Jaeger 1959) for arbitrary river stage fluctuations

$$\left.\begin{aligned} \varphi(x,z,t) &= \int_0^t u_\varphi(x,z,t-\tau)H(\tau)d\tau \\ Q(t) &= \int_0^t u_Q(t-\tau)H(\tau)d\tau \\ V(t) &= \int_0^t u_V(t-\tau)H(\tau)d\tau \end{aligned}\right\} \tag{9.45a}$$

where u_φ [T^{-1}], u_Q [LT^{-2}], and u_V [LT^{-1}], respectively, are the instantaneous unit impulse response functions for φ, Q, and V. The convolution process involves translation of the instantaneous unit response function $u(\tau)$ to $\tau = t$ and folding the function about this new origin and integration from $\tau = 0$ to $\tau = t$ of the product of the input function $H(\tau)$ and the corresponding value of the instantaneous unit response function (after folding), i.e., $u(t-\tau)$. Each of the convolution integrals above can be derived by linear superposition up to time t of solutions each corresponding to incremental change in $H(\tau)$, ΔH, occurring over the incremental time interval $\Delta\tau$, and taking the limit $\Delta\tau \rightarrow 0$ of the summed incremental contributions.

Quick inspection of the first two equations in Equation (9.45a) and bearing in mind Equation (9.43) reveal that $u_Q(t) = -K\int_0^b \frac{\partial}{\partial x}u_\varphi(0,z,t)dz$. Further, it can be shown that bank storage $V(t)$ *impulse response* function u_V is equal to the *unit step response* function corresponding to $Q(t)$, g_Q,: $u_V(t) = g_Q(t) = \int_0^t u_Q(\tau)d\tau$ (Hantush et al. 2002).

Hall and Moench (1972) present unit response functions and derive the corresponding impulse response functions for linear unconfined flow in finite or semi-infinite aquifer with and without a thin semi-permeable layer separating the stream from the aquifer. They express the convolution integral for Q in discrete form to simulate the effect of a hypothetical stream stage hydrograph

$$Q(n)=\sum_{n=1}^{n\le N} H(n)u_Q(N-n+1)\Delta t_n \tag{9.45b}$$

where $Q(n)$ and $H(n)$, respectively, Q and H at discrete time t_n; N corresponds to discrete point in time when stage hydrograph ends; and Δt_n is the nth time increment. Discrete forms have a great appeal to real applications, where flood hydrographs are often reported at discrete points in time. Furthermore, they can be used to extend solutions for sudden rise or fall in the river stage to general flood stage hydrographs. Rather than convoluting with the impulse response function, Moench et al. (1974) considers the alternative convolution integral for $Q(t)$, by convoluting the time derivative of the stream stage with the unit step response function instead: $Q(t)=\int_0^t g_Q(t-\tau)(dH(\tau)/d\tau)d\tau$. They apply the discrete form of the convolution integral to compute stream losses and gains caused by arbitrary stage fluctuations that resulted from a reservoir release on the North Canadian River in central Oklahoma. They showed that the routed hydrograph was significantly modified by bank storage.

Harmonic tidal fluctuations propagate inland from the shore through coastal aquifers that have submarine outcrops. This motion can be described by Equations (9.13) or (9.23) depending on whether the aquifer is horizontal or has an inclined bed. The solution of Equation (9.13) subject to a simple harmonic tidal fluctuation of the form $h(0,t) = h_0 \sin(\omega t)$ for a semi-infinite aquifer, $h(\infty,t) = 0$, after a sufficiently large time is (Todd 1980, citing, among others, Werner and Noren 1951)

$$h=h_0 exp\left(-\sqrt{\pi S/Tt_0}\right)sin\left(2\pi t/t_0-\sqrt{\pi S/Tt_0}\,x\right),\quad t \gg t_0 \tag{9.46}$$

where h is water table elevation (or piezometric head for a confined aquifer) [L]; T is aquifer transmissivity [L^2T^{-1}]; S is the aquifer storativity, $= S_y$ for unconfined aquifers; $t_0 = 2\pi/\omega$ is the tidal period [T]; h_0 is the wave amplitude [L]. This equation describes a diminishing sinusoidal wave whose amplitude at distance x from the shore is $h_0\exp\left(-(\omega S/2T)^{1/2}x\right)$. At a distance x from the shore, the time lag, t_L, of a given maximum or minimum after it occurs in the ocean is given by $t_L=(S/2\omega T)^{1/2}x$. Thus, the wave travels with a velocity $v_w = x/t_L=(2\omega T/S)^{1/2}$, with the wave length inland given by $L_w = v_w t_0 = 2\pi(2T/\omega S)^{1/2}$.

The rate of flow into the aquifer per unit length of coast is given by

$$Q(t)=h_0\sqrt{\omega ST}cos(\omega t-\pi/4) \tag{9.47}$$

from which, the volume of flow stored in the aquifer per half-cycle per unit length of coast is the integral of $Q(t)$ from $t = 3t_0/8$ to $t = 7t_0/8$, which is

$$V = 2h_0\sqrt{ST/\omega} \tag{9.48}$$

Both equations show that flow into the aquifer per unit coast length and volume stored per half-cycle are functions of aquifer storativity and transmissivity and tidal wave characteristics.

The impact of tidal-induced stream stage fluctuations on two-dimensional groundwater flow in coastal aquifers was analyzed by Kim et al. (2007). The stream is assumed to be at right angle with the coastline boundaries and partially penetrates the aquifer. The aquifer below the stream is confined and overlain by semi-permeable bed sediments, and unconfined elsewhere. Both aquifers are of semi-infinite extent. Stream stage fluctuation is assumed to be a function of time only and uniform throughout the channel length. No explicit relationship between inland stream stage fluctuations and tidal fluctuations was provided. The solution was obtained for an arbitrary stream stage function and sinusoidal tidal fluctuations - similar to the one given above but with a phase shift - by successive application of Laplace and Fourier transforms and inverting numerically back to the time and spatial domains. Fourier inversion was performed via Romberg integration and the Laplace inversion was obtained by means of the so called de Hoog algorithm (de Hoog et al. 1982). However, the solution is limited by the fact that a tidal wave moving inland through the channel will diffuse, subside, and eventually dissipates farther away from the coast. An average, uniformly fluctuating water table over a long stretch of the channel hardly captures such dynamics.

In deviation from the commonly used analytical models, the effect of partial penetration on stream-aquifer interactions was modeled analytically by Zlotnik and Huang (1999) in a semi-infinite aquifer, with a semi-permeable bed separating the stream from the aquifer. They couple one dimensional, leaky confined aquifer flow under the stream to one-dimensional, linear unconfined flow, as governed by Equation (9.13), in the aquifer section beyond. An analytical solution describing head fluctuations in response to a unit step rise in stream stage was obtained by means of Laplace transformation for the specific case of negligible storage (i.e., $S^* = 0$) in the aquifer section under the stream, and the solution for the more general case of aquifer storage was obtained using the Stehfest (1970) algorithm for the numerical Laplace transform inversion. However, the models are based on Dupuit aquifer flow, consequently vertical flow in the vicinity of the stream was neglected.

Moench and Barlow (2000) provided semi-analytical solutions (analytical in the Laplace domain) for various stream-aquifer flow systems, with and without semi-pervious streambank material. They coupled one-dimensional transient flow in a confined aquifer to time-dependent vertical flow in an aquitard with a constant head at the top or with an impermeable layer overlying the aquitard. Further, the case of water table conditions in the aquitard was considered, and two-dimensional, cross-sectional flow in an anisotropic water table aquifer was also treated. Drainage into/from pore space due to rise/fall of the water table in the aquitard and unconfined aquifer is described by Equation (9.4), after dropping nonlinear terms, evaluated at the initial saturated thickness of the aquifer system. Analytical expressions in the Laplace domain are obtained for the piezometric head or water table level and seepage between the stream and aquifer for the finite/semi-infinite leaky-confined and unconfined flow scenarios in response to an instantaneous step change of the water level in the stream. Solutions in the time domain were obtained by numerical inversions of the Laplace domain analytical solutions with the Stehfest (1970) algorithm. Numerical examples show that the salient features of the response, particularly the delayed response phenomenon, of water table aquitards and unconfined aquifers to changes in the water level of an adjacent stream are similar to those that occur in response to the withdrawal/injection of groundwater from/into a well tapping a water table aquifer or leaky confined aquifer. Extension of the step-response solutions to arbitrary, continuously-varying stage hydrograph was demonstrated by Barlow et al. (2000) in two case studies, one for the Blackstone River stream-aquifer system, Massachusetts, and the other Cedar River stream-aquifer system, Iowa, using convolution integral to calibrate for aquifer hydraulic parameters and to estimate groundwater head, seepage rate, and bank storage. The convolution integrals were carried in discrete forms as described in the previous section. Due to partial penetration of the aquifer by the Cedar River and resistance to flow at the river-aquifer interface, the streambank-leakance parameter was lumped and obtained from calibration. Stream-aquifer analysis of the Blackstone River revealed that due to the shallow water table depth the aquifer behaves like a confined aquifer, with a negligible specific yield.

Limitations and Numerical Modeling: The foregoing solutions of bank-storage problem apply to ideal stream-aquifer systems and are based on common assumptions, such as a homogeneous aquifer, Dupuit-Forchheimer conditions, partially penetrating streams, and simplified flow domain geometry. For example, as we have discussed in a previous section, flow in the vicinity of partially penetrating streams is predominantly vertical, and vertical velocities can also be significant near the water table and in the vicinity of hydraulic divides. Anisotropic aquifers of low vertical hydraulic conductivity can significantly reduce the rate of flow into or out of the aquifer and limit the size of the storage zone (Chen and Chen 2003). In major alluvial aquifers, many of these assumptions are questionable (Sharp 1977), and the degree to which invalid assumptions reduce the accuracy of prediction will depend on the degree of divergence of the assumptions from the real world and will likely remain unknown. The limitations of analytical models therefore should be fully recognized when attempting to understand hydrogeologic processes or in making decisions based on model output.

Recognizing the importance of the vertical component of seepage velocity in a significantly sloped water table near the banks, especially in thick and finite porous media, Whiting and Pomeranets (1997) simulated bank storage by means of a numerical model in a two-dimensional cross section of a channel-floodplain system. They solved the transient saturated groundwater flow equation, with a free moving boundary after dropping the nonlinear terms in Equation (9.4), using the finite element method and the Neuman and Witherspoon (1971) method to account for the seepage face. Floodplain drainage and bank storage per unit channel length were simulated for different values of floodplain width to channel width, for clay, silt, and sand/gravel floodplains. Results showed that drainage of water from the floodplain due to a drop in channel water level occurs over a period of days in gravel, weeks to a few years in sand, years in silt, and decades in clay. Floodplains composed of silt appeared to be ideal for sustained baseflow, albeit at relatively slow drainage rates. Due to the simulated lower storage capacity and very small discharge from clay soils, clay floodplains make very poor sources of significant baseflow. On the other hand, wide sandy floodplains provide a good yield of water and duration of flow for sustaining baseflow, but they may not be capable to sustain streamflows during extended drought periods.

The three-dimensional nature of groundwater flow in the vicinity of a partially penetrating stream and the variation of bank storage flux both into and out of the channel wetted perimeter in response to a 2-meter rise in the river stage was evident in a study conducted by Saquillace (1996) on bank storage in an alluvial aquifer adjacent to the Cedar River in Iowa. Analysis of water chemistry and MODFLOW model results indicated significant horizontal water movement into the aquifer and vertical seepage beneath the river. The flux exchange between the river and the aquifer was largest near the river's edge and decreased substantially with distance from the edge to the center of the river even though hydraulic conductivity was simulated as uniform along the river bottom. The model also showed that bank storage sustained baseflow for a period of five weeks after the peak river stage. Chen and Chen (2003) applied MODFLOW and MODPATH to determine the size of bank storage zone and investigated various hydrogeological factors and climate conditions on bank storage due to stream stage fluctuations.

Finally, stream-aquifer interactions have primarily focused on single-phase flow in aquifers and the models presented above are not suited for dual-porosity porous media such as fractured aquifers. Therein, the porous medium is composed of highly conductive fractures with low volumetric fraction within low permeability matrix of primary porosity, with the fluid being exchanged between the fractures and porous matrix. This phenomenon also applies to alluvial aquifers with extensive silty and/or clay lenses dispersed throughout. The transfer of water between fractures and porous blocks may be assumed to be at pseudo-steady state and proportional to the difference in hydraulic head between both media. Önder (1998) derived analytical solutions for the piezometric head and groundwater flow per unit length of the stream into (or from) the surrounding fractured aquifer, due to a sudden step rise in stream stage. He also obtained formulas of limiting cases for both piezometric head fluctuations and stream-to-aquifer flow rate. The solutions may be useful in the analysis of recession hydrographs in streams cutting through fractured formations.

9.2.5 Coupled River Flow and Aquifer Interactions

The analyses presented in the previous section were concerned with aquifer behavior exclusively in response to *a priori* known river stage fluctuations. In simulating groundwater flow the hydraulic head in the channel is predefined and the effect of groundwater movement on stream elevation is neglected. The decoupling of groundwater and surface water systems would not have been possible without availability of measurements describing stage variations in time. Decoupling also is likely to be attributed to the belief that groundwater movement has a much larger timescale than that of open channel flow and that complexities associated with modeling their interactions have often limited the implementation of an integrated modeling approach. A broader approach therefore is to consider channel flow dynamics as well as groundwater movement in an interactive mode.

Hydraulic Approach: This approach entails an explicit formulation of the flow continuity equation and momentum equations. Flow rate and depth within a stream channel are functions of the reach inflow hydrograph, lateral inflow, bank and bed drainage, channel morphologic and hydraulic characteristics, and climate conditions.

The conservation form of a streamflow continuity equation is (Chow et al. 1988)

$$\frac{\partial A}{\partial t}+\frac{\partial Q}{\partial x}=q_l+q_g \tag{9.49}$$

where Q is channel flow rate or discharge [L^3T^{-1}]; A is cross-sectional area of flow [L^2]; q_l is the lateral inflow per unit channel length [$L^3T^{-1}L^{-1}$] on account of runoff, subsurface storm runoff, and direct rainfall; q_g is flow from aquifer into the channel per unit length through its wetted perimeter [$L^3T^{-1}L^{-1}$]; t is time; and x is longitudinal distance along the channel. q_g describes the movement of water between the two systems and is given by Darcy's law

$$q_g=\frac{K'}{b'}w(\varphi-\varphi_0) \tag{9.50}$$

where φ is hydraulic head at the bank [L]; φ_0 is stream water level [L]; K' [LT^{-1}] and b' [L] are, respectively, the hydraulic conductivity and thickness of a semi-permeable sediment layer; and w [L] is the wetted perimeter of channel ≈ stream width for a wide channel.

The non-conservation form of the continuity equation is usually expressed in terms of channel flow velocity v instead of Q, i.e., expressed for a unit width of channel,

$$\frac{\partial y}{\partial t}+v\frac{\partial y}{\partial x}+y\frac{\partial v}{\partial x}=\frac{q_l+q_g}{W} \tag{9.51}$$

where y is channel flow depth [L]; $v = Q/A$ is average channel flow velocity [LT^{-1}]; and W is channel width [L]. Equation (9.51) applies to prismatic channels, in which the cross-sectional shape does not vary along the channel and the bed slope is constant.

The conservation form of the momentum equation (also known as the Saint-Venant equation) is

$$\frac{\partial Q}{\partial t}+\frac{\partial}{\partial x}\left(\frac{Q^2}{A}\right)+Ag\frac{\partial y}{\partial x}=Ag\left(S_0-S_f\right) \tag{9.52}$$

where g is gravitational acceleration [LT^{-2}]; S_0 is channel bed slope [LL^{-1}]; and S_f is friction slope [LL^{-1}] derived from resistance equations such as the Manning or Chézy equation. The effects of abrupt contraction and expansion of the channel and the effect of wind shear are neglected in Equation (9.52). Further, inactive (off-channel storage) such as the case in floodplains are also neglected in both the continuity and momentum conservation equations (see, e.g., Gunduz and Aral 2005b). Equation (9.52) is based on lateral flow with negligible velocity component in the x direction, i.e., the arriving flow has no momentum in the direction of the channel flow (e.g., Henderson 1966, p. 269). Although not shown, a correction coefficient – also called *Boussinesq coefficient* – is usually introduced to account for the nonuniform distribution of velocity at a channel cross section in computing the momentum (Chow et al. 1988, Equation 9.1.37).

The nonconservative form of the momentum for a unit width of the channel equation is given by

$$\frac{\partial v}{\partial t}+v\frac{\partial v}{\partial x}+g\frac{\partial y}{\partial x}+v\frac{q_l+q_g}{Wy}=g\left(S_0-S_f\right) \tag{9.53}$$

Equation (9.53) describes a full dynamic wave, with the first two terms on the left, respectively, local acceleration and convective acceleration, representing inertial effects, the third accounts for pressure force effect (free surface slope), the fourth is lateral flow inertial effect, and the first and second terms on the right side, respectively, represent gravity (bed slope) and friction resistance (friction slope) effects. Routing flows in channels using the continuity Equation (9.49), or its nonconservative form of Equation (9.51), and the momentum Equation (9.52) or (9.53) and its kinematic and diffusive simplifications below is referred to as hydraulic routing or distributed flow routing.

Pinder and Sauer (1971) coupled Equations (9.51) and (9.53) to depth-averaged, two-dimensional aquifer flow equation to simulate the effect of floodplain bank storage on flood wave modification. Figure 9.12 depicts a conceptual river-aquifer system. Flow below the channel is confined and unconfined elsewhere. Leakage through the wetted perimeter of the channel (banks and river bed) was lumped into a leakage term in

Equation (9.50) distributed over the confined section of the aquifer, below the channel.

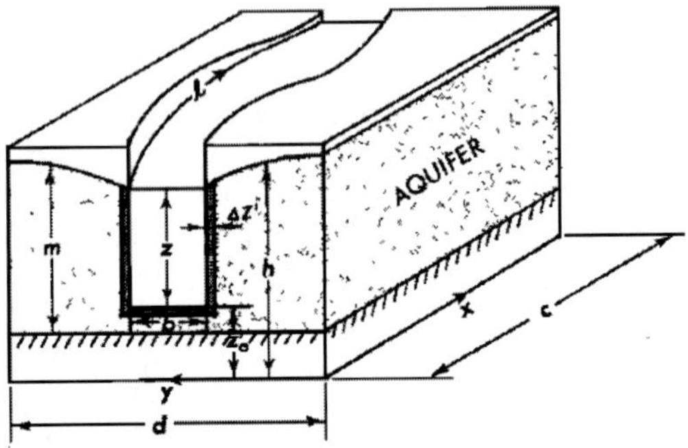

Figure 9.12: Diagramatic representation of aquifer-stream system. From Pinder and Sauer (1971), used with permission of the American Geophysical Union.

An iterative, alternating direction, implicit finite difference technique method was used to solve confined and unconfined flow equations, while an explicit, finite difference staggered-net method was deemed sufficiently accurate to solve the channel flow continuity and momentum Equations (9.51) and (9.53) with q_l =0. An iterative scheme was implemented to solve coupled flow dynamics, for each finite difference time step, the open channel flow Equations (9.51) and (9.53) are solved first, with the groundwater inflow term $q_g = 0$; the groundwater equations are then solved during the same time step using the calculated stream elevations. q_g value can now be updated by means of Equation (9.50) and the computed stream elevations and groundwater hydraulic heads. The open channel flow equations must be solved once again for the updated q_g. This process is repeated until the change in q_g between successive calculations is less than or equal to a predetermined error tolerance. Simulation results showed that bank storage can have significant effect on attenuating and dispersing flood waves in the lower segment of a long reach, and that the degree of flood wave modification is significantly influenced by the channel reach length and the hydraulic conductivity of the floodplain aquifer.

Zitta and Wiggert (1971) considered the case of an idealized channel fully penetrating a floodplain aquifer and coupled Equations (9.51) and (9.53), with $q_l = 0$ and $q_g = 2q$, to the nonlinear Boussinesq Equation (9.9) governing one-dimensional, unsteady flow in an unconfined aquifer. q is groundwater flow to the channel per unit length per side. The stream is perfectly hydraulically connected to the aquifer (i.e., without a semi-pervious bank). The Boussinesq equation was solved by an explicit finite difference method using the same time interval used for solving Equations (9.51) and (9.53) by the explicit characteristic method. The coupled open channel and groundwater flow equations are solved for a prescribed stage flood hydrograph by means of the same iterative method described above. Uniform flow conditions are assumed at the downstream channel boundary and initially uniform water table height in the floodplain and the channel. Numerical simulation showed that bank storage causes marked attenuation of the wave and that this attenuation is characterized by a reduction in the peak channel flow and a sustained baseflow recession. Flood wave modification was minimal for wide channels and relatively low permeability of the floodplain.

If we neglect inertia terms, Equation (9.53) reduces to

$$\frac{\partial y}{\partial x} = \left(S_0 - S_f\right) \tag{9.54}$$

which applies to diffusive waves observed in a wide range of natural floods. Combining the diffusive wave approximation given in Equation (9.54) with Equation (9.51) yields the following diffusion wave equation (Cunge 1969)

$$\frac{\partial Q}{\partial t} = D\frac{\partial^2 Q}{\partial x^2} - c_k\frac{\partial Q}{\partial x} + c_k q \ ; \quad D = Q/2BS_f \quad ; \quad c_k = \partial Q/\partial A \tag{9.55}$$

where D is the wave diffusivity [L^2T^{-1}]; c_k is the kinematic wave celerity [LT^{-1}]; and q here is total lateral flow rate (runoff and groundwater seepage) per unit channel length [L^2T^{-1}]. For the specific case $\partial y/\partial x \approx 0$, we have $S_f \approx S_0$, and Equation (9.55) reduces to the well known kinematic wave approximation in the presence of a distributed lateral flow

$$\frac{\partial Q}{\partial t} = c_k\frac{\partial Q}{\partial x} + c_k q \tag{9.56}$$

The bank storage effect was examined by Hunt (1990) by coupling the kinematic wave equation approximation Equation (9.56) in a rectangular channel with one-dimensional linearized aquifer flow normal to channel length as governed by Equation (9.13). He obtains an analytical solution for an infinitely wide aquifer and a first-order approximation to the case of a finite aquifer breadth. In both cases the aquifer is considered to be in perfect hydraulic connection with the channel. Application results showed that bank storage delays the arrival of the flood wave, reduces the peak, and produces extended tailing of the hydrograph as the flood receded.

Perkins and Koussis (1996) replaced the MODFLOW's STREAM package by the Muskingum-Cunge method to simulate diffusive wave routing in channels interacting with aquifers. The Muskingum-Cunge method allows for introducing the diffusive term in Equation (9.55) by numerical diffusion arising from the finite difference approximation of the kinematic wave Equation (9.56). Simulation results showed that wave diffusion is as important as bank storage on modification of flood waves. Accurate simulation results demand streamflow routing time steps much smaller than a typical aquifer time step, and shorter aquifer time steps yielded numerical results by the proposed diffusive wave routing technique more accurate than the MODFLOW's STREAM module. Stronger stream-aquifer coupling required equal aquifer and flood routing time steps. The rather expected effects of bank storage such as flood peak attenuation and retardation, extended tailing of hydrograph recession, and shortening of its leading edge were observed in the simulations.

Hydrologic Approach: An alternative approach to the distributed or hydraulic flow routing is hydrologic routing or lumped system routing. Hydrologic routing employs an integrated (lumped) form of the continuity equation and a linear or nonlinear relation between discharge into or out of the reach or reservoir. Although hydraulic routing approach more adequately describes the dynamics of flow than does the hydrologic routing methods, the latter has been widely used to rout flow in natural streams because of its simplicity. Furthermore, hydrologic routing with linear storage models allows for the development of linear response functions by means of analytical techniques and numerical schemes for flow in stream/river reaches with a significant groundwater component. The linear response functions presented below furnish a framework to obtain numerically efficient solutions for optimization problems of conjunctive surface water and groundwater management.

In level pool routing channel storage is a function of the discharge into the reach, whereas in general storage can be expressed as a linear combination of both flow into and flow out of the reach. In the well known Muskingum method, storage, S [L^3], in a river reach is expressed as a linear combination of flow into the stream reach, I [L^3T^{-1}], and flow out of the reach, O [L^3T^{-1}],

$$S(t) = \eta\left[\xi\, I(t) + (1-\xi)\, O(t)\right] \tag{9.57}$$

in which η is the storage time constant for the reach [T]; and ξ is a weighting factor varying usually from 0 to 0.5, with a mean value of 0.2 in natural streams. In natural streams ξ varies from 0 to 0.3 (Chow et al. 1988). The storage time constant, η, represents approximately the time for a wave to travel the channel length. The specific case of $\xi = 0$, $S(t) = \eta\, O(t)$, corresponds to level pool routing. According to Equation (9.57), channel storage, S, is proportional to the linear combination of a wedge storage, $\eta\,\xi\,(I - O)$, and a prizm storage, $\eta\, O$.

The integrated form of the equation of continuity for a stream segment interacting with the surrounding alluvial bank is

$$\frac{dS(t)}{dt} = I(t) - O(t) + Q(t) + r(t) \tag{9.58}$$

where $Q(t)$ is groundwater discharge to stream segment [$L^3\ T^{-1}$]; and $r(t)$ is contribution of quickflow and direct channel precipitation adjusted for evaporation from channel surface [$L^3\ T^{-1}$]. During a storm event $r(t) > 0$ is primarily the sum of hillslope runoff, interflow, and direct precipitation on the channel ($r(t) > 0$), whereas between storm events $r(t) < 0$ accounts for channel surface evaporation rate and plant transpiration rate right on the stream-bank interface. Assuming that channel flow is initially at steady state, $dS(0)/dt = 0$, and that $r = 0$ at time 0, i.e., $I_0 = O_0 - Q_0$, Equation (9.58) would still retain the same form, but with $S(t)$, $I(t)$, $O(t)$, and $Q(t)$ redefined as deviations from their initial values at time 0.

Aquifer discharge integrated along the river reach length is

$$Q = \Gamma(\sigma - s) \tag{9.59}$$

where σ is stream stage drawdown measured relative to a high datum [L]; s is drawdown relative to the same datum of the water table adjacent to the stream [L]; and $\Gamma \approx wl(K'/b')$ is referred to as reach transmissivity (Illangasekare and Morel-Seytoux 1982) [L^2T^{-1}], which is the product of reach length L [L] and the so called river coefficient, $w(K'/b')$; and w is the wetted perimeter [L].

Rushton (2007) refer to Equation (9.59) as the concept of the river coefficient that is widely used in regional groundwater models (e.g., MODFLOW) wherein the size of a numerical cell is larger than the stream width. A detailed two-dimensional numerical groundwater flow analysis in the vicinity of partially penetrating rivers shows that the river coefficient should not be exclusively based on the properties of the riverbed deposits (i.e., streambed conductance K'/b'), but also on the hydraulic conductivity of the aquifer. Illangasekare and Morel-Seytoux (1982) provide a relationship for the reach transmissivity, but apparently neglect the resistance of bed deposits: $\Gamma = (T/e)L^{(w/2+e)} / \left(5w + (e/2)\right)$, where T is the transmissivity of the of the aquifer underlying the reach [L^2T^{-1}]; and e is the aquifer saturated thickness [L].

In departure from the traditional viewpoint that a hydrologic model is an end in itself, Morel-Seytoux and Daly (1975) reinforced the more broader view of Young (1970) that responses to alternative management strategies should be evaluated in economic terms (e.g., cost and/or benefit) rather than on the basis of physical terms (e.g., flow rates or volumes of water). In this context, a stream-aquifer model necessarily becomes an intermediate component of a more complex system. In the lexicon of systems analysis, the stream-aquifer model is embedded as a constraint within a broader mathematical optimization framework for which an optimum set of values for the decision variables (or management control variables) are obtained for an optimal value of the objective function (e.g., cost/benefit).

For a nonprismatic channel and when stream and/or aquifer hydraulic properties and conditions vary along the channel length, the stream can be divided into a set of R homogeneous reaches and flow can be routed from one reach to the other sequentially, with groundwater flow into or out of each reach computed on the basis of aquifer hydraulic properties and flow conditions within the reach. In other words, Equations (9.57) and (9.58) can also be used for semi-distributed flow routing in gaining or losing streams.

Drawdown, s, in an aquifer intersected by R stream reaches and subjected to groundwater pumping from P wells can be described by the basic linear aquifer flow equation $S_y \, \partial s / \partial t = \nabla \mathbf{T} \nabla s - \sum_{w=1}^{P} Q_w \delta_w - \sum_{r=1}^{R} Q_r \delta_r$, where $\mathbf{T}$ is the two-dimensional transmissivity tensor assuming x and y are along the principal directions; Q_w is the volumetric pumping rate of well w [L^3T^{-1}]; δ_w is the Dirac delta function singular at the point of coordinates x_w and y_w for well w; P number of pumping wells; Q_r is the aquifer discharge rate to the stream per unit reach length [$L^3T^{-1}L^{-1}$]; δ_r is the Dirac delta function singular along the rth reach of the stream; and R is number of stream reaches. The solution of this equation at well w is given by (Morel-Seytoux and Daly 1975; Morel-Seytoux 1975)

$$s_w(t) = \sum_{p=1}^{P} \int_0^t Q_p(\tau) k_{wp}(t-\tau) d\tau + \sum_{r=1}^{R} \int_0^t Q_r(\tau) k_{wr}(t-\tau) d\tau \tag{9.60}$$

in which Q_r = the aquifer discharge to a stream reach r per unit length of the reach [$L^3T^{-1}L^{-1}$] is related to the drawdown to the water level in reach r, σ_r, to the natural drawdown to the aquifer water table, ν_r, and to the pumping rates in the aquifer, Q_p, by this integral equation

$$Q_r(t) + \Gamma_r \sum_{m=1}^{R} \int_0^t Q_m(\tau) k_{rm}(t-\tau) d\tau = \Gamma_r \left[\sigma_r(t) - \vartheta_r(t) \right] - \Gamma_r \sum_{p=1}^{P} \int_0^t Q_p(\tau) k_{rp}(t-\tau) d\tau \tag{9.61}$$

where Γ_r is the transmissivity of reach r [L^2T^{-1}]; R is the total number of reaches; P is the number of pumping wells, k_{rm} and k_{rp} are, respectively, the "reach" and "well" kernels; and t is time.

For a uniform drawdown in reach r, the relationship between storage S_r (relative to its value at $t = 0$) and change in stage is

$$S_r = A_r \left(\sigma_r^0 - \sigma_r \right) \tag{9.62}$$

where A_r is the reach surface area [L^2]; and σ_r^0 is the initial river drawdown [L]. The solution of the linear system Equation (9.57) and (9.58) with (9.62) is (Morel-Seytoux 1975)

$$\sigma_r(t) = \sigma_r^0 - \frac{1}{A_r} \left\{ \sum_{m=1}^{r} \int_0^t \kappa_{rm}(t-\tau) \left[a_m Q_m(\tau) + b_m I(\tau) \right] d\tau \right\} \tag{9.63}$$

where κ_{rm} is the mth reach routing kernel; and $I(t)$ is the inflow to the first [L^3T^{-1}], upstream reach. Expressions for κ_{rm} and coefficients a_m and b_m are given by Morel-Seytoux (1975).

Equation (9.63) shows explicitly the river drawdown dependence on the inflow to the first reach $I(t)$ and on flow to the aquifer $Q_m(t)$ in each reach

In Equation (9.61) $\varphi_r(t)$ is the aquifer drawdown natural redistribution that otherwise would take place if there was no stream in the system and no pumping from the wells. It represents the redistribution of the water table as a result of a nonuniform initial drawdown in the aquifer. Morel-Seytoux (1975) and Illangasekare and Morel-Seytoux (1982) solve this problem by interpreting the initial drawdown conditions in the aquifer s_π^0 as the result of pumping during the period immediately preceding the zero time at the points π and letting Q_π^0 be these artificial pumping rates; it follows that

$$\vartheta_r(t) = \sum_{\pi=1}^{\Pi} \int_0^{t+1} Q_\pi^0 k_{r\pi}(t+1-\tau)\, d\tau \tag{9.64}$$

where $k_{r\pi}$ is aquifer drawdown natural redistribution kernels; and Π is number of grid points.

The general equation describing return flow Q_m to reach m can be obtained by substituting Equation (9.63) and (9.64) into Equation (9.61). The resulting equation relates the unknown return flows and the initial aquifer drawdowns, upstream river inflows, and pumping rates. The sequence for calculating the discrete form of the continuous kernels presented above is described by Morel-Seytoux and Daly (1975) and Morel-Seytoux (1975). The discrete form of the kernels (or impulse response functions) k_{rm}, k_{rp}, κ_{rm}, and $k_{r\pi}$ can be generated by a finite difference model or by other methods using a standard groundwater simulator, from which the remaining unknown influence coefficients are computed from linear algebraic system of equations. The main advantage of this approach is that discrete kernels are calculated only once and set aside for use in subsequent manipulations, only the influence coefficients (Morel-Seytoux and Daly 1975; Morel-Seytoux 1975; Illangasekare and Morel-Seytoux 1982) need to be recalculated if the position of the stream channel or surface water body is modified. This property has a great appeal to solving conjunctive surface water and groundwater management problems by means of mathematical optimization, which otherwise requires repeated (e.g., finite difference model simulations).

The problem of river flow interactions with their alluvial aquifers and the impact of bank storage in the riparian floodplains on flood wave modification and baseflow sustenance can be solved analytically by making use of the linear response theory. Figure 9.13 is a schematic of a prismatic channel and the surrounding floodplain aquifer and a typical cross section of the aquifer on one side of the channel. Assuming groundwater flow is one-dimensional and normal to channel flow direction, a mathematical system can be formulated by combining the Muskingum relation given in Equation (9.57) and the continuity Equation (9.58) with the linearized approximation of the Boussinesq equation. The coupled stream-aquifer system can be solved analytically for various boundary conditions (Hantush et al. 2002; Hantush 2005).

The boundary value problem for the floodplain aquifer is

$$S_y \frac{\partial h(x,t)}{\partial t} = T \frac{\partial^2 h(x,t)}{\partial x^2} + N(x,t) \tag{9.65}$$

and

$$h(x,0) = 0, \tag{9.66a}$$

$$Q/2 = T \frac{\partial h(0,t)}{\partial x} = PK' \frac{[h(0,t) - H(t)]}{b'}, \tag{9.66b}$$

$$h(\infty,t) = 0, \text{ or } h(l,t) = \phi(t), \text{ or } \partial h(l,t)/\partial x = 0 \tag{9.66c}$$

in which $h(x,t)$ is water table elevation relative to its initial value at time 0 (initially at steady state); x is distance from the stream; $N(x,t)$ is deviation of net groundwater recharge rate per unit area (negative when evapotranspiration exceeds infiltration) relative to its initial value at $t = 0$; Q is aquifer discharge from both sides of the floodplain (negative values implies flow to the aquifer); l is aquifer width or distance to an observation well; $\phi(t)$ describes water table fluctuations at distance l from streambank; T is average aquifer transmissivity; S_y is the specific yield of the alluvial sediments; P is half the wetted perimeter of the channel; K' is the hydraulic conductivity of the streambed [L T^{-1}]; and b' is the thickness of the streambed [L]. The distance l in Equation (9.66c) can be the width of a riparian buffer strip where head might be controlled by drainage from upper lands in the watershed, or the distance to an observation well tapping the alluvial sediments. The head at this location might also be controlled by pumping wells. The impervious, no-flow condition, third expression in Equation (9.66c), accounts for hydraulic divides, bedrocks or mountain ranges. Some average of water level in the stream $H(t)$ is assumed throughout a reach of length L, and channel storage $S(t)$ therefore is $S(t) = W \cdot L \cdot H(t)$, where W is the channel width. Again, to better account for additional head losses due to partial penetration of the aquifer by the stream, P might as well be treated as an empirical adjustment, rather than half the wetted perimeter (recall, the one-dimensional approximation Equation (9.65) neglects the effect of partial penetration). As a crude estimate, P can vary anywhere between half the wetted perimeter and some average value of the saturated thickness at the interface, $h(0,t)$. In addition to commonly made assumptions, the system is assumed homogeneous and symmetric around the channel axis, and water table fluctuations are much smaller than some average saturated thickness of the aquifer.

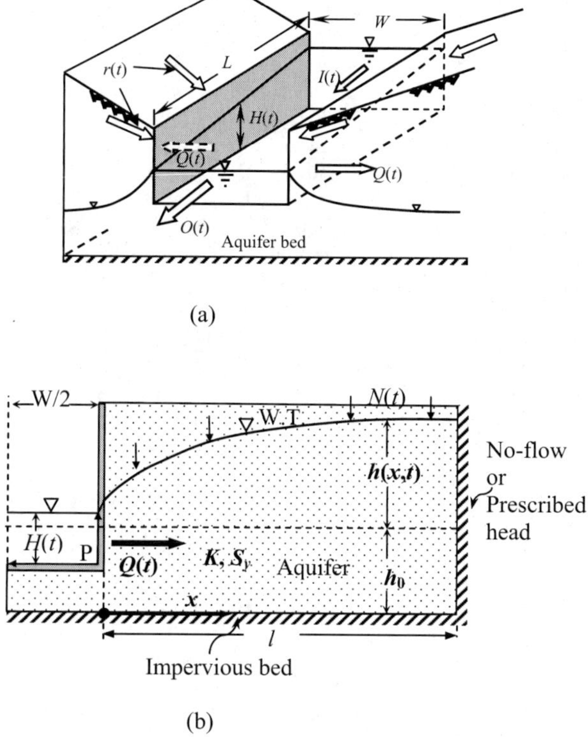

Figure 9.13: Schematic illustration of a stream-aquifer system: (a) surface and subsurface flow routing components; and (b) cross-sectional view of unconfined aquifer separated from the stream by a semi-permeable layer. From Hantush (2005), used with permission of Elsevier.

The response and interactions of the stream-aquifer system Equations (9.57), (9.58), (9.65), and (9.66) are driven by four system excitations, namely, the inflow to the channel, $I(t)$, net recharge rate to the water table, $N(t)$, net lateral flow to the channel, $r(t)$, and drawdown at some distance l from the bank, $\phi(t)$. For a semi-infinite aquifer width or the case of impermeable bedrock at $x = l$, the last forcing term is irrelevant. During a storm event $r(t)$ is primarily the sum of direct runoff, interflow, and direct precipitation on the channel ($e(t) > 0$), whereas during no-rain periods $r(t) = L \cdot W \cdot e(t)$ is the channel surface evaporation rate and, thus, assumes negative values ($e(t) < 0$). During dry periods, $N(t)$ is related to riparian evapotranspiration losses and thus has values less than 0.

By virtue of linearity of the stream-aquifer system, Hantush (2005) expresses the channel outflow rate $O(t)$ and aquifer discharge $Q(t)$ in terms of convolution integrals of the various system excitations and derives analytical expressions for their impulse response and unit-step response functions (also, see Hantush et al. 2002). They obtain convolution integrals for the channel outflow volume $V_O(t)$ and bank storage $V_Q(t)$ in terms of the unit-step response functions and alternative forms in terms of the impulse response functions. For example, the impulse response function relating the channel outflow rate $O(t)$ to the inflow hydrograph, $u_s(t)$, gives the channel outflow rate distribution in t in response to a unit amount of water applied instantly at $t = 0$ at the channel inlet $[T^{-1}]$. The impulse response function relating the channel outflow rate to the aquifer net recharge $[L^2\ T^{-1}]$, $u_n(t)$, is the channel outflow rate response to a unit amount of recharge per unit area applied instantly at $t = 0$. Impulse response functions for $\phi(t)$ and $r(t)$, $u_\phi(t)$ and $u_r(t)$, respectively, and those associated with $Q(t)$, $u_a^*(t)$; $u_n^*(t)$, $u_\phi^*(t)$, and $u_r^*(t)$, are defined similarly.

During drought periods, $u_r^*(t)$ relates withdrawal of groundwater from the banks to direct evaporation from the channel surface. If we assume evaporation from channel surface exceeds that from the water table, the falling stage relative to the aquifer head produces hydraulic gradient in the direction of the stream, thereby driving the water from alluvial banks to the stream.

The unit-step response functions relate system excitations to volumes rather than rates; they are the integral in time of the corresponding impulse response functions. For example,

$$g_s(t) = \int_0^t u_s(\tau)\, d\tau \tag{9.67}$$

$g_n(t)$, $g_\phi(t)$, and $g_r(t)$, respectively, are similarly defined as the integrals from $\tau = 0$ to $\tau = t$ of $u_n(\tau)$, $u_\phi(\tau)$, and $u_r(\tau)$ (Hantush 2005, Table 1 and 2).

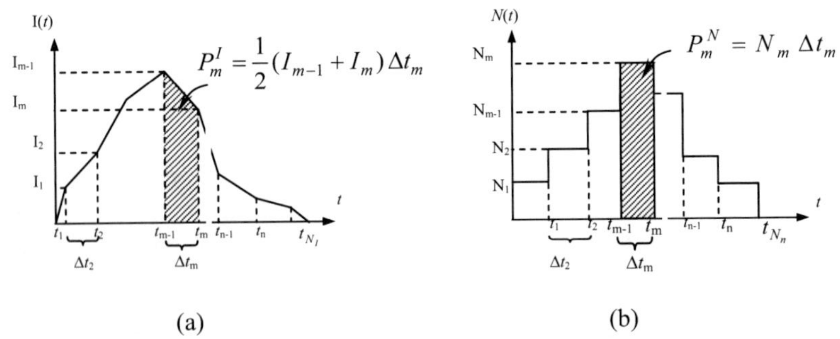

Figure 9.14: Discrete time inputs: (a) discrete time inflow hydrograph; and (b) piecewise uniform aquifer recharge rate. From Hantush (2005), used with permission of Elsevier.

Figures 9.14a and 9.14b illustrate, respectively, discrete time measurements of a typical inflow hydrograph and piecewise uniform estimates of water table accretion rate, all measured relative to initial values at $t = 0$. In general, these quantities are sampled at discrete points and nonuniformly in time. Figure 9.14a may also resemble head fluctuations in an observation well, with $\phi(t)$ as the ordinate, or discrete measurements of net lateral channel flow, with $r(t)$ as the ordinate.

The inflow hydrograph in Figure 9.14a shows discrete inflow rates I_m at the discrete time points t_m, $m = 1, 2, \ldots N_I$. The time period between successive measurements $\Delta t_m = t_m - t_{m\text{-}1}$ is not necessarily uniform over the record. This also applies to sampled water table levels at $x = l$. Stepwise uniform recharge is considered in the sampling period (Figure 9.14b), N_m is the recharge rate in time period Δt_m. The discrete-time forms of the channel outflow rate $O(t)$, flow rate to the aquifer $Q(t)$, and channel outflow volume and bank storage $V_O(t)$ and $V_Q(t)$, respectively, can be obtained following Hantush et al. (2002):

$$O_n = \sum_{m=1}^{n\le N_I} P_m^I U_{n,m}^I + \sum_{m=1}^{n\le N_n} P_m^N U_{n,m}^N + \sum_{m=1}^{n\le N_\phi} P_m^\phi U_{n,m}^\phi + \sum_{m=1}^{n\le N_r} P_m^r U_{n,m}^r \tag{9.68}$$

where

$$\begin{aligned} &P_m^I = (I_{m-1} + I_m)/(2\Delta t_m), \quad P_m^N = N_m \, \Delta t_m \,, \quad P_m^\phi = (\phi_{m-1} + \phi_m)/2, \\ &P_m^r = (r_{m-1} + r_m)/(2\Delta t_m), \quad \Delta t_m = t_m - t_{m-1} \end{aligned} \tag{9.69}$$

and

$$U_{n,m}^I = \left[g_s(t_n - t_{m-1}) - g_s(t_n - t_m)\right]/\Delta t_m \tag{9.70}$$

$$U_{n,m}^N = \left[g_n(t_n - t_{m-1}) - g_n(t_n - t_m)\right]/\Delta t_m \tag{9.71}$$

$$U_{n,m}^\phi = \left[g_\phi(t_n - t_{m-1}) - g_\phi(t_n - t_m)\right]/\Delta t_m \tag{9.72}$$

$$U_{n,m}^r = \left[g_r(t_n - t_{m-1}) - g_r(t_n - t_m)\right]/\Delta t_m \tag{9.73}$$

wherein N_I, N_n, N_ϕ, and N_r are the number of sampling time intervals for $I(t)$, $N(t)$, $\phi(t)$, and $r(t)$, respectively. I_n, N_n ϕ_n, and r_n equal to 0 for $n \ge N_I$, N_n, N_ϕ, N_r. $U_{n,m}^I$, $U_{n,m}^N$, $U_{n,m}^\phi$, and $U_{n,m}^r$ are, respectively, the corresponding discrete unit pulse response functions. If a water divide (i.e., no-flow) exits at $x = l$, all but the third summation on the right of Equation (9.68) are retained.

Similarly, the discrete form of $Q(t)$, Q_n, is

$$Q_n = \sum_{m=1}^{n \le N_I} P_m^I U_{n,m}^{I*} + \sum_{m=1}^{n \le N_n} P_m^N U_{n,m}^{N*} + \sum_{m=1}^{n \le N_\phi} P_m^\phi U_{n,m}^{\phi*} + \sum_{m=1}^{n \le N_r} P_m^r U_{n,m}^{r*} \tag{9.74}$$

where

$$U_{n,m}^{I*} = \left[g_a^*(t_n - t_{m-1}) - g_a^*(t_n - t_m) \right] / \Delta t_m \tag{9.75}$$

$$U_{n,m}^{N*} = \left[g_n^*(t_n - t_{m-1}) - g_n^*(t_n - t_m) \right] / \Delta t_m \tag{9.76}$$

$$U_{n,m}^{\phi*} = \left[g_\phi^*(t_n - t_{m-1}) - g_\phi^*(t_n - t_m) \right] / \Delta t_m \tag{9.77}$$

$$U_{n,m}^{r*} = \left[g_r^*(t_n - t_{m-1}) - g_r^*(t_n - t_m) \right] / \Delta t_m \tag{9.78}$$

and $g_a^*(t)$, $g_n^*(t)$, $g_\phi^*(t)$, and $g_r^*(t)$, are the unit-step response functions associated with Q and corresponding to I, N, ϕ, and r, respectively (Hantush 2005, Table 1 and 2).

The discrete form of $V_O(t)$ is (Hantush 2005)

$$V_{O_n} = \sum_{m=1}^{n} P_m^\nu U_{n,m}^I + \sum_{m=1}^{n} P_m^\psi U_{n,m}^N + \sum_{m=1}^{n} P_m^\varphi U_{n,m}^\phi + \sum_{m=1}^{n} P_m^\upsilon U_{n,m}^r \tag{9.79}$$

where

$$P_m^\nu = (\nu_{m-1} + \nu_m)/(2\Delta t_m), \quad P_m^\psi = (\psi_{m-1} + \psi_m)/(2\Delta t_m), \quad P_m^\varphi = (\varphi_{m-1} + \varphi_m)/(2\Delta t_m)$$
$$P_m^\upsilon = (\upsilon_{m-1} + \upsilon_m)/(2\Delta t_m) \tag{9.80}$$

and

$$\nu_m = \sum_{k=1}^{m \le N_I} P_k^I, \qquad \psi_m = \sum_{k=1}^{m \le N_n} P_k^N, \qquad \varphi_m = \sum_{k=1}^{m \le N_\phi} P_k^\phi, \qquad \upsilon_m = \sum_{k=1}^{m \le N_r} P_k^r \tag{9.81}$$

wherein V_{O_n} is outflow volume at discrete time t_n; and P_k^I, P_k^N, P_k^ϕ, and P_k^r are given in Equation (9.69).

Alternative to evaluation of further integrals, the bank storage $V_Q(t)$ can be obtained directly by integrating the continuity Equation (9.58) from $\tau = 0$ to $\tau = t$, and making use of Equation (9.57)

$$\begin{aligned} V_Q(t) &= \int_0^t Q(\tau)\, d\tau \\ &= v(t) - \eta\left[\xi I(t) + (1-\xi) O(t)\right] - V_O(t) + \upsilon(t) \end{aligned} \tag{9.82}$$

which has the following discrete form

$$V_{Q_n} = v_n - \eta\{\xi I_n + (1-\xi) O_n\} - V_{On} + \upsilon_n \tag{9.83}$$

where V_{Q_n} is bank storage at t_n.

It should be noted that the discrete time relationships (9.68), (9.74), and (9.79) can be modified for stepwise linear representation of $I(t)$, $N(t)$, $\phi(t)$, $v(t)$, $\psi(t)$, $\varphi(t)$, and $\upsilon(t)$ in sampled records by a straightforward extension of the results of Hantush et al. (2002), Equations (9.39a)-(9.39e) and (9.45)-(9.46c), to this analysis and integration of $g_s(t)$, $g_n(t)$, $g_\phi(t)$, and $g_r(t)$ from $\tau = 0$ to $\tau = t$. Alternatively, a more simplified approach is the trapezoidal rule where V_{O_n} and V_{Q_n} are computed directly from O_n and Q_n values, i.e., Equations (9.41) and (9.42) of Hantush et al. (2002).

Figure 9.15a shows the effect of aquifer hydraulic conductivity K on modification of a hypothetical flood wave (Hantush 2002). These results illustrate the effect of bank storage on peak flood reduction, increasingly sustained flow (i.e., extended tailing), and somewhat delayed peak arrival with increasing K. The effect of retardation or streambed leakance R on flow to the aquifer is depicted in Figure 9.15b.

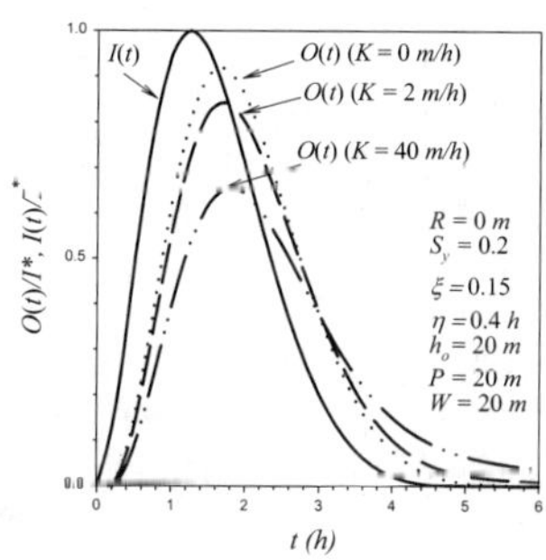

(a)

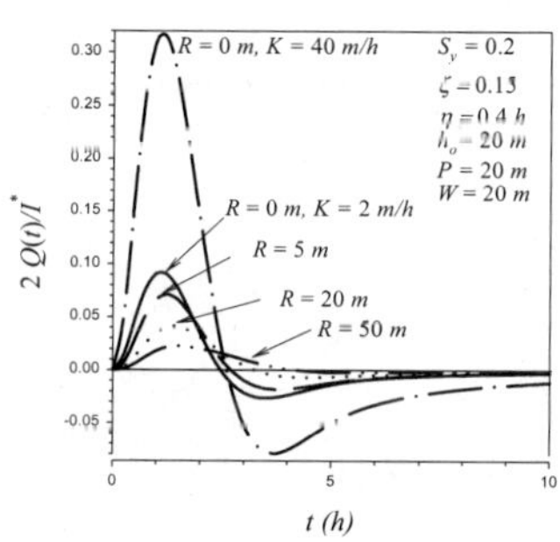

(b)

Figure 9.15: Sensitivity results: (a) scaled stream discharge rate variation with aquifer hydraulic conductivity; and (b) scaled stream to aquifer flow rate variation with the retardation factor. I^* is the peak channel inflow rate.

Figure 9.16 illustrates the effects of a hypothetically simulated regulated reservoir releases. The figure shows the decline in streamflow due to reduction of reservoir releases over a period of 10 hrs, with no water table replenishment, $N = 0$, and with water table accretion at a rate of $N = 0.05$ cm/hr per unit aquifer width per unit channel length (Hantush 2005). Despite the complete shutdown of the reservoir releases, the stream continues to be supplied by groundwater discharge, albeit, at much slower baseflow rates. Even when $N = 0$ cm/hr, baseflow persists for a period exceeding 200 hrs owing to bank storage releases.

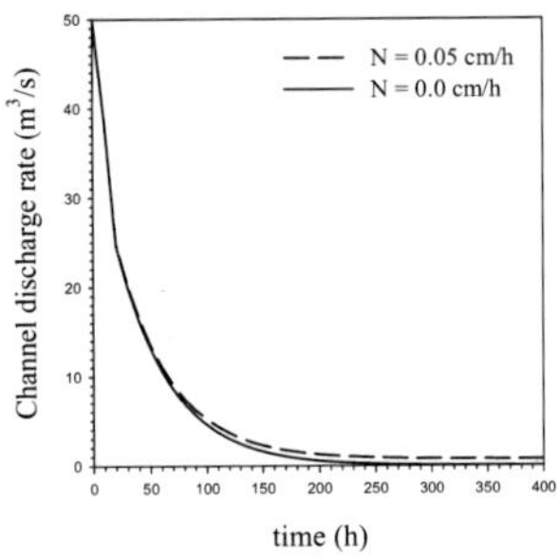

Figure 9.16: Simulated channel discharge rates for the hypothetical scenario of channel inflow rate reduction to 0 cm/h, with 0 and 0.05 cm/hr aquifer recharge. From Hantush (2005), used with permission of Elsevier.

The analytical discrete kernels are robust because they can be applied in a semi-distributed manner to heterogeneous stream-aquifer systems by dividing a river reach into piecewise homogeneous segments and routing flow from one segment to the other. Channel and aquifer hydraulic properties and hydrologic conditions may be assumed uniform in each segment. The outflow from one segment becomes the inflow to the following segment. However, apart from the simplifying assumptions that were made, the dependence of the above derived relationships on the essentially empirical Muskingum parameters η and ξ, and whether these parameters are unique to the channel are additional limitations. As stated earlier, ξ varies between 0 and 0.3 with a mean value of order 0.2 (Chow et al. 1988). While in practice the two parameters are inferred from flow measurements by curve fitting, estimates of η can nevertheless be computed directly from flow measurements and channel geometry (recall, it is the average reach travel time). Its value may also be estimated as the observed time of travel of peak flow through the reach. Cunge (1969) showed if the parameters ξ and η are appropriately defined, the finite-difference form of the Muskingum routing equation can be considered an approximate solution to the kinematic wave or diffusive wave equation (also, Chow et al. 1988; Viessman and Lewis 1996; Perkins and Koussis 1996).

While the analytical framework described above brings forward in rather unambiguous fashion the essential features of the problem, numerical techniques can accommodate more general flow scenarios and less restrictive assumptions. A finite-difference approach that utilizes the nonlinear Muskingum storage relationship for river routing with bank storage was presented by Birkhead and James (2002). In their approach the storage within the reach is given by $S=(b/a)\left[\xi I^{m/n}+(1-\xi)O^{m/n}\right]$, where constants a and n reflect the stage-discharge rating function, and b and m mirror the stage-volume rating function. The discrete form of the flow from the river reach to the aquifer was derived from integral convolution; its discrete form is given by Hall and Moench (1972) and Moench et al. (1974) for unconfined flow Equation (9.13) without semi-pervious streambank. Rather than accounting for aquifer flow in the continuity equation, Equation (9.58), they add bank storage into channel storage to complete mass balance.

9.2.6 Stream Depletion

When pumping groundwater from a well near a stream, a cone of depression induced by water table drawdown develops around the well and the volume of the cone represents a volume of water that has been taken from storage in the aquifer. The cone of depression continues to expand until its perimeter reaches the stream where further lowering of the water table in the vicinity capture some of the groundwater flow that would have, without pumping, discharged as baseflow to the stream (Sophocleous 2002). The induced hydraulic gradient toward the aquifer causes water to flow from the stream to the surrounding porous banks and through the streambed to the aquifer below. This phenomenon is referred to as stream depletion or induced infiltration (e.g., Hantush 1965). Flow in small streams can be substantially reduced or may even cease by excessive pumping and poorly managed groundwater development. In addition to diminishing water supplies, stream depletion can have undesirable ecological consequences and can cause problems involving adjudication or litigation over the legal rights of communities whose main water supply depends primarily on a sustained downstream flow. While in this section we have chosen to focus on stream depletion, poorly developed aquifers could also lead to the drying of springs and wetlands that constitute natural discharge areas of groundwater systems (Sophocleous 1997, 2000).

The most fundamental result of stream depletion is that obtained by Theis (1941) and Glover and Balmer (1954) for a well near a stream and pumping groundwater at a constant rate Q_w from an extensive aquifer. Figure 9.17a depicts a stream partially penetrating an extensive unconfined aquifer and a well fully taping the aquifer at a distance l from the stream. A semi-permeable bed sediment layer separates the stream from the aquifer. Figure 9.17b illustrates a hypothetical, equivalent stream-aquifer system with the stream fully penetrating and separated from the aquifer by a vertical semi-pervious layer of similar thickness and hydraulic conductivity as in the actual system. The stream in the equivalent system is situated at a distance x_0 from the well larger than l. The stream and aquifer extend indefinitely in the direction normal to the figure and the aquifer is of a semi-infinite extent in the x direction (Figure 9.17c). The boundary value problem describing this situation can be cast in radial (r,θ) or in the Cartesian (x,y) coordinate system. In an extensive and homogeneous aquifer without a stream, radial coordinates are convenient because of the radially-symmetric nature of flow toward the well.

The boundary-value problem describing drawdown in a homogeneous and isotropic leaky aquifer near a fully penetrating stream and pumped by a well at distance l from a stream is

$$\frac{1}{\nu}\frac{\partial Z}{\partial t}=\frac{\partial^2 Z}{\partial x^2}+\frac{\partial^2 Z}{\partial y^2}-\frac{2}{\lambda^2(1+\beta)}Z+\frac{2Q_w}{K}\delta(x-x_0)\delta(y-y_0) \tag{9.84}$$

$$Z(x,y,0)=0;\quad Z(x,\pm\infty,t)=0;\quad Z(\infty,y,t)=0 \tag{9.85a}$$

$$\left.\begin{aligned}&\frac{\partial Z(-l,y,t)}{\partial x}=\frac{1}{R}Z(-l,y,t); && \text{with a clogging layer}\\ &Z(-l,y,t)=0; && \text{without a clogging layer}\end{aligned}\right\} \tag{9.85b}$$

where $Z=h_0^2-h^2$; h is height of water table above the horizontal base of the aquifer [L]; h_0 is initial depth of saturation [L]; $\lambda=\sqrt{(b_a/K_a)\bar{h}K}$ is the leakage factor [L]; b_a [L] and K_a [LT^{-1}], respectively, the thickness and hydraulic conductivity of the aquitard; $\bar{h}\approx h_0$ is some average of aquifer saturated thickness; $\beta=h_0/\bar{h}$ (≈ 1 for $h - h_0 << h_0$); Q_w is the constant discharge of the well [L^{3}T^{-1}]; δ( ) is the Dirac delta function; $\nu=K\bar{h}/S_y$ is the aquifer diffusivity [L^{2}T^{-1}]; K is the aquifer hydraulic conductivity [LT^{-1}]; S_y is the specific yield of the aquifer; $R=Kb'/K'$ is the retardation factor [L]; K' [LT^{-1}] and b' [L], respectively, are the hydraulic conductivity and thickness of the semi-pervious layer of the streambed; l is distance between the pumped well and the stream [L]; and x_0 and y_0 are the x and y coordinates of the well, respectively.

This boundary value problem in Equations (9.84)-(9.85) also describes flow toward a well in a confined aquifer in contact with a stream. In this case, $Z=h_0-h$ corresponds to the drawdown s in the piezometric head [L]; $\nu=T/S$, where $T=KB$ is aquifer transmissivity; $S = S^*B$ is the aquifer storativity; S^* is the specific storativity; B is thickness of the aquifer; $\lambda=\sqrt{(b'/K')T}$; $\beta=1$; and $1/T$ replacing $2/K$ in the last term on the right of Equation (9.84) [e.g., Equation (9.10) of Zlotnik (2004)]. Note that the stream stage and head in the source bed are held constant at h_0. The specific case of $\lambda \to \infty$ corresponds to no leakage across the bed and purely unconfined/confined aquifer flow. Worth noting also is that $Z=2h_0s'$, where $s'=s-s/2h_0$ is called the *corrected drawdown* (i.e., the drawdown that would occur in an equivalent confined aquifer); and $s=h_0-h$ is water table drawdown. As expected, $s'\approx s$ when $s << h_0$.

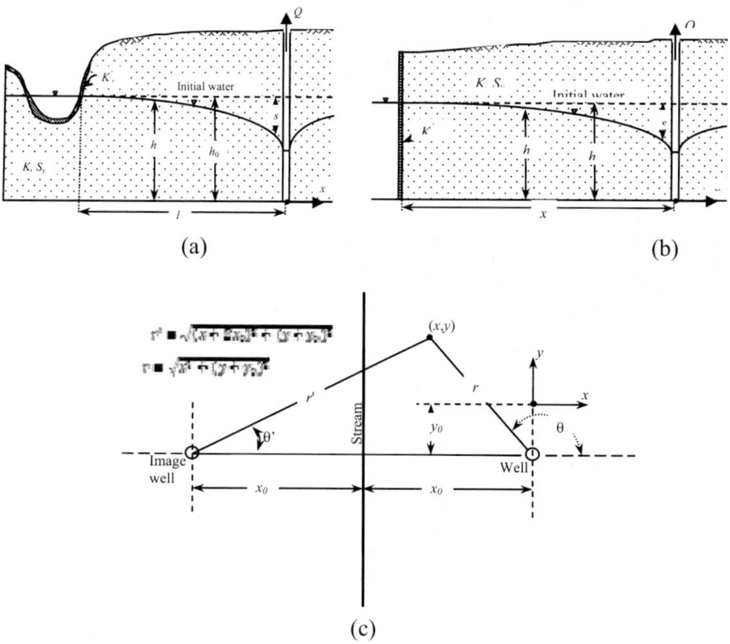

Figure 9.17: Conceptual stream-aquifer system with a well pumping near a stream: (a) actual system (partial incision by the stream); (b) an equivalent stream-aquifer system (fully incised aquifer); and (c) image well theory for solutions of drawdown near a stream, actual well and its image on the opposite side from the stream.

Equation (9.84) subject to the initial and boundary conditions (9.85a)-(9.85b) describes asymmetric aquifer flow and for the specific case of perfect hydraulic connection with the stream (i.e., without a clogged streambed) can be solved readily using the method of images. In general, solutions for symmetric radial flow are much simpler to obtain than asymmetrical radial flow caused by the presence of a water body or an impermeable boundary for which case a Cartesian coordinate system may be more convenient. However, for the specific case of an unclogged streambed and by virtue of linearity of confined flow systems (or unconfined flow with water table drawdown much smaller than the average saturated thickness of the aquifer) the method of images offers a powerful tool for solving asymmetric flow toward a pumping well using the rather simpler radial flow solutions and superposition of the responses to the actual and their image wells. In the method of image wells, actual boundaries and the semi-infinite aquifer are replaced by an equivalent hydraulic system composed of an extensive aquifer and actual wells and their images across the boundaries (see Figure 9.17c). The drawdown and stream depletion rate are thus computed from radial solutions derived from well(s) pumping from and/or recharging an extensive aquifer. For example, for a fixed-head boundary, such as a stream with a constant stage, the image of a pumping well located at a distance l from the stream is located on the opposite side at the same distance from the stream (hence, a mirror image of the actual pumping well) but recharging, instead of discharging, at a rate of Q_w (Figure 9.17c). Of course, this method is applicable if the actual boundaries form simple geometries; for more details refer to Todd (1980, pp. 139-146).

The stream depletion rate, Q_r [$L^3T^{-1}L^{-1}$], assuming the river is infinitely long, is the integral of the flow rate to the aquifer per unit river length from $y = -\infty$ to $y = +\infty$

$$Q_r = -K \int_{-\infty}^{+\infty} h(-l,y,t)\frac{\partial}{\partial y}h(-l,y,t)dy = \frac{K}{2}\int_{-\infty}^{+\infty}\frac{\partial}{\partial y}Z(-l,y,t)dy \tag{9.86}$$

The rate of river depletion at time t since pumping started in a semi-infinite nonleaky aquifer fully penetrated by a river with a pervious bed can be obtained by applying the method of images to Theis (1935) solution (Glover and Balmer 1954; see, also, Theis 1941)

$$Q_r = Q_w erf(u), \quad u = l/\sqrt{4\nu t} \tag{9.87}$$

For the case of a fully-penetrating stream and semi-permeable bank, the solution is obtained by solving the boundary-value problem (9.84), (9.85a) and the first part of (9.85b) by means of Laplace transform and the method of separation of variables (Hantush 1965)

$$Q_r = Q_w \left\{ erfc(u) - e^{-u^2+(u+w)^2} erfc(u+w) \right\}; \quad w = \sqrt{\nu t}/R \tag{9.88}$$

In order to extend the utility of Equation (9.87) to a partially penetrating stream with or without a clogging layer, Jacob (1950) hypothesized the use of an artificial distance x_0 larger than the actual distance between the well and the stream, l (i.e., horizontal prolongation of the aquifer). That is, the flow length in the aquifer section between the well and the stream is extended and the streambank therefore is pushed farther away to the left in Figure 9.17a so that, though artificially, resistances due to partial penetration and/or semi-pervious layer are accounted for. Methods for estimating extended flow lengths x_0 are presented by several investigators (e.g., Kazmann 1948; Rorabaugh 1956; Hantush 1959). The modification provided by Equation (9.88) is based on the conjecture that the effect of a semi-pervious riverbed can still be accounted for by the resistance of a hypothetical vertical semi-permeable layer and that the effect of partial penetration is made up for by an effective distance between the well and the stream x_0' smaller than that proposed by Jacob (1950), i.e., x_0, but still larger than the actual distance l. In addition to underestimating the drawdown distribution, ignoring completely the effect of the semi-perviousness of the clogging layer will result in significantly overestimating the rate of river depletion (Figure 9.18).

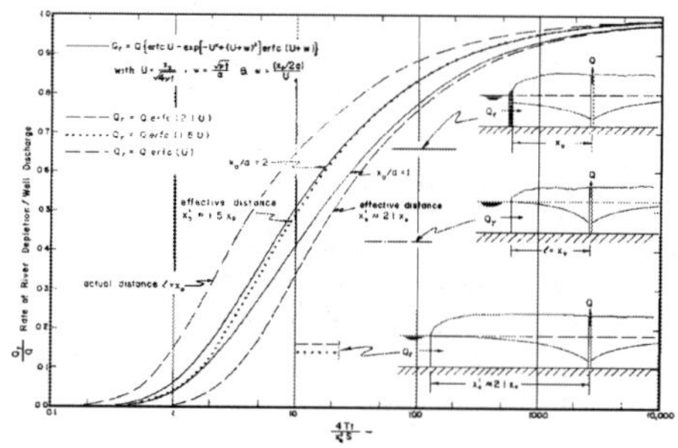

Figure 9.18: Comparison of rate of stream depletion; pervious and semi-pervious streambed theories; $a = Kb'/K'$ denotes the retardation coefficient. From Hantush (1965), used with permission of the American Geophysical Union.

The smaller extended distance proposed by Hantush (1965) helps alleviate the problem of artificially added storage capacity of the aquifer that comes with the additional length of the aquifer, while the added semi-permeable layer accounts for the streambed resistance to flow. Larkin and Sharp (1992) reported that streambed conductivity is typically as much as one to three orders of magnitude lower than aquifer conductivity. The neglect of conductivity contrasts and stream partial penetration in analytical models can lead to significant errors (Sophocleous et al. 1995; Conrad and Beljin 1996).

Spalding and Khaleel (1991) examined the limitations of the above stream depletion formulas and associated solutions for drawdown and developed a technique to estimate the additional seepage length (i.e., x_0 and $x_0^{'}$) and the retardation coefficient (R) utilized to account for the effects of partially penetrating streams and resistance of semi-pervious beds for a given stream-aquifer system. They compared numerical results obtained form a two-dimensional groundwater flow model AQUIFEM (Townley and Wilson 1980) to the above stream depletion and associated drawdown analytical solutions. Among others, they showed that neglecting the effects of aquifer storage on the side of the stream opposite to the well caused errors as large as 21%, and that analytical solutions tend to overestimate stream depletion under common stream-aquifer systems.

Rather than hypothesizing about the extended aquifer length, Hunt (1999) obtains exact analytical solutions for drawdown and stream depletion rate for the case of a slightly penetrating and narrow stream with a clogging bed by solving the nonleaky form of Equation (9.84) (i.e., dropping distributed leakage term) expressed in terms of corrected drawdown s' or drawdown s and accounting for the stream as a line source. The solution is

$$Q_r = Q_w \left\{ erfc(u) - exp\left(\frac{\gamma^2 t}{4S_y T} + \frac{\gamma l}{2T} \right) erfc(u + \sqrt{\frac{\gamma^2 t}{4S_y T}}) \right\} \tag{9.89}$$

where $u = l / \sqrt{4vt}$; $\gamma = WK'/b'$; W is wetted perimeter; $v = T/S$; $T = K\bar{h}$ is average aquifer transmissivity; and l is distance from pumping well to the stream. Comparison of Equation (8.89) with (8.88) reveals that if the Hantush retardation coefficient is redefined as $R = 2T/\gamma$, instead of Kb'/K', the two equations are equivalent and the Hantush solution for a fully-penetrating streambed and a semi-infinite aquifer can be used to described flow depletion from a slightly penetrating streambed in an infinite aquifer by modifying the retardation coefficient [referred to as *streambed leakance* by Hunt (1999) or *streambed leakage coefficient* (Zlotnik 2004)]. Rather than extending aquifer length (Hantush 1965) to account for flow resistance due to partial penetration, the result above suggests a modified retardation factor. However, the Hunt solution is limited to a stream cross section that has width and depth small compared to the saturated aquifer thickness.

For a shallow stream of finite width W, stream depletion caused by pumping a well tapping a homogeneous aquifer and at distance l from the stream is (Zlotnik 2004)

$$Q_r = Q_w \left\{ erfc(u) - exp\left[(d/(2B_s u))^2 + d/B_s \right] erfc\left[u + (d/(2B_s u)) \right] \right\} \tag{9.90}$$

in which

$$B_s = \lambda coth \frac{W}{2\lambda}; \qquad \lambda = \sqrt{b'T/K'} \tag{9.91}$$

where $u = l/\sqrt{4vt}$; and λ is defined above as the leakage factor.

It is interesting to note that Equations (9.90) and (9.89) are identical to the Hantush functional form in Equation (9.88) but with modified parameters. In Equation (9.90), the streambed leakage coefficient is a more realistic representation of Hantush (1965) retardation parameter (Kb'/K'), and for the specific case of a small stream width ($W \ll 2\lambda$), the expression for B_s reduced to that derived by Hunt (1999): $B_s \approx 2T/\lambda$.

Equations (9.87)-(9.90) share a common feature in that the river is the only source of water to sustain a long pumping period. These solutions predict that after aquifer storage is exhausted the well flow rate, Q_w, is entirely derived from streambed filtration: $\lim_{t \to \infty} Q_r / Q_w = 1$. This is possible if the river constitutes a limitless (infinite) source of water and river stage response to drawdown in the aquifer is insignificant.

The effect of leakage from an overlying bed as a source was first considered by Hunt (2003). The aquitard has a free-surface drawdown and overlies a leaky confined aquifer. A narrow stream slightly penetrates the aquitard below the water table and whose bed is situated at some distance above the upper boundary of the confined aquifer. The stream furnishes recharge to the aquitard and together with the main aquifer storage are the source of water supply for a well fully penetrating the confined aquifer at distance l from the stream and pumping at a constant rate, Q_w. The aquifer and the aquitard are extensive and have uniform properties. The solution for stream depletion is Equation (9.89) modified for leakage from the aquitard

$$Q_r = Q_w \left\{ erfc(u) - exp\left(\frac{\gamma^2 t}{4ST} + \frac{\gamma l}{2T} \right) erfc\left(u + \sqrt{\frac{\gamma^2 t}{4ST}} \right) - \frac{\gamma l}{T} \int_0^1 F\left(u, \frac{tT}{Sl^2} \right) G\left(u, \frac{tT}{Sl^2} \right) du \right\} \quad (9.92)$$

where S is aquifer storativity; $F(u,v)$ and $G(u,v)$ are functions provided by Hunt (2003); and all remaining parameters are as defined. The typical temporal behavior of the stream-depletion curve for a leaky water table aquitard is characterized by two inflection points often observed in the delayed yield phenomenon (Figure 9.19). Equation (9.92) predicts smaller stream depletion at earlier times than the case of a fully confined aquifer. As the dimensionless parameter $K^* = K'l^2/(B'T)$ increases, Q_r becomes less at small times, because increasing K^* causes a larger portion of water pumped from the well at early times to be drawn from water stored in the overlying aquitard. However, after a long pumping period the finite volume of water in the aquifer will be exhausted at which time drawdown in the aquifer and aquitard approach each other and the well discharge is derived entirely from the stream.

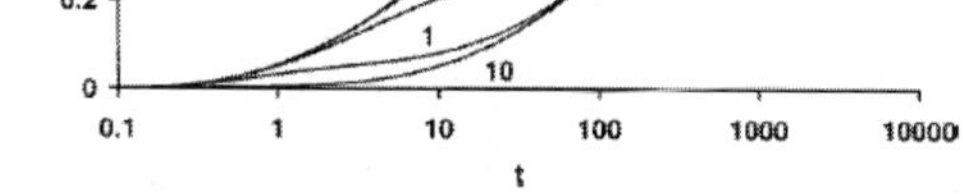

Figure 9.19: Typical stream-depletion behavior; $\lambda = K'Wl/(b'T)$; W is stream width; K' and b' are hydraulic conductivity and thickness of aquitard below streambed, respectively; $\varepsilon = S/S'_y$; S is storativity of the leaky confined aquifer; S'_y is specific yield of the aquitard (Hunt 2003).

Zlotnik and Tartakovsky (2008) considered a similar problem, but with a constant-head source bed underlying a water table aquifer. A well pumping at a constant rate Q_w is fully penetrating the unconfined aquifer and located at distance l from a narrow stream slightly penetrating the upper aquifer. Leakage from the source bed occurs through an aquitard with negligible storage. The aquifer system is of infinite horizontal extent and has uniform properties. Stream depletion rate in this case is given by (Zlotnik and Tartakovsky 2008)

$$Q_r = Q_w \left\{ \frac{a_1}{2} e^{-1/B_d} erfc\left(u - \sqrt{\frac{K_a t}{b_a S_y}} \right) - \frac{a_2}{2} e^{-1/B_d} erfc\left(u + \sqrt{\frac{K_a t}{b_a S_y}} \right) + a_1 a_2 exp\left(\lambda_d / 2 + \lambda_d^2 t_d / 4 - t_d / B_d^2 \right) erfc\left(u + \sqrt{\frac{\gamma^2 t}{4 S_y T}} \right) \right\} \tag{9.93}$$

in which

$$a_1 = \frac{B_d}{2/\lambda_d + B_d}; \quad a_2 = \frac{B_d}{2/\lambda_d - B_d}; \quad B_d^2 = \frac{b_a T}{K_a l^2}; \quad \lambda_d = \frac{\gamma l}{T}; \quad t_d = \frac{Tt}{S_y l^2} \tag{9.94}$$

where K_a [LT^{-1}] and b_a [L] are, respectively, the hydraulic conductivity and thickness of the aquitard; and u and all other variables are defined earlier. Taking the limit as $t_d \to \infty$ and recalling that $\gamma = WK'/b'$ is the streambed retardation coefficient (or streambed leakance), one obtains this result

$$\frac{Q_r}{Q_w} = \frac{1}{(2/\gamma)\sqrt{K_a T / b_a} + 1} exp\left(-l\sqrt{K_a / (b_a T)} \right) \tag{9.95}$$

which describes the maximum stream depletion rate. Note that the expression on the right is less than unity. After a long pumping period, well discharge is no longer sustained entirely by streambed filtration; only a fraction of the discharge is derived from stream depletion, whose magnitude depends on the stream-aquifer system parameters and geometry. The remaining fraction is supplied by source bed leakage across the aquitard. Zlotnik and Tartakovsky (2008) refer to this formula as the *maximum stream depletion rate*. It is obvious from Equation (9.95) that leakage causes the steady-state depletion rate to decrease exponentially with the distance between the well and the stream. This decrease is controlled by a complex interplay among the streambed leakance coefficient (γ), aquifer transmissivity (T), and aquitard parameters, K_a and b_a. This analysis was preceded by one (Zlotnik 2004) in which the stream fully penetrates an unconfined aquifer underlain by a constant-head source bed. Stream depletion rate due to a well pumping at a constant rate beside the stream was obtained for three cases: (a) a semi-infinite aquifer, (b) aquifer squeezed between the stream and an impervious boundary, and (c) an aquifer bounded by two parallel streams. In all cases, the stream is assumed to be perfectly connected to the aquifer and the stream-aquifer system is underlain by an aquitard and source bed. The drawdown in the unconfined aquifer is assumed to have no influence on the hydraulic head in the underlying aquifer and the source bed therefore becomes a limitless source of water, which in addition to streambed filtration sustains the well discharge.

In contrast, Hunt (2003) conceptualizes that volume of water stored in the aquitard is finite and leakage ceases when storage is exhausted and heads in the main aquifer and the aquitard approach equilibrium after a long pumping period.

The case of a relatively wide stream slightly penetrating a finite alluvial valley with sections on the left, below, and on the right of the stream having varying values of aquifer transmissivity and storativity/specific yield, was analyzed by Butler et al. (2001). The final solution, however, was obtained by numerical inversion of the Laplace transform. Comparison of the semi-analytical solution to MODFLOW revealed that neglect of vertical flow component can cause errors in stream depletion estimates by greater than 20% when streambed leakance ($K'W^2/(b'T)$) is large (> 1) and the distance between the well and the stream relative to aquifer width (l/W) is small. Large aquifer anisotropy (K_z/K_x) can also be important factor in stream-aquifer interactions (Sophocleous et al. 1995). The closer the well to the stream, the more important the impact of anisotropy becomes. Butler et al. (2007) extended a similar analysis to the case of a finite alluvial aquifer separated from a constant-head aquifer below by an aquitard. An analytical solution, which extends Theis (1935) and Hunt (1999) solutions to a relatively wide river, was developed by Fox et al. (2002) for drawdown caused by a well pumping at a constant rate near the river. The river slightly penetrates the aquifer and has a semi-pervious bed. The aquifer is homogeneous and extensive. They derive an analytical solution for drawdown in the aquifer section between the well and the stream, but stop short of deriving a solution for drawdown elsewhere and the formula for stream depletion rate.

In all of the solutions presented above for the cases of narrow and relatively wide rivers, flow beneath a slightly penetrating stream was assumed to be horizontal, and for the case of a wide river flow was represented by the vertically-averaged confined flow equation with a distributed leakage term describing seepage into or out of the streambed. As was discussed earlier groundwater can have significant vertical components in the vicinity of and below streams, somewhat similar to the case of flow toward partially penetrating wells. While these analytical solutions have successfully addressed the issue of aquifer storage on the side of the stream opposite to the well, the extent of the error incurred by the solutions, especially for cases of high anisotropy and wells relatively close to a stream, and limitations are not fully understood. Three-dimensional analytical solutions or numerical experiments (e.g., Spalding and Khaleel 1991; Sophocleous et al. 1995; Butler et al. 2001) can be helpful in quantifying such limitations.

Finally, it should be noted that the so-called semi-analytical solutions often require numerical inversion of complex functions in the Laplace domain (e.g., Zlotnik and Huang 1999; Moench and Barlow 2000; Butler et al. 2001, 2007; Kim et al. 2007; Intaraprasong and Zhan 2009), and the accuracy of the final numerical results therefore depends on efficient numerical inversion of the solutions in the Laplace domain. For example, Tseng and Lee (1998) conducted a detailed comparative analysis and concluded that the popular Stehfest inverse Laplace transform technique could yield errors much larger than that based on numerical evaluation of the Theis well function $W(u)$, and for large values of u, the Laplace inversion technique showed numerical oscillations between positive and negative values.

9.3 Surface-Subsurface Interactions at the Hillslope Scale

When rain falls on the land surface, a part of it is trapped in the vegetation and foliage as interception. Depending on soil properties and antecedent moisture conditions, some or all of the rain water effectively reaching the ground may enter the soil through the process of infiltration. A part of this infiltrated water moves downward in response to gravity and reaches the groundwater table, from where it moves essentially horizontally and eventually contributes to the stream as baseflow. If the soil is layered and has restrictive geologic strata, some of the infiltrated water moves parallel to the soil surface as interflow and contributes to the stream as subsurface storm flow. Subsurface storm flow is often aided by preferential flow paths (earthworm activity, decayed root channels, etc.) that conduct the rain water to the stream fairly rapidly. If the rainfall has sufficient intensity and duration, the soil is not able to absorb all the water supplied at the surface, and the excess water moves as Hortonian runoff on the soil surface. This surface water moves very rapidly to the stream (within minutes to hours) and is often the cause of flashy response in a stream to rainfall events. Depending on the nature of the slope and the geologic layering of the soil, some of the subsurface water may reappear on the surface as return flow. Thus, surface and subsurface waters interact in complex ways to generate streamflow. The pathways of water movement are schematically shown in Figure 9.20. The various flow mechanisms that contribute to streamflow are described below.

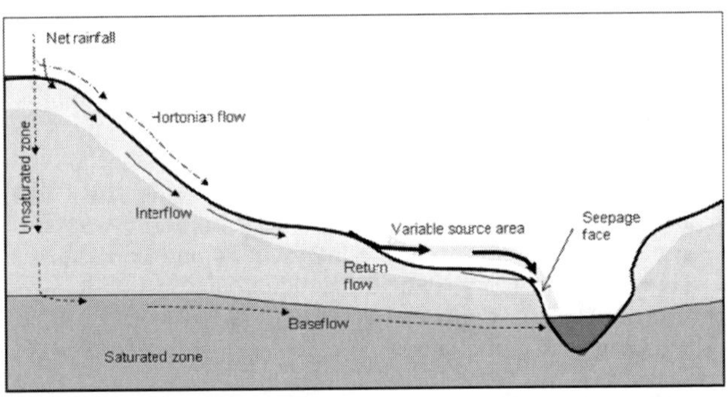

Figure 9.20: Schematic showing the various runoff mechanisms that contribute to streamflow at the hillslope scale.

9.3.1 Streamflow generation mechanisms

There are two primary mechanisms that produce overland flow from rain storms: (i) Hortonian overland flow that occurs when the rainfall intensity exceeds the infiltration rate of the soil (Horton 1933; Freeze 1972a, 1972b; Dunne 1983); and (ii) saturation overland flow that occurs when rising water tables intersect the soil surface, generating exfiltration. These mechanisms are part of a continuum of processes and may operate individually, but more commonly in combination (Freeze 1972a). In both cases, precipitation falling directly on the saturated regions at the soil surface produces direct surface runoff or overland flow. These surface flow zones may occupy only a portion of a catchment and may vary in size depending upon soil properties such as saturated hydraulic conductivity, the moisture release curve, organic content, depth to restricting layers, antecedent soil water content, and topography.

Hortonian Overland Flow: Hortonian overland flow occurs from a catchment when rainfall exceeds the ability of the soil to allow water to infiltrate (*aka* infiltration-excess overland flow). Infiltration into soils depends on several soil hydraulic properties - notably the conductivity and the porosity. Two conditions must be met for Hortonian overland flow to occur: (i) the rainfall intensity must be greater than the saturated hydraulic conductivity of the soil, and (ii) the duration of the rainfall event must be long enough to achieve surface saturation. Vegetated soils typically have high infiltration capacities because of the propensity of the roots to have porous soil material surrounding them. Vegetation also intercepts rainfall thus protecting the soil from compaction and crust-formation. Infiltration-excess overland flow is not a significant runoff-generating process in such catchments. Conversely, bare soil surfaces tend to develop a crust with relatively low hydraulic conductivity promoting Hortonian overland flow. Thus, Hortonian overland flow is an important streamflow response mechanism in (1) semi-arid to arid regions where rainfalls tend to be intense and natural surface conductivities are low, (2) areas where surface conductivity has been reduced by compaction, and (3) impermeable areas. Horton's (1933, 1945) view was modified by Betson (1964), who proposed the partial-area concept according to which surface water may originate as Hortonian overland flow on a limited contributing area that varies from basin to basin but, except for the possibility of some expansion during extreme events, remains fairly constant on a given basin (Dunne and Black 1970).

Saturation Overland Flow: In many forested catchments, rainfall intensity rarely exceeds the saturated conductivity of the surface soil, and overland flow develops because rain falls on temporarily or permanently saturated areas (wetlands) with no storage for water to infiltrate. Flow developing under these conditions is known as saturation-excess overland flow. This overland flow occurs due to saturation from below – i.e., it consists of direct water input to the saturated area plus the return flow contributed by the emergence of groundwater from upslope regions.

In humid areas, the water table is usually coincident with the stream surface and is fairly shallow in near-stream areas. When rainfall occurs, the water deficit in regions neighboring streams is rapidly satisfied and the water table rises to the ground surface. Continued supply of water from rainfall on these saturated areas leads to generation of overland flow even if the rain rate is less than the saturated hydraulic conductivity of the soil. Saturation overland flow is liable to be a dominant runoff mechanism in drainage basins with concave hillslope profiles and wide, flat valleys. Saturation from below can also occur where subsurface flow lines converge and water arrives faster than it can be transmitted away, and where the soil is layered and conductivity decreases with depth leading to perched water. When subsurface water flowing downslope to the stream enters the saturated area near the stream, some of the water is forced to reemerge onto the ground surface because the capacity of the soils and rocks to transmit all of the incoming water downslope is insufficient. This reemerging subsurface water is known as return flow.

The areas of a catchment that are prone to saturation tend to be near the stream channels, or where groundwater discharges to the surface. These areas grow in size during a storm and shrink during extended dry periods (Ragan 1966; Dunne and Black 1970; Dunne 1978; Ward 1984; Dunne et al. 1975; Beven 1978). The areas on which saturation-excess overland flow develops expand and shrink during a storm in response to rainfall reflecting the overall wetness of the watershed. This mechanism of runoff generation is often referred to as the variable source area concept and was modeled by Govindaraju and Kavvas (1991). Variable source areas exert a very strong influence on the nature of the streamflow hydrograph for a storm event.

The location and size of zones of surface saturation that generate saturation overland flow can be predicted based on the distributed topographic attributes and soil properties of a catchment as reflected in TOPMODEL (Beven and Kirkby 1979; Beven and Wood 1993). TOPMODEL uses the equation for conservation of mass for several reservoirs in a catchment – for example, an interception reservoir and a soil reservoir. Rainfall provides the input to the interception reservoir, which is assigned a capacity depending on the vegetation type. The outputs from the interception reservoir are evaporation, calculated using an evaporation formula and throughfall, which then forms the input to the soil reservoir. The conservation of mass equation provides a method for calculating the water balance for the soil reservoir. By linking together the water balance equations for all of the hypothetical reservoirs in the catchment, a routing computation can be completed. TOPMODEL assumes that the subsurface flow rate per unit width of contour q_{si} [L^2T^{-1}] at any point 'i' in the landscape can be presented by:

$$q_{si} = T_i \tan\beta_i e^{-S_i/M} \tag{9.96}$$

where T_i is the soil transmissivity when the soil profile is saturated [L^2T^{-1}]; β_i is the local slope angle; S_i is the local deficit of readily drainable soil water [L^{-1}]; and M is related to the exponential variation of hydraulic conductivity with depth [L^{-1}]. From this equation the following expression for the local saturation deficit can be derived that is independent of the infiltration rate or rainfall intensity.

$$S_i = S_t + M\delta - M\ln\left(\frac{A_i}{\omega_i \tan\beta_i}\right) \tag{9.97}$$

where

$$\delta = \frac{1}{A_t} \int_{A_t} ln\left(\frac{A_i}{\omega_i \tan \beta_i}\right) dA \tag{9.98}$$

and A_i is the upslope contributing area [L^2] draining across a length of contour ω_i [L]; A_t is the total catchment area [L^2]; and S_t is the mean storage deficit of the entire catchment [L^{-1}]. A point in the landscape is saturated and generates saturation overland flow when $S_i < 0$. If $S_i <$ infiltration volume during a time period, that point becomes saturated during the time interval. A continuous accounting of S_t in Equation (9.97) allows simulation of the dynamic expansion and contraction of zones of surface saturation in the landscape. The parameter δ is an area weighted topographic and soil attribute that allows comparisons between catchments of the potential to generate zones of surface saturation.

Interflow: Water that has infiltrated the soil surface and is impeded from downward movement due to stratification (abrupt or gradual) tends to move laterally in shallow soil horizons as subsurface storm flow or interflow. While this subsurface water moves slowly to the stream and contributes to baseflow, it may be aided by the presence of preferential flow pathways (e.g., soil cracks, old animal burrows, decayed root channels, etc.) leading to quickflow response in the stream.

Groundwater flow: In general, groundwater flow results in the longest travel time (days, weeks, to years) for the water that fell on the soil surface to eventually reach the stream. Baseflow in low-flow periods is comprised almost entirely of groundwater discharge. Consequently, baseflow tends to vary quite slowly and over long time periods in response to changing inputs of water through net recharge.

Flow in the Saturated Zone: Groundwater model studies suggest that basin-wide regional groundwater flow cannot respond quickly enough to rainfall events (Freeze 1974). However, some tracer studies revealed that groundwater can be a significant component of event response when mechanisms that quickly produce steep hydraulic gradients in materials of high conductivity in near-stream areas are activated.

Macropore Flow: The response of interflow to rainfall events would be quite sluggish if it is not aided by the presence of macropores where Darcian flow through the soil matrix is largely short-circuited by water moving in conduits. Macropores are typically on the order of 3 to 100 mm in diameter and are interconnected to varying degrees, thus they can allow water to bypass the soil matrix and move rapidly at speeds much greater than those predicted by Darcy's Law (Kirkby 1988). Stillman et al. (2006) show that the effective conductivity of soils increased by several orders of magnitudes in the presence of macropores. The combination of macropores and tile-drains was shown to generate Hortonian-like streamflow responses even when no surface flow was observed. It is generally difficult to assess the importance of macropores or to simulate their effects in catchment-scale models because their number, orientation, size, and inter-connectedness are highly site-specific and macroscopic properties have to be obtained through calibration.

Flow in the Unsaturated Zone: The unsaturated portion of the soil holds water at negative gage pressures (i.e., water pressure is less than atmospheric pressure), and water in this region primarily moves vertically down and contributes to the water table. The hydraulic resistance offered by the soil is high resulting in low effective velocities. Flow through the unsaturated zone is one of the primary mechanisms of replenishing the aquifer through recharge. In special cases, unsaturated flow may contribute to baseflow in a stream (Hewlett and Hibbert 1963; Weyman 1970; Nutter 1975).

9.3.2 Mathematical Formulation of the Hillslope Hydrology Problem

A physics-based study of stream-aquifer interaction will require the dynamic modeling of the three basic flow components occurring on a hillslope: overland flows, streamflows, and subsurface saturated-unsaturated flows (see Figure 9. 20). These flow processes are in a state of continuous interaction through their common boundaries. In investigating the stream hydrograph, the response time of any water in the system is controlled by how far it has to travel to get to the stream (slope length) and the mechanics of its transfer (pathway). Methods of predicting expansion and shrinkage of surface flow regions during rainfall events are useful since overland flow routes from the near-channel contributing areas have practically zero time lag (as compared to baseflow contributions) in reaching the receiving waters downstream. Accurate modeling of this phenomenon necessitates the study of overland flow on variable domains. In regions adjacent to streams, wetter conditions are maintained owing to shallow water tables. These regions are more responsive to precipitation events and are likely to develop overland flow even when the rainfall intensity is less than the saturated hydraulic conductivity of the soil. Under prolonged rainfall the saturated regions adjacent to streams, contributing to overland flow, will expand in areal extent. Similarly, this source area diminishes when rain stops.

The mathematical representation of the flow processes over a hillslope leads to a set of nonlinear partial differential equations. We present the governing equations for describing the flow processes.

Streamflows: The equations of continuity and momentum in a channel have been covered in Section 9.2.5 and are given by Equations (9.51) and (9.53), respectively, (also, see Gonwa and Kavvas 1986).

Overland Flows: Most hydrologic models treat the overland flow process as fully turbulent, broad sheet flow, which may be satisfactory for computing runoff rates. However, overland flow can occur over large parts of the landscape and the depths and velocities of flow can be extremely variable. The flow may be laminar or fully turbulent. It can exist as broad sheet flow or as flow in microchannels such as rills and ephemeral gullies so that the hydraulics of overland flow is very complex. In hydrologic analyses, the flow as a whole is often treated as broad sheet flow, with the result that computed flow velocities are lower than true values. Consequently, parameter values for the broad sheet flow assumption needed for application of a model to field conditions can be very different from values determined in idealized laboratory studies and may require calibration. Nevertheless, for many purposes, hydrologic models based on the broad sheet flow assumption are useful.

Overland flows may be treated as flows per unit width on very wide, rectangular channels (Woolhiser 1974). Since the width is very much greater than the depth, the hydraulic radius is approximately equal to the depth of flow. Using a similar procedure (as the analysis for streamflows discussed above), results in the following one dimensional equation for overland flows

$$\frac{\partial}{\partial x}(V_o h)+\frac{\partial h}{\partial t}=R-f \tag{9.99}$$

$$S_f = S_0 - \frac{\partial h}{\partial x} - \frac{V_o}{g}\frac{\partial V_o}{\partial x} - \frac{1}{g}\frac{\partial V_o}{\partial t} - \frac{RV_o}{gh} \tag{9.100}$$

with the generalized friction law for overland flows given as

$$V_o = \alpha\, h^m S_f^j \tag{9.101}$$

where h is the depth of flow [L]; V_o is the depth averaged overland flow velocity [LT^{-1}]; S_0 is the slope of the overland flow bed [LL^{-1}]; x is the coordinate along the horizontal direction [L]; t is time [T]; S_f is the slope of the total energy line [LL^{-1}]; R [LT^{-1}] is rainfall rate [LT^{-1}]; and f is infiltration [LT^{-1}]. In Equation (9.101), α, m, j are constants of the friction relation (e.g., Manning or Chezy equation).

Subsurface Saturated-Unsaturated Flows: The subsurface component is complicated due to the presence of two regions, the saturated and the unsaturated zones. These two zones are separated by the water table, which is a theoretical surface at which the water pressure in the soil equals the atmospheric pressure. Within usual limitations, subsurface flow may be described by a continuity equation expressed as

$$\frac{\partial}{\partial t}(\rho\theta)+\nabla\bullet(\rho\mathbf{q})=s \tag{9.102}$$

where $\mathbf{q}$ is the vector of specific discharge of water [LT^{-1}]; ρ is the density of water [ML^{-3}]; θ is the moisture content by volume of the soil [L^3L^{-3}]; s is a source/sink term (incorporating the effects of evapotranspiration, lumped well discharges, etc.) [$ML^{-3}T^{-1}$]; and $(\nabla\cdot)$ is the divergence operator. The Darcy-Buckingham law states that

$$\mathbf{q} = -K(\theta)\nabla\phi \tag{9.103}$$

where $K(\theta)$ is the hydraulic conductivity of the soil, which varies with the moisture content in some nonlinear fashion [LT^{-1}]; and ϕ is the piezometric head [L]. Combining Equations (9.102) and (9.103) and simplifying yields the following governing equation (Winter 1983)

$$\left(S_w S_s + \eta \frac{dS_w}{d\psi}\right)\frac{\partial \varphi}{\partial t} = \frac{\partial}{\partial x}\left[K_x \frac{\partial \varphi}{\partial x}\right] + \frac{\partial}{\partial y}\left[K_y \frac{\partial \varphi}{\partial y}\right] + \frac{\partial}{\partial z}\left[K_z \frac{\partial \varphi}{\partial z}\right] + J \tag{9.104}$$

where S_w is the water saturation [L^3L^{-3}]; S_s is the specific storativity [L^{-1}]; η is the soil porosity [L^3L^{-3}]; x and y are horizontal directions [L]; t = time [T]; J = sources/sinks [$L^3L^{-3}T^{-1}$]; and the piezometric head may be expressed as $\phi = \psi + z$ (ψ = pressure head [L] and z = elevation head [L]). Under isotropic conditions, two functional relationships are required: $K(\theta)$ and $\psi(\theta)$. These functions are called characteristic curves that, in general, are highly nonlinear and hysteretic so that nonunique specifications for $K(\theta)$ and $\psi(\theta)$ may result for a given θ depending on the previous wetting or drying history of the soil. Either of these relationships can be determined experimentally for individual soils, or typical models can be used to describe them in mathematical form. Some popular examples of these models (when hysteresis is neglected) are provided in Table 9.1. Given the nature of these relationships, the governing equations tend to be highly nonlinear, requiring a high degree of skill and computational effort for numerical solutions. Hysteresis may be avoided by considering only drying or wetting cycle, and is often neglected in practical applications.

Table 9.1: Some popular relationships for soil characteristics.

Soil water retention	Hydraulic conductivity	Parameters
Brooks and Corey (1964)		
$S_w = \frac{\theta - \theta_r}{\eta - \theta_r} = \left(\frac{\psi_b}{\psi}\right)^{\lambda}$	$K_r = \frac{K(\theta)}{K_s} = (S_w)^n$	λ = pore size index ψ_b = air entry pressure $n = 3 + 2/\lambda$
Campbell (1974)		
$\frac{\theta}{\eta} = \left(\frac{H_b}{\psi}\right)^{1/b}$	$\frac{K(\theta)}{K_s} = \left(\frac{\theta}{\eta}\right)^n$	b = constant H_b = scaling parameter $n = 3 + 2b$
Van Genuchten (1980)		
$S_w = \left[\frac{1}{1+(\alpha\lvert\psi\rvert)^n}\right]^m$	$\frac{K(\theta)}{K_s} = S_w^{1/2}\left\{1-\left[1-S_w^{1/m}\right]^m\right\}^2$	n = constant m = constant

K_s = saturated hydraulic conductivity ($\theta = \eta$), θ_r = residual water content, η = porosity

Initial and Boundary Conditions: For an initially dry soil, a specified mass flux is employed over the soil surface. After the soil surface is saturated, the rate of influx is controlled by both the surface and subsurface flow conditions, and the soil surface boundary condition is changed from a specified flux to a specified head condition. The moving water boundary on the soil surface (see Figure 9.20) needs special attention because its location is not known beforehand. In transient problems the location of this boundary is time dependent. The extent of the variable source area therefore determines where the boundary conditions of the subsurface change from one of specified flux to one of the specified head.

The upper boundary condition of the channel is typically specified as a flux from upstream sections, while uniform flow is assumed at the lower boundary. The upstream moving boundary condition for overland flow is taken as the zero-depth condition (Govindaraju et al. 1988) and a zero-depth gradient lower boundary is often used (Morris 1979).

The lateral inflow into the stream at any section is the sum of the rainfall, the contribution from overland flow sections on the sides, and baseflow. This lateral inflow is expressed as

$$Q = R(t) + Q_{ov} \pm Q_s \tag{9.105}$$

where $R(t)$ is the time-varying rain falling over the width of the stream section (positive for rainfall and negative for evaporation) [L^2T^{-1}]; Q_{ov} is the overland flow contribution [L^2T^{-1}]; and Q_s is the quantity infiltrating into (or exfiltrating from) the soil over the width of the channel section [L^2T^{-1}].

Similarly, the lateral inflow into the overland flow nodes is determined by the algebraic sum of precipitation and infiltration into the soil. Mathematically, this is represented as

$$Q_O = R(t) - K(\psi)\left[\frac{\partial \psi}{\partial z} + 1\right], \quad Q_O > 0$$

$$Q_O = 0, \quad Q_O \le 0 \tag{9.106}$$

where $R(t)$ represents the rainfall intensity contribution to the lateral inflow [LT^{-1}]; ψ represents the pressure head on the subsurface boundary nodes [L]; and z is vertical coordinate (positive upwards) [L].

Because of the sparsity of data in most hydrological situations, two sets of physically meaningful initial conditions for the subsurface component are in common use: the static or no-flow condition and the steady-state condition. Under the static initial condition the total hydraulic head ϕ is constant for all spatial locations with a horizontal water table at level with the stream depth. The moisture contents and pressure heads above the water table are at equilibrium. For steady-state initial conditions, the time derivatives of the flow equations disappear and the equations may then be solved subject to appropriate boundary conditions.

9.4 Watershed-Scale Groundwater and Surface Water Interactions

At the watershed scale, groundwater flow patterns and their interactions with surface water are influenced by topography, geology, and climate of the region. These features influence the water table configuration below the surface, which in turn influences the distribution of recharge and discharge conditions. In a typical hilly upland area in a temperate region, underlain by a water table aquifer, streamlines are vertically downward at the surface divide, horizontal and parallel to the impermeable base over most of the aquifer, and vertically upward underneath a draining stream. The flow patterns and water table configurations depend on the balance between net vertical recharge (i.e., water table accretion) and ability of the aquifer to transmit lateral groundwater flow toward the stream.

In a nested channel network, groundwater flow systems are more complex than the localized flow system described above with local, intermediate, and regional flow systems existing simultaneously. In local flow systems, water moves from a recharge area to the immediately adjacent discharge area (stream, wetland, or lake). In regional flow systems, flow is from the recharge area farthest from the main valley to the draining main river. Bridging the two systems, intermediate flow systems prevail when flow paths are longer than local but shorter than the regional case (Tóth 1963; Dingman 2002, p. 338). The development and dissipation of local flow systems depends on the magnitude of recharge events and permeability of the porous media (Winter 1983). Hydraulic heads, water table distribution, and flow velocities can be obtained by solving Equation (9.1), with effective values of both hydraulic conductivity and effective storativity depending on the resolution of interest, subject to free-surface condition given in Equation (9.4) and other appropriate boundary and initial conditions. Topographic variations and complexities arising from aquifer heterogeneity and anisotropy and complex geometry of interfacial boundaries with surface-water bodies reveal the limited scope of analytical techniques and the need for numerical models as appropriate tools to obtain solutions for such complex flow systems (Winter 1976; Anderson and Munter 1981). In the most general case, variably saturated (i.e., coupled unsaturated-saturated) numerical models are used to delineate seepage face development and seepage to and from surface-water bodies (Winter 1983). Later in this section, we will present modeling approaches and some of the most widely used models for regional-scale groundwater and surface-water interactions.

Watershed-scale, event-based rainfall-runoff models and continuous-time hydrologic and water quality models are rarely applied without recession analysis and its use in separating quickflow (surface runoff and subsurface return flow) from baseflow (or groundwater runoff). Knowledge of surface runoff is not only important for water resources management and flood hazards mitigation, but also imperative to evaluation of urban and agricultural storm runoff control strategies. Urban and/or agricultural runoff is the primary nonpoint source of surface-water pollution in many watersheds. In agricultural watersheds, subsurface flow is the primary pathway for nitrate transport to surface water. Other agricultural chemicals, such as pesticides, leach from soils down to groundwater and find their way to surface water bodies during baseflow periods. Knowledge of baseflow is therefore useful for waste dilutions and pollution control especially during low-flow periods where streamflows are entirely derived from groundwater discharge. Recession analysis, among others, is also important in frequency analysis for estimating low flow statistics and in regional low flow studies for indexing the storage capacity of the catchment (Tallaksen 1995).

9.4.1 Baseflow Analysis

Baseflow is the portion of streamflow contributed by groundwater discharge; it is derived from groundwater storage and from persistent, slowly varying sources, such as lake and wetland drainage and snowmelt percolation. During prolonged drought periods and in the absence of reservoir releases, streamflow is almost entirely sustained by baseflow. Streamflow recession is the withdrawal of water from watershed storage with no inflow; it is represented by the falling limb of a hydrograph to the right of the point of peak flow rate (Figure 9.21). At some point in time beyond the point of inflection on the recession curve, streamflow is maintained entirely by baseflow. The baseflow recession curve represents withdrawal from groundwater storage only and describes the relationship between declining aquifer discharge and time. Following the end of a precipitation event, baseflow may be composed of old groundwater (in storage and water table accretion prior to the beginning of the event), event and post-event groundwater recharge, and bank storage seepage.

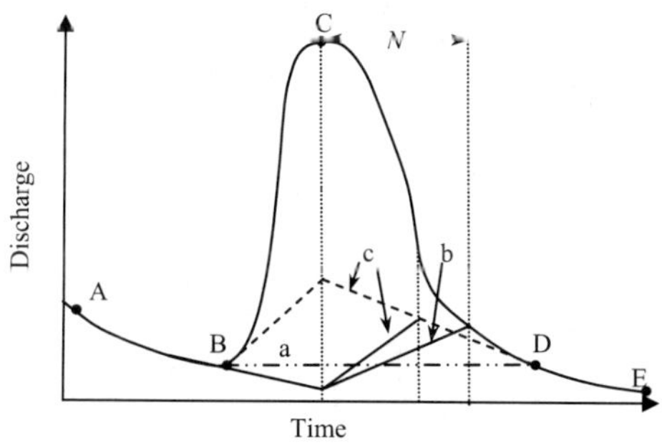

Figure 9.21: Typical storm hydrograph and graphical baseflow separation: straight line method (a); area method (b); and variable slope method (c). $N = A^{0.2}$; A is the watershed area in square miles; and N is the time in days from the hydrograph peak.

Physical models of baseflow have been covered in previous sections for ideal stream-aquifer flow situations; e.g., Equations (9.11), (9.12), (9.14), (9.18), (9.19), (9.25), (9.26), (9.38), (9.39), and (9.74). Among these equations, Equation (9.12) corresponds to the familiar empirical hyperbolic baseflow recession curve, whereas (9.14) and (9.38) correspond to the widely used empirical exponential baseflow recession curve described below. In fact, the correspondence between the empirical recession relationships and the physically-based counterparts sets the stage for scaling-up aquifer parameters to the watershed scale (Brutsaert and Nieber 1977).

A practical relationship between outflow from a basin, Q [L^3T^{-1}], to the volume of water stored in the basin, S [L], is the nonlinear storage model

$$Q = aS^n \tag{9.107}$$

where a and n are constants; n = 1 for the specific case of a linear storage model in which the discharge is directly proportional to the storage in the groundwater reservoir. Under baseflow conditions, long after direct runoff and recharge cease, streamflow is derived entirely from groundwater storage with no inflow to the aquifer. The differential equation describing baseflow recession is

$$dS = -Q dt \tag{9.108}$$

which with Equation (9.107) leads to

$$\frac{dS}{S^n} = -a\,dt \tag{9.109}$$

For $n = 1$, we have

$$Q = Q_0 e^{-a(t-t_0)}; \quad (\text{linear storage model}) \tag{9.110}$$

If $n > 1$, then

$$Q = Q_0 \left[1 + c(t - t_0)\right]^{n/(1-n)} \tag{9.111a}$$

in which

$$c = (n-1)a^{1/n} / Q_0^{(1-n)/n} \tag{9.111b}$$

where Q_0 [L^3T^{-1}]is the outflow at the initial time t_0.

For the linear storage model, Equation (9.110), the constant a is the slope of the straight line on a semi-logarithmic plot of Q versus t, regardless of the choice of t_0

and magnitude of the corresponding flow rate Q_0. Obviously, this is not the case for the nonlinear storage model, Equation (9.111a), where the slope of the strait line on a plot of $Q^{(1-n)/n}$ versus t is dependent on Q_0. Because it is practically impossible to determine the beginning of recession for each rainfall event from a continuous river flow record with intermittent dry and wet periods, the problem of choosing t_0 can be avoided by noting that Equations (9.108) and (9.109) can be combined with (9.107) to yield the following alternative differential balance.

$$dQ = -\alpha Q^{\beta}\, dt \tag{9.112}$$

and plotting $-dQ/dt$ versus Q on logarithmic graph paper, then determining α and β from the slope and intercept of the line that best fits the plot (Brutsaert and Nieber 1977), with $\alpha = na^{1/n}$ and $\beta = 2 - 1/n$. Constant α can be related to coefficient c in Equation (9.111a) by making use of Equation (9.111b). Equations (9.110) and (9.111) can be verified as the solutions of Equation (9.112) for $n = 1$ and $n > 1$, respectively.

Assuming some geometric similarity of the drainage pattern within the catchment, an equivalent lateral inflow, q [$L^3T^{-1}L^{-1}$], can be defined as (Brutsaert and Nieber 1977)

$$q = \frac{Q}{2L} \tag{9.113}$$

where L [L] is the total length of all the tributary and main channel sections upstream from the gauging station at which the flow is Q, as well as an equivalent aquifer width (on both side of the channel) is

$$l = \frac{\gamma A}{2L} \tag{9.114}$$

where A is the drainage area of the catchment [L^2]; and γ is the fraction of the catchment underlain by aquifers contributing to streamflow.

The substitution of Equations (9.113) and (9.114) into (9.11), (9.12), and (9.14), or any other exponential and hyperbolic, physically-based baseflow equations and comparison with (9.112) should yield unique values for β and expressions relating α to catchment characteristics and aquifer geometric and hydraulic properties. For example, it can be shown that β = 3, 3/2, and 1, for Equations (9.11), (9.12), and (9.14), respectively (Brutsaert and Nieber 1977; Brutsaert and Lopez 1998). A detailed procedure is outlined by these authors for the estimation of constant α from measured catchment outflow rates and the estimation of catchment-scale aquifer hydraulic conductivity, specific yield, hydraulic diffusivity, and hydraulic desorptivity. A methodology for estimating scale dependence of the hydraulic properties and their distributions is also presented. On the basis of a numerical model experiment, Szilagyi et al. (1998) showed that complex aquifer geometry, a gently sloping impervious layer, and spatially variable hydraulic conductivities have no impact on the accuracy of catchment-scale hydraulic conductivity and the mean aquifer depth estimated using Brutsaert and Nieber (1977) recession flow analysis.

In spite of their widespread use, methods for separating baseflow contributions from total streamflow remain mired in subjectivity and inconsistency. Even those methods that are physically-based and rely on analytical solutions of the Boussinesq equation (e.g., Szilagyi and Parlange 1998; Szilagyi 1999; Huyck et al. 2005) are designed for recession periods and limited by the conditions required for the applicability of the Boussinesq equation and fully penetrating streams. Despite their subjectivity, methods for graphical baseflow separation are widely - and perhaps will continue to be - used primarily due to their simplicity. More recently, filtering techniques (e.g., Wahl and Wahl 1995; Piggott et al. 2005) and recursive digital filters (Nathan and McMahon 1990; Szilagyi et al. 2003; Szilagyi 2004; Aksoy et al. 2009) have been suggested to disaggregate the daily streamflow into quickflow and baseflow components. Other approaches involve fitting first-order or second-order autoregressive process with a random error term (e.g., Tallaksen 1995). All of these approaches and methods, however, share one common trait: they are impossible to verify without direct and adequate groundwater discharge measurements throughout the catchment. Direct observations of baseflow – the groundwater portion of streamflow - are usually unavailable during wet periods and sometime after cessation of precipitation events.

Figure 9.21 depicts a typical hydrograph during a storm event and commonly used graphical baseflow separation techniques. The hydrograph is composed of a rising limb (segment B-C) followed by a receding limb (C-D). The rising limb is usually preceded by baseflow recession before direct runoff begins (A-B). A new baseflow recession begins immediately after direct runoff ends (D-E). Segment A-B represents gradually diminishing baseflow prior to the time of intense rainfall. Point D on the declining limb may be approximated as the point on a semi-log plot where log Q starts to deviate from the linear relationship between log Q and time.

The straight line method of separating baseflow from direct runoff involves drawing a straight line from point B, where the rise commences, to the intersection with the falling limb. A more realistic approach is to connect point B where direct runoff starts with point D on the receding limb of the hydrograph where direct runoff ends by an inclined straight line.

The area method assumes direct runoff ends a fixed time N from the time to peak flow, where $N = A^{0.2}$; A is the watershed area in square miles; and N is the time in days from the hydrograph peak. Point B is projected forward to the time of the peak. A straight line is used to connect this projection at the peak to the point on the recession limb at time N after the peak.

In the variable slope method, point B is projected ahead to the time of peak. Point D is extrapolated backward to the time of the point of inflection on the recession limb. A straight line is used to connect the endpoints of the extrapolated curves. Another approach involves extrapolating point D backward in time to the point directly under the peak and connecting this point to point B by a straight line.

The three-component separation method involves a two-step procedure separating surface runoff, interflow, and baseflow (Singh 1992). First, direct or quick runoff is separated from baseflow using any of the variable slope methods. This process is repeated to separate interflow from surface runoff using the values of direct runoff estimated in the first step.

9.4.2 Regional-Scale Modeling

Watersheds are complex systems comprised of subsystems continuously interacting with each other through a variety of hydrological processes such as infiltration, unsaturated flow, subsurface storm flow, groundwater flow, overland flow, groundwater recharge, baseflow, exfiltration, run-off and run-on, channel flow, evaporation, transpiration, plant uptake, transmission loss, precipitation as rainfall or snow, and snow melt. These processes comprise basic components of the hydrologic cycle and some or all of the processes might be present or occurring at any watershed at any given time. There is a constant flux of water between the atmosphere, the surface, and the subsurface. However, due to the large differences in time scales associated with the flow of water within these three flow domains, the hydrologic cycle is partitioned into three individual components (Furman 2008).

A watershed can be broadly divided into two compartments: surface and subsurface compartments. The surface compartment contains overland planes and channels. The subsurface compartment can similarly be separated into two partitions: the unsaturated (vadose zone) and the saturated medium (groundwater). All these compartments/mediums could interact with each other. Sometime this interaction is unilateral, sometimes mutual. In watersheds groundwater– surface water interactions consist of two parts: stream-aquifer interaction and overland-subsurface interaction. The former is extensively studied by coupling one-dimensional streamflow with saturated subsurface flow, whereas overland-subsurface interaction is generally studied under the context of runoff generation (Huang and Yeh 2009). Groundwater can receive water from the overland planes through infiltration and recharge. If the groundwater table rises up to the ground surface in response to a rainfall event, it can provide water flow to the overland plane (saturation overland flow or exfiltration). Similarly groundwater can replenish streamflows through baseflow or receive water from the channels depending on the head difference between the groundwater table and the flow stage in the stream.

Depending on the problem of interest and the desired level of detail, watershed discretization and the extent of the watershed processes to be considered can vary. Watershed models are developed to serve different audiences. They can vary from very simple screening models to complex research oriented models. As such, their conceptualizations of the groundwater-surface water interactions vary. Some watershed models don't have groundwater components at all. In this section we will provide a summary of different methods used in models for handling surface water-groundwater interactions, varying from simple empirical relationships to physically based, one-dimensional to three-dimensional. We will also provide a summary of different levels of widely used surface water-groundwater coupling mechanisms utilized in practice and research.

Coupling Surface and Subsurface Processes: Numerous methods are available for coupling surface and subsurface processes. Gunduz (2006) classified such methods under four categories (actually five but one of the categories is only applicable to solve water quality problems). In theory as the level of coupling increases the accuracy is expected to increase along with the computational burden. From the simplest to the most complex, these coupling methods are:

I. Sink/source function type coupling or no coupling (e.g., Tayfur et al. 1993; Esteves et al. 2000; Kazezyilmaz et al. 2007)
II. Non-iterative (external) coupling (e.g., Smith and Wollhiser 1971; Motha and Wingham 1995)
III. Iterative (internal) coupling (e.g., Pinder and Sauer 1971; Van der Kwaak 1999; Morita and Yen 2002; Panday and Huyakorn 2004)
IV. Simultaneous or full coupling (e.g., Gunduz and Aral 2005b)

In all except I. there is some degree of coupling. Type I coupling does not involve solving any kind of differential equations. The flux term that links the surface and subsurface components is solved independently. Depending on the direction of the flux, it becomes either a source or a sink for the surface component (exfiltration or infiltration, respectively). In case of infiltration the flux term becomes a source for the subsurface. For example, Kazezyilmaz et al. (2007) developed a wetland model where the wetland is assumed to have a shallow part exhibiting overland flow characteristics and a lake component. The former is solved using the diffusion wave approach. The wetland is assumed to interact with groundwater through groundwater discharge or recharge terms, which are computed using Darcy's law. The exchange between surface and groundwater is calculated in lateral direction at the banks of the wetland. Flow in the detention area of the wetland is computed with a storage routing method. Here, the exchange between groundwater and the lake is calculated in the vertical direction. In both parts, the groundwater head needs to be known and model needs groundwater level data as input. Therefore, the surface water/groundwater interaction is really represented by source/sink term in the surface-water balance equation.

In Type II coupling, the surface component is typically solved first, independent of the subsurface component. Results from that computation are then used to calculate the flux term that links the surface and subsurface. The subsurface component is then solved using the interacting flux terms as source/sink. Once the subsurface terms are calculated the computation is complete. In other words, no iteration is involved. Although it is fast, it may results in large errors. Some call this type of coupling as no-coupling due lack of feedback between the two systems (e.g., Furman 2008). The name *external coupling* was first used by Freeze (1972a).

Type III coupling is an improvement over Type II and involves feedback between the two systems (Furman 2008). The surface and subsurface components are solved independently similar to Type I. However, at each time step both systems are solved iteratively until a convergence criterion is met. Since both systems are solved independently, different time steps can be used for each, typically smaller time steps for the surface component and larger time step for the subsurface component. Although one gets the impression that this method is computationally very demanding, in solving one-dimensional dynamic open channel flow coupled with a two-dimensional transient groundwater flow, Pinder and Sauer (1971) reported convergence after only 2 to 5 iterations. *Alternating coupling* is another name used for Type III coupling (Morita and Yen 2002). Harada et al. (2005) coupled a two-dimensional groundwater flow model to a steady-state channel flow wherein channel discharge is related to stream stage using Manning's equation. Type III coupling was implemented to simulate exchange with aquifer in the alluvial fan of the Yasu River in Japan.

The final and most complicated coupling, Type IV, requires solving the surface and subsurface flow equations simultaneously. This results in a large set of equations. A key point in implementation of this type of coupling is the fact that the same time interval must be used in discretization of both surface and subsurface processes. Since surface processes require much smaller time steps, they dictate the size of the computational time step. As a result, the computational time and power required in this type of coupling is substantial higher than any other coupling. In a recent paper, Huang and Yeh (2009) argued that the fully coupled approach (Type IV) is a special case of all coupling combinations. Based on the physical nature of the interface, Huang and Yeh (2009) describe two cases of surface/subsurface coupling: continuous and discontinuous. If a sediment layer is present at the interface then a discontinuous interface may be justified. From a numerical perspective they define three cases: time lagged (decoupled), iterative, and simultaneous solutions.

Groundwater-Surface Water Interaction Modeling Approaches: The MODFLOW (Harbaugh and McDonald 1996) computer package is by far one of the most widely used groundwater modeling packages; it has been linked to surface water modules to simulate stream-aquifer interactions, lakes/ponds, and wetlands (e.g., Cheng and Anderson 1994; Restrepo et al. 1998, Bauer et al. 2006), and many other applications, from small stream-reach scale to the regional scale. Acknowledging the existence of myriad groundwater flow simulation models, we discuss below MODFLOW's basic concepts for simulating groundwater and surface water interactions and linkages with surface channels flow modules and watershed models. A brief discussion of the emerging analytic element method (AEM) and modeling the phenomenon at the watershed-scale are also provided in the following sections.

MODFLOW

The Modular Finite Difference Ground Water Flow model (MODFLOW) is probably the most widely used and tested groundwater model around the world. To some degree, it is the de facto standard in groundwater modeling (Furman 2008). It is often used in comparing new methodologies and/or solutions (mostly analytical) in the literature. MODFLOW (McDonald and Harbaugh 1988; Harbaugh and McDonald 1996; Harbaugh et al. 2000; Harbaugh 2005) is a three-dimensional finite difference based groundwater flow model [see Equation (9.104) with $S_w = 1$ and $dS_w / d\psi = 0$]. It incorporates the effects of many steady-state and transient processes, such as areal recharge, rivers, drains, evapotranspiration, and pumping. MODFLOW-2005 can simulate stream-groundwater interactions and lake-groundwater interactions. Both versions have some limitations that provided the impetus for development of many sub-modules. The theory and detailed description of MODFLOW are beyond our scope. Only a brief description will be given. Interested readers can refer to above references.

MODFLOW solves the three-dimensional groundwater equations utilizing a finite difference technique. Therefore, the aquifer is divided into cells with uniform properties. At each time step, the head is calculated at the center of each cell. Two types of boundary conditions apply to cells used to simulate boundary conditions: *specified-head* and *no-flow cells*. With the former, a piezometric head is specified that may vary with time or stress periods. As it is obvious from its name, in no-flow cells flux is zero across one or more sides. Cells not listed under any of these categories are called *variable-head cells* and are computed by the model. Interaction of groundwater with streams and lakes can be modeled in MODFLOW using a head-dependent flow boundary condition, similar in form to Equation (9.7).

The River package included in MODFLOW-2005 uses a head-dependent boundary condition to handle stream-groundwater interactions. The head or stage along the stream is determined from the depth of flow in the channel. Streams are divided into reaches and segments, where a reach is a section of a stream associated with a particular finite-difference cell. A particular finite-difference cell can have more than one stream reach, but only one finite-difference cell can be assigned to a single reach. The group of reaches between connections with another stream/tributary, a lake, or a watershed boundary is called a segment. Streambed properties can vary among reaches within a segment, but not within a reach.

A low permeability riverbed material often separates the open water of a river from the groundwater system. Under this condition, MODFLOW computes the flow rate between a river and aquifer from the following relationship,

$$\left.\begin{aligned} Q &= c(h_r - h) & h &> h_b \\ Q &= c(h_r - h_b) & h &\le h_b \end{aligned}\right\} \tag{9.115}$$

where Q is flow between the river and the aquifer (taken positive if it is directed into aquifer) [L^3/T]; h_r is water level in the river [L]; h is the head at the computational node [L]; h_b is height of the river bed [L]; and c is riverbed conductance [L^2/T] defined as KLw_b / d_b, in which K is hydraulic conductivity of the river bed [L/T]; L is the length of the reach; w_b is width of the river [L]; and d_b is the thickness of the river bed layer.

Several researchers have noted that when the water table falls below the riverbed, the water in the stream and the water table are disconnected and unsaturated flow conditions develop between the riverbed and the water table. Such conditions usually arise due to low-permeable clogging layers on the riverbed. Figure 9.22 shows the stream–aquifer relationships for the case of a clogged streambed. From the top to the bottom panel, the figure shows a connected gaining stream, connected losing stream, disconnected stream with a shallow water table, and disconnected stream with a deep water table. Since MODFLOW does not consider the effect of disconnected clogged streams adequately, it underestimates flow from the river to the aquifer when the water table falls below the river bed (Osman and Bruen 2002; Fox and Durnford 2003; Fox and Gordji 2007).

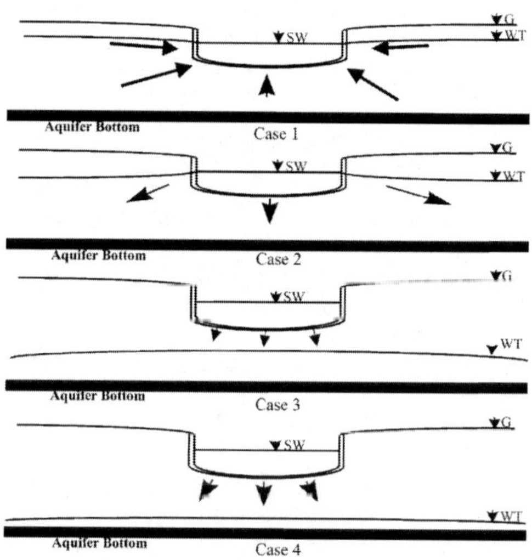

Figure 9.22: Stream–aquifer relationships for a partially penetrating stream. From Osman and Bruen (2002), used with permission of Elsevier.

Several modifications to Equation (9.115) are proposed to improve the simulation of water movement in the unsaturated zone beneath the streambed. Rovey (1975) modified Equation (9.115) for the case $h \le h_b$ by replacing h_b with $h_b + h_a$ where h_a is the air entry head (bubbling pressure) of clogging layer. The Bouwer (1969) model is similar to Rovey's but uses h_{cr} instead of h_a, where h_{cr} is defined as critical pressure head and is computed by integrating the relative hydraulic conductivity ($K(\psi)/K_s$) from zero to a pressure head at which moisture content and hydraulic conductivity become practically irreducible (Bouwer 1969). Osman and Bruen (2002) argued that there is a limiting shallow water elevation (h_c) below the stream bed that controls the seepage rate from the river to the aquifer. They modified Equation (9.115) to

$$\left.\begin{array}{ll} Q = c(h_r - h) \ ; & h > h_b \\ Q = c(h_r - h_b + h_{suc}) \ ; & h_c < h \le h_b \\ Q = c(h_r - h_c) \ ; & h \le h_c \end{array}\right\} \tag{9.116}$$

where h_{suc} is the suction at the base of the clogging layer approximated by h_b - h [L]. The limiting shallow water table is approximated by $h_c = h_r - h_{smax}$ with h_{smax} being the maximum suction head [L] given by Equation (10) in Osman and Bruen (2002). Fox and Gordji (2007) expanded Osman and Bruen's (2002) two-regime model, where the saturated stream/aquifer boundary is immediately converted to an unsaturated flow boundary, to a three-regime model by introducing a transition zone.

Figure 9.23 depicts the change in water flux from a river reach as a function of the head, h, in the cell containing the reach, as calculated using above equations. Flow is zero when h is equal to the water level in the river, h_r. For values of h higher than h_r, flow is into the river, represented as a negative inflow to the aquifer. For values of h lower than h_r, flow is positive into the aquifer. Based on MODFLOW Equation (9.115), this positive flow increases linearly as h decreases, until it reaches h_b; thereafter the flow remains constant. The Osman and Bruen (2002) modification, Equation (9.116), for unsaturated flow condition below riverbed shows a constant flux to the aquifer only after the water table has dropped below elevation h_c, and not h_b. The three-regime proposal of Fox and Gordji (2007) shows a smooth transition, rather than abrupt change, to a constant flux at some depth below h_c.

There are numerous external packages developed for MODFLOW. Furman (2008) provides a valuable overview and summary of MODFLOW packages developed for surface water-groundwater interaction. Citing Alazmi and Vafai (2001), Furman (2008) lists five approaches for treating boundary conditions at the surface-subsurface water interface. The boundary conditions are based on conservation of mass and continuity of momentum which are typically represented by velocity and its normal gradient at the interface.

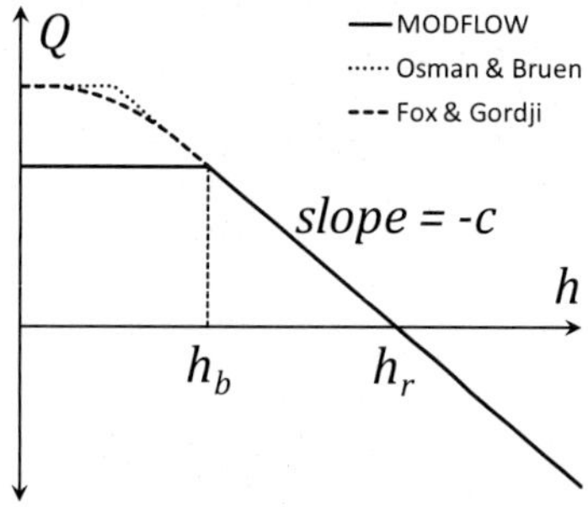

Figure 9.23: Comparison among three theories of the relationship between flow, Q, from a river to the underlying aquifer and head, h, where h_b is the elevation of the bottom of the riverbed, h_r is the head in the river, and c is the riverbed conductance. Positive Q indicates flow into aquifer, negative Q indicates flow into river.

Below we summarize some of the efforts on coupling MODFLOW with other models.

- GSFLOW: The Coupled Ground-Water and Surface-Water Flow model is based on the integration of the Precipitation-Runoff Modeling System (PRMS) and MODFLOW-2005 (Markstrom et al. 2008). The purpose behind the development of GSFLOW was to simulate groundwater and surface water flow in watersheds by simultaneously simulating flow across the land surface and within subsurface saturated and unsaturated media. Simulations in GSFLOW take place within and among three regions. The first region is bounded on top by the plant canopy and on the bottom by the lower limit of the soil zone; the second region consists of all streams and lakes; and the third region is the subsurface zone beneath the soil zone (Markstrom et al. 2008). The hydrology in the first region is simulated with PRMS, whereas hydrologic processes within the second and third regions are simulated with MODFLOW-2005.

- DAFLOW-MODFLOW: The surface-water flow model DAFLOW and MODFLOW are coupled by Jobson and Harbaugh (1999). The DAFLOW model developed by Jobson (1989) simulates one-dimensional flow through a system of interconnected channels by solving the diffusive-wave form of the Saint-Venant flow equations. The stream system is subdivided into a series of branches, with each branch further divided into a number of subreaches. DAFLOW is designed to simulate flow in upland stream systems where flow reversals do not occur and backwater conditions are not severe (Jobson 1989). Because the response time of surface-water systems are usually much smaller than those of groundwater systems, the appropriate time step of a surface water model should be smaller than the appropriate time step of the groundwater model. Considering this, DAFLOW was structured so that multiple DAFLOW time steps can occur within a MODFLOW time step. During each iteration of a MODFLOW time step, a subroutine is called to calculate seepage to or from the stream. DAFLOW must also be run to route the surface water flow for the MODFLOW time step. For seepage calculations, groundwater head can either be assumed to remain constant or vary linearly during the groundwater (GW) time step. The heads are assumed to remain constant during a single surface water (SW) time step. The SW flow conditions at the beginning of the GW time step are stored so that the iterative process can be repeated. Leakage to or from a stream subreach is computed using Darcy's law.

- SWATMOD: This model is developed by linking MODFLOW (McDonald and Harbaugh 1988) to the Soil and Water Assessment Tool SWAT model (Arnold et al. 1993). It is a comprehensive surface- and ground-water modeling tool for evaluating options for water-resources management (Sophocleous et al. 1999).

- GMS: The Groundwater Modeling System (GMS), developed and maintained by Aquaveo (2009), is a graphical interface for numerous models including MODFLOW-2000, MODPATH, MT3DMS/RT3D, SEAM3D, ART3D, UTCHEM, FEMWATER, PEST, UCODE, MODAEM and SEEP2D. It provides tools for groundwater simulation including site characterization, model development, calibration, post-processing, and visualization. The interface offer GIS capabilities, making it possible to build a conceptual model using GIS feature objects and conduct groundwater simulations rather seamlessly at the watershed scale. The conceptual model defines the boundary conditions, sources/sinks, and material property zones for a model. The model data can then be automatically discretized to the model grid. An example watershed-scale application of GMS is the study of Hantush and Wang (2003) who applied GMS to estimate long-term nitrate baseflow loading from two agricultural watersheds.

Analytic Element Method

The Analytical Element Method (AEM) is a computational method based on superposition of hundreds of analytical functions to solve regional groundwater flow problems (Strack 1989; Haitjema 1995). We mention it because it was motivated by a groundwater-surface water interaction problem, i.e., modeling the effect of the Tennessee-Tombigbee Waterway on the surrounding aquifers (Strack and Haitjema 1981a, 1981b). Each particular feature in an aquifer system is represented by an analytical function within the AEM framework. For instance line sinks are used for stream sections, areal sink distributions are used to represent ponds and lakes, and line doublets are used to model discontinuities in aquifer thickness or hydraulic conductivity (Bruker and Haitjema 1996). Readers are referred to two books written by the original developers Strack (1989) and Haitjema (1995) for details.

In groundwater-surface water interaction problems, AEM represents recharge to groundwater as area sinks and streams are modeled by line elements. The line elements are characterized by a jump in the stream function, which is linked directly with the amount of water captured by the stream (Strack 2003). An example application can be found in Bruker and Haitjema (1996). They developed a conjunctive surface water and groundwater flow modeling approach called GFLOW1 to prevent over infiltration of streams or lakes. They represented changing boundary conditions by adding or removing streams from the groundwater domain based on availability of water in the stream. Baseflow contribution to streams and lakes (line sinks) are computed by accumulating all groundwater inflows and outflows. Hunt et al. (1996) presented application of AEM to simulate groundwater-lake interactions.

Sometimes groundwater flux has a significant component parallel to the river channel in the same direction as the streamflow. Such stream-aquifer systems are called *underflow-component dominated* (Larkin and Sharp 1992). Unlike almost all of the analytical models, the AEM is capable of simulating underflow component of groundwater flow, e.g., near a meandering stream (e.g., Bakker 2007). However, it should be noted that underflow can be computed using any of extant two- and three-dimensional numerical models, including MODFLOW. Bakker (2007) presented an analytical element approach for simulating the steady-state interactions between groundwater and surface water features with leaky beds.

In spite of the advantages of AEM (reliance on basic principles, scale-independence, and computational efficiency), and maturity of the method as a research program for groundwater flow modeling (Kraemer 2004), AEM applications have been generally limited to two-dimensional and steady-state groundwater flow applications. An overview of the methodology and areas of applications, including groundwater-surface water interactions, has been provided by Hunt (2006). He cites the two limitations above as the areas where more method development is needed.

Watershed Models: In a broad sense, watershed models can be classified into physically-based models and process-based/empirical models that combine mass balance with empirical relationships. With a few exceptions, current models are mostly a hybrid of physically-based and empirical processes. A new generation of physically-based watershed models is also emerging that is based on scaling up point flow and momentum conservation equations by means of stochastic ensemble-averaging methods.

Physically-Based Models

Physically-based watershed models vary in complexity in handling the surface water-groundwater interactions. Few models use full dynamic coupling. However, such models are not practical to use and are limited to research studies. The Integrated Hydrology Model (InHM) (Van der Kwaak 1999), MODHMS (Panday and Huyakorn 2004), and HydroGeoSphere (Therrien et al 2005; Goderniaux et al. 2009) are some examples. For example, rather than specifying water bodies as prescribed head boundary condition, the InHM model is capable of determining lake location and stage as part of the simulations (Smerdon et al. 2007).

Physically-based watershed models like MIKE SHE (Graham and Butts 2005) and GSSHA (Downer et al. 2006) typically link the two-dimensional form of the Saint-Venant equation (or its simplifications like diffusive or kinematic wave) for overland flow with the one-dimensional Richards equation. Simplifications of Richard's equation such as the Green-Ampt equation can be used too. For each time step the upper boundary condition in the unsaturated zone is either a constant flux such as rainfall rate or evaporation rate (no ponding), or a constant head (e.g., depth of ponded water). The lower boundary condition is typically a pressure boundary. If the groundwater table is too deep, then a zero pressure head gradient is implemented at the lower boundary to simulate gravity drainage

The flux rate computed from the last cell of the unsaturated media becomes recharge for groundwater. Interaction of groundwater and stream is handled either by head or flux boundaries. In the former, the elevation of the water surface of the stream is used as the specified head in the solution of groundwater. The flux boundary condition is similar to the one used in MODFLOW RIVER package, where flux through the river bed is calculated based on the elevation difference of groundwater and stream surfaces using Darcy's equation, when they are connected. When the groundwater surface elevation is below the river bed, that is the river and the groundwater systems are disconnected, then the river is assumed to leak to groundwater at a rate equal to the saturated conductivity.

There are numerous studies in the literature that rely on physically-based watershed models to study effects of various management strategies and land use scenarios on groundwater levels, recharge baseflow, etc., mostly MIKE SHE applications. Smerdon et al. (2008) studied recharge variations in a mountainous watershed in Canada by using a combination of MIKE SHE and a simpler seasonal water budget model, with the former utilized in data rich areas. Recharge estimates are then used to develop a regional groundwater model with MODLFOW. Hammersmark et al. (2008) studied the pre- and post-stream restoration periods in a northern California meadow using MIKE SHE and showed how such restoration scenarios affect groundwater levels, subsurface storage and baseflow. Schroder and Rosbjerg (2004) used MIKE SHE to explore how the groundwater feedback to evapotranspiration (capillary raise) influences the water balance of a small watershed in Denmark.

Process-Based/Empirical Models

These models are based on empirical parameterizations of different processes and rarely utilize momentum equations or simplifications thereof. In general, empirical models rely on empirical parameters that rarely have any physical meaning and are derived primarily from regression analyses of observed data. They are usually suitable for conditions under which the relationships have been developed. In other words, they become less reliable under the conditions outside the limit of the original environment and generally are not suitable for predictions under different conditions (Kalin and Hantush 2003). Therefore, empirical models almost always have to be calibrated with observed data before applying to a specific area/case.

These types of models have no direct coupling and are of the Type I coupling described earlier. Soil Water and Assessment Tool - SWAT (Neitsch et al. 2005), Hydrologic Simulation Program Fortran - HSPF (Bicknell et al. 2001), Generalized Watershed Loading Function - GWLF (Haith and Shoemaker 1987) are some examples of such models. The portion of rainfall reaching the ground is first partitioned into its surface runoff and infiltrating water components. If operated at daily time scale the Soil Conservation Service (SCS) curve number method is typically used (SCS 1986). At smaller time scales other infiltration equations, such as Green-Ampt, Horton or Philips' equations can be used to compute infiltration. The unsaturated zone can be divided into soil layers and routing of water flux through these zones is achieved through empirical relationships. For instance, SWAT uses the following empirical relationship to compute flux between two successive layers

$$f = (SW - FC) \cdot \left\{ 1 - e^{-\frac{\Delta t}{TT}} \right\} \qquad (9.117)$$

where f [L] is amount of water that moves from a layer to the next layer on a day; SW is available soil water [L]; FC is field capacity [L]; Δt is time step [T]; TT is travel time given by $(SAT\text{-}FC)/K_{sat}$ with SAT being amount of water in the soil layer when completely saturated; and K_{sat} being the saturated hydraulic conductivity. Once the flux of water from the last soil layer is computed, the movement of water from that point all the way to the groundwater is simply handled with a delay function. The source of water to the groundwater is the recharge received from unsaturated zone. The equation used in SWAT to compute recharge is

$$w_{rech,i} = \left\{ 1 - e^{-\frac{1}{\delta_w}} \right\} w_{seep} + e^{-\frac{1}{\delta_w}} w_{rech,i-1} \qquad (9.118)$$

where $w_{rech,i}$ is recharge on day i; δ_w is groundwater delay parameter (found by calibration); and w_{seep} is the amount of water leaving the bottom of the soil profile to recharge water tables.

The linkage between groundwater and the river in such empirical models is unilateral. Streams can receive water from the groundwater in form of baseflow but the reverse mechanism, i.e., streams feeding the groundwater system, is not permitted. In SWAT, baseflow occurs only if the amount of water available in the soil profiles exceeds a user defined threshold, which is a parameter with no physical meaning and has to be calibrated. HSPF's conceptualization of the interactions is similar with most processes being represented through empirical relationships. Compared to HSPF and SWAT, GWLF's modeling of groundwater and its interaction with surface water and unsaturated zone is overly simplistic. Groundwater discharge and deep seepage from the unconfined aquifer are merely related to groundwater storage (conceptualized as a reservoir) with recession and seepage constants, respectively.

Watershed models of this kind have typically been utilized in addressing surface water quality and quantity issues due to their weak groundwater component. Several researchers tried to couple HSPF and SWAT with MODLFOW to improve the groundwater component of these models. For instance Cho et al. (2009) studied land use impacts on groundwater levels and streamflow in a Virginia watershed by using HSPF and MODLFOW. They run HSPF to obtain areal-average recharge estimates with varying land use practices. The recharge estimates are then fed into MODLFOW as inputs to study changes in groundwater levels owing to land use variations. Sophocleous et al. (1999) linked SWAT with MODLFOW (SWATMOD) to study long-term management studies in a Kansas watershed. Kim et al. (2008) also developed a SWAT-MODLFOW linkage for improved estimation of spatio-temporal distribution of groundwater recharge rates, aquifer evapotranspiration and groundwater levels.

Models Based on Up-Scaling Point-Scale Conservation Equations

The governing partial differential equations (PDEs) of the hydrologic processes used in watershed models are essentially developed for and applicable at the point scale. The Freeze and Harlan (1969) blueprint has been the foundation of many distributed, physically-based watershed models (e.g., MIKE SHE, Graham and Butts 2005; InHM, Van Der Kwaak 1999; CATHY, Paniconi et al. 2003; GSSHA, Downer et al. 2006; etc.), where a series of point-scale governing equations from different continua are solved. Yet, they are assumed to hold true at the scale of the computational units utilized in models, such as grid cells, small subwatersheds or hydrologic response units with homogenous soil, land use and topographic characteristics. In reality heterogeneities always exist in nature. In theory, one can start from the point scale PDEs and obtain the up-scaled form of those PDEs, which would then be applicable at the scale of the models computational unit and make use of not only the average values of the physical model parameters but also other statistical measures such as variance, covariance, etc.

In a series of papers Reggiani et al. (1998, 1999, 2000) presented an alternative approach to the Freeze and Harlan (1969) blueprint for watershed modeling. They derived up-scaled forms of the global balance laws of mass, momentum, and energy by integrating point-scale equations over specifically chosen regions, called *representative elementary watersheds* (REW). Due to the wide spectrum of time scales characterizing various hydrologic processes, they also integrated the same equations over a characteristic time interval. Each REW is divided into five different subregions: unsaturated zone, saturated zone, concentrated overland flow, saturated overland flow, and channel reach. The subregions interact with each other through exchange terms of mass, momentum, and energy. The unsaturated zone interacts with the saturated zone through recharge or capillary rise terms and with the concentrated overland flow domain through infiltration and evaporation processes. The saturated zone is linked to saturated overland flow with a flux term through a seepage face, and exchanges mass with the channel reach through either lateral channel inflow or groundwater recharge if there are channel losses. The mass exchange between the channel reach and the saturated zone is based on hydraulic head differences in each zone (Reggiani and Rientjes 2005), which is similar to MODLFOW's approach, i.e.,

$$e_{sr} = \frac{plK}{\Lambda}(h_r - h_s) \tag{9.119}$$

where p is channel wetted perimeter [L]; l is channel reach length [L]; K is channel bottom hydraulic conductivity [LT^{-1}]; Λ is a length scale characterizing the thickness of the channel bed transition zone [L]; and h_r and h_s are heads in the channel reach and the saturated zone, respectively [L]. The REW approach has not established itself yet within the modeling community. Some applications found in the literature include Fenicia et al. (2005), Reggiani and Rientjes (2005), Zhang et al. (2005, 2006), Zehe et al. (2006), and Mou et al. (2008).

Although the REW model is based on up-scaled equations of physics and thermodynamics laws, it is deterministic in nature. It is not suited for handling heterogeneities in model parameters at the sub-REW scale. The Watershed Environmental Hydrology (WEHY) model, recently developed by Kavvas et al. (2004, 2006), is a comprehensive watershed-scale model based on upscaling of point-scale conservation equations. The ensemble averaged forms of those equations are applicable at the computation grid scales, and are used to simulate the hydrology and pollutant transport in small watersheds. Sub-computational unit heterogeneities in model parameters are represented through their means, variances and covariances. Since the model is based on upscaled equations, it is scale independent, and applicable to both very small and very large watersheds.

In the WEHY model, the vertical unsaturated flow component is modeled using upscaled Green-Ampt conservation equations (Chen et al. 1994a, 1994b). Overland flow component is based on upscaled form of the two-dimensional kinematic wave approximation of the Saint-Venant equations (Yoon and Kavvas 2000). Subsurface storm flow (also called interflow) is represented by the upscaled equations developed by Dogrul et al. (1998) for transient subsurface flow on sloping impermeable beds in heterogeneous hillslopes. Groundwater flow of WEHY is based on two-dimensional transient, saturated, horizontal flow under Dupuit-Forcheimer assumptions (Equation 5-90 in Bear 1979). The one-dimensional diffusive wave equation combined with Chezy equation is used to solve channel flow. The WEHY model can simulate both infiltration excess (Hortonian type) and saturation excess (Dunne type) overland flow, which define the interaction of the overland flow with the soil layer below. The model is also capable of simulating subsurface storm flow, which can generate return flow from the soil subsurface to the land surface. Infiltrated water from the unsaturated zone replenishes the groundwater zone as groundwater recharge. Groundwater can receive from or lose to channels depending on head differences. Although detailed information on the coupling of subsurface and surface water processes is not provided in Kavvas et al. (2004), the explicit finite difference numerical schemes used in numerical modeling of the component processes leads us to believe that coupling is more likely to be of Type II, i.e., noniterative. The model is proprietary and solely developed for research purposes by the developers. Chen et al. (2004) and Yoshitani et al. (2009) tested the flow component of WEHY model on small, mountainous watersheds.

Watershed-Scale Interactions (Rattlesnake Creek Basin Case-Study): Understanding of stream-aquifer interaction is essential for sustained usefulness of aquifers and watersheds in many arid to sub-humid regions of the world. As an example, Figure 9.24 shows the Rattlesnake Creek basin in south central Kansas where water management in the form of pumping from wells and surface water withdrawals have been an important concern. During the mid-1960's to mid-1980's, appropriated irrigation amounts had increased 10-fold with alarming declines in water table levels and streamflows (Sophocleous et al. 1999). Associated deterioration of water quality through saltwater intrusion from underlying aquifers was also an important concern.

The 145-km-long by 25-km-wide Rattlesnake Creek basin is covered with a veneer of loess deposits and sand dunes, and the main aquifer underneath is comprised of Pleistocene alluvium (Fader and Stullken 1978). The watershed is mostly farmland with wheat, grain, sorghum, corn, soybeans, and cattle being the important land use categories. The mean annual precipitation is estimated to be 617 mm, mostly in the months of April through September, with light snowfall (< 50 mm). Over 60% of the watershed does not contribute to streamflow because of flat topography and isolated regions of water (see Figure 9.25).

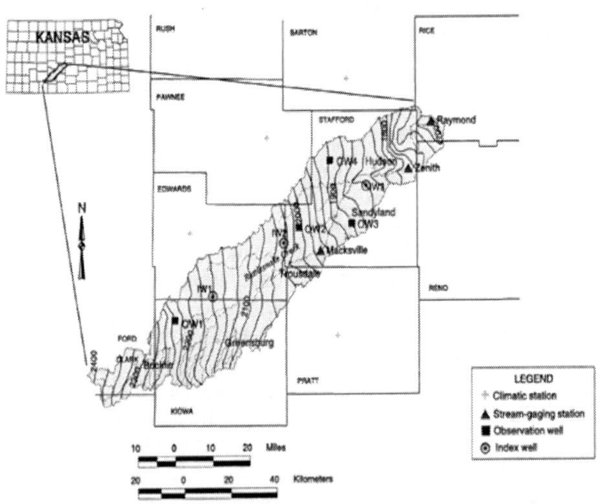

Figure 9.24: The Rattlesnake Creek Basin in Kansas with pre-development water table contours (m) and gauging stations. From Sophocleous et al. (1999), used with permission of Elsevier.

Figure 9.25 depicts the conceptualization of the basin hydrologic balance adopted by Sophocleous et al. (1999). The important water balances (see also Figure 9.26) are as follows:

- Surface runoff (SRO) becomes lateral inflow to a stream or pond.
- Recharge (R) to the relatively shallow aquifer occurs from infiltration (INF) that percolates downward beyond the root zone (PERC), along with pond seepage (POND SEEP), channel transmission losses (TL), and subsurface lateral flow (SLF) that does not flow into ponds.
- Groundwater pumped from the aquifer (Q) is applied as irrigation at the surface (IRR).
- Water in ponds evaporates (E), and water in the soil zone within reach of plant roots is removed by evapotranspiration (ET).
- The stream and aquifer interact via leakage driven by hydraulic gradients across the streambed (q).

Percolation below the root zone was distributed over the aquifer without explicitly accounting for the time lag and attenuation that would be associated with the unsaturated subsurface flow, given the long-term simulation horizon, the relatively high infiltration rates, low available water capacity of the basin soils, and the relatively shallow depth to water table over most of the basin. The authors combined MODFLOW (McDonald and Harbaugh 1988) and the watershed model SWAT (Arnold et al. 1993) to create SWATMOD, a comprehensive surface- and ground-water
modeling tool for evaluating options for water-resources management.

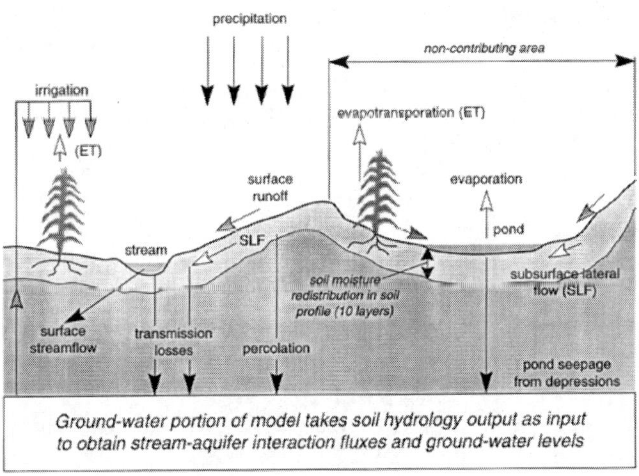

Figure 9.25: Conceptualization of noncontributing ponds and their interaction with underlying aquifer for the Rattlesnake Creek basin.

Both SWAT and MODFLOW were modified to implement the linkages shown in Figure 9.26. Specifically, two subroutines were developed by the authors: HYDBAL, to be called by SWAT, and MODSWB, to be called by MODFLOW. HYDBAL's functions are called at the end of each simulated aquifer time step to pass data back and forth between SWAT and MODFLOW, and to write SWAT's results to a "hydrologic balance" data file for each aquifer time step. MODSWB associates the geographic domain of the subbasins represented by SWAT with the aquifer grid domain and stream network defined in MODFLOW, and converts the hydrologic fluxes calculated by SWAT for each time step into flow rates for recharge, tributary flow, and both surface and groundwater diversions. In this manner, SWAT and MODFLOW were externally coupled by the authors. A time step of one day was implemented for the SWAT model, whereas a monthly time step was utilized for MODFLOW simulations to deal with the computational constraints and also because of the difference in the flow rates in the surface and subsurface systems. Readers are referred to Sophocleous et al. (1999) for more details and for evaluating the capabilities of the SWATMOD model. The authors show examples highlighting the important role of stream-aquifer interaction in a watershed-scale setting.

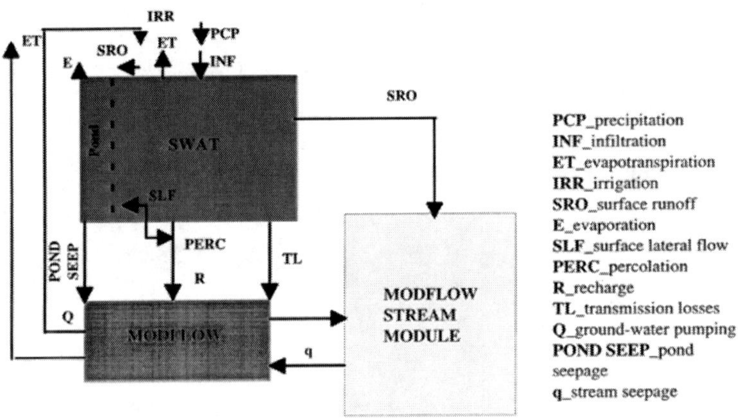

Figure 9.26: Water balance components and connections between the MODFLOW and SWAT models. From Sophocleous et al. (1999), used with permission of Elsevier.

9.5 References

Aksoy, H., Kurt, I., and Eris, E. (2009). "Filtered smoothed minima baseflow separation method." *J. Hydrol.*, 372, 94-101.

Akylas, E., and Koussis, A. (2007). "Response of sloping unconfined aquifer to stage changes in adjacent stream: 1. Theoretical analysis and derivation of system response functions." *J. Hydrol.*, 338, 85-95.

Alazmi, B., and Vafai, K. (2001). "Analysis of fluid flow and heat transfer interfacial conditions between a porous medium and a fluid layer." *Int. Commun. Heat Mass.*, 44, 1735–1749.

Anderson, M. P., and Munter, J. A. (1981). "Seasonal reversals of groundwater flow around lakes and the relevance to stagnation points and lake budgets." *Water Resour. Res.*, 17(4), 1139-1150.

Aquaveo (2009). *Groundwater Modeling System (GMS)*, Version 6.5, Aquaveo LLC, Provo, Utah, <http://www.aquaveo.com/gms> (Jul. 9, 2009).

Arnold, J. G., Allen, P. M., and Bernhardt, G. (1993). "A comprehensive surface-groundwater flow model." *J. Hydrol.*, 142, 47-69.

Bahr, A. K., and Mishra, G. C. (1997). "One-dimensional spring flow model for time variant recharge." *Hydrol. Sci. J.*, 42(3), 381-390.

Bakker, M. (2007). "Simulating groundwater flow to surface water features with leaky beds using analytic elements." *Adv. Water Resour.*, 30, 399-407.

Barlow, P. M., DeSimone, L. A., and Moench, A. F. (2000). "Aquifer response to stream-stage and recharge variations: II. Convolution method and applications." *J. Hydrol.*, 230, 211-229.

Barnes, B. S. (1939). "The structure of discharge-recession curves." *Trans. Am. Geophys. Union*, 20, 721-725.

Bauer, P., Gumbricht, T., and Kinzelbach, W. (2006). "A regional coupled surface water/groundwater model of the Okavango Delta, Botswana." *Water Resour. Res.*, W04403, doi:10.1029/2005WR004234.

Bear, J. (1972). *Dynamics of fluids in porous media*, Elsevier, New York.

Bear, J. (1979). *Hydraulics of groundwater,* McGraw-Hill, New York.

Bear, J., and Verruijt, A. (1987). "Modeling Groundwater Flow and Pollution." *Theory and applications of transport in porous media*, D. Reidel Publishing Company, Dordrecht, Holland.

Betson, R. P. (1964). "What is watershed runoff?" *J. Geophys. Res.*, 69, 1541-1551.

Beven, K. (1978). "The hydrological response of headwater and sideslope areas." *Hydrol. Sci. Bull.*, 23, 419-437.

Beven, K. and Kirkby, M. J. (1979). "A physically based, variable contributing area model of basin hydrology." *Hydrol. Sci. Bull.*, 24, 43-69.

Beven, K. and Wood, E. F. (1993). "Flow Routing and Hydrologic Response of Channel Networks." *Channel network hydrology*, K. Beven and M. J. Kirkby, eds., John Wiley, New York.

Bicknell, B. R., Imhoff, J. C., Kittle, J. L., Jr., Jobes, T. H., and Donigian, A. S., Jr. (2001). *Hydrological simulation program – FORTRAN, version 12, user's manual,* Aqua Terra Consultants, Mountain View, Calif.

Birkhead, A. L., and C. S. James. (2002). "Muskingum river routing with dynamic storage." *J. Hydrol.*, 264, 113-132.

Boussinesq, J. (1877). "Essai sur la theorie des eaux courantes." *Mem. Acad. Sci. Inst. France*, 23(1), 252-260. [Cited by Brutsaert (2005).]

Boussinesq, J. (1904). "Recherches theoriques sur l'ecoulement des nappes d'eau infiltrees dans le sol et sur le debit des sources." *J. Math. Pures Appl.*, 5me ser, 10, 5-78. [Cited by Brutsaert (2005).]

Bouwer, H. (1969). "Theory of seepage from open channels." *Advances in hydroscience*, V. T. Chow, ed., Academic Press, New York, 121-172.

Brooks, R. H., and Corey, A. T. (1964). "Hydraulic properties of porous media." *Hydrology Paper No. 3*, Colorado State University, Fort Collins, Colo.

Bruggeman, G. A. (1999). *Analytical solutions of geohydrological problems: Developments in water science*, 64. Elsevier, Amsterdam.

Bruker, S. M., and Haitjema, H. M. (1996). "Modeling steady state conjunctive groundwater and surface water flow with analytic elements." *Water Resour. Res.*, 32(9), 2725-2732.

Brutsaert, W., and Nieber, J. H. (1977). "Regionalized drought flow hydrographs from a mature glaciated plateau." *Water Resour. Res.*, 13(3), 637-643.

Brutsaert, W. (1994). "The unit response of groundwater outflow from a hillslope." *Water Resour. Res.*, 30(10), 2759-2763.

Brutsaert, W., and Lopez, J. P. (1998). "Basin-scale geohydrologic drought flow features of riparian aquifers in the southern great plains." *Water Resour. Res.*, 34(2), 233-240.

Brutsaert, W. (2005). *Hydrology: An Introduction*, Cambridge University Press, Cambridge.

Bryan, K. (1919). "Classification of springs." *J. Geolog.*, 27, 522-561.

Butler, J. J., Jr., Zlotnik, V. A., and Tsou, M-S (2001). "Drawdown and stream depletion produced by pumping in the vicinity of a partially penetrating stream." *Ground Water*, 39(5), 651-659.

Butler, J. J., Jr., Zhang, X., and Zlotnik, V. A. (2007). "Pumping-induced drawdown and stream depletion in a leaky aquifer system." *Ground Water*, 45(2), 178-186.

Campbell, G. S. (1974). "A simple method for determining unsaturated conductivity from moisture retention data." *Soil Sci.*, 117, 311-314.

Carslaw, H. S., and Jaeger, J.C. (1959). *Conduction of heat in solids*, 2nd Ed., Oxford Univ. Press, London, U.K.

Cheng, X., and Anderson, M. P. (1994). "Simulating the influence of lake position on groundwater fluxes." *Water Resour. Res.*, 30(7), 2041-2049.

Chen, X., and Chen, X. (2003). "Stream water infiltration, bank storage, and storage zone changes due to stream-stage fluctuations." *J. Hydrol.*, 280, 246-264.

Chen, Z. Q., Govindaraju, R. S., and Kavvas, M. L. (1994a). "Spatial averaging of unsaturated flow equations for areally heterogeneous fields: Development of models." *Water Resour. Res.*, 30(2), 523–533.

Chen, Z. Q., Govindaraju, R. S., and Kavvas, M. L. (1994b). "Spatial averaging of unsaturated flow equations for areally heterogeneous fields: Numerical simulations." *Water Resour. Res.*, 30(2), 534–544.

Chen, Z. Q., Kavvas, M. L., Fukami, K., Yoshitani, J., and Matsuura, T. (2004). "Watershed environmental hydrology (WEHY) model: Model application." *J. Hydrol. Eng.*, 9(6), 480-490.

Childs, E. C. (1971). "Drainage of groundwater resting on a sloping bed." *Water Resour. Res.*, 7(5), 1256-1263.

Cho, J., Barone, V. A., and Mostaghimi, S. (2009). "Simulation of land use impacts on groundwater levels and streamflow in a Virginia watershed." *Agr. Water Manage.*, 96(1), 1-11.

Chow, V.T., Maidment, D. R., and Mays, L. W. (1988). *Applied hydrology*, McGraw-Hill, New York.

Conrad, L. P., and Beljin, M. S. (1996). "Evaluation of an induced infiltration model as applied to glacial aquifer systems." *Water Resour. Bull.*, 32(6), 1209-1220.

Cooper, H. H., and Rorabaugh, M.I. (1963). *Ground-water movements and bank storage due to flood stages in surface streams*, U.S. Geological Survey Water-Supply Paper 1536-J, 343-366.

Cunge, J. A. (1969). "On the subject of a flood propagation method (Muskingum method)." *J. Hydraulics Research*, Intern. Assoc. Hydraul. Res., 7(2), 205-230.

de Hoog, F. R., Knight, J. H., and Stokes, A. N. (1982). "An improved method for numerical inversion of Laplace transforms." *S.I.A.M. J. Sci. Stat. Comp.*, 3, 357-366.

Dever, R. J., and Cleary, R. W. (1979). "Unsteady-state, two-dimensional response of leaky aquifers to stream stage fluctuations." *Advances in Water Resources*, 2, 13-18.

Dingman, S. L. (2002). *Physical hydrology*, Prentice Hall, New Jersey.

Dogrul, E. C., Kavvas, M. L., and Chen, Z. Q. (1998). "Prediction of subsurface storm flow in heterogeneous sloping aquifers." *J. Hydrol. Eng.*, 3(4), 258–267.

Donnan, W. W. (1946). "Model tests of a tile-spacing formula." *Soil Sci. Soc. Am. Proc.*, 11, 131-136.

Downer, C. W., Ogden, F. L., Niedzialek, J. M., and Liu, S. (2006). "Gridded surface/subsurface hydrologic analysis (GSSHA) model: A model for simulating diverse streamflow producing processes." *Watershed models,* V. P. Singh and D. Frevert, eds., CRC Press, Boca Raton, Fla., 131-159.

Drummond, D. D. (1998). *Hydrogeology, simulation of ground-water flow, and ground-water quality of the Upper Coastal Plain aquifers in Kent County, Maryland*, Maryland Geological Survey, Report of Investigations No. 68.

Dunne, T., and Black, R. D. (1970). "Partial area contributions to storm runoff in a small New England watershed." *Water Resour. Res.*, 6, 1296-1311.

Dunne, T., Moore, T. R., and Taylor, C. H. (1975). "Recognition and prediction of runoff-producing zones in humid regions," *Hydrological Sciences Bull.*, 20, 305-327.

Dunne, T. (1978). "Field studies of hillslope flow processes." *Hillslope hydrology*, M. J. Kirkby, ed., John Wiley, New York, 227-293.

Dunne, T. (1983). "Relation of field studies and modeling in the prediction of runoff." *J. Hydrol.*, 65, 25-48.

Dupuit, J. (1863). *Etudes theoriques et pratiques sur le mouvement des eaux dans les canaux de couverts et a travers les terrains permeable*, 2nd Ed., Dunod, Paris. [Cited by Bear (1979).]

Emich, V. N. (1962). "On horizontal drains in layered soil." *PMTF*, 4, 131-133 (in Russian). [Cited by Bear (1979).]

Esteves, M., Faucher, X., Galle, S., and Vauclin, M. (2000). "Overland flow and infiltration modeling for small plots during unsteady rain: Numerical results versus observed values." *J. Hydrol.*, 228, 265-282.

Fader, S. W., and Stullken, L. E. (1978). *Geohydrology of the Great Bend Prairie, south-central Kansas*, Kansas Geological Survey, Irrigation Series 4, 19.

Fenicia, F., Zhang, G. P., Rientjes, T., Hoffmann, L., Pfister, L., and Savenije, H. H. G. (2005). "Numerical simulations of runoff generation with surface water-groundwater interactions in the Alzette river alluvial plain (Luxembourg)." *Phy. Chem. Earth*, 30(4-5), 277-284.

Fox, G. A., DuChateau, P., and Durnford, D. S. (2002). "Analytical model for aquifer response incorporating distributed stream leakage." *Ground Water*, 40(4), 378-384.

Fox, G. A., and Durnford, D. S. (2003). "Unsaturated hyporheic zone flow in stream-aquifer conjunctive systems." *Adv. Water. Resour.*, 26, 989-1000.

Fox, G. A., and Gordji, L. (2007). "Consideration for unsaturated flow beneath a streambed during alluvial well depletion." *J. Hydrol. Engrng.*, 12(2), 139-145.

Freeze, R. A., and Harlan, R. L. (1969). "Blueprint for a physically-based, digitally-simulated hydrologic response model." *J. Hydrol.* 9, 237-258.

Freeze, R. A. (1972a). "Role of subsurface flow in generating surface runoff: 1. Baseflow contributions to channel flow." *Water Resour. Res.*, 8, 609-623.

Freeze, R. A. (1972b). "Role of subsurface flow in generating surface runoff: 1. Upstream source areas." *Water Resour. Res.*, 8, 1272-1283.

Freeze, R. A. (1974). "Streamflow generation." *Reviews of Geophysics and Space Physics*, 4, 627-647.

Furman, A. (2008). "Modeling coupled surface–subsurface flow processes: A review." *Vadose Zone J.*, 7(2), 741-756.

Gill, M. A. (1985). "Bank storage characteristics of a finite aquifer due to sudden rise and fall of river level." *J. Hydrol.*, 76, 133-142.

Glover, R. E., and Balmer, C. G. (1954). "River depletion from pumping a well near a river." *Trans. Am. Geophys. Union*, 35(3), 468-470.

Goderniaux, P., Brouyère, S., Fowler, H. J., Blenkinsop, S., Therrien, R., Orban, P., and Dassargues, A. (2009). "Large scale surface–subsurface hydrological model to assess climate change impacts on groundwater reserves." *J. Hydrol.*, 373, 122–138.

Gonwa, W. S., and Kavvas, M. L. (1986). "A modified diffusion wave equation for flood propagation in trapezoidal channels." *J. Hydrol.*, 83, 119-136.

Govindaraju, R. S., Jones, S. E., and Kavvas, M. L. (1988). "On the diffusion wave model for overland flow: 1. Solution for steep slopes." *Water Resour. Res.*, 24(5), 734-744.

Govindaraju, R. S., and Kavvas, M. L. (1991). "Dynamics of moving boundary overland flows over infiltrating surfaces at hillslopes." *Water Resour. Res.*, 27(8), 1885-1898.

Govindaraju, R. S., and Koelliker, J. K. (1994). "Applicability of linearized Boussinesq equation for modeling bank storage under uncertain aquifer parameters." *J. Hydrol.*, 157, 349-366.

Graham, D. N., and Butts, M. B. (2005). "Flexible, integrated watershed modeling with MIKE SHE." *Watershed models*, V. P. Singh and D. K. Frevert, eds., CRC Press, Boca Raton, Fla., 245-272.

Gunduz, O., and Aral, M. M. (2005a). "A Dirac-δ function notation for source/sink terms in groundwater flow." *J. Hydrol. Engrng.*, 420-427.

Gunduz, O., and Aral, M. M. (2005b). "River networks and groundwater flow: A simultaneous solution of a coupled system." *J. Hydrol.*, 310, 216-234.

Gunduz, O. (2006). "Surface/subsurface interactions: Coupling mechanisms and numerical solution procedures." *Groundwater and ecosystems*, A. Baba, K. W. F. Howard, and O. Gunduz, eds., NATO Sciences Series, Springer.

Gureghian, A. B. (1978). "Solutions of Boussinesq's equation for seepage flow." *Water Resour Res.*, 14(2), 231-236.

Haith, D. A., and Shoemaker, L. L. (1987). "Generalized watershed loading functions for stream flow nutrients." *Water Resour. Bull.*, 23(3), 471-478.

Haitjema, H. M. (1995). *Analytic element modeling of groundwater flow,* Academic Press, San Diego, Calif.

Hall, F. R., and Moench, A. F. (1972). "Application of the convolution equation to stream aquifer relations." *Water Resour. Res.,* 8(2), 487-493.

Hammersmark, C. T., Rains, M. C., and Mount, J. F. (2008). "Quantifying the hydrological effects of stream restoration in a montane meadow, northern California, USA." *River Res. Appl.*, 24(6), 735-753.

Hantush, M. M., and Mariño, M. A. (2001). "Analytical modeling of the influence of denitrifying sediments on nitrate transport in aquifers with sloping beds." *Water Resour. Res.*, 37(12), 3177-3192.

Hantush, M. M., Harada, M., and Mariño, M. A. (2002). "Hydraulics of stream flow routing with bank storage." *J. Hydrol. Engrng.*, 7(1), 76-89.

Hantush, M. M., and Wang, M. (2003). "Modeling long-term nitrate base-flow loading from two agricultural watersheds." *Proc., American Water Resources Association (AWRA) Spring Specialty Conference, Kansas City, Missouri, Mo., May 12-14.*

Hantush, M. M. (2005). "Modeling stream-aquifer interactions with linear response functions." *J. Hydrol.*, 113(1-4), 59-79.

Hantush, M. S. (1959). "Analysis of data from pumping wells near a river." *J. Gephys. Res.*, 64(11), 1921-1932.

Hantush, M. S. (1960). "Modification of the theory of leaky aquifers." *J. Geophys. Res.*, 65(11), 3713-3725.

Hantush, M. S. (1964). "Hydraulics of wells." *Advances in hydroscience*, V. T. Chow, ed., Academic Press, New York, 281-433.

Hantush, M. S. (1965). "Wells near streams with semipervious beds." *J. Geophys. Res.*, 70(12), 2829-2838.

Hantush, M. S. (1967). "Flow of groundwater in relatively thick leaky aquifers." *Water Resour. Res.*, 3(2), 583-590.

Harada, M., Yamada, T., and Hantush, M. M. (2005). "Interaction process between stream and aquifer in alluvial fan of the Yasu River." *J. Hydrosci. and Hydraul. Engrng., Japan Society of Civil Engineers*, 23(1).

Harbaugh, A. W., and McDonald, M. G. (1996). *User's documentation for MODFLOW-96: An update to the U.S. Geological Survey modular finite-difference ground-water flow model*, U.S. Geological Survey Open-File Report 96-485.

Harbaugh, A. W., Banta, E. R., Hill, M. C., and McDonald, M. G. (2000). *MODFLOW-2000, the U.S. Geological Survey modular ground-water model-*

modularization concepts and the ground-water flow process, U.S. Geological Survey Open-File Report 00-92.

Harbaugh, A. W. (2005). *MODFLOW-2005, the U.S. Geological Survey modular ground-water model: The ground-water flow process*. U.S. Geological Survey Techniques and Methods 6-A16.

Haushild, W., and Kruse, G. (1962). "Unsteady flow of ground water into a surface reservoir." *Transactions, American Society of Civil Engineers*, 127, 408-424.

Henderson, F. M. (1966). *Open channel flow*, Macmillan Publishing Co., Inc., New York.

Hewlett, J. D., and Hibbert, A. R. (1963). "Moisture and energy conditions within a sloping soil mass during drainage." *J. Geophysical Res.*, 68, 1081-1087.

Hogarth, W., Govindaraju, R. S., Parlange, J.-Y., and Koelliker, J. K. (1997). "Linearized Boussinesq equation for bank storage: A correction." *J. Hydrol.*, 198, 377-385.

Hooghoudt, S. B. (1940). "Algemeene beschouwing van het problem van de detailonwatering en de infiltratie door middle van parallel loopende drains, greppels, slooten, en kanalen." Versl. Landbouwk. Onderz, 46(14)B, Algemeene Landsdrukkerij, 's-Gravenhage. [Cited by Ritzema (2006).]

Horton, R. E. (1933). "The role of infiltration in the hydrologic cycle." *Trans. Am. Geophy. Union*, 14, 446-460.

Horton, R. E. (1945). "Erosional development of streams and their drainage basins: Hydrophysical approach to quantitative morphology." *Geol. Soc. Am. Bull.*, 56, 275-370.

Huang, G., and Yeh, G-T (2009). "Comparative study of coupling approaches for surface water and subsurface interactions." *J. Hydrol. Engrng.*, 14(5), 453-462.

Hunt, B. (1990). "An approximation for the bank storage effect." *Water Resour. Res.*, 26(11), 2769-2775.

Hunt, R. J., Haitjema, H. M., Krohelski, J. T., and Feinstein, D. T. (1996). "Simulating ground water-lake interactions: Approaches and insights." *Ground Water*, 41(2), 227-237.

Hunt, R. J., and Krohelski, J. T. (1996). "The application of an analytical element model to investigate groundwater-lake interactions at Pretty Lake, Wisconsin." *Lake and Reservoir Manage.*, 12, 487-495.

Hunt, B., (1999). "Unsteady stream depletion from ground water pumping." *Ground Water*, 37(1), 98-102.

Hunt, B. (2003). "Unsteady stream depletion when pumping from semiconfined aquifer." *J. Hydrol. Engrng.*, 8(1), 12-19.

Hunt, B. (2004). "Spring-depletion solution." *J. Hydrol. Engrng.*, 9(2), 144-149.

Hunt, R. J. (2006). "Ground water modeling applications using the analytic element method." *Ground Water*, 44(1), 5-15.

Huyakorn, P. S., and Pinder, G. F. (1983). *Computational methods in subsurface flow*, Academic Press, Orlando, Fla.

Huyck, A. A. O., Pauwels, V. R. N., and Verhoest, N. E. C. (2005). "A base flow separation algorithm based on the linearized Boussinesq equation for complex hillslopes." *Water Resour. Res.*, 41, W08415, doi:10.1029/2004WR003789.

Illangasekare, T., and Morel-Seytoux, H. J. (1982). "Stream-aquifer influence coefficients as tools for simulation and management." *Water Resour. Res.*, 18(1), 168-176.

Intaraprasong, T., and Zhan, H. (2009). "A general framework of stream-aquifer interaction caused by variable stream stages." *J. Hydrol.*, 373, 112-121.

Jacob, C. E. (1950). "Flow of ground water." *Engineering hydraulics*, H. Rouse, ed., John Wiley & Sons, New York.

Jobson, H. E. (1989). *Users manual for an open-channel streamflow model based on the diffusion analogy*, U.S. Geological Survey Water-Resources Investigations 89-4133.

Jobson, H. E., and Harbaugh, A. W. (1999). *Modifications to the diffusion analogy surface-water flow model (DAFLOW) for coupling to the modular finite-difference ground-water flow model (MODFLOW)*, U.S. Geological Survey Open-File Report 99-217.

Kalin, L., and Hantush, M. M. (2003). *Evaluation of sediment transport models and comparative application of two watershed models*, EPA/600/R-03/139, USEPA-NRMRL, Cincinnati, Ohio.

Kavvas, M. L., Chen, Z. Q., Dogrul, C., Yoon, J. Y., Ohara, N., Liang, L., Aksoy, H., Anderson, M. L., Yoshitani, J., Fukami, K., and Matsuura, T. (2004). "Watershed environmental hydrology (WEHY) model based on upscaled conservation equations: Hydrologic module." *J. Hydrol. Eng.*, 9(6), 450-464.

Kavvas, M. L., Yoon, J. Y., Chen, Z. Q., Liang, L., Dogrul, C., Ohara, N., Aksoy, H., Anderson, M. L., Yoshitani, J., Reuter, J., and Hackley, S. (2006). "Watershed environmental hydrology model: Environmental module and its application to a California watershed." *J. Hydrol. Eng.*, 11(3), 261-272.

Kazezyılmaz-Alhan, C. M., Medina, M. A., Jr., and Richardson, C. J. (2007). "A wetland hydrology and water quality model incorporating surface water/groundwater interactions." *Water Resour. Res.*, 43, W04434, doi:10.1029/2006WR005003.

Kazmann, R. G. (1948). "The induced infiltration of river water to wells." *Trans. Am. Gephys. Union*, 29, 85-92.

Kim, K-Y, Kim, T., Kim, Y., and Woo, N-C (2007). "A semi-analytical solution for groundwater response to stream-stage variations and tidal fluctuations in a coastal aquifer." *Hydrol. Process.*, 21, 665-674.

Kim, N. W., Chung, I. M., Won, Y. S., and Arnold, J. G. (2008). "Development and application of the integrated SWAT-MODFLOW model." *J. Hydrol.*, 356(1-2), 1-16.

Kirkby, M. J. (1988). "Hillslope runoff processes and models." *J. Hydrol.*, 100, 315-339.

Koussis, A. D., Smith, M. E., Akylas, E., and Tombrou, M. (1998). "Groundwater drainage flow in a soil layer resting on an inclined leaky aquifer." *Water Resour. Res.*, 34(11), 2879-2887.

Kraemer, S. R. (2004). "The maturity of analytic element groundwater flow modeling as a research program (1980-2003)." *Proc., IV International Conference on Analytic Element Modeling of Groundwater Flow*. Ecòle Nationale Supérieure des Mines de Saint-Etienne, St. Etienne, France.

Kresic, N. (2007). *Hydrogeology and groundwater modeling*, 2nd Ed., CRC Press, Boca Raton, Fla.

Larkin, R. G., and Sharp, J. M. (1992). "On the relationship between river-basin geomorphology, aquifer hydraulics, and groundwater flow direction in alluvial aquifers." *Geological Society of America Bulletin*, 104, 1608-1620.

Maillet, E. (1905). "Essai d'hydraulique souterraine et fluvial." *Libraire Sci.*, A. Herman, Paris. [Cited by Tallaksen (1995).]

Manga, M. (1996). "Hydrology of spring-dominated streams in the Oregon Cascades." *Water Resour. Res.*, 32(8), 2435-2439.

Mariño, M. A. (1975). "Digital simulation model of aquifer response to stream stage fluctuation." *J. Hydrol.*, 25, 51-58.

Markstrom, S. L., Niswonger, R. G., Regan, R. S., Prudic, D. E., and Barlow, P. M. (2008). *GSFLOW-Coupled Ground-water and Surface-water FLOW model based on the integration of the Precipitation-Runoff Modeling System (PRMS) and the Modular Ground-Water Flow Model (MODFLOW-2005)*, U.S. Geological Survey Techniques and Methods 6-D1.

McDonald, M. G., and Harbaugh, A. W. (1988). *A modular three-dimensional finite-difference ground-water flow model*, Techniques of Water-Resources Investigations of the U.S. Geological Survey, Book 6, Chapter A1.

Meisler, H. (1986). "Northern Atlantic coastal plain regional aquifer-system study." *Regional aquifer-system analysis program of the U.S. Geological Survey, Summary of projects, 1978-1984*, R. J. Sun, ed., U .S. Geological Survey Circular 1002, 168-194.

Moench, A. F., Sauer, V. B., and Jennings, M. E. (1974). "Modification of routed streamflow by channel loss and base flow." *Water Resour. Res.*, 10(5), 963-968.

Moench, A. F., and Barlow, P. M. (2000). "Aquifer response to stream-stage and recharge variations: I. Analytical step-response functions." *J. Hydrol.*, 230, 192-210.

Morel-Seytoux, H. J. (1975). "A combined model of water table and river stage evolution." *Water Resour. Res.*, 11(6), 968-972.

Morel-Seytoux, H. J., and Daly, C. J. (1975). "A discrete kernel generator for stream-aquifer studies." *Water Resour. Res.*, 11(2), 253-260.

Morita, M., and Yen, B. C. (2002). "Modeling of conjunctive two dimensional surface: Three dimensional subsurface flows." *J. Hydraul. Eng.*, 128(2), 184–200.

Morris, E. M. (1979). "The effect of small slope approximation and lower boundary conditions on solutions of the Saint-Venant equations." *J. Hydrol.*, 40, 31-47.

Motha, J. A., and Wigham, J. M. (1995). "Modeling overland flow with seepage." *J. Hydrol.*, 169, 265-280.

Mou, L., Tian, F., Hu, H., and Sivapalan, M. (2008). "Extension of the representative elementary watershed approach for cold regions: Constitutive relationships and an application." *Hydrol. Earth Syst. Sci.*, 12(2), 565-585.

Nathan, R. J., and McMahon, T. A. (1990). "Evaluation of automated techniques for base flow and recession analysis." *Water Resour. Res.*, 26(7), 1465-1473.

Neitsch, S. L., Arnold, J. G., Kiniry, J. R., and Williams, J. R. (2005). *Soil and water assessment tool theoretical documentation*, Version 2005, <http://www.brc.tamus.edu/swat/doc.html> (Jun. 2009).

Neuman, S. P., and Witherspoon, P. A. (1971). "Analyzing nonsteady flow with a free surface using the finite element method." *Water. Resour. Res.*, 7, 611-623.

Nutter, W. L. (1975). *Moisture and energy conditions in a draining soil mass*, Georgia Institute of Technology Environmental Resources Center, ERC 0875, Atlanta, Ga.

Önder, H. (1998). "One-dimensional transient flow in a finite fractured aquifer system." *Hydrol. Sci.*, 43(2), 243-265.

Osman, Y. Z., and Bruen, M. P. (2002). "Modelling stream-aquifer seepage in an alluvial aquifer: An improved loosing-stream package for MODFLOW." *J. Hydrol.*, 264, 69-86.

Panday, S., and Huyakorn, P. S. (2004). "A fully coupled physically-based spatially-distributed model for evaluating surface/subsurface flow." *Adv. Water Resour.*, 27, 361–382.

Paniconi, C., Marrocu, M., Putti, M., and Verbunt, M. (2003). "Newtonian nudging for a Richards equation-based distributed hydrological model." *Adv. Water Resour.*, 26(2), 161–178.

Parlange, J.-Y., Braddock, R. D., and Voss, G. (1985). "Similarity and iterative solution of the Boussinesq Equation: A Comment." *J. Hydrol.*, 80, 391-393.

Parlange, J.-Y., Hogarth, W. L., Parlange, M. B., Haverkamp, R., Barry, D. A., Ross, P. J., and Steenhuis, T. S. (1998). "Approximate analytical solution of the nonlinear diffusion equation for arbitrary boundary condition." *Trans. Porous Media*, 30, 45-55.

Perkins, S. P., and Koussis, A. D. (1996). "Stream-aquifer interaction model with diffusive water routing." *J. Hydraul. Engrng.*, 122(4), 210-218.

Piggott, A. R., Moin, S., and Southam, C. (2005). "A revised approach to the UKIH method for the calculation of baseflow." *Hydrol. Sci. J.*, 50(5), 911-920.

Pinder, G. F., and Sauer, S. P. (1971). "Numerical simulation of flood wave modification due to bank storage effects." *Water Resour. Res.*, 7(1), 63-70.

Polubarinova-Kochina (1962). *Theory of Groundwater Movement*, translated from Russian by J. M. R. De Wiest, Princeton University Press, Princeton, N.J.

Ragan, R. M. (1966). "An experimental investigation of partial area contributions." *Int. Assoc. Scientific Hydrol. Publ.*, 76, 241-249.

Reggiani, P., Sivapalan, M., and Hassanizadeh, S. M. (1998). "A unifying framework of watershed thermodynamics: 1. Balance equations for mass, momentum, energy and entropy and the second law of thermodynamics." *Adv. Water Resour.*, 22(4), 367–398.

Reggiani, P., Sivapalan, M., Hassanizadeh, S. M., and Gray, W. G. (1999). "A unifying framework of watershed thermodynamics: 2. Constitutive relationships." *Adv. Water Resour.*, 23(1), 15–39.

Reggiani, P., Sivapalan, M., and Hassanizadeh, S. M. (2000). "Conservation equations governing hillslope responses." *Water Resour. Res.*, 38(7), 1845–1863.

Reggiani, P., and Rientjes, T. H. M. (2005). "Flux parameterization in the representative elementary watershed approach: Application to a natural basin." *Water Resour. Res.*, 41(4), W04013.

Restrepo, J. I., Montoya, A. M., and Obeysekera, J. (1998). "A wetland simulation module for the MODFLOW ground water model." *Ground Water*, 36(5), 764-770.

Ritzema, H. P. (2006). "Subsurface flow to drains." *Drainage Principles and Applications*, 3rd Ed., H. P. Ritzema, ed., ILRI Publication No. 16, Wageningen, Alterra, 265-271.

Rorabaugh, M. I. (1956). *Ground water in northeastern Louisville, Kentucky, with reference to induced infiltration*, U.S. Geological Survey Water-Supply Paper 1360-B.

Rovey, C. E. K. (1975). "Numerical model of flow in stream aquifer system." *Hydrology Paper No. 4*, Colorado State University, Fort Collins, Colo.

Rushton. K. (2007). "Representation in regional models of saturated river-aquifer interaction for gaining/losing rivers." *J. Hydrol.*, 334, 262-281.

Saquillace, P. J. (1996). "Observed and simulated movement of bank-storage water." *Ground Water*, 34(1), 121-134.

Schroder, M. S., and Rosbjerg, D. (2004). "Groundwater recharge and capillary rise in a clayey catchment: Modulation by topography and the Arctic Oscillation." *Hydrol. Earth Syst Sc.* 8(6), 1090-1102.

Serrano, S. E., and Workman, S. R. (1998). "Modeling transient stream/aquifer interaction with the non-linear Boussinesq equation and its analytical solution." *J. Hydrol.*, 206, 245-255.

Sharp, J. M., Jr. (1977). "Limitations of bank-storage model assumptions." *J. Hydrol.*, 35, 31-47.

Singh, K. P. (1969). "Theoretical baseflow curves." *J. Hydraul. Div., Proc. Amer. Soc. Civil Engineers*, 95(6), 2029-2048.

Singh, V. J. (1992). *Elementary hydrology*, Prentice Hall, N.J.

Smerdon, B. D., Mendoza, C. A., and Devito, K. J. (2007). "Simulations of fully coupled lake-groundwater exchange in a subhumid climate with an integrated hydrologic model." *Water Resour. Res.*, 43, W01416.

Smerdon, B. D., Allen, D. M., Grasby, S. E., and Berg. M. A. (2008). "An approach for predicting groundwater recharge in mountainous watersheds." *J. Hydrol.*, 365(3-4), 156-172.

Smith, R. E., and Woolhiser, D. A. (1971). "Overland flow on an infiltrating surface." *Water Resour. Res.*, 7(4), 899–913.

Soil Conservation Service (1986). *Technical Release 55: Urban Hydrology for Small Watersheds,* U.S. Department of Agriculture.

Sophocleous, M. A., Koussis, A. D., Martin, J. L., and Perkins, S. P. (1995). "Evaluation of simplified stream-aquifer depletion models for water rights administration." *Ground Water*, 33(4), 579-588.

Sophocleous, M. A. (1997). "Managing water resources systems: Why safe yield is not sustainable." *Ground Water*, 35(4), 561

Sophocleous, M. A., Koelliker, J. K., Govindaraju, R. S., Birdie, T., Ramireddygari, S. R., and Perkins, S. P. (1999). "Integrated numerical modeling for basin-wide water management: The case of the Rattlesnake Creek Basin in south-central Kansas." *J. Hydrol.*, 214, 179-196.

Sophocleous, M. A. (2000). "From safe yield to sustainable development of water resources: The Kansas experience." *J. Hydrol.*, 235, 27-43.

Sophocleous, M. A. (2002). "Interactions between groundwater and surface water: The state of the science." *Hydrogeology J.*, 10, 52-67.

Spalding, C. H., and Khaleel, R. (1991). "An evaluation of analytical solutions to estimate drawdowns and stream depletions by wells." *Water Resour. Res.*, 27(4), 597-609.

Stehfest, H. (1970). "Numerical inversion of Laplace transforms." *Commun. ACM*, 13(1), 47-49.

Stillman, J. S., Haws, N. W., Govindaraju, R. S., and Rao, P. S. C. (2006). "A model for transient flow to a subsurface tile drain under macropore-dominated flow conditions." *J. Hydrol.*, 317, 49-62.

Strack, O. D. L., and Haitjema, H. M. (1981a). "Modeling double aquifer flow using a comprehensive potential and distributed singularities: 1. Solution for homogeneous permeability." *Water Resour. Res.*, 17(5), 1535–1549.

Strack, O. D. L., and Haitjema, H. M. (1981b). "Modeling double aquifer flow using a comprehensive potential and distributed singularities: 2. Solution for inhomogeneous permeability." *Water Resour. Res.*, 17(5), 1551–1560.

Strack, O. D. L. (1989). *Groundwater mechanics*, Prentice-Hall, Englewood Cliffs, N.J.

Strack, O. D. L. (2003). "Theory and applications of the analytic element method." *Rev. Geophys.*, 41(2), 1-16.

Swanson, S. K., and Bahr, J. M. (2004). "Analytical and numerical models to explain steady rates of spring flow." *Ground Water*, 42(5), 747-759.

Szilagyi, J., and Parlange, M. B. (1998). "Baseflow separation based on analytical solutions of the Boussinesq equation." *J. Hydrol.*, 204, 251-260.

Szilagyi, J., Parlange, M. B., and Albertson, J. D. (1998). "Recession flow analysis for aquifer parameter determination." *Water Resour. Res.*, 34(7), 1851-1857.

Szilagyi, J. (1999). "On the use of semi-logarithmic plots for baseflow separation." *Ground Water*, 37(5), 660-663.

Szilagyi, J., Harvey, F. E., and Ayers, J. F. (2003). "Regional estimation of base recharge to ground water using water balance and base-flow index." *Ground Water*, 41(4), 504-513.

Szilagyi, J. (2004). "Heuristic continuous base flow separation." *J. Hydrol. Engrng.*, 9(4), 311-318.

Tallaksen, L. M. (1995). "A review of baseflow recession analysis." *J. Hydrol.*, 165, 349-370.

Tayfur, G., Kavvas, M. L., Govindaraju, R. S., and Storm, D. E. (1993). "Applicability of St. Venant equations for two-dimensional overland flows over rough infiltrating surfaces." *J. Hydraul. Eng*, 119, 51-63.

Theis, C. V. (1935). "The relationship between the lowering of the piezometric surface and the rate and duration of discharge from a well using ground-water storage." *Trans. Am. Geophys. Union*, 2, 519-524.

Theis, C. V. (1941). "The effect of a well on the flow of a nearby stream." *Trans. Am. Geophys. Union*, 22(3), 734-738.

Therrien, R., McLaren, R. G., Sudicky, E. A., and Panday, S. M. (2005). *HydroGeoSphere: A three-dimensional numerical model describing fully-integrated subsurface and surface flow and solute transport*, Groundwater Simulations Group, University of Waterloo, Waterloo, Ont., Canada.

Todd, D. K. (1980). *Groundwater hydrology,* 2nd Ed., John Wiley & Sons, New York.

Tolikas P., Damaskinidou, A., and Sidiropoulos, E. (1990). "Series representation of flux for the Boussinesq equation." *Water Resour. Res*., 26(12), 2881-2886.

Tóth, J. (1963). "A theoretical analysis of groundwater flow in small drainage basins." *J. Geophy. Res*., 68, 4795-4812.

Townley, L. R., and Wilson, J. L. (1980). *AQUIFEM: Description of and user's guide for a finite-element aquifer flow model*, Tech. Rep. 252, Mass. Inst. of Technol., Cambridge, Mass.

Tseng, P-H, and Lee, T-C. (1998). "Numerical evaluation of exponential integral: Theis well function approximation." *J. Hydrol*., 38-51.

Upadhyaya, A., and Chauhan, H. S. (2001). "Interaction of stream and sloping aquifer receiving constant recharge." *J. Irrig. Drain. Engrng*., 127(5), 295-301.

Van Genuchten, M. Th. (1980). "A closed-form equation for predicting the hydraulic conductivity of unsaturated soils." *Soil Sci. Soc. Am. J*., 44, 892-898.

Van der Kwaak, J. E. (1999). "Numerical simulation of flow and chemical transport in integrated surface–subsurface hydrologic systems." PhD thesis, Department of Earth Sciences, University of Waterloo, Ont., Canada.

Viessman, W., Jr., and Lewis, G. L. (1996). *Introduction to hydrology*, 4th Ed., Harper Collins College Publishers, New York.

Wahl, K. L., and Wahl, T. L. (1995). "Effects of regional ground-water declines on streamflows in the Oklahoma Panhandle." *Symposium on Water-Use Data for Water Resources Management*, AWRA, Tucson, Arizona, 239-249.

Ward, R. C. (1984). "On the response to precipitation of headwater streams in humid areas." *J. Hydrol*., 74, 171-189.

Werner, P. W., and Noren, D. (1951). "Progressive waves in non-artesian aquifers." *Trans. Am. Geophys. Union*, 32, 238-244.

Weyman, D. R. (1970). "Throughflow on hillslopes and its relation to the stream hydrograph." *Int. Assoc. Scientific Hydrol. Bull*., 15(2), 25-32.

Whiting, P. J., and Pomeranets, M. (1997). "A numerical study of bank storage and its contribution to streamflow." *J. Hydrol*., 202, 121-136.

Winter, T. C. (1976). *Numerical simulation analysis of the interaction of lakes and ground water*, U.S. Geological Survey Professional Paper 1001.

Winter, T. C. (1983). "The interaction of lakes with variably saturated porous media." *Water Resour. Res*., 19(5), 1203-1218.

Wooding, R. A., and Chapman, T. G. (1966). "Groundwater flow over a sloping impermeable layer: 1. Application of the Dupuit-Forchheimer assumption." *J. Geophys. Res*., 71(12), 2895-2902.

Woolhiser, D. A. (1974). "Unsteady free surface flow problems." *Proc. of the Institute on Unsteady Flow in Open Channels*, K. Mahmood and V. Yevjevich, eds., Colorado State University, Fort Collins, Colo., 195-213.

Workman, S. R., Serrano, S. E., and Liberty, K. (1997). "Development and application of an analytical model of stream/aquifer interaction." *J. Hydrol.*, 200, 149-164.

Yeh, W. W.-G. (1970). "Nonsteady flow to a surface reservoir." *J. Hydraul. Div., Proc. Amer. Soc. Civil Engineers*, 96(3), 609-618.

Yoon, J., and Kavvas, M. L. (2000). "Spatial averaging of overland flow equations for the hillslope with rill-interrill configuration." *Proc., ICHE 2000.*

Yoshitani J., Chen, Z. Q., Kavvas, M. L., and Fukami, K.. (2009). "Atmospheric model-based streamflow forecasting at small, mountainous watersheds by a distributed hydrologic model: Application to a watershed in Japan." *J. Hydrol. Eng.*, 14(10), 1107-1118.

Young, R. A. (1970). "Safe yield of aquifers: An economic reformulation." *J. Irrig. Drain. Div. Amer. Soc. Civil Eng.*, 96(IR4), 377-385.

Zehe, E., Lee, H., and Sivapalan, M. (2006). "Dynamical process upscaling for deriving catchment scale state variables and constitutive relations for meso-scale process models." *Hydrol. Earth Syst. Sci.*, 10(6), 981-996.

Zhang G. P., Fenicia, F., Rientjes, T. H. M., Reggiani, P., and Savenije, H. H. G. (2005). "Modeling runoff generation in the Geer river basin with improved model parameterizations to the REW approach." *Phy. Chem. Earth.*, 30(4-5), 285-296.

Zhang, G. P., Savenije, H. H. G., Fenicia, F., and Pfister, L. (2006). "Modelling subsurface storm flow with the representative elementary watershed (REW) approach: Application to the Alzette River Basin." *Hydrol. Earth Syst. Sci.*, 10(6), 937-955.

Zitta, V. L., and Wiggert, J. M. (1971). "Flood routing in channels with bank storage." *Water Resour. Res.*, 7(5), 1341-1345.

Zlotnik, V. A., and Huang, H. (1999). "Effect of shallow penetration and streambed sediments on aquifer response to stream stage fluctuations (analytical model)." *Ground Water*, 37(4), 599-605.

Zlotnik, V. A. (2004). "A concept of maximum stream depletion rate for leaky aquifers in alluvial valleys." *Water Resour. Res.*, 40, W06507, doi: 10.1029/2003WR002932.

Zlotnik, V. A., and Tartakovsky, D. M. (2008). "Stream depletion by groundwater pumping in leaky aquifers." *J. Hydrol. Engrng.*, 13(2), 43-50.

CHAPTER 10

DENSITY DEPENDENT FLOWS, SALTWATER INTRUSION AND MANAGEMENT

Bithin Datta[1] and **Anirban Dhar**[2]

[1]James Cook University, Townsville, QLD 4811, Australia
[2]Indian Institute of Technology Kharagpur, Kharagpur, W.B. 721302, India

10.1 Introduction

Intrusion of saltwater in coastal aquifer hinders its beneficial use as a source of water. The specific aim of coastal aquifer management may involve the design of a spatiotemporal pumping strategy to meet demands, while controlling the intrusion process, also through planned pumping. Many coastal aquifers, i.e., aquifers hydraulically connected with the ocean, can be used as major operational reservoirs in water resources systems. Development of such reservoirs for supply of water is very common in many parts of the world. The exploitation of these reservoirs is restricted in many cases, because of seawater intrusion into the coastal aquifers. Saltwater intrusion in coastal aquifers is global problem. Over the years different studies have been reported worldwide, e.g., Madras, India (Rouve and Stoessinger 1980), the Mediterranean coast of Israel (Shamir et al. 1984), the Waialae aquifer of southern Oahu, Hawaii (Essaid 1986; Emch and Yeh 1998), in southern Oahu, Hawaii (Souza and Voss 1987), the Nile delta aquifer in Egypt (Sherif et al. 1988), in Hallandale, Florida (Andersen et al. 1988), the Yun Lin Basin in southwestern Taiwan (Willis and Finney 1988), the Northern Guam aquifer (Contractor and Srivastava 1990), the Soquel-Aptos Basin, Santa Cruz County, California (Essaid 1990a, 1990b), the Jakarta Basin (Finney et al. 1992), the southwest region of Bangladesh (Nobi and Gupta 1997), Hernando County, Florida (Guvanasen et al. 2000), Island of Texel, The Netherlands (Oude Essnik 2001), the Jahe River Basin, Shandong Province, China (Cheng and Chen 2001), the Lei-Qiong Depression Zone, near the Leizhou Peninsula in southern China (Zhou et al. 2003), Savannah, Georgia (Kentel et al. 2005), the Llobregat delta, south of Barcelona, Spain (Abarca et al. 2006), and the coastal karstic aquifer in Crere, Greece (Karterakis et al. 2007). A comprehensive review of saltwater intrusion modeling, management and monitoring methods can be found in Bear et al. (1999) and Cheng and Ouazar (2004). Some of the works focused on prediction, while others stressed management.

Saltwater intrusion is a phenomenon that occurs due to movement of seawater towards freshwater aquifers and creation of a brackish environment. Heavier saltwater tries to underlie freshwater due to differences in buoyancy. A brackish environment or mixing zone of varying density is created. In this transition zone (diffuse interface) a gradual increase in density of mixed fluid is observed from freshwater to seawater. Figure 10.1a shows the transition zone in an unconfined coastal aquifer. Variation in transition zone thickness is observed depending on the coastal aquifer environment. In presence of thin transition zone, a coastal aquifer can be idealized by considering freshwater and saline water as immiscible fluids. The transition zone can be replaced by a sharp interface (Figure 10.1b). These idealizations can be used to facilitate modeling of the saltwater intrusion phenomenon in coastal aquifers. Although, sharp interface models are popular due to their simplicity, they inherit the possibility of oversimplifying the actual situation. In reality, a transition zone should be considered in coastal aquifer modeling. Incorporation of a transition zone causes the saltwater intrusion model to be highly nonlinear.

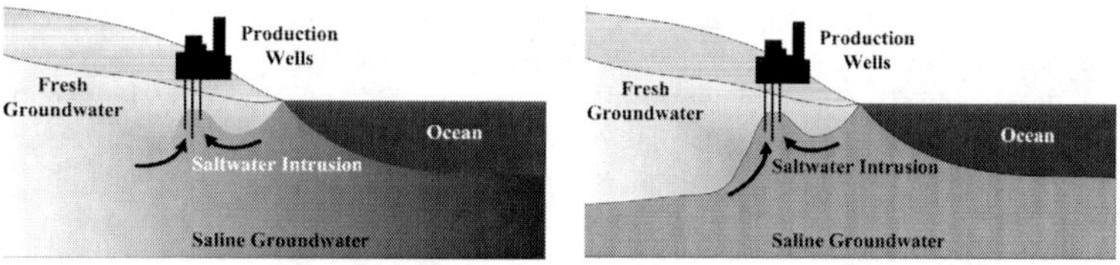

(a) (b)

Figure 10.1: (a) Diffuse interface in an unconfined coastal aquifer, and (b) sharp interface in an unconfined coastal aquifer.

Seawater is comprised of more than seventy elements, with six ions contributing greater than 99% of the total dissolved solids (TDS) concentration. Table 10.1 provides a typical seawater composition and corresponding concentrations (Younos 2005). The TDS of seawater varies largely between 500 mg/l to 50,000 mg/l. Lower range values indicate brackish water while upper range represents seawater. The TDS level in brackish water system fluctuates spatiotemporally. Variations may be due to tide, freshwater inflow (rain or river flows), and evaporation rate (Younos 2005). Although TDS concentration of 500 mg/l is considered as Secondary Maximum Contaminant Level (SMCL) by U.S. EPA (1992), a much lower value is desirable for drinking purpose.

Table 10.1: Composition of seawater (Younos 2005).

Element	Mass Fraction (%)	Concentration (mg/l)
Chloride (Cl)	55.04	19,400
Sulfate (SO_4)	7.68	940
Calcium (Ca)	1.16	411
Sodium (Na)	30.61	10,800
Magnesium (Mg)	3.69	1290
Potassium (K)	1.10	392

10.1.1 Density Dependent Flow

Saltwater intrusion simulation modeling approaches available for dealing with the freshwater and saltwater in coastal aquifer are: (a) sharp interface approach, and (b) diffuse interface approach. Saltwater intrusion is generally modeled considering two immiscible fluids with a sharp interface, or a single fluid approach with density dependence. Density-dependent models are highly nonlinear due to the coupled nature of the flow and transport equations. Variation of fluid density is the prominent feature of the density-dependent flow. Density dependence necessarily changes the flow pattern compared to common groundwater problems where variation of fluid density is not prominent. A historical perspective of saltwater-freshwater interaction is presented by Reilly and Goodman (1985). Segol (1994), Holzbecher (1998), and Bear et al. (1999) provide insight to the subject. Bobba (1993) provided a review of mathematical models for saltwater intrusion in coastal aquifer. Many approximate analytical and numerical models are available for simulating the physical processes in coastal aquifers.

Analytical Models: Most analytical models are based on Ghyben-Herzberg approach, which uses sharp interface assumption along with simple hydrostatic conditions to describe the saltwater intrusion phenomenon in a homogeneous, unconfined coastal aquifer. This approach underestimates depth to the saltwater interface and fails when there is freshwater flow towards sea (Freeze and Cherry 1979). The sharp interface assumption facilitates the derivation of analytical solutions. Different studies have used the sharp interface approach, e.g., Henry (1959), Bear and Dagan (1964), Hantush (1968), Schmorak and Mercado (1969), Strack (1976), van Dam and Sikkema (1982), Sikkema and van Dam (1982), Huppert and Woods (1995), Dagan and Zeitoun (1998), and Kacimov and Obnosov (2001). Recently, Dentz et al. (2006) derived analytical solutions for Henry's problem using a coupled density approach. A detailed review of analytical methods can be found in Cheng and Ouazar (1999). However, analytical solutions are available for only simplified field conditions.

Numerical Models: Numerical models are effective in representing physical processes irrespective of their complexity. A number of algorithms are available for obtaining numerical solutions of the governing partial differential equations. Usually, stability criteria dictate the choice of spatial and temporal discretization. The approaches and challenges for modeling density-dependent flow are reviewed by Simmons et al. (2001), Diersch and Koditz (2002), Post (2005), and Simmons (2005). Different numerical techniques have been applied for simulating the saltwater intrusion phenomenon, including the finite difference method (Mercer et al. 1980; Polo and Ramis 1983; Ledoux et al. 1990; Essaid 1990b; Das and Datta 2000), the finite element method (Lee and Cheng 1974; Segol et al. 1975; Voss 1984; Huyakorn et al. 1987; Contractor and Srivastava 1990; Huyakorn et al. 1996; Lin et al. 1997; Cheng et al. 1998; Guvanasen et al. 2000), the boundary element method (Volker and Rushton 1982; Taigbenu et al. 1984; Cabral and Wrobel 1993; Naji et al. 1998), the finite volume method (Liu et al. 2001), the mixed finite element-finite volume approach (Mazzia and Putti 2006), Picard and Newton linearization approaches (Putti and Paniconi 1995), and the network simulation method (Meca et al. 2007).

Many standard models are available for simulating the saltwater intrusion process. A review of such codes can be found in Sorek and Pinder (1999). Commonly used codes include SUTRA (Voss 1984; Voss and Provost 2002), FEMWATER (Lin et al. 1997), FEFLOW (Diersch 2002), HST3D (Kipp 1986, 1997), MOCDENS3D (Oude Essnik 1998), SWIFT (HSI GeoTrans 2000), SEAWAT (Guo and Langevin 2002), MODHMS (HydroGeoLogic Inc 2002), and SWI (Bakker 2003; Bakker and Schaars 2005).

10.1.2 Saltwater Intrusion and Management

Management of saltwater intrusion in coastal aquifers is a critical issue of modern times. Simulation models can provide necessary future management guidelines only if repetitively used in an exhaustive manner. Simulation models can only answer the question: what if a specific management strategy is implemented. Therefore, an effective mechanism is needed that can combine the optimization and physical processes simultaneously to efficiently evolve management strategies. Linking of the process simulation model with an optimal decision model can address this requirement.

***Objectives of Management Model*:** Identification of a proper objective function is a challenging task in management modeling. Different methods can be utilized for controlling saltwater intrusion in coastal aquifer. In some of the methods, surrogate objectives are used to quantify the management goals.

Different management objectives reported in the literature include maximizing the total pumping rate (Shamir et al. 1984; Hallaji and Yazicigil 1996; Cheng et al. 2000; Bhattacharjya 2003; Mantoglou 2003; Abarca et al. 2006; Karterakis et al. 2007), minimizing the pumped water salinity (Das and Datta 1999a, 1999b), minimizing the volume of saltwater intrusion into the aquifer (Finney et al. 1992; Emch and Yeh 1998), minimizing the drawdown (Hallaji and Yazicigil 1996), minimizing the deviation from the target concentration, and pumping and recharge rate (Abarca et al. 2006), and minimizing the distance of the stagnation point (Strack 1976) from coast (Park and Aral 2004). Pumping or injection costs are also included in different studies, e.g., Emch and Yeh (1998), Gordon et al. (2001), Reichard and Johnson (2005), and Ferreia da Silva and Haie (2007). However, only a few works considered multiple objectives of operation at a time (Shamir et al. 1984; Emch and Yeh 1998; Das and Datta 1999a; Park and Aral 2004).

Surrogate objectives are sometimes expressed as constraints. Constraints on maximum and minimum pumping rates (Hallaji and Yazicigil 1996; Emch and Yeh 1998; Das and Datta 1999a; Cheng et al. 2000; Zhou et al. 2003; Abarca et al. 2006), toe location (Cheng et al. 2000; Mantoglou 2003; Mantoglou et al. 2004; Park and Aral 2004; Ferreia da Silva and Haie 2007), interface elevation at a particular location

(Rao et al. 2003, 2004a), groundwater heads (Hallaji and Yazicigil 1996; Emch and Yeh 1998; Zhou et al. 2003; Reichard and Johnson 2005; Abarca et al. 2006), flow potential (Mantoglou 2003; Mantoglou et al. 2004), and salt concentration of the pumped water (Gordon et al. 2000; Das and Datta 1999a, 1999b) are more common.

***Incorporation of Physical Processes in Management Model*:** Accuracy of management strategy depends on the level of approximation and the way in which they are incorporated. In broader sense, physical processes can be incorporated using (a) an analytical model, (b) a numerical simulation model, (c) an embedding approach, or (d) a meta-model-based approach.

The way physical processes are included vary, e.g., analytical solutions (Cheng et al. 2000; Mantoglou 2003; Cheng et al. 2004; Park and Aral 2004) or numerical simulation using the sharp interface approach (Willis and Finney 1988; Finney et al. 1992; Emch and Yeh 1998; Rao et al. 2004a; Mantoglou et al. 2004). Gordon et al. (2000) used a chloride transport model. Only a few studies have used density-dependent flow and transport within the management framework, e.g., Das and Datta (1999a, 1999b) and Qahman et al. (2005). Das and Datta (1999a, 1999b) used the embedding technique for this purpose. Computational complexity increases when density variations are taken into account, which results in a highly nonlinear and nonconvex problem (Finney et al. 1992).

A good number of work used meta-models (approximation-based models) to reduce the computational complexity of physical process-based models. These models are useful approximators of the highly nonlinear nature of the saltwater intrusion process. The most primitive version of a meta-model is the "Response Matrix," which is generally linear in nature. This linear nature has been exploited also while formulating optimization models. Response matrix-based studies are reported in Hallaji and Yazicigil (1996), Zhou et al. (2003), Reichard, and Johnson (2005), Abarca et al. (2006), and Karterakis et al. (2007). Sometimes this approach yields incorrect results due to oversimplification of expressions for nonlinear processes. In recent years, a few studies concentrated on the use of complex meta-models, like Artificial Neural Network models, e.g., Bhattacharjya (2003), Rao et al. (2003, 2004b), and Bhattacharjya and Datta (2005).

***Different Optimization Algorithms in Management Models*:** In solving management problems, the choice of algorithm greatly influences the solution quality. Generally, complexity (linearity, nonlinearity and discontinuity) and nature of decision variables (discrete, continuous and integer) dictate the choice. Linearization of physical processes by response matrix approach is a widely used method, which facilitates the use of linear programming, e.g., Hallaji and Yazicigil (1996), Mantoglou (2003), Reichard and Johnson (2005), and Karterakis et al. (2007).

The saltwater intrusion problem is a necessarily nonlinear problem due to nonlinear physical processes involved. Applications of nonlinear algorithms include: sequential quadratic programming (Mantoglou 2003; Mantoglou et al. 2004) and the geostatistical inversion technique (Abarca et al. 2006). A large number of studies have used different versions of MINOS (Murtagh and Saunders 1993), which utilizes the reduced-gradient algorithm in conjunction with quasi-Newton algorithm, e.g., Willis and Finney (1988), Finney et al. (1992), Emch and Yeh (1998), and Das and Datta (1999a, 1999b). However, MINOS is "designed to find solutions that are locally optimal" (Murtagh and Saunders 1993). Non-differentiable formulations are the most difficult ones to solve using classical algorithms. Gordon et al. (2001) used Bundle-Trust nonsmooth optimization procedure to optimize discontinuous formulation.

Application of nontraditional algorithms (evolutionary algorithm, genetic algorithm, simulated annealing) has gained popularity in recent years. These algorithms can easily handle highly nonlinear, even non-smooth problems. In addition these algorithms are well suited for linking optimization algorithms with numerical simulation models. Over the years different genetic algorithms have been used in saltwater intrusion management, e.g., structured messy genetic algorithm (Cheng et al. 2000), real coded genetic algorithm (Bhattacharjya and Datta 2005), simple genetic algorithm (Cheng et al. 2004; Katsifarakis and Petala 2006, Qahman et al. 2005), and real coded progressive genetic algorithm (Park and Aral 2004). Application of other algorithms include the evolutionary algorithm (Mantoglou et al. 2004), simulated annealing (Rao et al. 2003, 2004a, 2004b), and differential evolution (Karterakis et al. 2007).

10.2 Density-Dependent Governing Equations

To simulate the density-dependent saltwater intrusion processes in coastal aquifers, the governing equations that represent the three-dimensional, advective-dispersive flow and transport processes are to be solved simultaneously. The three-dimensional advective-dispersive flow equation for a heterogeneous, anisotropic and fully saturated aquifer may be written as proposed by Huyakorn et al. (1987):

$$\frac{\partial}{\partial x_i}\left[K_{ij}\left(\frac{\partial h}{\partial x_j}+\eta c e_j\right)\right]=S_S\frac{\partial h}{\partial t}+\phi\eta\frac{\partial c}{\partial t}\frac{\rho}{\rho_o}\left(q_r-q_p\right) \qquad (10.1)$$

The reference hydraulic head is defined as

$$h=\frac{p}{\rho_0 g}+Y \qquad (10.2)$$

The density coupling coefficient is defined as

$$\eta=\frac{\varepsilon}{c_S} \qquad (10.3)$$

The density difference ratio (ε) is expressed as

$$\varepsilon = \left(\frac{\rho_S - \rho_o}{\rho_o} \right) \tag{10.4}$$

The actual hydraulic conductivity is defined as

$$K_{ij} = \frac{k_{ij} \rho g}{\mu} \tag{10.5}$$

The density of the mixed fluid is defined as

$$\rho = \rho_o \left(1 + \varepsilon \frac{c}{c_S} \right) \tag{10.6}$$

The advective-dispersive equation can be written as (Huyakorn et al. 1987)

$$\frac{\partial}{\partial x_i} \left(D_{ij} \frac{\partial c}{\partial x_j} \right) - V_i \frac{\partial c}{\partial x_i} = \phi \frac{\partial c}{\partial t} + q_r c \tag{10.7}$$

The Darcy velocity vector is expressed as:

$$V_i = K_{ij}^o \left[\frac{\partial h}{\partial x_j} + \eta c e_j \right] \tag{10.8}$$

The density of water primarily depends on concentration and temperature. It has been observed that chlorinity contributes a major portion of the seawater (Table 10.1). Figure 10.2 shows the variation of density with chlorinity and temperature. Coupling between flow and transport processes are described in terms of Equation (10.6).

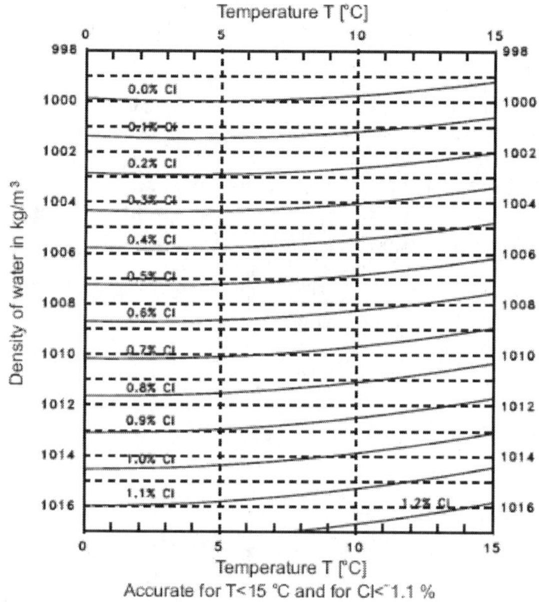

Figure 10.2: Density of water as a function of the chlorinity and temperature (Oude Essnik 2001).

10.2.1 Different Solution Approaches

The flow and transport equations described by Equations (10.1)-(10.8) can only be solved numerically. There are two approaches in numerical solution, viz: (a) forward modeling using discretized equations, and (b) optimization-based embedding technique. Different forward models for simulating the density-dependent flow and transport are SUTRA (Voss 1984; Voss and Provost 2002), FEMWATER (Lin et al. 1997), HST3D (Kipp 1986, 1997), SEAWAT (Guo and Langevin 2002), and FEFLOW (Diersch 2002).

Optimization-based embedding approaches are relatively rare in the literature. Das and Datta (2000) proposed an approach for solving the discretised finite difference form of the governing flow and transport equations for the transient saltwater intrusion process using nonlinear optimization. The resulting model serves as a simulation model utilizing classical nonlinear optimization technique. Equations (10.1)-(10.8) can be approximated by finite difference methods to form a set of discretized non-linear equations. They are non-linear due to the density coupling as represented in Equation (10.6). The discretized set of equations can be expressed in a functional form as

$$f_I(u_J) = 0 \quad , \forall I \in \{1, 2, \ldots, N_t\}, J \in \{1, 2, \ldots, 2 \times N_t\} \tag{10.9}$$

where u_j is a vector consisting of values of h, c, q_r, and q_p in the three-dimensional space and time domain of the problem. N_t is the total number of equality constraints. This system of equations can be solved using an appropriate numerical method described in Das (1995) and Das and Datta (2000). In the optimization-based embedded model, the discretized version of flow and simulation equations given by Equation (10.9) constitutes the constraints of optimization; the objective is to minimize the total error ε_I given by

$$Minimize: \sum_{I=1}^{N_t} |\tau_I| \tag{10.10}$$

subject to

$$f_I(u_J) = \tau_I \quad , \forall I \in \{1, 2, \ldots, N_t\}, J \in \{1, 2, \ldots, 2 \times N_t\} \tag{10.11}$$

Interestingly, solution of the above mentioned optimization problem provides the spatiotemporal distribution of head and concentration values.

10.2.2 Verification of Numerical Codes

Density-dependent numerical codes need to be verified against standard solutions such as the Henry problem (Henry 1964), the Elder problem (Elder 1967), the HYDROCOIN salt dome problem (Swedish Nuclear Power Inspectorate 1986), and the rotating immiscible fluids problem (Bakker et al. 2003). Das and Datta (2000) used Henry's seawater intrusion problem (Henry 1964) to evaluate the performance of the their non-linear, optimization-based simulation methodology. The two dimensional cases in the vertical plane were utilized in the evaluation.

Henry's problem is used as a standard problem for the verification of steady-state, density-dependent flow and transport models of seawater intrusion in coastal aquifers. In this problem, freshwater discharges to the vertical open sea boundary over a diffuse wedge of seawater that has intruded the aquifer. Several investigators have used different representation of the problem with different parameter values. A number of numerical models based on significantly different methods of solution give nearly identical results for Henry's problem.

Henry's seawater intrusion problem in a coastal aquifer is described schematically in Figure 10.3. The representation of the Henry problem, the parameter values and other data specified in the study are identical to the used by Huyakorn et al. (1987). The problem specifications are chosen so that the cases analyzed correspond to numerically solved by other researchers. The hypothetical two-dimensional confined aquifer is 100 m thick and 200 m long. Freshwater enters the aquifer through the inland face with a uniform velocity v_n and leaves the aquifer through the ocean face.

The heavier seawater enters the aquifer through the bottom portion of the aquifer and the lighter mixed fluid leaves the aquifer through a narrow outlet portion near the top end of the ocean face. The extent of the outlet portion was specified as identified by Huyakorn et al. (1987).

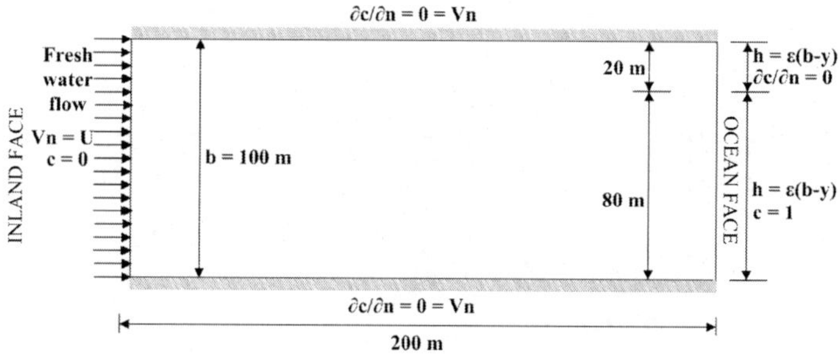

Figure 10.3: Schematic details of Henry problem in a coastal aquifer (Das 1995).

The results pertaining to a steady-state constant dispersion case is described in the following. The parameter values used for the steady state constant dispersion case are:

$$\alpha_L = \alpha_T = 0$$
$$d_o = 6.6\times10^{-2}\ \text{m}^2/\text{d}$$
$$U = 6.6\times10^{-3}\ \text{m/d}$$
$$\varepsilon = 0.025$$
$$\phi = 0.35$$
$$K^o_{xx} - K^o_{yy} - 1\ \text{m/d}$$
$$K^0_{zz} = 0$$
$$\Delta x = \Delta y = \Delta z = 10\ \text{m}$$

where α_L and α_T are the longitudinal dispersivities, respectively; d_o is the molecular diffusion coefficient; U is the freshwater influx through the inland face; K^o_{xx}, K^o_{yy} and K^0_{zz} are the freshwater hydraulic conductivities corresponding to x, y and z directions, respectively; and Δx, Δy, and Δz are the elemental step lengths. The flow and transport boundary conditions used were identical to that of Huyakorn et al. (1987). The boundary conditions at the inland face allow freshwater influx. The coastal boundary conditions allows convective mass transport out of the system over the top outlet portion 80 m $\leq y \leq$ 100 m. For this region, the normal concentration gradient is set equal to zero. Figure 10.4 compares the 0.5 isochlors for steady-state conditions and constant dispersion case computed using the embedded optimization methodology (Das 1995; Das and Datta 2000) with those obtained by other researchers (Huyakorn et al. 1987; Lee and Cheng 1974; Henry 1964; and the sharp interface solution of Henry 1959) as presented in Huyakorn et al. (1987). It can be seen that results by Das and Datta (2000) and Huyakorn et al. (1987) are in good agreement. The comparison is reasonably good with those of Lee and Cheng (1974) and Henry (1964).

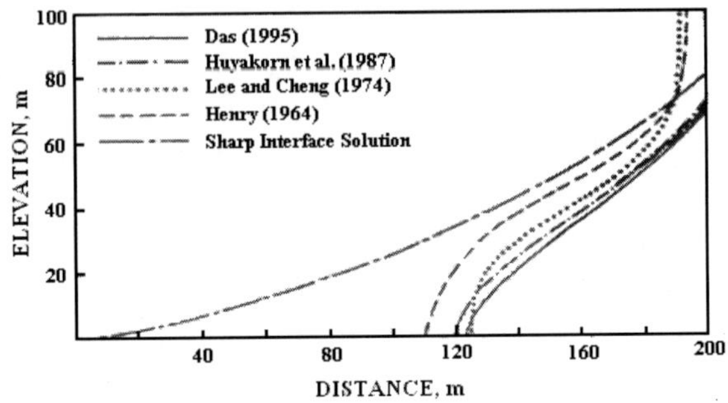

Figure 10.4: Comparison of 0.5 isochlors for steady-state conditions and constant dispersion (adapted from Das 1995).

10.3 Saltwater Intrusion and Management

Saltwater intrusion can be managed in combination with one or more than one of the following alternatives (Maimone 2002):

- *Demand management* - reduction in pumping by lowering the demand of water.
- *Non-potable water reuse* - reduction in demand by reuse of treated water.
- *Injection barrier* - creating hydraulic barriers by injecting water.
- *Extraction barrier* - extraction of saline water to protect wells further inland.
- *Tapping alternative aquifers* - transferring pumping stress to aquifers below or above the impacted one.
- *Well relocation* - relocation of wells to an area where fresh water head is higher or less prone to saltwater intrusion.
- *Plugging abandoned wells* - reduces risk of intermixing of saltwater and fresh aquifer.
- *Modified pumping rates* - space-time scheduling of pumping rates.
- *Pumping caps* - restricted pumping by imposing upper limits.
- *Physical barriers* - slurry walls or sheets piles to prevent intrusion at a small scale.
- *Scavenger wells* - specially designed wells that extract fresh water under hydrodynamic stabilization of saltwater-fresh mixing zone.
- *Controlled intrusion* - in case of a depleted source with adequate supply of substitute resource.
- *Conjunctive use* - use of surface water to relieve extra load on coastal aquifer.

- *Aquifer storage and recovery* - storage of potable surface water in saline aquifer for future use.
- *Intrusion with treatment* - desalination plant can be used to convert saline water to fresh water. However concentrate management is major issue in desalination process.
- *Suring pit* - passing freshwater with a certain velocity through pit.

Ideally, integrated management of saltwater intrusion in coastal aquifers should be able to incorporate any of these alternatives. The basic steps necessary for integrated management of saltwater intrusion is presented in Figure 10.5. Most common management strategies for coastal aquifers require careful planning of withdrawal strategies for control and remediation of saltwater intrusion in coastal aquifers. Such strategies can be evolved only if, the physical process involved in the coastal aquifers are simulated reliably and studied. Numerical simulation is the most direct way of identifying possible solutions of saltwater intrusion management problem, but sequential numerical solutions do not generate sensitivity and/or tradeoff information. Moreover, simulation methodology can examine only a finite number of alternatives. Also, the obtained solutions are only optimal with respect to alternatives considered during the analysis. In linked simulation-optimization, the optimization model performs the search process by iterating between the optimization and the simulation model. The simulation model provides the necessary information to the optimization model at every stage of iteration in order to reach the optimal solution. The advantage of this approach is that any type of simulation model can be incorporated in the search for optimal management strategies. The principal advantage of the linked simulation optimization is its applicability to highly nonlinear systems.

Management models are of three categories: (a) descriptive management model, decision is taken based on the simulation results; (b) prescriptive management model, optimization-based decision making; and (c) descriptive-prescriptive management model, combined simulation optimization framework. Ideally, the third category of models is important from actual management point of view.

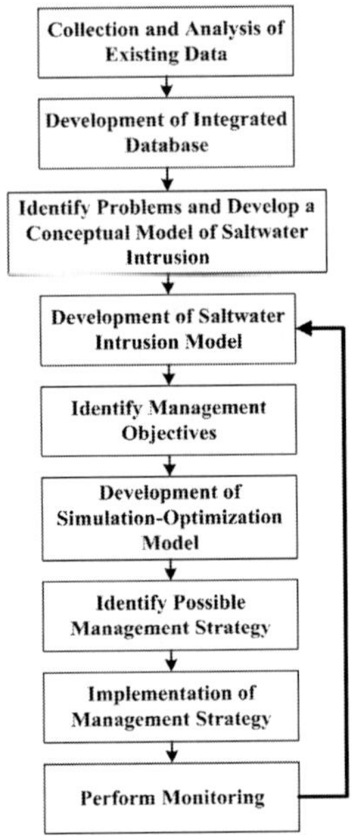

Figure 10.5: Integrated saltwater intrusion management (modified from Maimone 2002).

10.3.1 Descriptive-Prescriptive Management Model

This model basically combines simulation model with optimization model to give meaningful strategies for management. In this method, the search process of the optimization model is performed by iterating between the optimization model and the simulation model. The simulation model provides the necessary information to the optimization model at every stage of iteration, in order to reach the optimal solution. This approach is considerably better than the embedded technique as the optimization model needs to handle only few variables and explicit constraints compared to many variables required in the embedded technique. This methodology has the ability to easily incorporate complex simulation models as externally-linked simulation models. The main disadvantage of this approach is that repetitive iterations between the optimization model and the simulation model are required to arrive at optimal solutions. Therefore, the size of the optimization model in terms of number of decision variables would be much smaller compared to the one using the embedding technique approach. Multiple-objective management models with conflicting objectives and implemented in the descriptive-predictive framework are described below.

The first multiple-objective management model considers two conflicting objectives: (i) maximization of total pumping for beneficial use from wells at interior locations, and (ii) minimization of pumping from wells located along and in the vicinity of the ocean boundary. The second objective is applicable when pumping along the ocean boundary and in the vicinity of the ocean face is used to alter the hydraulic gradients closed to this boundary to control saltwater intrusion in land. In order to increase the extraction of water from the interior wells while maintaining the salinity within permissible limits, it may be required to increase pumping from the ocean face wells. Therefore, these two conflicting objectives constitute a multiple-objective management problem. This multiple-objective management model can be represented as:

$$Maximize: \sum_{t=1}^{T}\sum_{n=1}^{NI}\sum_{y=1}^{Y_l} Q_{ny}^{t} \tag{10.12}$$

$$Minimize: \sum_{t=1}^{T}\sum_{n=1}^{NB}\sum_{y=1}^{Y_l} Q_{ny}^{t} \tag{10.13}$$

subject to

$$\boldsymbol{C} = f(\boldsymbol{Q}) \tag{10.14}$$

$$C_{ny}^{t} \le C(u)_{ny}^{t} \quad ,\forall n \in \{1, \ldots, NI\},\ y \in \{1, \ldots, Y_t\},\ t \in \{1, \ldots, T\} \tag{10.15}$$

$$C_{ny}^{t} \ge 0 \quad ,\forall n \in \{1, \ldots, NI\},\ y \in \{1, \ldots, Y_t\},\ t \in \{1, \ldots, T\} \tag{10.16}$$

$$Q(l)_{n}^{t} \le \sum_{y=1}^{Y_l} Q_{ny}^{t} \quad ,\forall n \in \{1, \ldots, NI\},\ t \in \{1, \ldots, T\} \tag{10.17}$$

$$\sum_{y=1}^{Y_l} Q_{ny}^{t} \le Q(u)_{n}^{t} \quad ,\forall n \in \{1, \ldots, NI\},\ t \in \{1, \ldots, T\} \tag{10.18}$$

Here, NI is the total number of interior wells used for withdrawing water for beneficial use; NB is the total number of barrier wells in the vicinity of the ocean face used to pumped water for hydraulic extraction barrier; NT is the total number of wells in the aquifer, which is equal to $NI + NB$; Q_{ny}^{t} is the pumping from the nth well from layer y at time step t; T is the total number of time steps; Y_l is the total number of vertical layers; $\boldsymbol{C}$ represents the vector of simulated concentrations as obtained from solution of the simulation model; $\boldsymbol{Q}$ is a vector of spatial and temporal pumping rates from specified potential pumping locations; C_{ny}^{t} is the salt concentration of the pumped water at n^{th} well at vertical layer y in time step t; $C(u)_{ny}^{t}$ is the upper limit of salt concentration on the pumped water at n^{th} well at vertical layer y in time step t; $Q(u)_{n}^{t}$ is the upper limit on the total withdrawal rate from the nth well in time step t; and $Q(l)_{n}^{t}$ is the lower limit on the total withdrawal rate from the nth well in time step t.

The second multiple-objective management model considers the two conflicting objectives: (i) maximization of total pumping for beneficial use from wells at interior locations, and (ii) minimization of maximum salt concentration of the pumped water from the interior wells. Again, increase in pumping from the interior wells will result in increased flow from the ocean towards the aquifer, and therefore, would result in increase in the salinity in the aquifer. Therefore, these two objectives are conflicting in nature, and these constitute a multiple-objective management model. The other imposed constraints are same as in the previous two objectives model. This management model can be represented as:

$$Maximize: \sum_{t=1}^{T}\sum_{n=1}^{NI}\sum_{y=1}^{Y_l} Q_{ny}^{t} \tag{10.19}$$

$$Minimize: \; C^{*} = \max\left(C_{ny}^{t}\right), \forall n \in \{1, \ldots, NI\},\; y \in \{1, \ldots, Y_t\},\; t \in \{1, \ldots, T\} \tag{10.20}$$

subject to Equations (10.14)-(10.18).

The third multiple-objective management model considers three conflicting objectives. These objectives are: (i) maximization of total pumping for beneficial use from wells at interior locations, (ii) minimization of pumping from wells located along and in the vicinity of the ocean boundary, and (iii) minimization of maximum salt concentration of the pumped water from the interior wells. Here all the objectives are conflicting in nature. Therefore, these constitute a multiple-objective management model with three objectives. This management model can be represented as:

$$Maximize: \sum_{t=1}^{T}\sum_{n=1}^{NI}\sum_{y=1}^{Y_l} Q_{ny}^{t} \tag{10.21}$$

$$Minimize: \sum_{t=1}^{T}\sum_{n=1}^{NB}\sum_{y=1}^{Y_l} Q_{ny}^{t} \tag{10.22}$$

$$Minimize: \; C^{*} = \max\left(C_{ny}^{t}\right), \forall n \in \{1, \ldots, NI\},\; y \in \{1, \ldots, Y_t\},\; t \in \{1, \ldots, T\} \tag{10.23}$$

subject to Equations (10.14)-(10.18).

The simulation model represented by Equation (10.14) may be incorporated within the management model using the embedded technique, or by using the response matrix approach. Recently, Dhar and Datta (2009a) have shown the linking of the physically-based simulation model with the multiple-objective optimization algorithm.

***Performance Evaluation*:** Performance of the developed management models have been tested for an illustrative system. Computational efficiency of the simulation-optimization methodology can be enhanced by linking a sufficiently accurate approximate simulation model with the optimization model. Therefore, an approximate simulation model, a trained ANN, is linked as an external module with the optimization model. The purpose of the external simulation model is to simulate flow and transport processes in coastal aquifers. The optimization model uses simulation results from the external simulation model to obtain optimal management strategies (Figure 10.6).

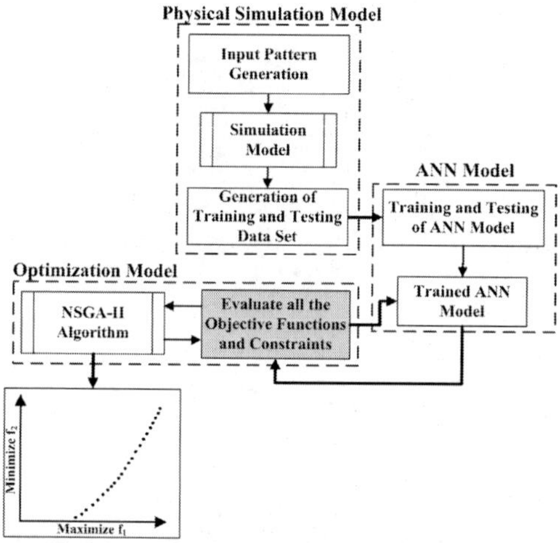

Figure 10.6: Linked simulation-optimization methodology.

The performance of the developed model is evaluated by applying the model to an illustrative study area as shown in the Figure 10.7. The area of the aquifer is 2.52 km^2 (1.8 km in length by 1.4 km in width), and the thickness of the aquifer is 100 m. The unconfined aquifer system is subjected to saltwater intrusion along the coastal side of the study area. The right hand face of the aquifer is the ocean face, which allows saltwater to enter into the aquifer through the bottom of the aquifer, and also allows the exit of the mixed water from the top of the aquifer. It is assumed that mixed water can exit the aquifer through the top 20% of the aquifer at the ocean face. The left hand side of the aquifers is inland face, which allows fresh water to enter in to the aquifer. The top of the aquifer is considered as phreatic. Uniformly distributed vertical discharges occur though the phreatic surface. The other three faces, the front, back and the bottom of the aquifer, are considered as impermeable. The three-dimensional unconfined hypothetical aquifer is assumed to be homogeneous and anisotropic with respect to freshwater hydraulic conductivity, molecular diffusion, and longitudinal and transverse dispersivity. Aquifer parameters values are specified in Table 10.2.

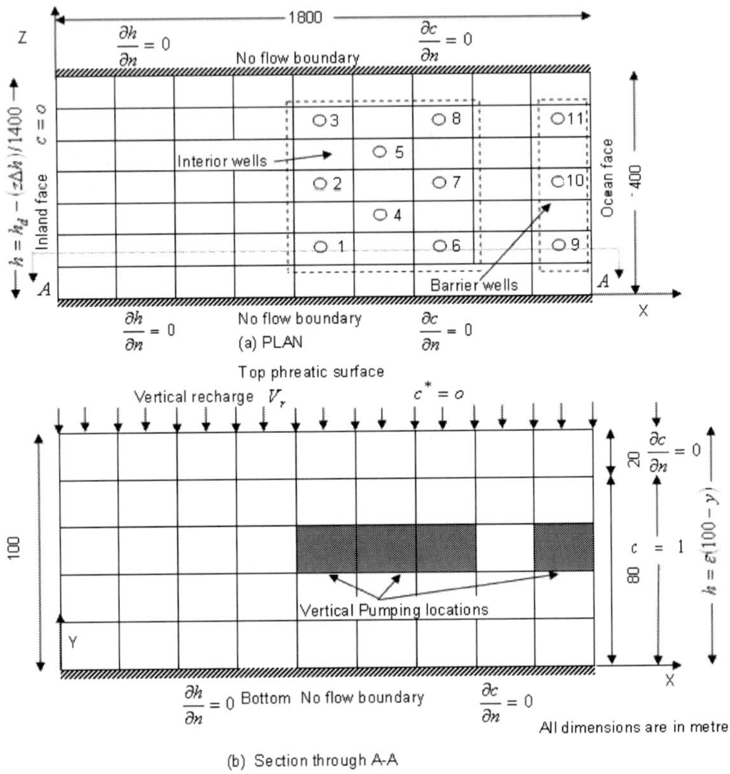

Figure 10.7: Illustrative study area.

The boundary condition in the aquifer (Figure 10.7) is considered as time invariant. The flow boundary condition on the ocean face is considered as hydrostatic in vertical direction. It is assumed to be constant throughout the ocean face. On the inland face, the reference hydraulic head is varying linearly along the length of the inland face. In this case, the value of h_f and Δh are assigned 1.66 m and 0.08 m, respectively. The boundary condition for the phreatic surface is specified as follows (Galeati et al. 1992):

$$K_{yy}\left(\frac{\partial h}{\partial y} + \varepsilon c\right) = v_r - S_y \frac{\partial h}{\partial t} \tag{10.24}$$

where K_{yy} is the actual hydraulic conductivity in y direction; S_y is the specific yield; v_r is the vertical recharge; c is the salt concentration; and h is the reference hydraulic head. The three other faces of the aquifer are impermeable. A zero-flux boundary condition is specified for these faces.

Table 10.2: Values of aquifer parameters for the study area.

Parameter	Values
Hydraulic Conductivity in x direction, K_{xx}^{o} (m/day)	25
Hydraulic Conductivity in y direction, K_{yy}^{o} (m/day)	0.25
Hydraulic Conductivity in z direction, K_{zz}^{o} (m/day)	25
Longitudinal dispersivity, α_L (m)	80
Lateral dispersivity, α_T (m)	25
Molecular diffusion, d_o (m^2/day)	0.69
Density difference ratio, ε	0.025
Soil porosity, ϕ	0.28
Vertical recharge rate, v_r (m/year)	0.20

The advective mixed outflux can exit through the top 20% portion of the aquifer. In this portion, the concentration gradient normal to the ocean face is equal to zero. In the rest 80% of the ocean face, the solute concentration is equal to one, as it allows influx of saltwater into the aquifer. Freshwater enters the aquifer through the inland face of the aquifer. Therefore, the solute concentration is equal to zero on the inland face. It is considered that zero concentration mass influx occurs through the top phreatic surface, and the advective component of the solute mass influx is equal to zero. Therefore, concentration gradient normal to the phreatic surface becomes zero. The other three faces, front, back, and bottom of the aquifer are impermeable. Hence the solute concentration gradient normal to these aquifer faces is set equal to zero. Steady state reference hydraulic head and concentration are assumed as the initial condition for transient flow and transport.

Figure 10.7 shows the locations of the pumping wells. The water is pumped from the middle layer of the aquifer. The pumping from the aquifer is considered as transient. A time step of 180 days is considered in the case of pumping. The upper limit of pumping rate is specified as 18,000 m^3/day and the lower limit is specified as 0 m^3/day. The pumping rate is considered constant for each time step of six months for a particular well. The pumping pattern is generated randomly over a period of three years, using a uniform distribution. The concentration values used are the normalized saltwater concentration with a range of (0,1), with a value of 1 corresponding to a concentration of 35,000 mg/l.

The training and testing patterns for this study area were generated using the embedded optimization-based numerical model developed by Das and Datta (2000). There are eight potential pumping locations to withdraw water for beneficial uses. Another three pumping locations are also considered near the coastal face. These pumping wells are used for controlling saltwater intrusion into the aquifer. Therefore, the total number of pumping wells for this study area is 11. The simulation is done over a period of 1.5 years with a time step of 0.5 years. The total number of inputs to the ANN model is 33. The input pumping patterns are randomly generated using a uniform distribution function. The pumping patterns are then fed to the numerical simulation model to generate spatial and temporal distribution of concentrations over the aquifer. The salt concentration observation locations are the same as the pumping locations used to withdraw water for beneficial use. Therefore, the outputs from the ANN model is the salt concentrations of the pumped water at each time step. The number of output from the ANN model is, therefore, equal to 24.

The training process involves finding of an optimal set of weights for the synaptic connections between the artificial neurons of the network from sufficient sets of input patterns. It has been observed that the testing error decreases monotonically to a minimum but then starts to increase, even as the training error continues to decrease (Burian et al. 2001). When noisy data are used to train the ANN, the ANN would initially learn the actual patterns; therefore, the testing error decreases initially along with the training error. After learning the actual patterns, the ANN may try to learn the noise also; thus, the testing error starts increasing even if the training error continues to decrease. It has been reported that partially-trained neural networks better approximate the unknown function (Hassoun 1999). In the present study, the testing error was carefully monitored during the training phase, and the training was stopped just before the testing error started to increase. The total set of generated patterns was divided into three sub sets, 10% of the total patterns were kept for testing, 10% of total patterns for validation or prediction, and remaining 80% were kept for training the neural network. The training pattern fixes the synaptic weights of the networks during the training phase. The testing set determines the stopping criteria for training. The training continues as long as the testing error goes on decreasing. The validation set determines the performance of the ANN on a new data set.

The performance of the developed ANN model was evaluated on the basis of relative error and coefficient of correlation criteria. The relative error, *RE*, shows the relative differences between actual and predicted salt concentration of the pumped water. The actual salt concentrations are those generated by the numerical simulation model. The predicted salt concentrations are those generated by the developed ANN model. The lesser the value of the relative error, the better the model performance. The relative errors ranged from 0.08% to 6.06%, with an average of 0.98%. These values are small and are in the acceptable range. Figure 10.8 shows the graphical representation of relative errors at different pumping locations. The coefficient of correlation between actual and predicted salt concentrations in the pumped water varied between 99.77% to 97.81% respectively, with an average of 99.41%. These values show the potential utility of the proposed approach.

The trained ANN model took only 2 seconds CPU time to calculate the salt concentration of the pumped water on a P-IV (1.7 MHz), 128MB RAM personal computer. Therefore, linking the ANN model with the optimization algorithm is computationally very efficient.

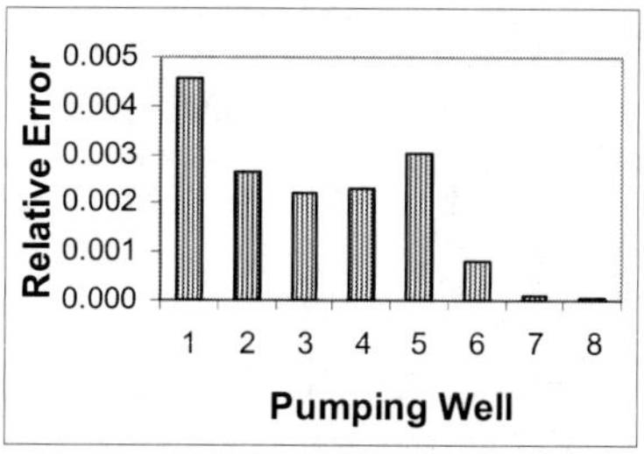

(a) For time step 1

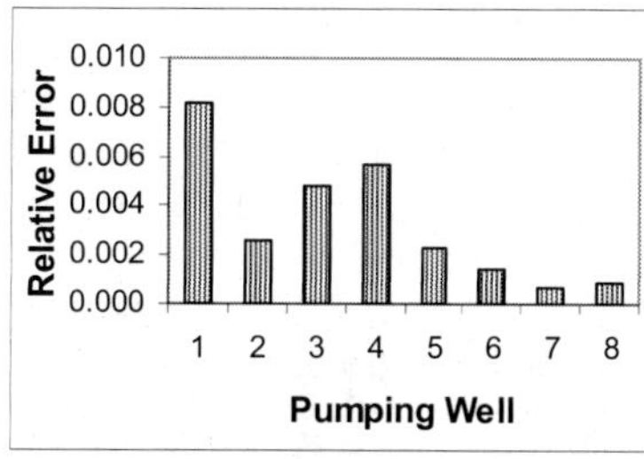

(b) For time step 2

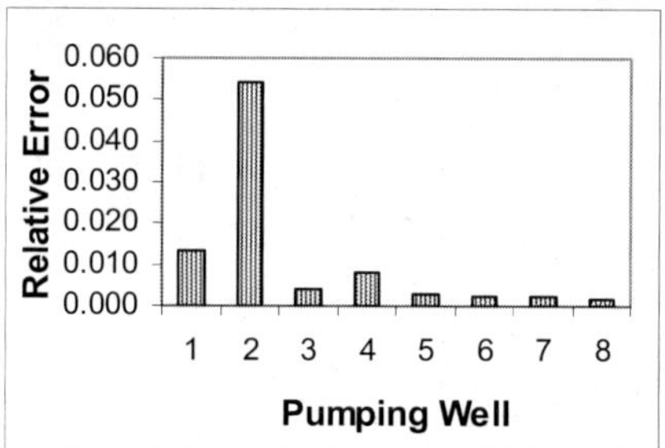

(c) For time step 3

Figure 10.8: Relative error between actual and predicted salt concentrations.

Pareto-optimal solutions: The multiple-objective management models developed above were applied to a hypothetical coastal aquifer to evaluate their performance. The Non-dominating Sorting Genetic Algorithm II (NSGA II) is used here to obtain the Pareto-optimal solutions. NSGA II is an elitist non-dominating sorting algorithm used to solve multiple-objective optimization problems. NSGA II searches for the Pareto optimal solutions by pushing the non-dominating solutions in a particular generation towards the Pareto-optimal solutions, while maintaining diversity among the solutions. The diversity of the solutions in the Pareto-optimal front is maintained by a crowding distance criterion. The efficiency of this algorithm is reported to be better than other multiple-objective evolutionary algorithms in maintaining diversity among the solutions of the Pareto-optimal front; however, this capability of maintaining diversity amongst the individual solutions in the Pareto-optimal front is highly dependent on the scaling of the different objective functions. The real coded NSGA II (Deb 2001) code was used in this study.

The first multiple-objective management model considers two conflicting objectives. These are: maximization of total pumping for beneficial use from wells at interior locations of the aquifer, and minimization of pumping from wells located along and in the vicinity of the ocean boundary. The second objective is applicable when pumping along the ocean boundary and in the vicinity of the ocean face is used to alter the hydraulic gradients near to this boundary to control saltwater intrusion. Therefore, in order to increase the extraction of water from the interior wells while maintaining the salinity within permissible limits, it may be required to increase pumping from the ocean face wells. The pumping from the wells located along, as well as in the vicinity of the ocean face, act as a hydraulic barrier. This barrier restricts the movement of saltwater into the aquifer. These two objectives are conflicting in nature, as the increase in pumping from the interior wells would also increase pumping from the barrier wells, so as to maintain pre-specified upper limit on salt concentration of the pumped water. This study considered eight interior potential pumping locations to withdraw of water for beneficial use. It also considered three pumping locations along and in the vicinity of the ocean boundary. All these wells are numbered 1 to 11 as shown in Figure 10.7. Locations 1 to 8 are for the interior wells, and locations 9 to 11 are for the barrier wells.

Pareto-optimal solutions for the two objectives optimization model obtained using non-dominating sorting Genetic Algorithm II (NSGA II) are shown in Figures 10.10 and 10.11. The x-axis of the Pareto-optimal front represents the total withdrawal rate of water for beneficial use from the interior wells. The y-axis represents the total pumping rate from the barrier wells. The solutions in Figure 10.9 correspond to a salt concentration upper limit of 0.2373. It can be observed from Figure 10.9 that for any improvement in one of the objectives, the other objective has to be sacrificed to a certain degree. The Pareto-optimal front touches the x-axis at a total pumping rate equal to 99,077 m^3/day. Therefore, to withdraw water up to 99,077 m^3/day from the interior wells, there is no requirement of barrier pumping to maintain the pre-specified salt concentration in the pumped water below the upper limit. The barrier pumping is necessary if the total withdrawal from the interior wells is more than 99,077 m^3/day. The amount of barrier pumping for different level of withdrawal from the interior wells can be obtained from the trade curve. For example, for withdrawal of 110,000 m^3/day from the interior wells, about 5,000 m^3/day of water has to be pumped from the barrier wells in order to maintain salt concentration in pumped water within the permissible upper limit.

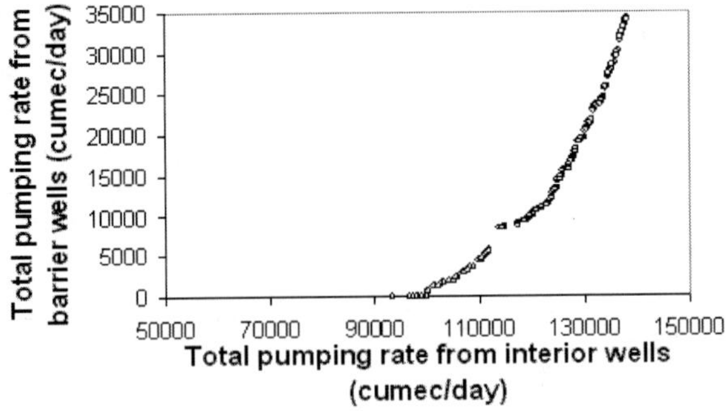

Figure 10.9: Pareto-optimal solutions for the first multiple-objective management model with salt concentration upper limit of 0.2373.

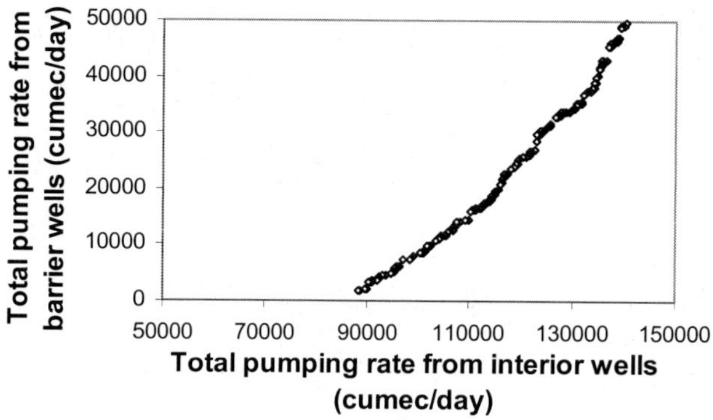

Figure 10.10: Pareto-optimal solutions for first multiple-objective management model with salt concentration upper limit of 0.2362.

The slope of the tradeoff curve represents the tradeoff between the two objectives (Datta and Perelta 1986). The tradeoff at any point on the tradeoff curve represents the amount of sacrifice needed in order to achieve unit amount of improvement on the other objective within the set of non-dominating solutions. In this case, the slope of the tradeoff curve represents the amount of sacrifice needed in the second objective (minimization of total barrier pumping) in order to achieve unit gain in the amount of withdrawal from the interior pumping locations. Figure 10.10 shows another tradeoff curve corresponding to an upper limit on salt concentration of 0.2362. In this case the tradeoff curve touches x-axis at about 89,467 m^3/day. That means, there is no necessity of barrier pumping for withdrawal of water up to 89,467 m^3/day from interior pumping locations in the aquifer. If the total pumping rate from the interior wells exceeds 89,467 m^3/day, the barrier pumping has to be induced to satisfy pre-specified upper limit of 0.2362 on salt concentration.

The second multiple-objective management model also considers two conflicting objectives. These are: maximization of total pumping for beneficial use from wells at interior locations, and minimization of maximum salt concentration in the water pumped from the interior wells. Again, increase in pumping from the interior wells will result in the increase of flow from the ocean towards the aquifer, and therefore, would result in increase in the salinity in the aquifer. Therefore, these two objectives are conflicting in nature. The tradeoff between the objectives is shown in Figure 10.11. The x-axis of the tradeoff curve represents total pumping rate from the interior wells, and the y-axis represents the maximum salt concentration of the pumped water. The increase in withdrawal of water from the interior wells lowers the hydraulic head in the aquifer. This increases the flow from the ocean boundary towards the wells, causing an increase in salt concentrations of the water pumped from the interior wells. This tradeoff between the two objectives is observed in Figure 10.11. It is evident that pumping from the coastal aquifer for beneficial use has to be restricted in order to limit the salinity of the pumped water within permissible limits. The amount of sacrifice needed in one of the objectives in order to achieve unit improvement in the other objective can be obtained from the slope of the tradeoff curve.

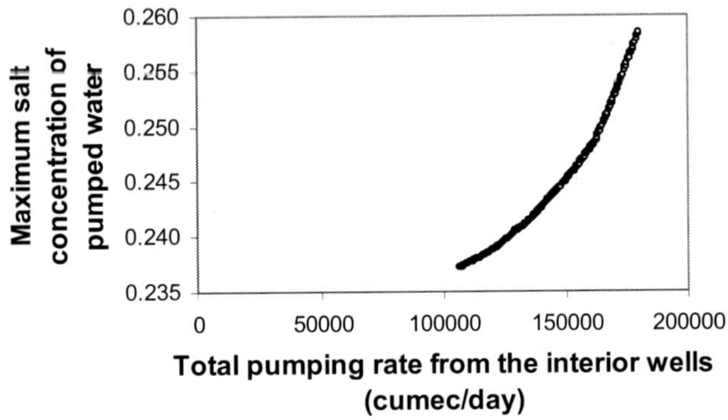

Figure 10.11: Pareto-optimal solutions for the second multiple-objective management model.

The third multiple-objective management model considers three conflicting objectives. These objectives are: maximization of total pumping for beneficial use from wells at interior locations, minimization of pumping from wells located along and in the vicinity of the ocean boundary, and minimization of maximum salt concentration of the pumped water from the interior wells. Here all the objectives are conflicting in nature. The non-inferior solutions are obtained again using the NSGA II algorithm. In this case, the non-inferior solutions represent a surface. The entire set of non-inferior solutions are generated by a single solution of the linked ANN-GA model. Figure 10.12 shows the Pareto-optimal solutions for different levels of the third objective, i.e., minimization of maximum permissible salt concentration in the pumped water. These tradeoff curves are shown for different specified levels of the third objective, in order to present these in two-dimensional space. It can be observed that the total amount of barrier pumping increases with the increase in total pumping from the interior wells. Also, the optimal total pumping rates extracted from the interior wells increase as the permissible maximum salt concentration increases. These evaluations show that solution results are as intuitively expected. Therefore, the methodology is capable of solving the multiple-objective problem. These evaluations demonstrate the potential applicability of the developed methodology.

The first multiple-objective optimization model with two objectives was solved using the embedded optimization technique (Das and Datta 1999a, 1999b) in order to validate the solutions obtain using the proposed methodology. The nonlinear optimization algorithm available in MINOS (Murtagh, and Saunders 1993) was used to solve the embedded optimization model. The multiple-objective management model was converted to a single-objective optimization model using the ε-constraint method. The single-objective optimization model was then solved using MINOS.

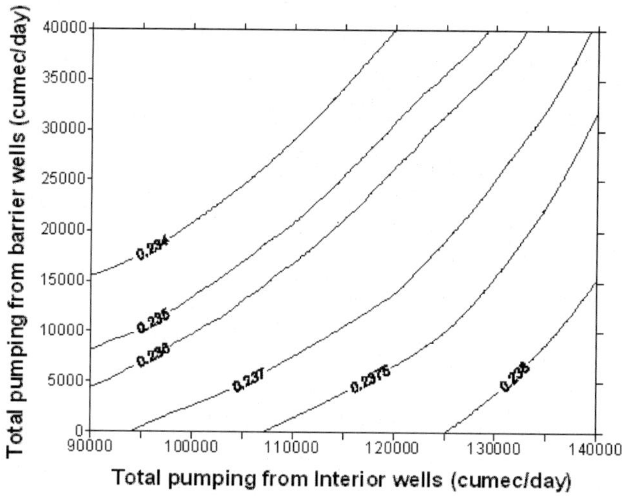

Figure 10.12: Pareto-optimal solutions for the third multiple-objective management model.

The Pareto-optimal solutions obtained using the ANN-GA approach and the embedded optimization technique are compared in Figure 10.13. The upper bound on salt concentration in pumped water was specified as 0.2373. The agreement between the ANN-GA model solutions and the embedded model solution is comparatively better in the lower ranges of total pumping rates from the interior wells. It can be noted that the permissible rates of total pumping from the interior wells are smaller for the embedded optimization solutions as compared to those obtained using the ANN-GA approach. One possible explanation is that the ANN-GA approach is more efficient in locating the global optimal solution compared to the classical optimization technique for the non-convex, nonlinear problem. However, this conclusion seems less convincing as the optimal objective function values are almost identical in the lower ranges of pumping. Again, these lower ranges of pumping represent more realistic values based on the physical constraints in the illustrative aquifer system. The embedded optimization model also considers the hydraulic and other physical constraints explicitly, and realistic bounds are specified on the hydraulic heads in this model. Such bounds cannot be explicitly incorporated in the ANN-GA model, unless the ANN is also trained extensively with input patterns satisfying such bounds. It is possible that the solutions obtained using the ANN-GA approach appear comparatively superior because, the ANN approximator is not adequate to fully consider the entire realm of the physical processes. Therefore, it specifies simulated scenarios that may not be adequate enough especially for ranges of stresses beyond the range of training patterns. The question of ANN's extrapolation capability is also relevant in this context.

In these limited solution results, it is also possible that the results from the embedded model deviate from the ANN-GA model results because of the bounds on hydraulic head specified in the embedded model. These bounds may constrain the solution at higher ranges of pumping, and such bounds are absent in the ANN-GA model. It is therefore reasonable to conclude that at least in these limited results, the solutions obtained using the ANN-GA model and the embedded model are comparable. However, our limited numerical experiment shows that even after further relaxing the upper and lower bounds on the hydraulic heads in the embedded optimization model, the optimal pumping values did not change. If this can be generalized, one possible conclusion could be that the GA is capable of locating improved optimal solutions. A definite answer to these issues related to the comparative performance on the ANN-GA model and the embedded model would need further extensive experimentation.

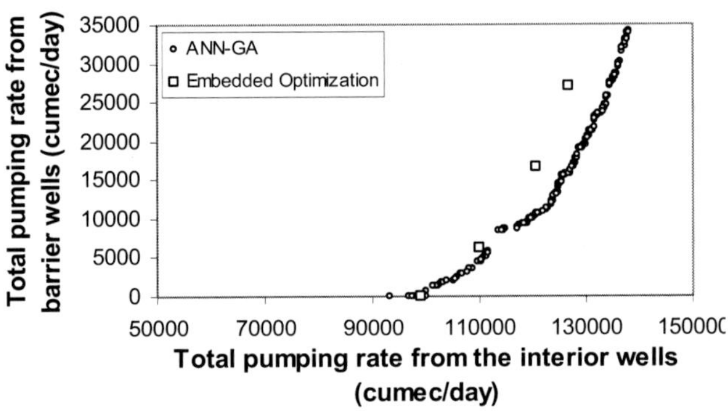

Figure 10.13: Pareto-optimal solutions for first multiple-objective management model with salt concentration upper limit of 0.2373 using ANN-GA and the embedded optimization technique.

Due to the nature of its formulation, the ANN-GA model solutions provide less information compared to the embedded model solutions. For example, the ANN-GA model provides only the spatial and temporal values of pumping rates, and salt concentrations at pre-selected observation locations. But the embedded optimization model generates special and temporal distribution of the hydraulic heads and salt concentrations for the coastal aquifer. The ANN-GA model however requires much less computational time for solution compared to the embedded optimization model. The first ANN-GA model has 2 objectives along with 24 decision variables, and 24 linear constraints. The Pareto-optimal solutions were obtained by the ANN-GA model after 2500 iterations. NSGA II took around 7 minutes of CPU time to complete 2,500 iterations, with a population size of 1,000 on a Sun E10000 UltrasparcII workstation. On the other hand, the embedded optimization model has 1924 decision variables along with 1890 nonlinear constraints, and 24 linear constraints. For the same problem, solution of the embedded model using MINOS took around 86 hours of CPU time to obtain a single point on the Pareto-optimal front. Possible deviations from prescribed optimal management strategies are important considerations in implementation of an optimal management strategy. Robust optimal solutions can ensure near optimal performance of a prescribed strategy even with limited deviations from the prescribed values, as well as due to parameter uncertainties. Dhar and Datta (2009b) have developed a robust strategy for management of saltwater intrusion in coastal aquifer.

10.3.2 Data Worth for Management Model

An integrated management framework considers combined operation and monitoring (Figure 10.14). The initial management of a coastal aquifer is generally based on existing information. Initial control of saltwater intrusion in a coastal aquifer would be based on information available at the time of development of the management strategy. The management objectives may be single or multiple. Multiple-objective management would require the selection of a particular management strategy from the set of non-dominated management alternatives. The judgment of the decision maker is used to select single optimal strategy. Once a strategy is chose, it can be implemented in the field.

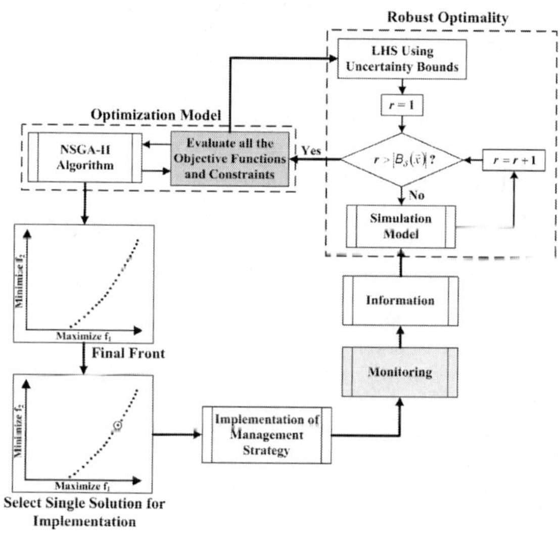

Figure 10.14: Integrated saltwater intrusion management framework.

These non-dominated management alternatives can be generated by using a linked simulation optimization model, where the physical processes are simulated either by a numerical simulation model or a ANN based model. However, there are operational uncertainties in implementing a selected optimal management strategy. Therefore, compliance monitoring of the implemented management strategies should be an essential part of the integrated management framework. The role of compliance monitoring is to obtain field data, e.g., spatial concentrations to provide feedback information on the impact of the implemented strategy. This feedback information is also utilized for intermediate modification of the earlier developed strategy which covered the entire planning horizon. This collected information can be used also as a new initial condition to develop an improved management strategy based on actual field data. This process can be implemented iteratively for efficient management of the aquifer.

Monitoring is an important component of integrated management framework. It is unlikely that the implemented pumping pattern will be in exact conformity with the prescribed one. Even if a pumping strategy is implemented for a definite goal, it is difficult to ensure that there will be no deviation from the prescribed pumping pattern in the field. Implementation of a prescribed management strategy is always associated with operational uncertainty. The focus of the monitoring network design for a coastal aquifer should address the consequences of operational uncertainty and provide feedback information for modification of the management strategy. A methodology combining a spatial interpolation scheme and an optimization model is needed for monitoring network design in coastal aquifers to serve this purpose. Dhar and Datta (2009b) have developed an integrated monitoring and management framework for saltwater intrusion in coastal aquifers.

10.4 Summary and Future directions

A comprehensive framework for modeling and management of density-dependent flow in coastal aquifers has been discussed. The linked simulation-optimization methodology has the potential to solve any management problem with single and multiple objectives. As a simplification, necessary for large study areas, the three-dimensional, transient, density-dependent flow and transport processes in coastal aquifer were approximated using a trained ANN model. The trained ANN model was then linked with a GA-based optimization model. Multiple-objective optimization problems were solved using NSGA-II algorithm. The ANN-GA model is simple in concept and takes considerably less CPU time for solution compared to the embedding technique approach. The performance evaluation of the developed ANN-GA model shows its potential applicability to solve saltwater intrusion management problems in coastal aquifers. If this methodology is established after more rigorous evaluations, it would be possible to solve large-scale coastal aquifer management problems that have been computational infeasible until now.

Extensions of this study should address some of the inherent limitations of the present work. Coastal aquifer management methodologies can be extended to consider other management alternatives. The performance can be evaluated incorporating parameter uncertainty, and variable uncertainty at the same time. Other types of approximate models can be tested for improvement of coastal process simulation. The developed methodologies for coastal aquifer management and monitoring need to be evaluated for real world systems. Further rigorous performance evaluation of the developed methodologies may be required before actual implementation of some of these methodologies.

Common management approaches for coastal aquifers require careful planning of withdrawal strategies in production and extraction barrier wells for control and remediation of saltwater intrusion in coastal aquifers. Such strategies can be evolved only if the physical processes involved in the coastal aquifers are simulated reliably. Numerical simulation would the most direct way of identifying possible solutions of saltwater intrusion management problem, but sequential use of numerical simulation model does not generate sensitivity and/or tradeoff information. Moreover, simulation methodology examines only a finite number of alternatives. Also, the obtained solutions are optimal with respect to alternatives considered during the analysis. In the linked simulation-optimization, the optimization model performs the search process by iterating between the optimization and simulation model. The simulation model provides the necessary information to the optimization model at every stage of iteration, in order to reach the optimal solution. The advantage of this approach is that any type of complex simulation model can be incorporated in the search of optimal management strategies.

The computational feasibility of using a linked simulation model is addressed by developing a meta-model-based (here ANN) simulation-optimization model for the multiple-objective management of saltwater intrusion. Performance evaluation of the meta-model-based methodology shows partial applicability of this approach and satisfactory performance in identifying feasible optimal solutions over an appreciable range of decision variables. Evaluations show that use of a partially-trained, ANN-based meta-model as a screening model for specifying an initial solution for physically-based, simulation-optimization has several computational advantages in terms of efficiency

Success of any management strategy greatly depends on the proper operation of the system. In field it is difficult to accurately maintain the prescribed pumping or other specified strategies. Development of management strategies based on operation uncertainty is necessary for integrated management. A robust optimal solution facilitates the field application of prescribed management strategy with some allowable bounds. These solutions can ensure near optimal performance of a prescribed strategy even with limited deviations from the prescribed values, as well as due to parameter uncertainties. Thus, robust optimization models need to be developed and solved for saltwater intrusion in costal aquifer. These models are described in Dhar and Datta (20009a, 2009b).

In reality feedback information is needed for effective implementation of any operation policy. Compliance monitoring is required to judge the effectiveness of the prescribed management strategy. However, monitoring has to be cost effective.

Optimal design of monitoring networks is necessary due to uncertainties in predicting the movement of pollutants in the groundwater system, and budgetary limitations. Monitoring network design problems are integer or mixed-integer in nature. In contrast to real variables there is a definite probability of reaching the global optima in such situations. Optimization algorithms like Evolutionary Algorithms (EA) and simulated annealing (SA) can assure solution of the posed problems.

In coastal aquifers, the ocean boundary acts as a continuous pollution source. Monitoring networks in coastal aquifers can be of divided into two categories: (a) detection of saltwater intrusion, and (b) compliance monitoring for implemented management strategies to control saltwater intrusion. Monitoring in coastal aquifer becomes an important issue as the aquifer remains under stress conditions (pumping). Variation in pumping pattern or deviation from optimal management strategy may lead to deterioration of coastal aquifer. Compliance monitoring strategy is required to curb the unplanned pumping. This issue is elaborated in Dhar and Datta (2009b).

A number of methodologies have been developed to address optimal management and monitoring of saltwater intrusion in costal aquifers and other general groundwater contamination problems. Most of these developed methodologies are generic in nature and should be applicable to various types of groundwater contamination management problems. This review addresses some of the approaches related to available methodologies for saltwater intrusion management, that can be applied to a variety of scenarios with appropriate modifications.

List of Notation

c solute concentration $(0 \le c \le 1)$ (ML^{-3}),
c_S solute concentration (ML^{-3}) corresponding to the maximum density ρ_S (ML^{-3}),
$\tilde{D}_{ij}$ dispersion tensor (L^2T^{-1})
e_j *jth* component of gravitational unit vector,
g gravitational acceleration (LT^{-2}),
h reference hydraulic head (L)
K_{ij} hydraulic conductivity tensor (LT^{-1}),
K_{ij}^o hydraulic conductivity at the reference (freshwater) condition (LT^{-1})
k_{ij} intrinsic permeability tensor (L^2),
p fluid pressure ($ML^{-1}T^{-2}$),
q_p volumetric flow rate of sinks per unit volume of the porous medium ($L^3T^{-1}L^{-3}$),
q_r volumetric flow rate of sources per unit volume of the porous medium ($L^3T^{-1}L^{-3}$),
S_S specific storage (L^{-1}),

t	time (T),
V_i	darcy velocity vector (LT^{-1}).
x_j	are Cartesian coordinates (L),
Y	elevation above datum (L)
ε	density difference ratio,
η	density coupling coefficient,
μ	dynamic viscosity of fluid $(\mu = \mu_o)$($ML^{-1}T^{-1}$)
μ_o	viscosity of the freshwater ($ML^{-1}T^{-1}$)
ϕ	porosity,
ρ	density of mixed fluid (freshwater and saltwater) (ML^{-3}),
ρ_o	reference (freshwater) density (ML^{-3}).

10.5 References

Abarca, E., Vázquez-Suñé, E., Carrera, J., Capino, B., Gámez, D., and Batlle, F. (2006). "Optimal design measures to correct seawater intrusion." *Water Resources Research*, 42, W09415, doi:10.1029/2005WR004524.

Andersen, P. F., Mercer, J. W., and White, H. O. (1988). "Numerical modeling of saltwater intrusion at Hallandale, Florida." *Ground Water*, 26(5), 619-630.

Bakker, M. (2003). "A Dupuit formulation for modeling seawater intrusion in regional aquifer systems." *Water Resources Research*, 39(5), SBH12.

Bakker, M., Oude Essink, G. H. P., and Langevin, C. D. (2003) "The rotating movement of three immiscible fluids-a benchmark problem" *Journal of Hydrology*, 287 (1-4), 270-278.

Bakker, M., and Schaars, F. (2005). "The sea water intrusion (SWI) package manual Part I." <http://bakkerhydro.org/swi/swimanpart1.pdf> (Dec. 20070.

Bear, J., and Dagan, G. (1964) "Some exact solutions of interface problems by means of the hodograph method," *Journal of Geophysical Research*, 69(2), 1563-1572.

Bear, J., Cheng, A. H.-D., Sorek, S., Ouazar, D., and Herrera, I. (1999). *Seawater intrusion in coastal aquifers: Concepts, methods and practices*, Kluwer Academic Publishers, London.

Bhattacharjya, R. K. (2003). "Management of saltwater intrusion in coastal aquifers using ANN-GA based simulation optimization approach", PhD thesis, I.I.T. Kanpur, India.

Bhattacharjya, R. K., and Datta, B. (2005). "Optimal management of coastal aquifer using linked simulation optimization approach." *Water Resources Management*, 19(3), 295-320.

Bobba, A. G. (1993). "Mathematical models for saltwater intrusion in coastal aquifers." *Water Resources Management*, 7, 3–37.

Burian, S. J., Durrans, S. R., Nix, S. J., and Pitt, R. E. (2001). "Training artificial neural networks to perform rainfall disaggregation." *Journal of Hydrologic Engineering*, 6(1), 43-51.

Cabral, J. J. S. P., and Wrobel, L.C. (1993). "Numerical analysis of saltwater intrusion using B-spline boundary elements." *International Journal for Numerical Methods in Fluids*, 16(11), 989–1005.

Cheng, A. H.-D., and Ouazar, D. (1999), "Analytical solutions." *Seawater intrusion in coastal aquifers: Concepts, methods and practices*, J. Bear, A. H.-D. Cheng, S. Sorek, D. Ouazar, and I. Herrera, eds., Kluwer Academic Publishers, Dordrecht.

Cheng, A. H.-D., Halhal, D., Naji, A., and Ouazar, D. (2000). "Pumping optimization in saltwater-intruded coastal aquifers." *Water Resources Resources*, 36(8), 2155-2165.

Cheng, A. H.-D., Benhachmi, M. K., Halhal, D., Ouazar, D., Naji, A., and EL Harrouni, K. (2004). "Pumping optimization in saltwater-intruded aquifers." *Coastal aquifer management, monitoring, modeling, and case studies*, A. H.-D. Cheng and D. Ouazar, eds., Lewis Publishers, 233-256.

Cheng, A. H.-D., and Ouazar, D., (2004). *Coastal aquifer management: Monitoring, modeling, and case studies*, Lewis Publishers, London.

Cheng, J. M., and Chen, C.X. (2001). "Three-dimensional modeling of density dependent saltwater intrusion in multilayered coastal aquifers in Jahe River basin, Shandong Province, China." *Ground Water*, 39(1), 137-143.

Cheng, J.-R., Strobl, R. O., Yeh, G.-T., Lin, H.-C., and Choi, W.H. (1998). "Modelling of 2D density-dependent flow and transport in the subsurface." *Journal of Hydrologic Engineering*, 3(4), 248–257.

Contractor, D. N., and Srivastava, R. (1990). "Simulation of saltwater intrusion in the Northern Guam Lens using a microcomputer." *Journal of Hydrology*, 118(1-4), 87-106.

Dagan, G., and Zeitoun, D. G. (1998). "Seawater–freshwater interface in a stratified aquifer of random permeability distribution." *Journal of Contaminant Hydrology*, 29(3), 185-203.

Das, A., and Datta, B. (1999a). "Development of multiobjective management models for coastal aquifers." *Journal of Water Resources Planning and Management*, 125(2), 76-87.

Das, A., and Datta, B. (1999b). "Development of management models for sustainable use of coastal aquifers." *Journal of Irrigation and Drainage Engineering*, 125(3), 112-121.

Das, A., and Datta, B. (2000). "Optimization based solution of density dependent seawater intrusion in coastal aquifers." *Journal of Hydrologic Engineering*, 5(1), 82-89.

Deb, K. (2001). *Multi-objective optimization using evolutionary algorithms*, John Wiley & Sons, Ltd., Chichester, U.K.

Dentz, M., Tartakovsky, D. M., Abarca, E., Guadagnini, A., Sanchez-Vila, X., and Carrera, J. (2006). "Variable-density flow in porous media." *Journal of Fluid Mechanics*, 561, 209-235.

Dhar, A., and Datta, B. (2009a). "Saltwater intrusion management of coastal aquifers. I: Linked simulation-optimization." *Journal of Hydrologic Engineering*, 14(12), 1263-1272.

Dhar, A., and Datta, B. (2009b). "Saltwater intrusion management of coastal aquifers. II: Operation uncertainty and monitoring." *Journal of Hydrologic Engineering*, 14(12), 1273-1282.

Diersch, H.-J. G. (2002). *FEFLOW finite element subsurface flow and transport simulation system reference manual*, WASY Inst. for Water Resour. Plann. and Syst. Res., Berlin.

Diersch, H.-J. G., and Kolditz, O. (2002). "Variable-density flow and transport in porous media: approach and challenges." *Advances in Water Resources*, 25(8-12), 899-944.

Emch, P. G., and Yeh, W. W.-G. (1998). "Management model for conjunctive use of coastal surface water and ground water." *Journal of Water Resources Planning Management*, 124(3), 129–139.

Essaid, H. I. (1986). "A comparison of the coupled freshwater-saltwater flow and the Ghyben-Herzbergh sharp interface approaches to modeling of transient behavior in coastal aquifer systems." *Journal of Hydrology*, 86(1-2), 169-193.

Essaid, H. I. (1990a). "A multilayered sharp interface model of coupled freshwater and saltwater flow in coastal systems: model development and application." *Water Resources Research*, 26(7), 1431–1454.

Essaid, H. I. (1990b). *The computer model SHARP, a quasi-three dimensional finite-difference model to simulate freshwater and saltwater flow in layered coastal aquifer system: Model development and applications*, Water-Resources Investigations Report 90-4130, U.S. Geological Survey.

Ferreia da Silva, J. F., and Haie, N. (2007). "Optimal locations of groundwater extractions in coastal aquifers." *Water Resources Management*, 21(8), 1299-1311.

Finney, B. A., Samsuhadi, S., and Willis, R. (1992). "Quasi-three-dimensional optimization model of Jakarta basin." *Journal of Water Resources Planning and Management*, 118(1), 18-31.

Freeze, R. A., and Cherry, J. A. (1979). *Groundwater*, Prentice Hall Inc., N.J.

Gordon, E., Shamir, U., and Bensabat, J. (2001). "Optimal extraction of water from regional aquifer under salinization." *Journal of Water Resources Planning and Management*, 127(2), 71-77.

Guo, W., Langevin, C. D. (2002). *User's guide to SEAWAT: A computer program for simulation of three-dimensional variable-density ground-water flow*, Open File Report 01-434, U.S. Geological Survey.

Guvanasen, V., Wade, S. C., and Barcelo, M. D. (2000). "Simulation of regional groundwater flow and saltwater intrusion in Hernando County, Florida." *Ground Water*, 38(5), 772-783.

Hallaji, K., and Yazicigil, H. (1996). "Optimal management of a coastal aquifer in southern Turkey." *Journal of Water Resources Planning and Management*, 122(4), 233-244.

Hantush, M. S. (1968). "Unsteady movement of fresh water in thick unconfined saline aquifers." *Bull. Int. Assoc. Sci. Hydrol.*, 13, 40-60.

Henry, H. R. (1959). "Salt intrusion into freshwater aquifers." *Journal of Geophysical Research*, 64, 1911-1919.

Holzbecher, E. (1998). *Modelling density-driven flow in porous media*, Springer, Berlin.

HSI GeoTrans (2000). "Theory and implementation for SWIFT for windows the Sandia waste-isolation flow and transport model for fractured media." <http://www.geotransinc.com/Swift_Theory.PDF> (Dec. 2007).

Huppert, H. E., and Woods, A. W. (1995). "Gravity-driven flows in porous layers." *Journal of Fluid Mechanics*, 292, 55-69.

Huyakorn, P. S., Anderson, P. F., Mercer, J. W., and White, H. O. (1987). "Saltwater intrusion in aquifers: Development and testing of a three-dimensional finite element model." *Water Resources Research*, 23(2), 293–312.

Huyakorn, P. S., Wu, Y. S., and Park, N. S. (1996). "Multiphase approach to the numerical solution of a sharp interface saltwater intrusion problem." *Water Resources Research*, 32(1), 93–102.

HydroGeoLogic Inc. (2002). *MODHMS-MODFLOW-based hydrologic modeling system: Documentation and user's guide*, Herndon, Virg.

Kacimov, A. R., and Obnosov, Y. V. (2001). "Analytical Solution for a Sharp Interface Problem in Sea Water Intrusion into a Coastal Aquifer." *Proc., Mathematical, Physical and Engineering Sciences*, 457(2016), 3023-3038.

Karterakis, S. M., Karatzas, G. P., Nikolos, I. K., and Papadopoulou, M. P. (2007). "Application of linear programming and differential evolutionary optimization methodologies for the solution of coastal subsurface water management problems subject to environmental criteria." *Journal of Hydrology*, 342(3-4), 270-282.

Katsifarakis, K. L., and Petala, Z. (2006). "Combining genetic algorithms and boundary elements to optimize coastal aquifers' management." *Journal of Hydrology*, 327(1-2), 200-207.

Kentel, E., Gill, H. and Aral, M. M. (2005). *Evaluation of groundwater resources potential of Savannah Georgia region*, Report No. MESL-01-05, Multimedia Environmental Simulations Laboratory, Georgia Institute of Technology, Atlanta, Ga.

Kipp, K. L. (1986). *HST3D, a computer code for simulation of heat and solute transport in three-dimensional groundwater flow systems*, Water Resources Investigations Report 86-4095, U.S. Geological Survey.

Kipp, K. L., Jr. (1997). *Guide to the revised heat and solute transport simulator, HST3D-Version 2*, Water Resources Investigations Report 97-4157, U.S. Geological Survey.

Ledoux, E., Sauvagnac, S., and Rivera, A. (1990). "A compatible single-phase/two-phase numerical model: 1. Modeling the transient salt-water/fresh-water interface motion." *Ground Water*, 28(1), 215-223.

Lee, C.-H., and Cheng, T.-S. (1974). "On seawater encroachment in coastal aquifers." *Water Resources Research*, 10(5), 1039–1043.

Lin, H.-C. J., Richards, D. R., Talbot, C. A., Yeh, G.-T., Cheng, J.-R., Cheng, H.-P., and Jones, N. L. (1997), "A three-dimensional finite element computer model for simulating density-dependent flow and transport in variable saturated media: Version 3.0", U.S Army Engineering Research and Development Center, Vicksburg, Miss.

Liu, F., Turner, I., and Anh, V. (2001). "A finite volume unstructured mesh method for modeling saltwater intrusion into aquifer systems." *Proc., First International*

Conference on Saltwater Intrusion and Coastal Aquifers: Monitoring, Modeling and Management, Morocco.

Maimone, M. (2002). "Developing an effective coastal aquifer management program." *Proc., 17th Salt Water Intrusion Meeting*, Delft, The Netherlands, 327-336.

Mantoglou, A. (2003). "Pumping management of coastal aquifers using analytical models of saltwater intrusion." *Water Resources Research*, 39(12), doi:10.1029/2002WR001891.

Mantoglou, A., Papantoniou, M., and Giannoulopoulos, P. (2004). "Management of coastal aquifers based on nonlinear optimization and evolutionary algorithms." *Journal of Hydrology*, 297(1-4), 209–228.

Mazzia, A., and Putti, M. (2006). "Three-dimensional mixed finite element-finite volume approach for the solution of density-dependent flow in porous media." *Journal of Computational and Applied Mathematics*, 185(2), 347-359.

Meca, A. S., Alhama, F., and Fernández , C. F. G. (2007). "An efficient model for solving density driven groundwater flow problems based on the network simulation method." *Journal of Hydrology*, 339(1-2), 39-53.

Mercer, J. W., Larson, S. P., and Faust, C. R. (1980). *Finite-difference model to simulate the areal flow of saltwater and freshwater separated by an interface,* Open-File Report 80-407, U.S. Geological Survey.

Murtagh, B. A., and Saunders, M. A. (1993). *MINOS 5.4 User's guide*, Technical Report SOL 83-20R, Dept. of Operation Research, Stanford University, Stanford, Calif.

Naji, A., Ouazar, D., and Cheng, A. H.-D. (1998). "Locating the saltwater-freshwater interface using nonlinear programming and h-adaptive BEM," *Engineering Analysis with Boundary Elements*, 21(3), 253-259.

Nobi, N. and Gupta, A. D. (1997). "Simulation of regional flow and salinity intrusion in an integrated stream aquifer system in a coastal region: Southwest region of Bangladesh." *Ground Water*, 35(5), 786-796.

Oude Essnik, G. H. P. (1998). "MOC3D adapted to simulate 3D density-dependent groundwater flow." *Proc., Modflow'98 Conference*, Golden, Colo., 291-303.

Oude Essnik, G. H. P. (2001). "Density dependent groundwater flow: Salt water intrusion and heat transport." Lecture notes for Hydrological Transport Processes/Groundwater Modelling II, Utrecht University.

Park, C., and Aral, M. (2004). "Multi-objective optimization of pumping rates and well placement in coastal aquifers." *Journal of Hydrology*, 290(1-2), 80-99.

Polo, J. F. and Ramis, F. J. R. (1983). "Simulation of salt water-fresh water interface motion." *Water Resources Research*, 19(1), 61–68.

Post, V. E. A. (2005). "Fresh and saline groundwater interaction in coastal aquifers: Is our technology ready for the problems ahead?" *Hydrogeology Journal*, 13,120-123.

Putti, M., and Paniconi, C. (1995). "Picard and Newton linearization for the coupled model of saltwater intrusion in aquifers." *Advances in Water Resources*, 18(3), 159-170.

Qahman, K., Larabi, A., Ouazar, D., Naji, A., and Cheng, A. H.-D. (2005). "Optimal and sustainable extraction of groundwater in coastal aquifers." *Stochastic Environmental Research and Risk Assessment*, 19(2), 99-110.

Rao, S., Thandaveswara, B. S., Bhallamudi, S. M., and Srivivasulu, V. (2003). "Optimal groundwater management in deltaic regions using simulated annealing and neural networks." *Water Resources Management*, 17(6), 409–428.

Rao, S., Bhallamudi, S., Thandaveswara, B., and Mishra, G. (2004a). "Conjunctive use of surface and groundwater for coastal and deltaic systems." *Journal of Water Resources Planning and Management*, 130(3), 255–267.

Rao, S. V. N., Sreenivasulu, V., Bhallamudi, S. M., Thandaveswara, B. S., and Sudheer, K. P. (2004b). "Planning groundwater development in coastal aquifers." *Hydrological Sciences Journal*, 49(1), 155-170.

Reilly, T. E., and Goodman, A. S. (1985). "Quantitative analysis of saltwater–freshwater relationships in groundwater systems: A historical perspective." *Journal of Hydrology*, 80(1-2), 125–160.

Reichard, E., and Johnson, T. (2005). "Assessment of regional management strategies for controlling seawater intrusion." *Journal of Water Resources. Planning Management*, 131(4), 280–291.

Rouve, G., and Stoessinger, W. (1980). "Simulation of the transient position of the saltwater intrusion in the coastal aquifer near Madras coast." *Proc., 3rd Internat. Conf. Finite Elements in Water Resources*, Univ. of Miss., Oxford, Miss.

Schmorak, S., and Mercado, A. (1969). "Upconing of fresh water - sea water interface below pumping wells, field study." *Water Resources Research*, 5, 1290-1311.

Segol, G., Pinder, G. F., and Gray, W. G. (1975). "A Galerkin-finite element technique for calculating the transient position of saltwater front." *Water Resources Research*, 11(2), 343-347.

Segol, G. (1994). *Classic groundwater simulations proving and improving numerical models*, Prentice-Hall, Old Tappan, N.J.

Shamir, U., Bear, J., and Gamliel, A. (1984). "Optimal annual operation of a coastal aquifer." *Water Resources Research*, 20(4), 435-444.

Sherif, M. M., Singh, V. P, and Amer, A. M. (1988). "A two-dimensional finite element model for dispersion (2D-FED) in coastal aquifer." *Journal of Hydrology*, 103(1-2), 11-36.

Sikkema, P. C., and van Dam, J. C. (1982). "Analytical formulae for the shape of the interface in a semi-confined aquifer." *Journal of Hydrology*, 56(3-4), 201-220.

Simmons, C. (2005). "Variable density groundwater flow: From current changes to future possibilities." *Hydrogeology Journal*, 13, 116–119.

Simmons, T. C., Fenstemaker, T. R., and Sharp, J. M., Jr. (2001). "Variable-density groundwater flow and solute transport in heterogeneous porous media: approaches, resolutions and future challenges." *Journal of Contaminant Hydrology*, 52(1-4), 245-275.

Sorek, S., and Pinder, G. F. (1999). "Survey of Computer Codes and Case Histories." *Seawater intrusion in coastal aquifers: Concepts, methods and practices*, J. Bear, A. H.-D. Cheng, S. Sorek, D. Ouazar, and I. Herrera, eds., Kluwer Academic Publishers, Dordrecht.

Souza, W. R., and Voss, C. I. (1987). "Analysis of anisotropic coastal aquifer system using variable-density flow and solute transport simulation." *Journal of Hydrology*, 92(1-2), 17-41.

Strack, O. D. L. (1976). "A single-potential solution for regional interface problems in coastal aquifers." *Water Resources Research*, 12(6), 1165-1174.

Swedish Nuclear Power Inspectorate (1986). *HYDROCOIN - An international project for studying groundwater hydrology modeling strategies*, Level 1 Final Report, Stockholm.

Taigbenu, A. E., Liggett, J. A., and Cheng, A. H.-D. (1984). "Boundary integral solution to seawater intrusion into coastal aquifers." *Water Resources Research*, 20(8), 1150–1158.

U.S. Environmental Protection Agency (USEPA) (1992). "Secondary drinking water regulations: guidance for nuisance chemicals." *Groundwater & drinking water*, <http://www.epa.gov/safewater/consumer/2ndstandards.html> (May 2007).

van Dam, J. C., and Sikkema, P. C. (1982). "Approximate solution of the problem of the shape of the interface in a semi-confined aquifer." *Journal of Hydrology*, 56(3-4), 221–237.

Volker, R., and Rushton, K. (1982). "An assessment of the importance of some parameters for seawater intrusion in aquifers and a comparison of dispersive and sharp-interface modeling approaches." *Journal of Hydrology*, 56(3-4), 239–250.

Voss, C. I. (1984). *A finite-element simulation model for saturated-unsaturated, fluid-density-dependent ground-water flow with energy transport or chemically-reactive single-species solute transport,* Water-Resources Investigations Report 84-4369, U.S. Geological Survey.

Voss, C. I., and Provost, A. M. (2002). *SUTRA: A model for saturated-unsaturated variable-density ground-water flow with solute or energy transport*, Water-Resources Investigations Report 02-4231, U.S. Geological Survey

Willis, R., and Finney, B. A. (1988). "Planning model for optimal control of saltwater intrusion." *Journal of Water Resources Planning and Management*, 114 (2), 163–178.

Younos, T. (2005). "Desalination: Supplementing freshwater supplies approaches and challenges." *Journal of Contemporary Water Research & Education*, 132, 1-2.

Zhou, X., Chen, M., and Liang, C. (2003). "Optimal schemes of groundwater exploitation for prevention of seawater intrusion in the Leizhou peninsula in southern China." *Environmental Geology*, 43(8), 978–985.

CHAPTER 11

IN-SITU AIR SPARGING AND THERMALLY-ENHANCED VENTING IN GROUNDWATER REMEDIATION

Wonyong Jang and Mustafa M. Aral
Multimedia Environmental Simulations Laboratory
School of Civil and Environmental Engineering
Georgia Institute of Technology, Atlanta, GA 30332

11.1 Introduction

The development of remedial technologies for contaminated sites has been driven by increased environmental awareness and national hazardous waste cleanup programs managed by the U. S. Environmental Protection Agency (U.S. EPA). The programs are regulated by the Comprehensive Environmental Response, Compensation, and Liability Act (CERCLA or Superfund), Resource Conversation and Recovery Act (RCRA), Oil Pollution Act (OPA), and Underground Storage Tank (UST) program. Under these programs, a variety of remediation technologies have been used or recommended for the cleanup of contaminated sites. These remediation technologies include groundwater pump-and-treatment, soil vapor extraction (SVE), bioventing, chemical and biological treatment, bioremediation, in-situ air sparging (IAS), thermally-enhanced venting (TEV), and monitored natural attenuation.

IAS is a cost-effective remedial technology applied to remove volatile organic carbons (VOCs) dissolved in the groundwater and to remove nonaqueous-phase-liquid contaminants (NAPLs) from above and below the water table (Kirtland and Aelion 2000; Unger et al. 1995). IAS has gained broad attention because it appears simple to implement, and has been used at many sites contaminated by chlorinated solvents and petroleum hydrocarbons (Bass et al. 2000).

TEV applies heated, humid air to the contaminated zone in order to enhance removal or recovery of less-volatile residual hydrocarbon contaminants such as naphthalene and dodecane in the unsaturated zone. As heat energy is introduced into the ground, the physicochemical properties of fluids and VOCs change under nonisothermal conditions. For example, the vapor pressure of semi-volatile organic compound (or the maximum concentration in the gas phase) can increase by three- or five-fold with a temperature increase of 20-30°C (Kaluarachchi and Islam 1995). Thermal energy plays an important role in increasing remediation efficiency and in reducing remediation time of the contaminated unsaturated zone.

11.2 In-Situ Air Sparging

IAS uses clean, pressurized air to remediate VOC-contaminated groundwater and to remove NAPL contaminants trapped within the saturated and unsaturated zones (Jang and Aral 2006; Thomson and Johnson 2000). As illustrated in Figure 11.1, IAS injects clean air below the water table. As the air rises through the saturated zone due to buoyancy and capillary forces, it contacts aqueous phase and NAPL contaminants, which diffuse into the air due to the concentration gradient and are then carried out into the unsaturated zone above the water table (Roosevelt and Corapcioglu 1998). In the unsaturated zone, the air containing the contaminants is usually withdrawn by SVE wells, and the contaminants are treated by post-treatment units. As IAS supplies oxygen into the contaminated zone, it can enhance the aerobic biodegradation of organic contaminants in the subsurface (Aelion et al. 1997; Kirtland et al. 2001). Aerobic biodegradation of organic compounds is verified and evaluated by monitoring the concentration variation of CO_2, O_2 and contaminants over time in the subsurface (Aelion et al. 1997; Hinchee and Arthur 1991; Lee et al. 2002; Wood et al. 1993).

IAS can provide permeable barriers to contaminant plume migration, where remedial activities occur. Contaminant removal by IAS commonly relies on two primary mechanisms, volatilization and biodegradation of contaminants as described above. IAS is readily applicable for contaminants with relatively high volatility, such as benzene, xylene, toluene, and tricholorethene (TCE) (U.S.EPA 2004). IAS works best in uniform coarse-grained soils, such as sand and gravel, and it is effective when intrinsic permeability is greater than or equal to 10^{-9} cm^2 (Sellers 1999; U.S.EPA 2004).

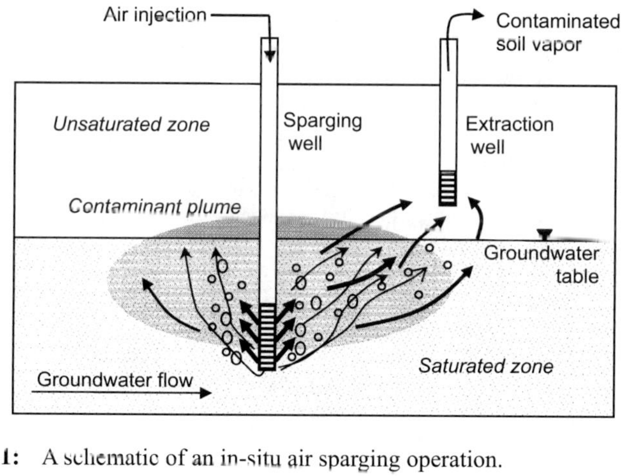

Figure 11.1: A schematic of an in-situ air sparging operation.

In IAS, air injection causes the dynamic movement of the groundwater and injected air by displacing water with air in the soil matrix. The lateral and upward movements of the injected air are important factors in determining the radius of influence (ROI) and the degree of contact between the air and contaminated groundwater, which controls the remedial performance of IAS (McCray 2000). Before a full-scale IAS system is installed, pilot tests for IAS should be performed to obtain field data about the feasibility of IAS, the ROI, and mechanical requirements, such as pressure and flow rates. Based on case studies of IAS, Bass et al. (2000) identified several difficulties in IAS application, which include the layout and density of sparging wells, rebound of contaminant concentrations after treatment, and remediation times.

11.2.1 Applicability of Air Sparging System

The applicability and performance of IAS are determined by a variety of factors, such as contaminant properties, geological and hydrogeologic conditions, and contaminant distribution (Wilson 1995). Contaminant properties include vapor pressure, solubility, and biodegradability. Geological and hydrogeologic conditions include aquifer depth, soil texture, porosity, permeability, heterogeneity, and stratigraphy. Contaminant distribution includes the horizontal and vertical extent of contamination, contaminant concentrations, and the nature of the contamination (e.g., dissolved phase versus NAPL).

11.2.1.1 Target Contaminants

Air sparging is applicable for the treatment of volatile/less volatile and/or tightly sorbed chemicals that cannot be remediated using vapor extraction alone (Miller 1996). Target contaminants for IAS include various fuels (e.g., gasoline, diesel, jet fuel) and fuel components such as benzene, toluene, ethylbenzene, and xylenes (BTEX); oils and greases; and chlorinated solvents (e.g., tetrachloroethene, trichloroethene, dichloroethenes, etc.). Methyl *tert*-butyl ether (MTBE), another fuel-related contaminant, has been used as a gasoline additive since the late 1970s as an octane enhancer and/or as a component of oxyfuel. MTBE is a volatile, flammable and colorless liquid that dissolves rather easily in water. Its use was highly increased by the implementation of the Clean Air Act Amendments of 1990, which require the use of oxygenates, such as MTBE and ethanol, for gasoline in areas with unhealthy levels of air pollution to reduce harmful emissions from motor vehicles. Due to the wide use of MTBE, it has been frequently detected in contaminated groundwater (Johnson et al. 2000).

An extensive review of IAS case studies is given by Bass et al. (2000). Table 11.1 lists the chemical properties of target compounds and describes representative case studies. The application of air sparging results in the treatment of groundwater contaminated by various volatile and semi-volatile organic compounds, the removal of immobile contaminant sources having NAPLs, and the creation of a barrier to dissolved contaminant plume migration(Heron et al. 2002; Johnson et al. 1999; NFESC 2001). If mobile light or dense NAPLs (i.e., LNAPL or DNAPL) are present at a particular site, air sparging operations may spread the NAPL beyond current contamination zones.

Volatilization is the primary mechanism for contaminant mass removal during air sparging; thus, the vapor pressure of contaminants is a key factor in determining the applicability of the technology. Contaminants with a high Henry's law constant are generally good candidates for IAS remediation. The Henry's law constant can be approximated by the ratio of a compound's vapor pressure to its aqueous solubility. Compounds with a relatively low Henry's law constant will take more time and incur higher operational costs due to the extended duration and air sparging flow rates needed to volatilize contaminant mass from the subsurface.

Table 11.1: List of some target contaminant properties for use in air sparging technology.

Chemical(s)	Vapor pressure* (mm Hg at 15°C)	Henry's law constant* (atm m^3/mol, 25°C)	Application cases
Benzene	58.38	5.55×10^{-3}	(Benner et al. 2000; Kirtland and Aelion 2000; Lundegard and Labrecque 1995)
Toluene	16.48	6.35×10^{-3}	
m-xylene	4.59	6.78×10^{-3}	
o-xylene	3.55	4.19×10^{-3}	
p-xylene	4.82	6.15×10^{-3}	
Ethyl benzene	5.26	8.14×10^{-3}	
Tetrachloroethene (PCE)	10.59	2.69×10^{-3}	(Brown et al. 1999; Rabideau et al. 1999)
Trichloroethene (TCE)	45.13	1.16×10^{-3}	
cis-1,2-dichloroethene	129.3	7.36×10^{-3}	
trans-1,2-dichloroethene	218.5	6.70×10^{-3}	
1,1-dichloroethene	406.7	2.28×10^{-3}	
Vinyl chloride (VC)	2179	2.24×10^{-3}	
Methyl-tert-butyl ether (MTBE)	161.2	5.41×10^{-4}	(CWA 2006; Lin and Cheng 2007)
Total petroleum hydrocarbons (TPH)	A mixture of petroleum-related chemicals.		(Aelion and Kirtland 2000; Aelion et al. 1997)

*Data from Yaws (1999).

11.2.1.2 Site Geological Conditions

The successful application of IAS technology depends on the ability of the IAS system to effectively deliver injected air to the treatment area and the ability of the subsurface materials to effectively transmit the air (Miller 1996). IAS is best suited to sites with relatively coarse-grained (moderate to high permeability) homogeneous soils and medium to shallow aquifer depths at less than 50 ft below ground surface (Miller 1996; NFESC 2001). Fine-grained, low permeability soils limit the migration of the injected air in the subsurface. In homogeneous soil media, high permeability soils allow the upward transmission of air, while low permeability soils retard the upward movement of injected air (Jang and Aral 2009). Site geological conditions such as stratification, heterogeneity, and anisotropy can prevent uniform airflow through soil media, thus reducing the effectiveness of air sparging. For IAS applications, good characterization of the geology in the area where an IAS system is to be installed is critical. The subsurface geology should be evaluated by collecting at least one continuous soil core from ground surface to the bottom of the contaminated aquifer in the area where air sparging is being considered (NFESC 2001). A hydrogeologist should evaluate the soil core to determine soil types and to identify the depth and location of distinct soil layers that may influence airflow.

11.2.1.3 Oxygen Supply and Biodegradation

Biodegradation of VOCs is one of major processes that naturally reduce VOC levels at contaminated sites (Lovley 1997). Within petroleum-contaminated zones, oxygen is depleted due to oxygen consumption by biotic and abiotic oxidation reactions. Oxygen can be supplied into the subsurface by oxygen-containing groundwater flow and air influx (soil breathing) into the ground without any human-induced activities (Andreoni and Gianfreda 2007; Jang and Aral 2007b; Lovley 1997); however, the oxygen supplied is usually much less than the oxygen consumed through aerobic biodegradation of organic compounds by indigenous microorganisms, which quickly depletes the dissolved oxygen within contaminated zones. This leads to low oxygen levels in the subsurface; thus, anaerobic conditions are often prevalent within and near the margins of groundwater plumes associated with contaminated sites. Under such situations, air sparging can become a significant source of oxygen (Jang and Aral 2009). As oxygen in the injected air dissolves into the groundwater, IAS can increase dissolved oxygen (DO) concentrations in the contaminated groundwater, which can then stimulate the growth of microorganisms and increase the aerobic biodegradation of organic contaminants. Aerobic biodegradation usually requires DO concentrations greater than 1 to 2 mg/L, while anaerobic bacteria generally cannot function at DO levels greater than 0.5 mg/L (Canada 2008; Wiedemeier 1998).

A variety of indigenous organisms exist in the soil, and many of them have the ability to biodegrade hydrocarbons under various environmental conditions. The microorganisms include bacteria, fungi, and protozoans, which utilize hydrocarbons as sources of energy and nutrients for cell growth (Reisinger 1995). The population of microorganisms changes based on the availability of nutrients, electron donors (e.g., petroleum hydrocarbons) and electron acceptors (e.g., oxygen, nitrate, sulfate), and changes in pH. Of the consortium of indigenous organisms, dominance depends on their adaptability to the environment they reside in. Hydrocarbons can biodegrade under either aerobic or anaerobic conditions. Aerobic biodegradation uses oxygen as the electron acceptor, while anaerobic biodegradation uses alternate electron acceptors, such as nitrate or sulfate (Reisinger 1995). For hydrocarbon oxidation under anaerobic conditions, nitrate, manganese (IV), iron (III), sulfate, and carbonate can be used as terminal electron acceptors in the order of reduction potentials (Lovley 1997; Wiedemeier 1998). BTEX, MTBE, and polycyclic aromatic hydrocarbons (PAHs) are biodegradable under aerobic conditions (Andreoni and Gianfreda 2007; Kane et al. 2001; Kirtland and Aelion 2000). Kirtland et al. (2001) showed that IAS with SVE accelerates the aerobic biodegradation of BTEX and petroleum hydrocarbons. MTBE is less biodegradable than benzene; thus, the effectiveness IAS on accelerating MTBE biodegradation is less than that for benzene (U.S.EPA 1998). Table 11.2 gives the stoichiometric relationships for aerobic and anaerobic bioreactions for various common organic compounds. Generally, the aerobic bioreaction of petroleum hydrocarbons is more rapid and efficient than the anaerobic bioreaction due to the lower free-energy requirement for initiation and higher energy production under the aerobic bioreaction (Cookson 1995).

Chlorinated ethenes (CEs), such as PCE and TCE, are among the most frequently detected VOCs at contaminated sites. Biodegradation of CEs depends on many physicochemical and biological parameters such as oxygen concentrations, availability of carbon sources, and indigenous microorganisms. The pathways for CEs biodegradation are typically divided into three main groups: dehalogenation, cometabolism, and oxidation under anaerobic or aerobic conditions (Figure 11.2). Under anaerobic conditions, PCE can be biologically transformed by sequential reductive dehalogenation as follows: PCE → TCE → dichloroethene (DCE) → vinyl chloride (VC) → ethene (Jang and Aral 2007a; Jang and Aral 2008a; Lawrence 2007; Vogel and McCarty 1985). Because PCE byproducts TCE, DCE and VC are toxic, dehalogenation of PCE results in the introduction of new contaminants into the subsurface. Under aerobic conditions, however, CEs can be biologically degraded via cometabolic processes, which usually do not generate harmful byproducts.

Table 11.2: Stoichiometric relationships for bioreactions of VOCs.

Aerobic bioreaction
Benzene oxidation/respiration: $C_6H_6 + 7.5O_2 \rightarrow 6CO_2 + 3H_2O$
Toluene oxidation/respiration: $C_6H_5CH_3 + 9O_2 \rightarrow 7CO_2 + 4H_2O$
Ethylbenzene oxidation/respiration: $C_6H_5C_2H_5 + 10.5O_2 \rightarrow 8CO_2 + 5H_2O$
m-Xylene oxidation/respiration: $C_6H_4(CH_3)_2 + 10.5O_2 \rightarrow 8CO_2 + 5H_2O$
PCE aerobic cometabolism: $C_2Cl_4 + O_2 + 2H_2O \rightarrow 2CO_2 + 4H^+ + 4Cl^-$
TCE aerobic cometabolism: $C_2HCl_3 + 1.5O_2 + H_2O \rightarrow 2CO_2 + 3H^+ + 3Cl^-$
DCE oxidation/respiration: $C_2H_2Cl_2 + 2O_2 \rightarrow 2CO_2 + 2H^+ + 2Cl^-$
VC oxidation/respiration: $C_2H_3Cl + 2.5O_2 \rightarrow 2CO_2 + H_2O + H^+ + Cl^-$
Anaerobic bioreaction
Benzene oxidation/denitrification: $C_6H_6 + 6NO_3^- + 6H^+ \rightarrow 6CO_2 + 6H_2O + 3N_2$
Benzene oxidation/iron reduction: $C_6H_6 + 30Fe(OH)_3 + 60H^+ \rightarrow 6CO_2 + 30Fe_3^{2+} + 78H_2O$
Benzene oxidation/sulfate reduction: $C_6H_6 + 3.75SO_4^{2-} + 7.5H^+ \rightarrow 6CO_2 + 3.75H_2S + 3H_2O$
Benzene oxidation/methanogenesis: $C_6H_6 + 4.5H_2O \rightarrow 2.25CO_2 + 3.75CH_4$
Toluene oxidation/denitrification: $C_6H_5CH_3 + 7.2NO_3^- + 7.2H^+ \rightarrow 7CO_2 + 7.6H_2O + 3.6N_2$

The aerobic and anaerobic biotransformation of CEs has been reported at many contaminated sites (Clement et al. 2000; Hinchee et al. 1995; Lenczewski et al. 2003). For example, field investigations at PCE-contaminated sites showed the appearance of TCE, DCE, and VC in the groundwater (Maslia et al. 2007). At TCE-contaminated sites, DCE and VC were also detected within both water and gas phases in the unsaturated zone as well as within the water in the saturated zone. These occurrences are due to anaerobic biotransformation and partitioning processes of the contaminants (Lorah and Olsen 1999; Wiedemeier 1998). In addition to anaerobic bioreactions of CEs, aerobic bioreactions of DCE and VC have been also documented at contaminated sites, especially at the aerobic fringe of contaminant plumes (Clement et al. 2000; Hinchee et al. 1995; Lenczewski et al. 2003; Verce et al. 2002). It is important to note that the anaerobic and aerobic bioreactions of CEs, coexisting at contaminated sites, generate totally different byproducts (e.g., toxic vs. benign byproducts in Figure 11.2), resulting in distinct environmental effects. Thus, a well-defined classification of target sites into co-existent aerobic or anaerobic regions is very important in accurately predicting the fate and transport of parent and daughter contaminants (Jang and Aral 2007b; Jang and Aral 2008b).

The subsurface is highly heterogeneous in terms of hydrogeological, physical, chemical, and biological conditions. Considering the heterogeneities, such as rate-limited dispersion of oxygen through porous media and locally distinct utilization of oxygen, one can expect the development of three-types of bioreaction blocks: pure aerobic block, pure anaerobic block, and aerobic-anaerobic-mixed block (Jang and Aral 2007b). The last block refers to a representative subsurface element in which both aerobic and anaerobic bioreactions occur simultaneously.

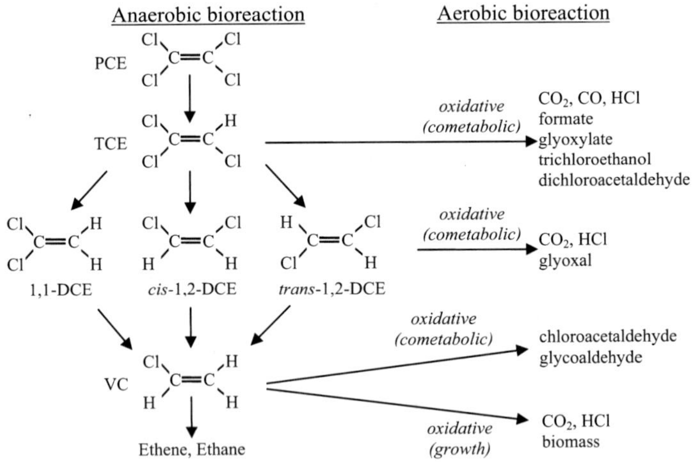

Figure 11.2: Anaerobic and aerobic biological degradations of chlorinated ethenes (van Hylckama Vlieg and Janssen 2001).

11.2.2 Theoretical, Experimental, and Numerical Approaches to IAS

Typically, airflow through porous media in the saturated zone occurs in the form of bubbles and/or discrete air channels that form in well-sorted, coarse-grained sediments. Many laboratory experiments have been conducted to determine airflow patterns for various air injection rates and porous media characteristics (Chen et al. 1996; Ji et al. 1993; Peterson et al. 1999; Reddy and Adams 2001; Roosevelt and Corapcioglu 1998). Ji et al. (1993) showed that an increase in air injection rate expands the size of overall air plume and increases the number of continuous air channels (i.e., a higher density of air channels in the saturated zone). They reported, however, that the size of each channel was almost independent of air injection rates. Peterson et al. (1999) conducted similar laboratory experiments to observe airflow and the zone of IAS influence as a function of grain size. They reported that airflow in discrete channels (i.e., channeled flow) was dominant for smaller grain sizes (average diameter < 1.3 mm), while air-bubbly flow was pervasive for larger grain sizes (average diameter ≥ 1.84 mm). Peterson et al. (1999) suggested that, in terms of a grain size, aquifers having sediments with diameters in the 2-3 mm range would be optimal for air sparging because the zone of influence for IAS is maximized under these conditions.

For the movement of distinct air bubbles through water-saturated porous media, Corapcioglu et al. (2004) theoretically estimated the rise velocity of air bubbles based on macroscopic balance equations for forces acting on the bubbles, accounting for inertia, gravity, buoyancy, surface tension, and drag. The theoretical velocity estimates showed good agreement with laboratory-measured velocities observed by Roosevelt and Corapcioglu (1998). Air channeling is ubiquitous at the pore scale and occurs at various scales during IAS (Clayton 1999). Heterogeneities of capillarity and relative permeability appear to determine the formation of air channels at macroscopic and larger scales (Ahlfeld et al. 1994; Ji et al. 1993). At smaller scales, air channels may be closely spaced relative to the representative elementary volume of a typical model, and air water flow can be assumed to be a continuum within relatively homogeneous porous media (McCray 2000). Air-water flow as a continuum has been applied for multiphase flow models.

The use of IAS technology is growing more rapidly than the theoretical and design knowledge associated with the technique, and mathematical models of air sparging are in the early stages of development (McCray 2000). Published air-sparging models became available in the early to mid-1990s. (McCray 2000) conducted an extensive review of the mathematical modeling of air sparging. In this review, air-sparging models were divided into two main groups: compartmentalized, lumped-parameter models and multiphase fluid-flow models. The former approach considers multiple, separate compartments (or blocks) for fluid flows and biological-physical-chemical processes. The compartments are linked in terms of mass transfer between them. The multiphase flow approach assumes that the water and gas phases are continuums. Such an assumption is often used to describe the simultaneous flow of water and air that occurs during IAS (Jang and Aral 2009). Numerical modeling of IAS is a valuable tool in predicting complex processes regarding fluid flow, contaminant transport, and nonequilibrium reactions and in solving problems occurring during IAS (Benner et al. 2002). Understanding the dynamic behavior of gas and water is essential to determine the feasibility of IAS and to design effective IAS systems.

11.2.2.1 Compartmentalized, Lumped-Parameter Model

In compartmentalized models for air sparging, porous media phases and contaminated/uncontaminated zones can be divided into multiple compartments (Rabideau et al. 1999; McCray 2000). The multiphase environment associated with IAS applications includes gas, water, soil, and NAPL phases that can be represented as compartments in which contaminants can be present in each (Rabideau and Blayden 1998) (Figure 11.3). The compartments are regarded as completely mixed reactors. The mass transfer between the compartments is expressed by equilibrium and/or nonequilibrium processes, such as dissolution, vaporization, and sorption. In order to consider the rate-limiting mechanism of contaminant diffusion, Rabideau et al. (1999) used two compartments to represent a "clean zone" and a "dirty zone" instead of aqueous and solid phases. The "clean" zone is in contact with air channels (i.e., gas phase compartment). Thus, contaminants in this zone are quickly removed by mass transfer into the gas phase and bioreactions. The contaminants in the "dirty" zone can be removed by relatively-limited contaminant diffusion from the "dirty" zone into the "clean" zone. Generally, the primary purpose of compartment models is to address the mass-transfer limiting diffusion of dissolved contaminants, and/or nonequilibrium interphase mass transfer between the air-water or NAPL-water phases (McCray 2000).

In compartmental models, the gas compartment is usually used to represent air channels in porous media. Compartmental models are useful to address the effect of air channeling on contaminant removal (Figure 11.3). Wilson (1992) used a well-mixed, two-compartment model (e.g., a water compartment and a reaction compartment between water and gas) to develop simple analytical air-sparging models that were used to estimate the removal of a dissolved contaminant under steady and unsteady conditions. Local equilibrium between water and gas phases was assumed in the analytical models. Wilson et al. (1994) developed a numerical model to simulate VOC transport into sparged air by dispersion and air-induced circulation of the water in the vicinity of a sparging well. Air was assumed to flow through discrete, persistent channels in the aquifer. By simplifying the air-sparging system into a few compartments and considering discrete fluid flows without considering the interaction between fluids, the compartmentalized, lumped-parameter models allow representation of many complex mass-transfer processes in a bulk volume of porous media. However, these models cannot describe the dynamic behavior of injected air and groundwater, the temporal- and spatial-distribution of the air, and the physicochemical processes that are associated with this dynamic behavior (Jang and Aral 2009; McCray 2000).

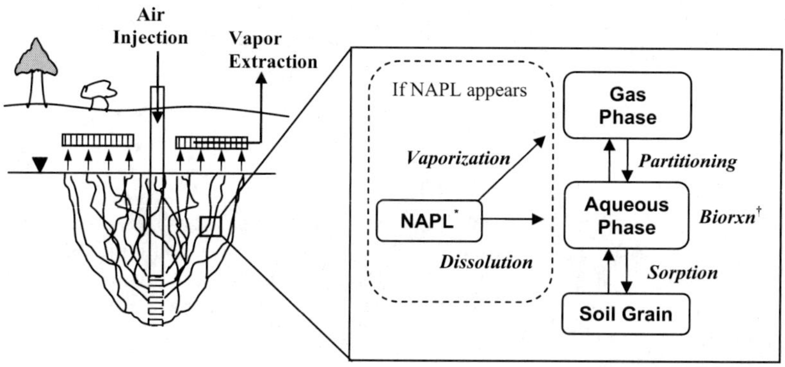

*NAPL = Nonaqueous phase liquid; †Biorxn= Aerobic and anaerobic Bioreactions.

Figure 11.3: A diagram of reactions and mass transfer between water, gas, and soil compartments.

The contaminant mass balance equations for gas, water, and NAPL phases can be written based on mass conservation in the subsurface system. Interphase mass transfer, dissolution, and vaporization of VOCs are often expressed by first-order kinetics (Rabideau and Blayden 1998). For the gas phase, the partitioning between gas and water and the vaporization from a NAPL source as well as influx and exflux of contaminants for a representative matrix element can be given as

$$V_G \frac{dC_G}{dt} = Q_G\left(C_G^{in} - C_G\right) + K_G A\left(HC_W - C_G\right) + V_G \lambda_V \left(C_G^{\max} - C_G\right) \quad (11.1)$$

where subscripts G and W indicate the gas and water phases, respectively; V_G is the volume of the gas phase [L^3], C_G and C_W are the contaminant concentrations in the gas and water phases [ML^{-3}]; Q_G is the flow of the gas [L^3T^{-1}]; C_G^{in} is the vaporized contaminant concentration in the gas coming into the compartment [ML^{-3}]; K_G is the gas-water mass transfer coefficient [LT^{-1}]; A is the interfacial area between gas and water phases [L^2]; H is the dimensionless Henry's constant; λ_V is the vaporization rate of the NAPL contaminant [T^{-1}]; and $C_G^{\max}$ is the maximum contaminant concentration in the gas phase [ML^{-3}]. For the water phase, the contaminant mass balance equation can be expressed as

$$V_W \frac{dC_W}{dt} = Q_W\left(C_W^{in} - C_W\right) - K_G A\left(HC_W - C_G\right) + V_W \lambda_D \left(C_W^{\max} - C_W\right) \quad (11.2)$$

where V_W is the volume of the water phase [L^3]; Q_W is the flow of the water [L^3T^{-1}]; C_W^{in} is the dissolved contaminant concentration in the water coming into the compartment [ML^{-3}]; λ_D is the dissolution rate of the NAPL contaminant [T^{-1}]; and $C_W^{\max}$ is the maximum contaminant concentration in the water phase [ML^{-3}]. The contaminant mass balance equation for the solid phase is

$$V_S \rho_S \frac{dC_S}{dt} = V_S \rho_S \lambda_S \left(K_S C_W - C_S\right) \tag{11.3}$$

where ρ_S is the bulk density of soil media [ML^{-3}]; λ_S is the first-order non-equilibrium sorption rate coefficient [T^{-1}]; C_S is the contaminant concentration on the soil media [$M_{contaminant}/M_{soil}$]; and K_S is the sorption coefficient under an equilibrium state [L/M_{soil}]. The dissolution and evaporation of a NAPL contaminant, which is trapped in pore spaces, result in its mass reduction. The NAPL-contaminant reduction can be expressed by

$$\rho_N \frac{dV_N}{dt} = -V_W \lambda_D \left(C_W^{\max} - C_W\right) - V_G \lambda_V \left(C_G^{\max} - C_G\right) \tag{11.4}$$

where subscript N indicates the NAPL phase; ρ_N is the density of the NAPL [ML^{-3}]; and V_N is the volume of the NAPL [L^3].

11.2.2.2 Multiphase Flow Model

When multiple immiscible fluids flow through porous media, the interfacial tension between the fluids is nonzero, and a distinct interface separates the fluids within each pore (Bear 1972). The existence of fluid interfaces makes multiphase flow problems much more difficult to analyze and model than single-phase flow such as the groundwater flow in the saturated zone (Celia et al. 1995). The ability of an interface to carry a nonzero stress is quantified by the interfacial tension for the fluid-fluid pair, and the pressure difference across an interface between two fluids is called the capillary pressure. For gas-water systems, the capillary pressure, P_c, can be written as

$$P_c = P_g - P_w \tag{11.5}$$

where P_g and P_w are gas and water pressures [$ML^{-1}T^{-2}$], respectively.

A major difficulty in modeling multiphase flow arises from the constitutive relationships governing multiphase fluid movement (Parker et al. 1987). The relationships are expressed as functional relationships of capillary pressures (P_c), saturations (s), and relative permeabilities (k_r) of coexisting phases. The relationships and their dependence on porous media and fluid characteristics have been widely studied (Ataie-Ashtiani et al. 2002; Brooks and Corey 1964; Demond and Roberts 1991; Miller et al. 1998; Mualem 1976; Parker et al. 1987; Stone 1970; van Genuchten 1980). Capillary pressure-saturation equations such as the Brooks and Corey equation (Brooks and Corey 1964) and van Genuchten equation (van Genuchten 1980) are widely used to describe the relationship between capillary pressure and effective saturation of a wetting fluid for multiphase flow (Leij et al. 1997). In air-water systems, the contact angle between the water-air interface and the solid phase is less than 90°; water is therefore the wetting fluid, which will preferentially wet the solid, and air then becomes the nonwetting fluid (Bear 1972).

In multiphase flow, each fluid is regarded as a continuum in the porous medium, and each fluid has its own flow pattern according to its saturation level. The permeability of each fluid within a porous medium is affected by the presence of the other phase. Darcy's law, originally describing the flow of a single-phase fluid completely saturating a porous medium, can be extended to describe multiphase flow (Bear 1972). Under the extension of Darcy's law, the permeability of each fluid is called an effective permeability. The ratio of the effective permeability of each fluid to the permeability of a single-phase fluid is defined as relative permeability. Relative permeability is generally expressed as a function of fluid saturation. In water-gas systems, the relative permeabilities of water and gas phases can be written using the van Genuchten model (van Genuchten 1980) as follows:

$$k_{rw} = s_{we}^{1/2}\left[1-\left(1-s_{we}^{1/m}\right)^{m}\right]^{2} \tag{11.6}$$

$$k_{rg} = c_{k}\left(1-s_{we}\right)^{1/2}\left(1-s_{we}^{1/m}\right)^{2m} \tag{11.7}$$

with

$$s_{we} = \frac{s_{w}-s_{wr}}{1-s_{wr}} = \begin{cases}\left[1+\left(\alpha P_{c}\right)^{n}\right]^{-m} & P_{c}>0\\ 1 & P_{c}\le 0\end{cases} \tag{11.8}$$

where k_{rw} and k_{rg} are the relative permeabilities of water and gas phases, respectively; c_k is the Klinkenberg factor, which accounts for the air slippage in air-water flow systems (Bear 1972); s_{we}, s_w, and s_{wr} are the effective, actual, and irreducible water saturations, respectively; and α, n, and m are the empirical parameters of the capillary pressure-saturation curve. Parker et al. (1987) extended the van Genuchten model to three fluid phases and suggested closed-form expressions for capillary pressure-saturation ($P_c - s$) and saturation-relative permeability ($s - k_r$) relationships in two- and three-phase porous media systems.

$P_c - s$ and $s - k_r$ relationships may be different depending on whether the porous medium is undergoing wetting or drying (Bear 1972). The phenomenon is called hysteresis, and is produced by pore scale effects associated with the difference in the contact angle between fluids and a solid, and nonwetting fluid entrapment during saturation path reversals (Lenhard et al. 1989). Typically, the relationships are highly nonlinear, and experimental measurement is often a difficult task. Ataie-Ashtiani et al. (2002)reported that the task can be more difficult in the presence of microheterogeneities that generate complex functional dependencies in the $P_c - s - k_r$ relationships. In both environmental remediation and petroleum reservoir engineering, nonhysteretic $P_c - s$ and $s - k_r$ relationships are usually accepted in multiphase flow modeling because the relationships are less computationally demanding and require fewer data (Miller et al. 1998).

Multiphase flow equations are complex and highly nonlinear because of the capillary pressure and relative permeability relationships incorporated into the equations. Most modeling of multiphase flows is based on numerical methods (Marulanda et al. 2000; McCray and Falta 1996; McCray and Falta 1997; Mei et al. 2002; Mohtar et al. 1994; van Dijke et al. 1995). A few investigators have developed analytical solutions (van Dijke et al. 1995; Philip 1998; van Dijke and van der Zee 1998). Due to the nonlinearity of the multiphase flow equations, analytical solutions are based on simplifying assumptions, such as steady-state air flow, incompressible air, constant air velocity, and stagnant water; relationships for relative permeability or capillary pressure have also been simplified to obtain solutions. van Dijke et al. (1995) developed an analytical solution to estimate air saturation levels and the ROI of IAS under steady-state conditions with a point source injection. The analytical solution showed good agreement with IAS simulation results. The analytical solution is a steady boundary-layer solution that neglects air flow due to vertical capillary pressure gradients. Philip (1998) presented a full air-flow solution that allows downward flow below an air-sparging point. The solution was given in terms of the dimensionless Kirchhoff potential and Stokes stream function. The former increases monotonically with capillary pressure and air saturation, and the latter maps the pattern of air flow.

The governing equation for multiphase flow can be derived by combining the mass conservation equation and Darcy's law (Bear 1972; Jang and Aral 2009) as

$$\frac{\partial\left(\phi s_f \rho_f\right)}{\partial t} = \nabla \cdot \left[\frac{\rho_f k_m k_{rf}}{\mu_f}\left(\nabla P_f + \rho_f g \nabla z\right)\right] + Q_f^* \tag{11.9}$$

where subscript f represents the mobile fluids; ϕ is the porosity; s is the fluid saturation; ρ is the fluid density [ML^{-3}]; μ is the dynamic viscosity [$ML^{-1}T^{-1}$]; k_r is the relative permeability; k_m is the intrinsic permeability tensor for soil media [L^2]; P is the fluid pressure [$ML^{-1}T^{-2}$]; g is the gravitational acceleration [LT^{-2}], z is the elevation [L]; and Q_f represent the strength of sources/sinks of mass [$ML^{-3}T^{-1}$]. For gas-water systems, the flow equations are coupled by capillary pressure term, P_c, as defined in Equation (11.5) (Parker and Lenhard 1987; Thomson et al. 1997). If a NAPL is present and is considered as an immobile residual, a change in NAPL saturation occurs only due to vaporization and/or dissolution. Also, if the saturation level of an immobile NAPL is small relative to that of water or gas and the pressures of gas and water phases are independent of the pressure of immobile NAPL, the presence of NAPL can be disregarded in solving water and gas flow equations (Sleep and Sykes 1989; Thomson et al. 1997).

The equation for the advective-dispersive transport of a contaminant in the water and gas phases can be written as (Jang and Aral 2009)

$$\frac{\partial(\phi s_f C_f)}{\partial t} = \nabla(\phi s_f D_f \nabla C_f) - \nabla(q_f C_f) + I_f + C_f^* Q_f' \quad ; f = w, g \quad (11.10)$$

where D and q are the dispersion tensor [L^2T^{-1}] and Darcy flux [LT^{-1}], respectively; I indicates the interphase mass transfer of a contaminant [$ML^{-3}T^{-1}$]; C^* is the specified concentrations of contaminants at a source/sink [ML^{-3}]; and Q' is the strength of the source/sink [T^{-1}]. The Darcy velocity for each fluid can be written as

$$q_f = -\frac{k_m k_{rf}}{\mu_f}(\nabla P_f + \rho_f g \nabla z) \quad (11.11)$$

Numerical analyses of multiphase flow have contributed significantly to the understanding of the theoretical aspects of IAS and to the design of IAS system (McCray 2000). Mohtar et al. (1994) developed a two-dimensional, steady-state model, SPARG, to simulate air sparging in porous media. The model, based on finite element method, was used to predict the distribution of capillary pressure during air sparging. Using a three-dimensional multiphase flow model developed by Huyakorn et al. (1994), Panday et al. (1994) depicted the distribution of steady-state gas pressure in the vicinity of a sparging well and the profiles of air velocity along a vertical cross-section and a horizontal plane. Transient air flows in a homogenous axially-symmetric porous medium were simulated by van Dijke et al. (1995), who developed dimensionless flow equations for water and gas based on the mixed form of Richard's equation. Their simulation results show that injected air induces water table mounding at the injection well, and that dynamic movements of air and water stabilized within 2 hours, leading to nearly steady-state air flow under conditions of hydrostatic water pressure. The phenomenon of water table mounding was also reported by Lundegard and Andersen (1996).

Using SVE with in conjunction with IAS, Unger et al. (1995) investigated the mechanisms controlling the removal of TCE as a DNAPL above and below the water table. At early times, TCE in the unsaturated zone was removed by rapid mass transfer from the NAPL to the gas phase. At later times, TCE removal was controlled by mass transfer between the NAPL and water phases in the saturated zone. For combined IAS-SVE applications, nonequilibrium effects can be associated with diffusion- and dispersion-limited mass transfer of VOCs between phases. A first-order mass transfer equation has been frequently used to represent nonequilibrium mass transfer of VOCs in porous media (Braida and Ong 2000; Sleep and Sykes 1989). In IAS modeling, Unger et al. (1995) used the CompFlow model developed by Forsyth and Shao (1991), and employed two horizontal sparging wells to improve the contact between the injected air and the DNAPL. The importance of contact time was also highlighted by Bass et al. (2000), who reviewed many field and experimental cases and provided predictive indicators for successful application of IAS. Benner et al. (2000) used the BIOVENTING^{plus} model to simulate IAS for sandy soils and shallow groundwater contaminated with LNAPL. Toluene, ethylbenzene, and xylene were the dominant LNAPL constituents. Mei et al. (2002) used a pseudo-transient method to investigate steady-state air flow by air venting and sparging in an axisymmetric domain. They noted that the compressibility of gas has significant effects on determining capillary pressure, the vertical velocity of gas, and ROI within some regions. Jang and Aral (2009) examined the remedial performance of IAS systems while investigating the flows of groundwater and injected air under different air injection rates. Their study better explained the phenomenon of groundwater and contaminant plume diversion around the IAS influence zones and the change in the ROI of IAS as a function of air injection rate.

11.2.3 Flows of Groundwater and Injected Air under IAS

Air flow in IAS plays a key role in determining the effectiveness of contaminant removal. Typically, a uniform distribution of air flow is expected in homogeneous soil media, while preferential air flow can be generated easily in heterogeneous media. In an IAS system, the injected gas interacts continuously with the groundwater while both fluid phases are subject to porous media flow. The gas flow pattern can be estimated from the relationships between the water and gas phase pressures. Ji et al. (1993) reported that a parabolic distribution of air flow can be obtained in homogeneous media, and an increase in air injection rates leads to an increase in the size of air sparging influence zone. Numerical approaches have been used to study the distribution of air flow (Jang and Aral 2009; Thomson and Johnson 2000; van Dijke et al. 1995). Air distribution in IAS depends on air injection rate and soil permeability in homogeneous soils. Air flow and gas saturation levels under IAS using a vertical injection well, developed by Jang and Aral (2009) using TechFlowMP code, are illustrated in Figure 11.4. In Figure 11.4(a), gas flow shows some lateral, downward movement around the injection screen, but overall, its upward movement is predominant within the air sparging zone. After air is introduced at the injection point, the size of the air plume increases and the air velocity decreases with its upward movement.

The distribution of gas saturation is determined by a variety parameters, such as air injection rate and soil permeability. In Figure 11.4(b), the distribution of gas saturation for an IAS system shows parabolic-shaped air plumes. Gas-saturation distributions can be used to estimate the zone of influence of IAS systems, which delineates a mixing zone where the injected air contacts with the groundwater. Because contaminant removal by IAS depends primarily on mass transfer of contaminants from the water phase to the gas phase, the size of the mixing zone should be large enough to span the width of the target contaminant plume.

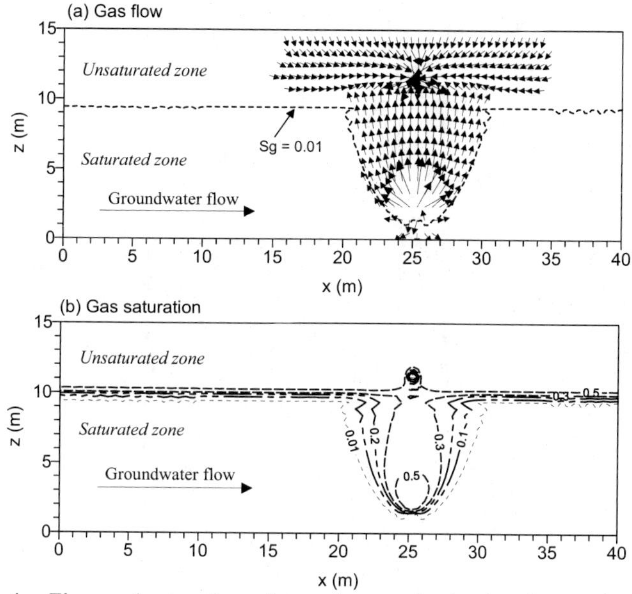

Figure 11.4: Flow and saturation of gas phase under *in-situ* air sparging (Q = 60 m^3/hr).

Air sparging reduces and diverts groundwater flow around an air-injection well within the zone influence of an IAS system, as seen in Figure 11.5. This figure presents the flow pathways and velocity profiles of groundwater in the horizontal and vertical planes. The reduction in the groundwater velocity is proportional to local gas saturation levels; the higher gas saturation induced by the greater injection rate results in greater diversion of the groundwater(Jang and Aral 2009). As pointed out by Nyer and Suthersan (1993), the lateral and vertical diversions around the air injection zone can result in the uncontrolled spreading of dissolved contaminants. The vertical diversion of groundwater should be taken into account in determining the depth of air-injection points, which should be sufficiently deep to allow remediation of the entire vertical extent of the contaminant plume.

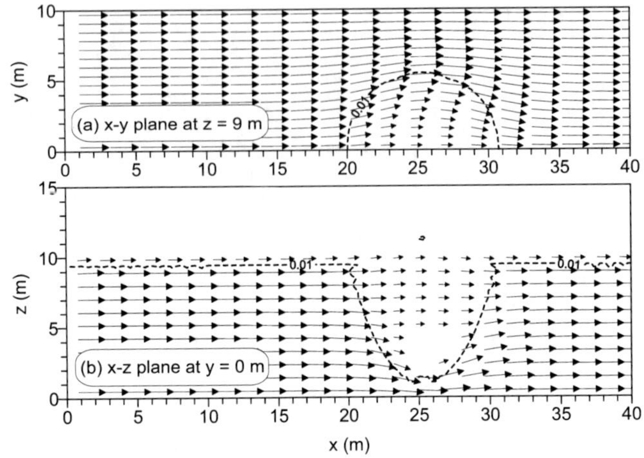

Figure 11.5: Groundwater flow on vertical and horizontal cross-sections under in-situ air sparging.

An IAS system with multiple injection wells can be implemented to expand the size of the IAS influence zone and also to provide a uniform distribution of injected air, which can improve the contact of the injected air and the contaminated groundwater. The distribution of gas saturation using two air injection wells is given in Figure 11.6. The IAS influence zone with a gas saturation of 0.1 extends up to 6.8 m in *y*-direction in the *x-y* plane in Figure 11.6(a). The gas saturation profile in the *x-z* plane indicates that influence zones of the two injection wells overlap, as shown in Figure 11.6(b). This region is important because gas saturation levels are low, thereby creating a pathway for a diverted contaminant plume. Even though the total air injection rate of this multi-injection-well system is half that of the one-injection-well system, shown in Figure 11.4, the former case produces wider and more uniform gas distribution than the latter case.

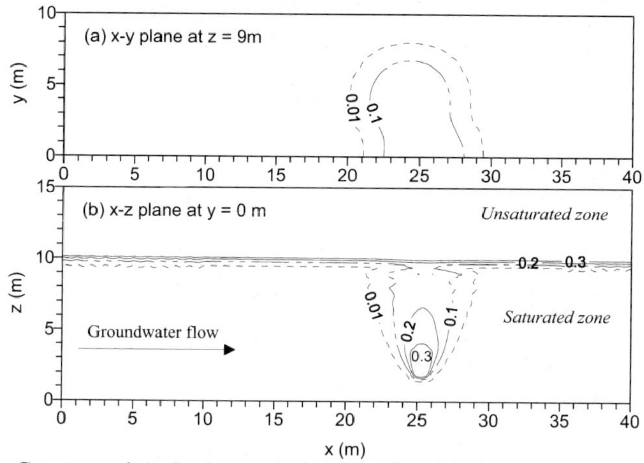

Figure 11.6: Gas saturations in the vertical and horizontal planes for a total in-situ air sparging rate of Q = 30 m^3/hr. Soil permeability k_{xx} = k_{yy}= 10^{-10} m^2 and k_{zz} = 5×10^{-11} m^2. The first and second injection wells are at (25.25 m, 0 m, 2.5 m) and (24.25 m, 6 m, 2.5 m) in (x, y, z), respectively.

11.2.4 Contaminant Transport

Normally, the size and shape of a contaminant plume is determined by the processes of advection and dispersion as well as physical, chemical, and biological processes. Under conditions of IAS wherein the presence of a gas phase locally reduces the water phase permeability, dissolved phase contaminants can be retarded by the reduced groundwater flow within the IAS influence zone as well as diverted around the IAS injection wells. Especially, the diversion of flow around air sparging wells prevents contaminated groundwater from entering the IAS zone of influence and interferes with the mixing of contaminated groundwater with the injected air, reducing the effectiveness of the remedial process.

In IAS operations, the reduction in the size and concentration of contaminant plumes will vary according to the air injection rate, with higher air injection rates resulting in the greater concentration reduction. To better understand how IAS affects contaminant transport, Jang and Aral (2009) simulated dissolved-phase TCE concentrations under IAS operations using one and two air injection wells. Results are illustrated in Figure 11.7. In Figure 11.7(a), a portion of the contaminant plume is diverted around the IAS zone of influence when using one air injection well; thus, a portion of the plume is untreated and migrates downstream beyond the zone of influence. In Figure 11.7(b), a second injection well (Injection well 2) is installed near the outer boundary of the TCE plume. In this case, the zone of influence created by the two wells fully envelops the contaminant plume in the y-direction, and the dissolved-phase contamination is successfully captured and remediated by the IAS system. Rather than using one injection well with a high injection rate, multiple injection wells with well-distributed injection rates are more effective in expanding the IAS zone of influence and intercepting contaminant plumes (Jang and Aral 2009). IAS using multiple injection wells helps to more uniformly distribute the contaminant loading per injected air volume, reduce the air injection rate at each well, and increase the overall remedial performance of the IAS system.

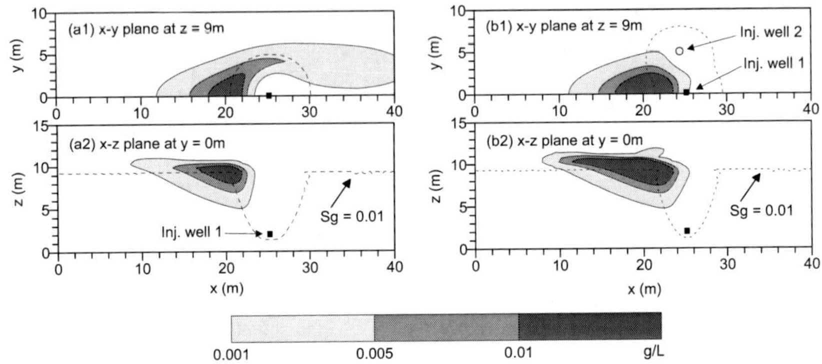

Figure 11.7: Contaminant transport under IAS operations using (a) one injection well, and (b) multiple injection wells. Total air injection $Q = 30\ m^3/hr$. Results from Jang and Aral (2009) using TechFlowMP code.

The design and operation of IAS systems with multiple injection wells is based on multiple factors, such as hydrogeologic parameters, spatial contaminant distributions, physicochemical properties of contaminants, and remedial goals. Especially, when the velocity of the groundwater flow or the concentration of contaminants is high, the contaminant loadings on each well will be high. Under such conditions, high flow rates or more injection wells may be required.

11.2.5 Radius of Influence

The ROI of an air sparging well is one of the most important parameters in the design of IAS systems, and is generally defined as the maximum radius from an injection well to the location at which remedial effects due to the injected air are observed, or as the radius from an injection well to a location at which the change in gas pressure due to air sparging is detected. The ROI can be estimated as a pre-selected gas saturation level, gas pressure, or gas-species concentration based on contaminated site conditions (Jang and Aral 2009; Lee et al. 2002). The most common method to estimate the ROI in field applications is to use the gas-phase saturation profiles (McCray and Falta 1996, 1997). In IAS simulations using the T2VOC code, Falta et al. (1995) showed that the anisotropy of the soil matrix has great importance in determining the ROI, and that the heterogeneities of the soil media also play an important role in determining airflow pathways. For example, horizontal, disk-shaped lenses of low permeability soil (such as a clay lens) contribute to the lateral movement of air injected below the lens. The effect of low-permeable layers has been demonstrated through experiments (Ji et al. 1993) and numerical studies (McCray and Falta 1996; van Dijke and van der Zee 1998). Philip (1998) and Mei et al. (2002) showed that the ROI in a fine-grained soil was larger than that in a coarse-grained soil because of stronger capillarity in the finer soil. Through air injection and extraction tests at a contaminated site, Lee et al. (2002) evaluated the ROI using two indicators: the variation of gas-phase pressure, and the propagation of DO concentrations in the groundwater. In their study, a pressure-based ROI was estimated to be the distance at which the gas-phase pressure was greater than or equal to 2.5 mm H_2O (U.S.ACE 1995), while an oxygen-based ROI was established to be the distance where dissolved oxygen concentrations were greater than or equal to 5% (U.S.EPA 1995). The pressure-based ROI was of similar magnitude as the oxygen-based ROI.

As seen in Figure 11.4(b), gas-saturation-based ROI varies with depth due to the change in the distributions of gas saturation with elevation. In Figure 11.4(b), the gas plume is nearly fully expanded at around $z = 9$ m before the injected air reaches the unsaturated zone, and thus the ROI at $z = 9$ m can be selected as a representative ROI value for a specific IAS operational condition. ROIs estimated under different IAS operational conditions (e.g., different air injection rates) are presented in Figure 11.8. The ratio of the ROI to the depth of the air injection point from the water table can be used to compare the zone of influence of IAS under different conditions. In Figure 11.8, the lower soil permeability yields a larger ROI at the same air injection rate, and the effect of air injection rate on ROI becomes greater as soil permeability decreases. The estimated ROIs at different soil permeabilities and injection rates can be used to estimate the effective remediation zones based on a certain level of gas saturation that could be obtained through pilot-scale field or laboratory experiments. However, because contaminant concentrations in the groundwater vary over time and location, it is difficult to predict the exact remedial performance of a specific gas saturation level.

11.2.6 Field Application of IAS Systems

The feasibility of air sparging technology at a particular site can be evaluated based on site characteristics and technical difficulties in implementation, remedial effectiveness, cost, and regulatory requirements (NFESC 2001; U.S.EPA 1991). The practical aspects of implementing IAS technology to remediate a particular contaminated site are discussed below.

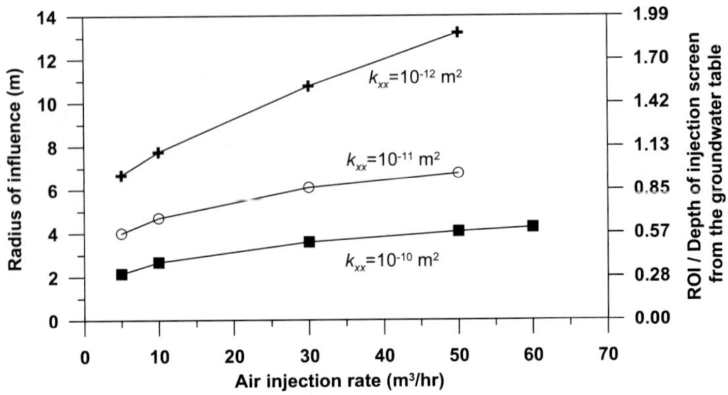

Figure 11.8: Radius of influence as a function of air injection rate at a gas saturation = 0.1, elevation of z = 9 m, and air injection rate of Q = 30 m^3/hr at a single injection well.

11.2.6.1 Subsurface Site Characterization

The site investigation and field data collection effort is a fundamental step for designing a remedial strategy for a contaminated site. Field data are critical in selecting the most appropriate remedial technique at a given site. Site investigation includes gathering information on the site history, geology, hydrogeology, nature and extent of contamination, geochemical and biological conditions, and potential hazards to human health and the environment. Table 11.3 summarizes the site characteristics that should be acquired for a particular contaminated site.

11.2.6.1 Air Sparging System Effectiveness and Feasibility

The effectiveness of IAS at a contaminated site should be evaluated based on site-specific data obtained through site characterization. The following categories can be used in evaluating the expected effectiveness of IAS for site remediation.

Contaminants of concern: IAS is typically effective at removing dissolved VOCs, and has the potential for treating NAPLs at low saturations. Section 11.2.1 discusses the characteristics of contaminants that are well-suited to IAS remediation. As contaminant removal by IAS is based primarily on vaporization and biodegradation, the contaminants of concern should be readily vaporized by the air sparging process. Contaminant mass reduction by aerobic biodegradation can also be expected when indigenous microbes as well as nutrients for microbial growth are present. If mobile NAPL is present at a site, it should be removed from the subsurface before implementing IAS operations. However, if NAPL is trapped in soil pore spaces and is immobile, IAS is an effective approach to remove NAPL from the subsurface

Table 11.3: Site characteristics/parameters needed to evaluate contaminated sites*.

Group	Characteristic/parameter	Comments
Site history	Site investigation/use records Chemical inventory records Contaminant release records	- Collect historical characterization data. - Identify potential contaminants due to previous site uses.
Site geology and hydrogeology	Subsurface geology Soil properties/stratification Groundwater depth and flow Hydraulic properties of subsurface materials Groundwater recharge and discharge Aquifer thickness	- Collect data within the target treatment area through field investigation. Includes collecting and analyzing soil cores, installing wells for groundwater and soil gas, performing aquifer tests, and monitoring groundwater elevations. - Conduct continuous groundwater monitoring to establish elevation and flow velocity and direction.
Contaminant	Contaminant sources and types Contaminants of concern Contaminant properties and concentration/distribution Volume of contaminated soil and groundwater Presence of NAPL	- Identify the extent of site contamination. Includes determining the horizontal and vertical distribution of contaminants (locations and phases), and locating zones having the highest contaminant concentrations. - Predict the migration of contaminants.
Physical, chemical, and biological conditions and processes	Dissolved oxygen (DO) Redox potential pH Temperature Nitrate, Fe(II), and Methane Microorganisms	- Collect data to predict the potential of chemical-biological reactions of contaminants. Includes characterizing the horizontal and vertical variation of these parameters within and around the contaminated zones. These parameters are optional at most air sparging sites.
Human health and the environment	Groundwater pumping wells Potential receptors of groundwater contamination and vapor migration Exposure pathways	- Identify human and environmental receptors, and resources that could be impacted by the contamination and remedial processes. - Evaluate the public acceptance of remedial processes.

*Modified from (NFESC 2001)

Geology and hydrogeology: IAS is applicable to relatively permeable soil media (e.g., sandy soils). IAS may not be suitable at sites with low permeability soils with high clay and silt contents and hydraulic conductivities less than 1×10^{-5} m/s (NFESC 2001). If the soil is homogeneous, the airflow distribution for vertical air injection wells will be the shape of an axisymmetric cone. However, if the soil is stratified and heterogeneous, preferential airflow pathways will develop, which may limit the contact between the injected air and contaminated groundwater, reducing the effectiveness of IAS. Thus, sites comprised of stratified and/or heterogeneous media require more detailed site characterization to identify the depth, location, and thickness of low permeable soil layers. Contaminants that accumulate on top of the low permeability lenses may be difficult to remove because the upward movement of injected air may be diverted around the lens. This means there would be no mixing of the contaminant and injected air. Thus, the presence and extents of low permeability lenses should be identified through site characterization. The stratification and heterogeneity of the subsurface also impacts groundwater flow. The groundwater flow direction and velocity are also important factors in determining the location and layout of air sparging wells.

Geochemical and biological reactions: The air injected during the IAS process can enhance the aerobic biodegradation of contaminants (Aelion and Kirtland 2000; Aelion et al. 1997; Bass et al. 2000; Kirtland et al. 2001; Lee et al. 2000; Travis and Rosenberg 1997), and the injection of additional nutrients can further accelerate biodegradation and biological growth(Blount et al. 2002; Brockman et al. 1995; Pfiffner et al. 1997). Brockman et al. (1995) investigated the effect of nutrients (nitrogen and phosphorus in gas phase) on in-situ bioremediation of a TCE-contaminated site through a staged experiment. In the first stage, only air was injected as a control experiment. In the second stage, 1-4% methane (by volume) was injected with the air. In the last stage, 4% methane with additional nitrous oxide (0.07% by volume) and triethyl phosphate (0.007% by volume) were injected. As compared to the control experiment, the methane injection in the second stage showed a rapid and large increase in the density of methanotrophic microorganisms. In the third stage, the addition of nitrogen and phosphorus increased the rate of TCE biodegradation by approximately three orders of magnitude.

11.2.6.1 System Design and Construction

IAS is typically used in conjunction with SVE, which removes the VOCs introduced into the gas phase of the unsaturated zone from the air sparging process (Johnson et al. 1993; Nyer and Suthersan 1993). In SVE pneumatic pressure gradients induced by vacuum pumps create a gas flow in the subsurface. IAS with SVE has been successfully applied to many sites contaminated with gasoline and chlorinated solvents (Aelion et al. 1997; Brockman et al. 1995; Hughes and Dacyk 1998; Kerfoot et al. 1998; Kirtland et al. 2001; Murray et al. 2000; Rabideau et al. 1999). Bass et al. (2000) summarized 44 air sparging cases, consisting of 8 sites for chlorinated solvents such as PCE and TCE, and 39 sites for petroleum hydrocarbons.

Air injection and soil vapor extraction system: An IAS system with SVE consists of primary air injection units and SVE units. The injection system includes an air compressor, a pressure vessel, an air filter, piping, values, a controller, and injection wells. The extraction system has a vacuum blower, a flow controller, a knockout tank, and extraction wells. A treatment unit for extracted soil vapor may be required at some sites. The capacity of the air compressor should be at least 20 cubic feet per minute (cfm) (34 m^3/hr) at pressure of 10 to 15 pounds per square inch (gage) (psig) above the calculated hydrostatic pressure at the top of the screened interval of air injection wells (NFESC 2001). In addition to air injection rates, the spacing of injection wells is very important to successful subsurface remediation. The recommended spacing is about 15 to 20 ft (4.5 to 6.1 m), with spacing not to exceed 30 ft (9.1 m) for successful IAS performance (Bass et al. 2000; Lesson et al. 2001; NFESC 2001). An example of the gas saturation distribution with about 5 m of injection well spacing is shown in Figure 11.6. The screen interval of vertical injection wells should be about 1 to 2 ft long, and the recommended depth of the screen is around 5 ft below the bottom of the contaminated zone (Lesson et al. 2001; NFESC 2001). Vertical injection wells are relatively easy to install with less cost, whereas horizontal injection wells installation cost are typically more expensive. Horizontal injection wells have the advantage of being able to generate long, relatively uniform influence zones in homogeneous soil media. However, the use of horizontal wells in heterogeneous soil media can cause the airflow distribution to be non-uniform. Under such conditions, it is very difficult to control aeration rates along the horizontal screen interval.

Monitoring network: An IAS system requires a monitoring network to measure the variation in the groundwater level, groundwater quality, and soil gas. The monitoring network consists of groundwater monitoring wells and multilevel monitoring points (NFESC 2001). Groundwater monitoring wells are typically 4-inch-diameter PVC with a relatively large screened intervals. Multilevel monitoring points (MMPs) are used for groundwater and soil vapor measurement, and are very important in identifying contaminant movement with depth. The screen interval of MMPs should be relatively small. MMPs should be installed in both the saturated and unsaturated zone. For more detailed information, refer to (NFESC 2001).

11.2.6.2 System Operation and Optimization

IAS systems can be operated in either continuous or pulsed modes. In the continuous mode, air is injected continuously via injection wells, while, in the pulsed mode, air is injected intermittently. The duration and frequency of air injection depend on site characteristics, such as hydraulic conductivity and groundwater velocity, as well as the concentration and distribution of contaminants in the subsurface. An advantage of pulsed IAS operation is that during periods of no air injection, the water saturation within an air sparging influence zone will increase as the gas phase is displaced by water below the water table, thus, facilitating the inflow of the dissolved-phase contaminant plume into the air sparging influence zone and enhancing the contaminant removal during the next cycle of air injection. The reduced operation time associated with the pulsed mode can also result in lower operation costs.

During IAS operations, remedial performance should be evaluated periodically, and the movement of contaminant plumes should be monitored. Especially, the diversion the plume around the IAS influence zone should be minimized through the adjustment of operation parameters, such as air injection flow rate, frequency, and duration. The optimization of IAS operations should be based on IAS performance, including the temporal variation of VOC concentration in the extracted soil vapor, cumulative contaminant removal, and air injection rate. The remedial efficiency and total contaminant mass removed can be estimated as follows:

$$r_{eff} = \frac{CQ_{Extraction}}{Q_{Injection}} \tag{11.12}$$

where r_{eff} is the contaminant mass removed per unit volume of injected air [ML^{-3} air]; C is the concentration of a contaminant in an extracted gas [ML^{-3}]; and $Q_{Extraction}$ and $Q_{Injection}$ are the volumetric flow rates of extracted soil gas and injected air [L^3T^{-1}], respectively. The total contaminant mass removed can be calculated as

$$M_{removed} = \sum_{i=1}^{n} CQ_{Extraction} \cdot \Delta t_i \tag{11.13}$$

where $M_{removed}$ is the total mass of a contaminant removed [M] for a period of time ($\sum \Delta t$); n is total number of time intervals; and Δt is the time interval for the measurement of extraction gas flow rates and contaminant concentrations in the extracted gas [T].

11.2.6.2 System Shutdown and Long-term Monitoring

An IAS operation may be terminated when cleanup goals are achieved or the operation is no longer effective in removing contaminant mass from the subsurface. In order to determine when it is appropriate to cease IAS operations, temporal variations of contaminant concentrations in the groundwater and extracted vapors must be evaluated. The rate of contaminant mass reduction will vary based site characteristics. If IAS operations reduce contaminant concentrations sufficiently to meet the remedial goals established for the site, a long-term monitoring plan should be implemented to confirm that the site has been successfully remediated. During the long-term monitoring period, further reduction in contaminant concentrations may occur due to natural processes in the subsurface. However, if IAS operations fail to achieve the cleanup goals, other remedial options or different a IAS application strategy would need to be considered. After successfully demonstrating attainment of cleanup goals through long-term monitoring work, the site should be closed based on a site closure plan.

11.3 Thermally-Enhanced Venting

Thermally-enhanced venting (TEV) is a heat-based, in-situ remedial technique used to enhance the removal or recovery of less-volatile residual hydrocarbon contaminants, such as naphthalene and dodecane, in the unsaturated zone. TEV applies heated air or steam instead of air at ambient conditions, used by normal venting, to cleanup the contaminated unsaturated zone. In addition to heated air and steam, electrical resistance heating, radio frequency heating, and thermal conduction can be implemented to introduce heat to the subsurface (U.S.EPA 2009). As heat energy is introduced into the ground, many physicochemical properties of fluids and contaminants change under nonisothermal conditions. A temperature increase of 20-30°C can increase in the maximum gas-phase concentration (or vapor pressure) of a less-volatile contaminant by three- to five-fold (Kaluarachchi and Islam 1995). The introduction of heated air into the contaminated zone increases the temperature of the local subsurface and this in turn increases the vaporization rates of residual NAPL contaminants. Therefore, TEV is a suitable method for enhancing the removal of medium- and less-volatile NAPLs in the unsaturated zone. Thermal energy plays an important role in increasing remediation efficiency and reducing remediation time. The application of soil heating can increase contaminant vapor pressure, contaminant diffusivity, the air-phase soil permeability through drying, and VOCs' vaporization, and can decrease NAPL viscosity which improves mobility. A schematic of a TEV system is shown in Figure 11.9.

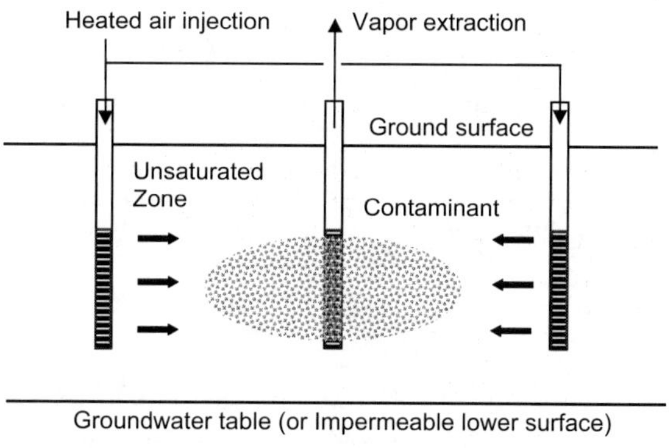

Figure 11.9: A schematic of a thermally-enhanced venting system.

11.3.1 Vapor Pressure of VOCs and Heat Sources of TEV

The physicochemical and thermal properties of VOCs depend on temperature. Especially, the vapor pressure of VOCs is significantly influenced by temperature. Figure 11.10 illustrates the relationship between vapor pressure and temperature for commonly-encountered VOCs.

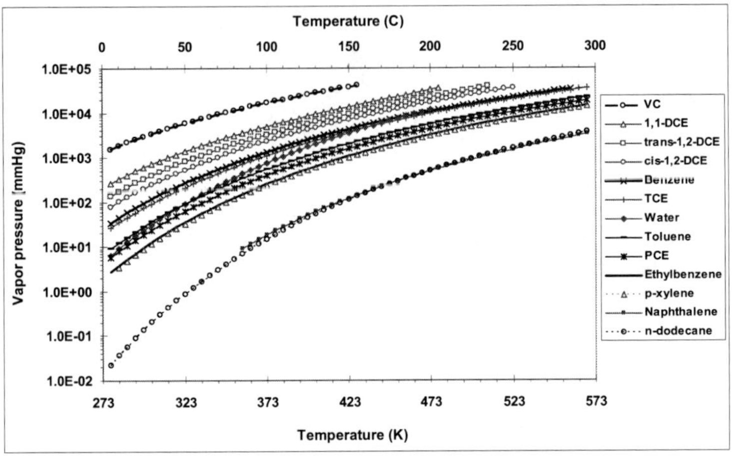

Figure 11.10: The variation of vapor pressure with temperature for common VOCs.

The addition of heat to the unsaturated zone can be accomplished by injection of hot gases (steam or heated air) or direct application of electrical heat energy into the ground (Davis 1997; FRTR 2009; U.S.EPA 2009). These methods are described in more detail below.

11.3.1.1 Injection of Hot Air or Steam

The introduction of hot air or steam below or near the contaminated zone can be used to heat contaminated soil and enhance the vaporization of VOCs. Islam and Kaluarachchi (1995) reported that the injection of hot air significantly increases the recovery of medium- to low-volatile compounds such as *n*-dodecane, naphthalene, and *n*-hexylbenzene. Fotinich et al. (1999) examined the effect of hot air on the removal of diesel fuel at two different flow rates and inlet temperatures (60 and 90°C) in a soil venting system. Their column experiments showed that a constant supply of hot air leads to a steady increase in temperature without any significant temperature drops due to the evaporation of diesel components. The increase in temperature from 60 to 90°C reduced the cleanup time by a factor of 4.5, while a threefold increase in air flow rate reduced the cleanup time by a factor of 3.8.

Field application of hot air for the purpose of unsaturated zone remediation is limited by the very low heat capacity of air (approximately 1 kJ/kg-°C). The heat capacity of steam is approximately four times that of air, and the heat of water evaporation is more than 2000 kJ/kg-°C. Thus, steam is more effective than air in transferring heat to the unsaturated zone (Davis 1997).

Steam injection has been applied extensively in removing NAPL contaminants in the variably saturated zone (Falta et al. 1992; Gudbjerg et al. 2004; Kaslusky and Udell

2002; Schmidt et al. 2002). Steam injection can enhance the release of contaminants from the soil matrix by decreasing viscosity and accelerating volatilization, and also by destroying some contaminants (U.S.EPA 2009). However, due to the condensation of steam (vaporized water), steam injection can leave behind residual water saturation, and dissolved contaminants may remain at high concentrations in the residual water (Kaluarachchi and Islam 1995; Davis 1997). Under such a situation, hot air can be injected to dry the soil and remove the dissolved contaminants. Steam injection may also cause the downward migration of DNAPL to a location below the steam zone (She and Sleep 1999). This downward migration occurs when the DNAPL that is volatilized by the steam zone condenses and accumulates in the cooler soils at the edge of the heated region. To avoid migration, the injection of a mixture of steam and air instead of pure steam has been proposed to remove DNAPL with the air being injected at the edge (Kaslusky and Udell 2002; Schmidt et al. 2002). The condensation of steam can also reduce the gas-phase permeability of the porous media (Kaluarachchi and Islam 1995). With deeper subsurface applications, hot air or steam should be introduced at relatively high pressure through injection wells.

11.3.1.2 Electrical Resistance Heating

In electrical resistance heating, an electric current is applied to the ground through arrays of electrodes. Resistance to the flow of electrical current through the soil generates temperatures greater than 100°C (U.S.EPA 2009), enhancing the vaporization of VOCs. Generally, the arrays of electrodes are installed around a central neutral electrode located in the contaminated region, creating a concentrated flow of electrical current toward the neutral electrode. For example, six-phase soil heating (SPSH) is a typical electrical resistance heating method that uses low-frequency electricity delivered to six electrodes in a circular array to heat soils.

Based on experimental studies using one-dimensional columns, Lingineni and Dhir (1992) reported that thermal venting using a resistance heater is highly effective when the thermal front (the location that soil temperature increases mainly due to heat conduction) move faster than the evaporation fronts (the location of minimum temperature due to volatilization). The rate of the thermal front movement depends mostly on the ratio of the energy contents of air and soil and air flow rate, while the rate of evaporation front movement depends on the residual saturation level, air flow rate, and volatility of contaminants.

Electrical heating has been proven effective in sandy media, and it also has a greater potential than steam to be effective in less permeable media such as clays and fine-grained sediments (Davis 1997; U.S.EPA 2009). The application of electrical resistance heating to less permeable soils can increase the permeability of the soils due to soil fractures generated while drying. Gauglitz et al. (1994) conducted field tests of SPSH with venting to remove TCE and PCE in clay soils. In their tests, electrical resistive heating effectively heated a clay zone to about 100°C. As the venting removed the vaporized soil water, air permeability increased. During heating, off-gas concentrations of TCE and PCE showed little change, but comparison of contaminant concentrations in soil samples between pre- and post-tests indicated substantial contaminant mass reduction. Through two-dimensional laboratory experiments, Heron et al. (1998) showed that resistive heating and SVE are very effective in removing dissolved TCE. In a review of rate-limited mechanisms, they pointed out that the major effect of heating on TCE removal was the increase in the volatility of TCE.

11.3.1.2 Radio Frequency/Electromagnetic Heating

The application of electromagnetic energy to the soil is an effective means of supplying the thermal energy required to heat soil at contaminated sites. Heating at power frequency (60 Hz power) is generally appropriate when the target soil temperature is less than the in-situ steam temperature (Vermeulen and McGee 2000). The rate of heating can be increased by raising the power frequency.

11.3.2 Heat Transport through Porous Media

A TEV system involves a variety of parameters because of the change in physical and chemical properties with temperature. The applicability and performance of TEV systems can be predicted on the basis of site-specific hydrogeological data and contaminant properties. To design an optimal TEV system, an understanding of thermal energy transport through porous media as well as fluid flow and contaminant transport in a nonisothermal subsurface environment is required. Because the fluid flow and contaminant transport aspects were addressed earlier for IAS, heat transport through porous media will be the focus of this section.

The transport of heat energy in the subsurface occurs by conductive and convective heat fluxes. During the TEV process, the conductive heat transport in the subsurface depends in a complex fashion on properties of the soil matrix (solid, water, gas, and NAPL), while convective heat transport occurs mainly by gas phase flow. The heat energy transport equations can be developed to describe temperature variations due to the injection of heated air, water vaporization and condensation, and contaminant vaporization. In numerical studies of TEV, heat energy transport is usually analyzed under the assumption of local thermal equilibrium between all phases and parallel heat conduction. The governing equation for thermal energy transport can be written as (Jang and Aral 2005)

$$\begin{aligned}&\frac{\partial}{\partial t}\left[S_h\left(T-T_{ref}\right)+\phi s_s\rho_v h_v+\phi s_g\rho_o h_o\right]+\\&+\nabla\cdot q_g\left[\rho_g\left(T-T_{ref}\right)+\rho_v h_v+\rho_o h_o\right]+\nabla\cdot\left[\phi s_g D_v\nabla\left(\rho_v h_v\right)+\phi s_g D_o\nabla\left(\rho_o h_o\right)\right]\\&-\nabla\cdot\left(K_A\nabla T\right)+Q_S=0\end{aligned} \tag{11.14}$$

where S_h is the overall heat capacity [J/m^3 °K]; T is the ambient temperature [°K]; T_{ref} is the reference temperature; h_f is the specific enthalpy of phases (subscript s = solid soil, w = liquid water, n = NAPL, g = air in gas, v = water vapor, and o = organic chemical vapor) [J/kg]; s is the saturation of phases; ρ_v and ρ_o are the densities of water vapor and organic vapor [kg/m^3], respectively; q_g is the Darcy velocity of gas phase [m/s]; K_A is the overall thermal conductivity [W/m °K]; ϕ is the porosity; and Q_S is the heat source/sink per unit volume due to heated air injection/extraction [J/m^3 s]. In multiphase systems, the volume-averaged overall heat capacity under thermodynamic equilibrium can be estimated as follows (Bear 1972; Kaluarachchi and Islam 1995; Kaviany 1995; Nield and Bejan 1999):

$$S_h = (1-\phi)\rho_s c_s + \phi \sum_{f=w,g,n} s_f \rho_f c_f \tag{11.15}$$

where ρ and c are the density [kg/m^3] and heat capacity [J/m^3 °K] of each phase, respectively; and subscript f indicates solid soil (s), water (w), gas (g), and NAPL (n) phases. The overall thermal conductivity depends on several parameters, such as the soil grain thermal conductivity, water thermal conductivity, gas thermal conductivity, NAPL thermal conductivity, porosity, saturation of phases, and geometry of media (Nield and Bejan 1999). The overall conductivity, K_A, can be obtained as a weighted arithmetic mean of the conductivities of phases by adopting a simple parallel conduction model as follows (Bear 1972; Nield and Bejan 1999):

$$K_A = (1-\phi)K_s + \phi \sum_{f=w,g,n} s_f K_f \tag{11.16}$$

where K_f is the heat conductivity of each phase [W/m °K]. The parallel method has been commonly used in nonisothermal subsurface simulations (Adenekan et al. 1993; Falta et al. 1992; Kaluarachchi and Islam 1995). Typically, the overall thermal conductivity is dominated by thermal conductivity of soil or rock, followed by that of liquids. Gas has the least effect on the overall conductivity, and the effect is negligible because the thermal conductivity of a liquid is 10 to 100 times larger than that of a gas (Adenekan et al. 1993).

11.3.3 Mathematical Modeling of TEV

Numerical modeling of multiphase flow, contaminant transport, and heat transport through porous media can be used to improve our understanding of the TEV process. Jang and Aral (2005) investigated the effects of TEV operations on the removal of a TCE as a NAPL and the transport of thermal energy in the contaminated unsaturated zone through numerical analysis using the TechFlowMP code (Jang and Aral 2005). In their study, a two-dimensional unsaturated domain shown in Figure 11.11 is used, which represents a contaminated region containing NAPL between injection and extraction wells. In Figure 11.11, the heated humidified air is injected into the unsaturated zone at the bottom of the domain, and the contaminated gas (the mixture of air, water vapor, and vaporized contaminant) is withdrawn from the top. While the heated air travels from the bottom to the top, the NAPL contaminant partitions into gas phase, and then the contaminant in gas phase is extracted from the subsurface. TCE, as an immobilized NAPL contaminant source in the pore space, is located in the unsaturated zone between 1 m $< r \leq$ 3 m and 1 m $\leq z \leq$ 2 m as shown in Figure 11.11.

The initial temperature of the domain is 10 °C. The heated air at constant temperature of 80 °C and a relative humidity of 50 % is injected from the bottom of the domain.

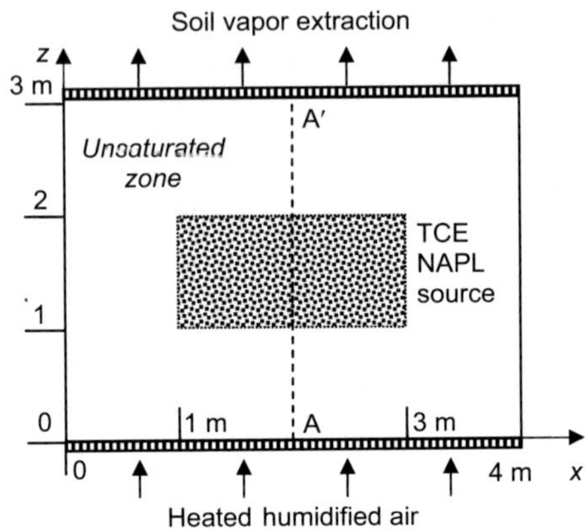

Figure 11.11: A two-dimensional domain for TEV modeling.

In TEV simulations, a set of nonlinear equations for gas flow, liquid water and NAPL saturations, chemical transport, and heat transport are solved sequentially and iterated until the solutions of all equations converge within specified criteria. The modeling results, including the gas phase Darcy velocity, temperature distributions, and NAPL saturations are given in Figure 11.12. The presence of an immobile NAPL contaminant at the contaminated zone lowers the effective saturation of the mobile fluid, the gas phase in this case, and thus decreases the relative permeability of the fluid. The reduction in the relative permeability of the mobile fluid can cause the fluid to be diverted around the low permeability zone (Frind et al. 1999; Jang and Aral 2009). As seen in Figure 11.12(a), the NAPL TCE source reduces the velocity of the gas phase within the source zone, resulting in a portion of gas bypassing the source zone, while the velocity outside the source zone increases. The magnitude of the bypassing decreases with time as the TEV operation removes the NAPL TCE.

The injection of the heated air raises the soil temperature in the domain as shown in Figure 11.12(b). The soil temperature within the source zone is lower than that outside the source zone at the same *z*-axis elevation, which is due to the evaporation of the NAPL TCE within the source zone. TCE vaporization utilizes a portion of thermal energy delivered to the source zone by the injected air; thus, the temperature increases within the source zone at a slower rate than outside the source zone. The temporal increase in soil temperature along line A-A' is shown in Figure 11.12(c). The increase in soil temperature plays an important role in enhancing the vaporization of TCE within the contaminant source zone. The TCE vaporization results in the reduction of NAPL saturation in the source zone with time, which is illustrated in Figure 11.12(d). The reduction in TCE NAPL saturation starts from the lower edge of the source zone (z = 1 m) because TCE evaporation is the greatest at that point, and the increase in soil temperature by heat transport occurs along with gas flow in the increasing z-direction.

In TEV applications at contaminated sites, the temperature of the injected air is an important factor for heat transfer in soil matrix and contaminant removal. Increasing the relative humidity and flow rate of the injected air will proportionally increase the heat capacity and the total thermal energy supplied to the domain, respectively.

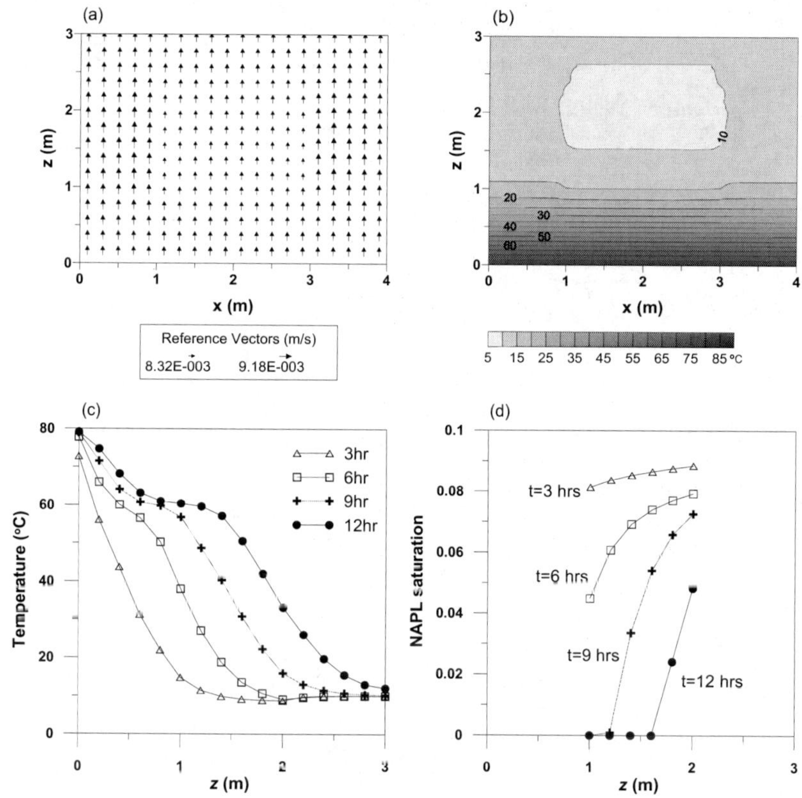

Figure 11.12: (a) Darcy velocity of gas phase at t = 2 hrs, (b) temperature distribution at t = 2 hrs, (c) temporal variation in soil temperature along line A-A', and (d) temporal reduction in TCE NAPL saturation along line A-A'.

11.3.4 Field Application Cases of TEV

In field applications of TEV, a combination of technologies, such as steam-air injection, electrical resistance heating, electromagnetic heating, and SVE, can be used to optimize remediation performance (LaChance et al. 2009; LLNL 1993; U.S.ACE 1999; U.S.DoE 2003; U.S.EPA 2003c; U.S.EPA 2003d; WSRC 2001). Table 11.4 summarizes a number of TEV projects.

Table 11.4: Summary of TEV projects.

Site (reference)	Contaminants/Media	Description
A.G. Communications Systems Site: Full scale system (U.S.EPA 2003d)	- TCE, cis-1,2-DCE, xylene and benzene - Media treated: soil and groundwater in the saturated and unsaturated zone.	- Steam-enhanced extraction using shallow vapor extraction wells, shallow and deep steam injection wells, vacuum-enhanced groundwater/vapor extraction wells, and deep groundwater extraction wells. - Steam supplied by a 294 kilowatt boiler at a pressure 3-7 psi. - Soil temperature: 84 to 140°F. - Removal: a total of 33,000 pounds of hydrocarbons. - Cost: $4.9 million.
Young-Rainy Star Center (U.S.DoE 2003)	TCE, cis-1,2-DCE, VC, methylene chloride, toluene, and NAPL. -Media treated: soil and groundwater.	- Steam-enhanced extraction and electro-thermal dynamic stripping process. - Temperature: greater than 84°C at a depth of 14-35 ft below ground surface. - Removal: an estimated 3,000 lbs of VOCs. - Cost: $3.8 million
Poleline Road Disposal Area: Pilot scale (U.S.ACE 1999)	1,1,2,2-PCA, TCE, and PCE -Media treated: soil and groundwater.	- SVE enhanced by six-phase soil heating. - Electrical power delivered to the soil by steel electrodes inserted vertically in a circular array. - Use of SPSH with SVE systems allowed more rapid treatment in colder environments (Alaska). - Treatment: 7,150 tons of solvent-contaminated soil. - Cost: $0.97 million
Lawrence Livermore National Laboratory Gasoline spill site: Field demonstration-Commercial scale (LLNL 1993)	Fuel hydrocarbons in gasoline: BTEX -Media treated: soil and groundwater	- Dynamic underground stripping: Combination of three technologies: steam injection-vacuum extraction, electrical heating of less permeable soils, and underground imaging to delineate heated areas. - Treatment: 100,000 cubic yards contaminated soil; over 7,600 gallons of gasoline removed. - Cost: $10.4 million
Savannah River Site 321-M Solvent Storage Tank Area (WSRC 2001)	NAPL PCE and TCE (NAPL: 90% PCE and 10% TCE).	- Dynamic underground stripping-hydrous pyrolysis oxidation: steam-injection well clusters, dual-phase groundwater/vapor extraction well, vadose zone SVE wells, steam and air injection. - Maximum design pressure of 40-60 psig at 143-152 °C for vadose zones. -Treatment: Total volume of 52,000 cubic yards.

Steam injection is the preferred method for heating the soil matrix (Baker and Heron 2009), and has been applied to many contaminated sites, including the A.G. Communications Systems site (AGCS), Savannah River Site 321-M, and the Young-Rainy Star Center site (U.S.DoE 2003; U.S.EPA 2003d; WSRC 2001). The AGCS site, located near Chicago, Illinois operated as a telecommunications manufacturing facility for over 40 years (U.S.EPA 2003d). The soil and groundwater at the site was contaminated with chlorinated solvents, including TCE and *cis*-1,2- DCE, and components of mineral spirits, including xylene and benzene. The source of the contamination was identified as an area in the vicinity of the former tank farm and beneath the manufacturing facility. The geologic strata at the site consists of three till layers overlying dolomite bedrock: Valparaiso Till consisting of a fine- to medium-grained sand layer at 36-38 ft below ground surface (bgs); a dense, over-consolidated, well-sorted laminated silt at 38-48 ft bgs; and Lemont Drift consisting of a thick, coarse-grained sand and gravel layer at 48-65 ft bgs. The depth of groundwater is about 38-40 ft bgs. A steam-enhanced extraction (SEE) system was pilot-tested at the site from January through July 1994. Full-scale SEE operations were performed from September 1995 to November 1999, which covered an area of about 250,000 ft^2 to a depth of about 50 ft at the AGCS site. The system included 65 steam injection wells (39 shallow wells screened to a depth of 35 ft bgs and 26 deep wells screened to a depth of 46 ft bgs), 186 shallow vapor extraction wells, 76 combination groundwater/vapor extraction wells, and two vapor extraction units operated at 150 to 250 scfm. Steam was supplied at pressures ranging from 3 to 7 psi; during operation, soil temperatures ranged from 84 ^{o}F to 140 ^{o}F, and groundwater temperatures ranged from 68 ^{o}F to 165 ^{o}F. The average groundwater concentration for TCE was reduced from approximately 20,000 μg/L to less than 1,000 μg/L over the period from September 1995 to September 1997. TCE and *cis*-1,2-DCE concentrations were reduced by more than 90% at individual wells during December 1995 to October 1997. The SEE system at the AGCS site removed more than 33,000 lbs of hydrocarbons from soil vapor and groundwater through November 1999 (U.S.EPA 2003d).

Electrical resistive heating has been used to remediate a variety of contaminated sites. Examples include the Charleston Naval Complex, North Charleston, South Carolina (NFESC 2007); the Poleline Road Disposal Area, Fort Richardson, Alaska (U.S.EPA 2003b); the Avery Dennison Site, Waukegan, Illinois (U.S.EPA 2003a); and Launch Complex 34, Cape Canaveral Air Station (CCAS), Florida (Battelle 2008). A site investigation at the CCAS site identified a large TCE NAPL source at Launch Complex 34, which was used as a launch site for Saturn rockets from 1960 to 1968 (Battelle 2008). Historical activities at the site included discharging wastes generated from rocket engine and cleaning operations into discharge pits. TCE was detected in the Upper Sand Unit (from ground surface to approximately 20 to 26 ft bgs), the Middle Fine-Grained Unit (thickness of 10-15 ft), Lower Sand Unit (thickness of 15-20 ft), and the Lower Clay Unit with thickness of 1.5-3 ft. The estimated mass of NAPL TCE was approximately 10,490 kg, based on a threshold TCE concentration of about 300 mg/kg in the soil. Electrical resistive heating was implemented to remove TCE DNAPL, and a field demonstration was performed from August 18, 1999 to July

12, 2000, with the post-demonstration assessment performed from August to December 2000. The electrical resistive heating system used 13 electrodes (each having two conductive intervals located at 25-30 ft bgs and 38-45 ft bgs) and 12 SVE wells with 2-ft screens to depths of 4-6 ft bgs. A total of 1.7 million kW-hrs of energy was applied to the subsurface through the electrodes. This field demonstration resulted in TCE mass reductions of 90% and 97% in the soil and NAPL, respectively. Resistive heating was found to be more efficient in the deeper portion of the aquifer, with less heating observed in the shallow portion. The greatest change in TCE NAPL mass was observed in the Lower Sand Unit, followed by the Middle Fine-Grained Unit. The Upper Sand Unit showed the least removal.

A conductive heating technique was implemented at a Confidential Chemical Manufacturing Facility (CCMF), Portland, Indiana (U.S.EPA 2003c). The CCMF site has been used since 1886, first as a lumber yard, and then for wheel manufacturing. From 1937 to the mid-1970's, the site was used to manufacture hard rubber products for automobiles and then to manufacture plastic exterior automobile parts. The two most heavily contaminated areas at the site were designated as GP-31 and GP-28. The GP-31 site covered an area of 150 ft by 50 ft to a depth of 18 ft and was contaminated with TCE (up to 79 mg/kg) and PCE (up to 3,500 mg/kg) present as DNAPL. The GP-28 site covered an area of 30 ft by 20 ft to a depth of 11 ft and was contaminated with 1,1-DCE at a maximum concentration of 0.65 mg/kg. For in-situ conductive heating, 130 and 18 heater/vacuum wells were installed at the GP-31 and GP-28 sites, respectively. The heater/vacuum wells were 4.5 inches in diameter and operated at 1,400 °F to 1,600 °F for heating, and soil gas was extracted using a vacuum. The maximum soil temperature in the treatment area at a depth of 13 ft was 212 °F to 500 °F. The TEV operation at the CCMF site lowered PCE and TCE concentrations at both sites below the cleanup goals, which were 8 mg/kg for PCE and 25 mg/kg for TCE (U.S.EPA 2003c).

11.4 Conclusions

IAS and TEV technologies are discussed in this chapter. Both technologies have advantages and limitations. For example, the advantages of IAS include relatively simple implementation, relatively short cleanup times, and the ability to use low cost (direct-push) wells. Its application is widely recognized by engineering and regulatory communities as an effective remediation technology for VOC removal. However, IAS limitations include the potential for migration of VOC vapors to human and/or ecological receptors at potential levels of concern, and the tendency for groundwater flow to be diverted around the treatment area due to air injection into the saturated zone and associated reductions in water phase permeability. For TEV, its primary advantage is its ability to enhance the removal of semi- or low-volatile NAPL contaminants through the addition of thermal energy into the subsurface. Its primary disadvantage is the additional cost associated with supplying this thermal energy.

As both IAS and TEV are in-situ remedial technologies, the contaminated soil and groundwater are treated without being excavated/withdrawn and transported, resulting in significant cost savings. However, a high degree of heterogeneity in soil and aquifer systems may make it difficult to treat contaminated sites to uniform levels. The field application of IAS and TEV requires consideration of a variety of site-specific characteristics regarding hydrology, geology, and physical-chemical-biological subsurface conditions. In addition, regulatory issues (e.g., permits for well installation, air pollution control devices, and underground injection control permits) and public acceptance of remedial technologies and operations should be taken into consideration when applying IAS and TEV.

11.5 References

Adenekan, A. E., Patzek, T. W., and Pruess, K. (1993). "Modeling of multiphase transport of multicomponent organic contaminants and heat in the subsurface - Numerical-model formulation." *Water Resources Research*, 29(11), 3727-3740.

Aelion, C. M., and Kirtland, B. C. (2000). "Physical versus biological hydrocarbon removal during air sparging and soil vapor extraction." *Environmental Science & Technology*, 34(15), 3167-3173.

Aelion, C. M., Kirtland, B. C., and Stone, P. A. (1997). "Radiocarbon assessment of aerobic petroleum bioremediation in the vadose zone and groundwater at an AS/SVE site." *Environmental Science & Technology*, 31(12), 3363-3370.

Ahlfeld, D. P., Dahmani, A., and Ji, W. (1994). "A Conceptual-Model of Field Behavior of Air Sparging and Its Implications for Application." *Ground Water Monitoring and Remediation*, 14(4), 132-139.

Andreoni, V., and Gianfreda, L. (2007). "Bioremediation and monitoring of aromatic-polluted habitats." *Applied Microbiology and Biotechnology*, 76(2), 287-308.

Ataie-Ashtiani, B., Hassanizadeh, S. M., and Celia, M. A. (2002). "Effects of heterogeneities on capillary pressure-saturation-relative permeability relationships." *Journal of Contaminant Hydrology*, 56(3-4), 175-192.

Baker, R. S., and Heron, G. (2009). "In-Situ Delivery of Heat by Thermal Conduction and Steam Injection for Improved DNAPL Remediation." *http://www.terratherm.com/resources/TechPapers/Terratherm%20Delivery%20of%20Heat%20Paper%20-%20Monterey.pdf.*

Bass, D. H., Hastings, N. A., and Brown, R. A. (2000). "Performance of air sparging systems: A review of case studies." *Journal of Hazardous Materials*, 72(2 3), 101-119.

Battelle. (2008). "Demonstration of Resistive Heating Treatment of DNAPL Source Zone at Launch Complex 34 in Cape Canaveral Air Force Station, Florida: Final Innovative Technology Evaluation Report, EPA/540/R-08/004." Battelle, Columbus, OH.

Bear, J. (1972). *Dynamics of Fluids in Porous Media*, American Elsevier, New York.

Benner, M. L., Mohtar, R. H., and Lee, L. S. (2002). "Factors affecting air sparging remediation systems using field data and numerical simulations." *Journal of Hazardous Materials*, 95(3), 305-329.

Benner, M. L., Stanford, S. M., Lee, L. S., and Mohtar, R. H. (2000). "Field and numerical analysis of in-situ air sparging: a case study." *Journal of Hazardous Materials*, 72(2-3), 217-236.

Blount, G. C., Caldwell, C. C., Cardoso-Neto, J. E., Conner, K. R., Jannik, G. T., Jr., C. E. M., Noffsinger, D. C., and Ross, J. A. (2002). "The Use of Natural Systems to Remediate Groundwater: Department of Energy Experience at the Savannah River Site." *Remediation*, 12(3), 43-61.

Braida, W., and Ong, S. K. (2000). "Influence of porous media and airflow rate on the fate of NAPLs under air sparging." *Transport in Porous Media*, 38(1-2), 29-42.

Brockman, F. J., Payne, W., Workman, D. J., Soong, A., Manley, S., and Hazen, T. C. (1995). "Effect of Gaseous Nitrogen and Phosphorus Injection on in-Situ Bioremediation of a Trichloroethylene-Contaminated Site." *Journal of Hazardous Materials*, 41(2-3), 287-298.

Brooks, R. H., and Corey, A. T. (1964). "Hydraulic Properties of Porous Media." *Hydrology Paper 3., 27 pp., Colorado State University, Fort Collins, Co.*

Brown, R., Herman, C., and Henry, E. (1999). "The use of aeration in environmental clean-ups." *Haztech International '91 conference proceedings, Pittsburgh, Pennsylvania.*, 1-42.

Canada, E. (2008). "Tabs on Contaminated Sites, Contaminated Sites Program - Federal Sites." *http://www.on.ec.gc.ca/pollution/ecnpd/tabs/tab21-e.html.*

Celia, M. A., Reeves, P. C., and Ferrand, L. A. (1995). "Recent Advances in Pore Scale Models for Multiphase Flow in Porous-Media." *Reviews of Geophysics*, 33, 1049-1057.

Chen, M. R., Hinkley, R. E., and Killough, J. E. (1996). "Computed tomography imaging of air sparging in porous media." *Water Resources Research*, 32(10), 3013-3024.

Clayton, W. S. (1999). "Effects of pore scale dead-end air fingers on relative permeabilities for air sparging in soils." *Water Resources Research*, 35(10), 2909-2919.

Clement, T. P., Johnson, C. D., Sun, Y. W., Klecka, G. M., and Bartlett, C. (2000). "Natural attenuation of chlorinated ethene compounds: model development and field-scale application at the Dover site." *Journal of Contaminant Hydrology*, 42(2-4), 113-140.

Cookson, J. T. (1995). *Bioremediation engineering : design and application*, McGraw-Hill, New York.

Corapcioglu, M. Y., Cihan, A., and Drazenovic, M. (2004). "Rise velocity of an air bubble in porous media: Theoretical studies." *Water Resources Research*, 40(4), doi:10.1029/2003WR002618.

CWA. (2006). "Removal of MTBE from Drinking Water Using Air Stripping: Case Studies." National Water Research Institute, California Water Agencies (CWA).

Davis, E. L. (1997). "How heat can enhance in-situ soil and aquifer remediation: Importnat chemical properties and guidance on choosing the appropriate technique, Ground Water Issue EPA/540/S-97/502." U.S. Environmental Protection Agency.

Demond, A. H., and Roberts, P. V. (1991). "Effect of Interfacial Forces on 2-Phase Capillary Pressure-Saturation Relationships." *Water Resources Research*, 27(3), 423-437.

Falta, R. W., Pruess, K., and Finsterle, S. (1995). "T2VOC User's Guide, Lawrence Berkeley Lab. Rep. LBL-36400 UC-400, Berkeley, CA."

Falta, R. W., Pruess, K., Javandel, I., and Witherspoon, P. A. (1992). "Numerical modeling of steam injection for the removal of nonaqueous phase liquids from the subsurface. 1. Numerical formulation." *Water Resources Research*, 28(2), 433-449.

Forsyth, P. A., and Shao, B. Y. (1991). "Numerical-Simulation of Gas Venting for Napl Site Remediation." *Advances in Water Resources*, 14(6), 354-367.

Fotinich, A., Dhir, V. K., and Lingineni, S. (1999). "Remediation of simulated soils contaminated with diesel." *Journal of Environmental Engineering-ASCE*, 125(1), 36-46.

Frind, E. O., Molson, J. W., Schirmer, M., and Guiguer, N. (1999). "Dissolution and mass transfer of multiple organics under field conditions: The Borden emplaced source." *Water Resources Research*, 35(3), 683-694.

FRTR. (2009). "Remediation Technologies Screening Matrix and Reference Guide, Version 4.0." U.S. Army Environmental Center, *http://www.frtr.gov/matrix2/top_page.html.*

Gauglitz, P. A., Roberts, J. S., Bergsman, T. M., Carley, S. M., Heath, W. O., Miller, M. C., Moss, R. W., Schalla, R., Schlender, M. H., Jarosch, T. R., Eddy-Dilek, C. A., and Looney, B. B. (1994). "Field test of six phase soil heating at the Savannah river site." Thirty-third Hanford symposium on health and the environment, G. W. Gee and N. R. Wing, eds., November 7-11, 1994, Richland, Washington.

Gudbjerg, J., Sonnenborg, T. O., and Jensen, K. H. (2004). "Remediation of NAPL below the water table by steam-induced heat conduction." *Journal of Contaminant Hydrology*, 72(1-4), 207-225.

Heron, G., Gierke, J. S., Faulkner, B., Mravik, S., Wood, L., and Enfield, C. G. (2002). "Pulsed air sparging in aquifers contaminated with dense nonaqueous phase liquids." *Ground Water Monitoring and Remediation*, 22(4), 73-82.

Heron, G., Van Zutphen, M., Christensen, T. H., and Enfield, C. G. (1998). "Soil heating for enhanced remediation of chlorinated solvents: A laboratory study on resistive heating and vapor extraction in a silty, low-permeable soil contaminated with trichloroethylene." *Environmental Science & Technology*, 32(10), 1474-1481.

Hinchee, R. E., and Arthur, M. (1991). "Bench Scale Studies of the Soil Aeration Process for Bioremediation of Petroleum-Hydrocarbons." *Applied Biochemistry and Biotechnology*, 28-9, 901-906.

Hinchee, R. E., Wilson, J. T., and Downey, D. C. (1995). *Intrinsic bioremediation*, Battelle Press, Columbus.

Hughes, W. D., and Dacyk, P. I. (1998). "Air sparging of chlorinated volatile organic compounds in a layered geologic setting." *International Conference on Remediation of Chlorinated Recalcitrant Compounds, Monterey California, May 18-21, 1998*, 279-284.

Huyakorn, P. S., Panday, S., and Wu, Y. S. (1994). "A 3-dimensional multiphase flow model for assessing NAPL contamination in porous and fractured media .1. Formulation." *Journal of Contaminant Hydrology*, 16(2), 109-130.

Islam, K. M. M., and Kaluarachchi, J. J. (1995). "Thermal venting to recover less-volatile hydrocarbons from the unsaturated zone. 2. Model applications." *Journal of Contaminant Hydrology*, 17(4), 313-331.

Jang, W., and Aral, M. M. (2005). "Three-dimensional Multiphase Flow and Multi-species Transport Model, TechFlowMP." *Report No. MESL-02-05*, Multimedia Environmental Simulations Laboratory, School of Civil and Environmental Engineering, Georgia Institute of Technology, Atlanta, GA.

Jang, W., and Aral, M. M. (2006). "Modeling of Multiphase Flow and Contaminant Removal under In-situ Air Sparging." Air & Waste Management Association's 99th Annual Conference & Exhibition, New Orleans, LA.

Jang, W., and Aral, M. M. (2007a). "Analyses of groundwater flow, contaminant fate and transport, and distribution of drinking water at Tarawa Terrace and vicinity, U.S. Marine Corps Base Camp Lejeune, North Carolina: Historical reconstruction and present-day conditions—Chapter G: Simulation of three-dimensional multispecies, multiphase mass transport of tetrachloroethylene (PCE) and associated degradation by-products." Agency of Toxic Substances and Disease Registry, Atlanta, GA.

Jang, W., and Aral, M. M. (2007b). "Modeling of co-existing anaerobic-aerobic biotransformations of chlorinated ethenes in the subsurface." The Third International Conference on Environmental Science and Technology 2007, American Academy of Science, Houston, TX.

Jang, W., and Aral, M. M. (2008a). "Effect of biotransformation on multispecies plume evolution and natural attenuation." *Transport in Porous Media*, 72(2), 207-226.

Jang, W., and Aral, M. M. (2008b). "The effect of oxygen transport on biotransformation of trichloroethylene in the subsurface." *World Environmental & Water Resources Congress 2008*, Honolulu, Hawaii.

Jang, W., and Aral, M. M. (2009). "Multiphase Flow Fields in In-situ Air Sparging and its Effect on Remediation." *Transport in Porous Media*, 76(1), 99-119.

Ji, W., Dahmani, A., Ahlfeld, D. P., Lin, J. D., and Hill, E. (1993). "Laboratory study of air sparging - air-flow visualization." *Ground Water Monitoring and Remediation*, 13(4), 115-126.

Johnson, P. C., Das, A., and Bruce, C. (1999). "Effect of flow rate changes and pulsing on the treatment of source zones by in situ air sparging." *Environmental Science & Technology*, 33(10), 1726-1731.

Johnson, R., Pankow, J., Bender, D., Price, C., and Zogorski, J. (2000). "MTBE - To what extent will past releases contaminate community water supply wells?" *Environmental Science & Technology*, 34(9), 210A-+.

Johnson, R. L., Johnson, P. C., Mcwhorter, D. B., Hinchee, R. E., and Goodman, I. (1993). "An Overview of in-Situ Air Sparging." *Ground Water Monitoring and Remediation*, 13(4), 127-135.

Kaluarachchi, J. J., and Islam, K. M. M. (1995). "Thermal venting to recover less-volatile hydrocarbons from the unsaturated zone. 1. Theory." *Journal of Contaminant Hydrology*, 17(4), 293-311.

Kane, S. R., Beller, H. R., Legler, T. C., Koester, C. J., Pinkart, H. C., Halden, R. U., and Happel, A. M. (2001). "Aerobic biodegradation of methyl tert-butyl ether by aquifer bacteria from leaking underground storage tank sites." *Applied and Environmental Microbiology*, 67(12), 5824-5829.

Kaslusky, S. F., and Udell, K. S. (2002). "A theoretical model of air and steam co-injection to prevent the downward migration of DNAPLs during steam-enhanced extraction." *Journal of Contaminant Hydrology*, 55(3-4), 213-232.

Kaviany, M. (1995). *Principles of heat transfer in porous media*, Springer-Verlag, New York.

Kerfoot, W. B., Schouten, C. J. J. M., and Engen-Beukeboom, V. C. M. v. (1998). "Kinetic analysis of pilot test results of the C-SPARGE Process." *International Conference on Remediation of Chlorinated Recalcitrant Compounds, Monterey California, May 18-21, 1998*, 271-277.

Kirtland, B. C., and Aelion, C. M. (2000). "Petroleum mass removal from low permeability sediment using air sparging/soil vapor extraction: impact of continuous or pulsed operation." *Journal of Contaminant Hydrology*, 41(3-4), 367-383.

Kirtland, B. C., Aelion, C. M., and Widdowson, M. A. (2001). "Long-term AS/SVE for petroleum removal in low-permeability Piedmont saprolite." *Journal of Environmental Engineering-ASCE*, 127(2), 134-144.

LaChance, J., Baker, R. S., Galligan, J. P., and Bierschenk, J. M. (2009). "Application of "Thermal Conductive Heating/In-Situ Thermal Desorption (ISTD)" to the Remediation of Chlorinated Volatile Organic Compounds in Saturated and Unsaturated Settings." *http://www.terratherm.com/resources/TechPapers/TerraTherm%20CVOC%20Paper%20Monterey.pdf.*

Lawrence, S. J. (2007). "Analyses of groundwater flow, contaminant fate and transport, and distribution of drinking water at Tarawa Terrace and vicinity, U.S. Marine Corps Base Camp Lejeune, North Carolina: Historical reconstruction and present-day conditions—Chapter D: Properties of Degradation Pathways of Common Organic Compounds in Groundwater. Agency of Toxic Substances and Disease Registry, Atlanta, GA."

Lee, C. H., Jang, W. Y., Jeon, Y. H., Lee, J., and Lee, K. K. (2000). "Study on air permeability and radius of influence in soil vapor extraction and bioventing." *Korean Society of Groundwater Environment*, 7(1), 24-31.

Lee, C. H., Lee, J. Y., Jang, W. Y., Jeon, Y. H., and Lee, K. K. (2002). "Evaluation of air injection and extraction tests at a petroleum contaminated site, Korea." *Water Air and Soil Pollution*, 135(1-4), 65-91.

Leij, F. J., Russell, W. B., and Lesch, S. M. (1997). "Closed-form expressions for water retention and conductivity data." *Ground Water*, 35(5), 848-858.

Lenczewski, M., Jardine, P., McKay, L., and Layton, A. (2003). "Natural attenuation of trichloroethylene in fractured shale bedrock." *Journal of Contaminant Hydrology*, 64(3-4), 151-168.

Lenhard, R. J., Parker, J. C., and Kaluarachchi, J. J. (1989). "A Model for Hysteretic Constitutive Relations Governing Multiphase Flow .3. Refinements and Numerical Simulations." *Water Resources Research*, 25(7), 1727-1736.

Lesson, A., Johnson, P. C., Johnson, R. L., Hinchee, R. E., and McWhorter, D. B. (2001). "Air sparging Design Paradigm." Prepared for Battelle, Columbus, OH.

Lin, C.-W., and Cheng, Y.-W. (2007). "Biodegradation kinetics of benzene, methyl tert-butyl ether, and toluene as a substrate under various substrate concentrations." *Journal of Chemical Technology & Biotechnology*, 82(1), 51-57.

Lingineni, S., and Dhir, V. K. (1992). "Modeling of soil venting processes to remediate unsaturated soils." *Journal of Environmental Engineering-ASCE*, 118(1), 135-152.

LLNL. (1993). "Dynamic Underground Stripping Demonstration at DOE's Lawrence Livermore National Laboratory, Gasoline Spill Site, Livermore, California." Lawrence Livermore National Laboratory.

Lorah, M. M., and Olsen, L. D. (1999). "Natural attenuation of chlorinated volatile organic compounds in a freshwater tidal wetland: Field evidence of anaerobic biodegradation." *Water Resources Research*, 35(12), 3811-3827.

Lovley, D. R. (1997). "Potential for anaerobic bioremediation of BTEX in petroleum-contaminated aquifers." *Journal of Industrial Microbiology & Biotechnology*, 18(2-3), 75-81.

Lundegard, P. D., and Andersen, G. (1996). "Multiphase numerical simulation of air sparging performance." *Ground Water*, 34(3), 451-460.

Lundegard, P. D., and Labrecque, D. (1995). "Air Sparging in a Sandy Aquifer (Florence, Oregon, USA) - Actual and Apparent Radius of Influence." *Journal of Contaminant Hydrology*, 19(1), 1-27.

Marulanda, C., Culligan, P. J., and Germaine, J. T. (2000). "Centrifuge modeling of air sparging -- a study of air flow through saturated porous media." *Journal of Hazardous Materials*, 72(2-3), 179-215.

Maslia, M. L., Sautner, J. B., Faye, R. E., Suárez-Soto, R. J., Aral, M. M., Grayman, W. M., Jang, W., Wang, J., Bove, F. J., Ruckart, P. Z., Valenzuela, C., Green, J. W. J., and Krueger, A. L. (2007). "Analyses of groundwater flow, contaminant fate and transport, and distribution of drinking water at Tarawa Terrace and vicinity, U.S. Marine Corps Base Camp Lejeune, North Carolina: Historical reconstruction and present-day conditions—Executive Summary." Agency of Toxic Substances and Disease Registry, Atlanta, GA.

McCray, J. E. (2000). "Mathematical modeling of air sparging for subsurface remediation: state of the art." *Journal of Hazardous Materials*, 72(2-3), 237-263.

McCray, J. E., and Falta, R. W. (1996). "Defining the air sparging radius of influence for groundwater remediation." *Journal of Contaminant Hydrology*, 24(1), 25-52.

McCray, J. E., and Falta, R. W. (1997). "Numerical simulation of air sparging for remediation of NAPL contamination." *Ground Water*, 35(1), 99-110.

Mei, C. C., Cheng, Z., and Ng, C. O. (2002). "A model for flow induced by steady air venting and air sparging." *Applied Mathematical Modelling*, 26(7), 727-750.

Miller, C. T., Christakos, G., Imhoff, P. T., McBride, J. F., Pedit, J. A., and Trangenstein, J. A. (1998). "Multiphase flow and transport modeling in heterogeneous porous media: challenges and approaches." *Advances in Water Resources*, 21(2), 77-120.

Miller, R. R. (1996). "Technology Overview Report: Air Sparging." Ground-Water Remediation Technologies Analysis Center.

Mohtar, R. H., Wallace, R. B., and Segerlind, L. J. (1994). "Finite Element Simulation of Oil Spill Cleanup Using Air Sparging." Computational Methods in Water Resource, A. Peter, ed., Kluwer Academic, Boston, 967-974.

Mualem, Y. (1976). "New Model for Predicting Hydraulic Conductivity of Unsaturated Porous-Media." *Water Resources Research*, 12(3), 513-522.

Murray, W. A., Lunardini Jr., R. C., Ullo Jr., F. J., and Davidson, M. E. (2000). "Site 5 air sparging pilot test, Naval Air Station Cecil Field, Jacksonville, Florida." *Journal of Hazardous Materials*, 72(2-3), 121-145.

NFESC. (2001). "Air Sparging Guidance Document." Naval Facilities Engineering Command (NFESC), Technical Report TR-2193-ENV.

NFESC. (2007). "Cost and Performance Review of Electrical Resistance Heating (ERH) for Source Treatment: Final Report, TR-2279-ENV." Naval Facilities Engineering Service Center (NFESE).

Nield, D. A., and Bejan, A. (1999). *Convection in Porous Media*, Springer, New York.

Nyer, E. K., and Suthersan, S. S. (1993). "Air sparging - Savior of ground-water remediations or just blowing bubbles in the bath tub." *Ground Water Monitoring and Remediation*, 13(4), 87-91.

Panday, S., Wu, Y. S., Huyakorn, P. S., and Springer, E. P. (1994). "A 3-dimensional multiphase flow model for assessing NAPL contamination in porous and fractured media. 2. Porous-medium simulation examples." *Journal of Contaminant Hydrology*, 16(2), 131-156.

Parker, J. C., and Lenhard, R. J. (1987). "A Model for Hysteretic Constitutive Relations Governing Multiphase Flow .1. Saturation-Pressure Relations." *Water Resources Research*, 23(12), 2187-2196.

Parker, J. C., Lenhard, R. J., and Kuppusamy, T. (1987). "A parametric model for constitutive properties governing multiphase flow in porous-media." *Water Resources Research*, 23(4), 618-624.

Peterson, J. W., Lepczyk, P. A., and Lake, K. L. (1999). "Effect of sediment size on area of influence during groundwater remediation by air sparging: a laboratory approach." *Environmental Geology*, 38(1), 1-6.

Pfiffner, S. M., Palumbo, A. V., Phelps, T. J., and Hazen, T. C. (1997). "Effects of nutrient dosing on subsurface methanotrophic populations and trichloroethylene degradation." *Journal of Industrial Microbiology & Biotechnology*, 18(2-3), 204-212.

Philip, J. R. (1998). "Full and boundary-layer solutions of the steady air sparging problem." *Journal of Contaminant Hydrology*, 33(3-4), 337-345.

Rabideau, A. J., and Blayden, J. M. (1998). "Analytical model for contaminant mass removal by air sparging." *Ground Water Monitoring and Remediation*, 18(4), 120-130.

Rabideau, A. J., Blayden, J. M., and Ganguly, C. (1999). "Field performance of air sparging system for removing TCE from groundwater." *Environmental Science & Technology*, 33(1), 157-162.

Reddy, K. R., and Adams, J. A. (2001). "Effects of soil heterogeneity on airflow patterns and hydrocarbon removal during in situ air sparging." *Journal of Geotechnical and Geoenvironmental Engineering*, 127(3), 234-247.

Reisinger, H. J. (1995). "Hydrocarbon Bioremediation- An Overview." Applied bioremediation of petroleum hydrocarbons, R. E. Hinchee, J. A. Kittel, and H. J. Reisinger, eds., Battelle Press, Columbus, xiii, 534.

Roosevelt, S. E., and Corapcioglu, M. Y. (1998). "Air bubble migration in a granular porous medium: Experimental studies." *Water Resources Research*, 34(5), 1131-1142.

Schmidt, R., Gudbjerg, J., Sonnenborg, T. O., and Jensen, K. H. (2002). "Removal of NAPLs from the unsaturated zone using steam: prevention of downward migration by injecting mixtures of steam and air." *Journal of Contaminant Hydrology*, 55(3-4), 233-260.

Sellers, K. (1999). *Fundamentals of hazardous waste site remediation*, Lewis Publishers, Boca Raton.

She, H. Y., and Sleep, B. E. (1999). "Removal of perchloroethylene from a layered soil system by steam flushing." *Ground Water Monitoring and Remediation*, 19(2), 70-77.

Sleep, B. E., and Sykes, J. F. (1989). "Modeling the Transport of Volatile Organics in Variably Saturated Media." *Water Resources Research*, 25(1), 81-92.

Stone, H. L. (1970). "Probability Model for Estimating 3-Phase Relative Permeability." *Journal of Petroleum Technology*, 22, 214-&.

Thomson, N. R., and Johnson, R. L. (2000). "Air distribution during in situ air sparging: An overview of mathematical modeling." *Journal of Hazardous Materials*, 72(2-3), 265-282.

Thomson, N. R., Sykes, J. F., and Van Vliet, D. (1997). "A numerical investigation into factors affecting gas and aqueous phase plumes in the subsurface." *Journal of Contaminant Hydrology*, 28(1-2), 39-70.

Travis, B. J., and Rosenberg, N. D. (1997). "Modeling in situ bioremediation of TCE at Savannah River: Effects of product toxicity and microbial interactions on TCE degradation." *Environmental Science & Technology*, 31(11), 3093-3102.

U.S.ACE. (1995). "Soil Vapor Extraction and Bioventing, EM 1110-1-4001." U.S. Army Corps of Engineers (U.S. ACE), Department of Army, Washington, DC.

U.S.ACE. (1999). "Soil Vapor Extraction Enhanced by Six-Phase Soil Heating at Poleline Road Disposal Area, OU-B, Ft. Richardson, Alaska." U.S. Army Corps of Engineers.

U.S.DoE. (2003). "Steam Enhanced Extraction and Electro-Thermal Dynamic Stripping Process (ET-DSP) at the Young-Rainy Star Center (formerly Pinellas) Northeast Area A, Largo, Florida, GJO–2003–482–TAC." U.S. Department of Energy (U.S. DoE).

U.S.EPA. (1991). "Site Characterization for Subsurface Remediation, EPA/625/4-91/026." U.S. Environmental Protection Agency (U.S. EPA).

U.S.EPA. (1995). "Bioventing Principles and Practices: Bioventing Desing, EPA/540/R-95/534a, 80pp." Office of Research and Development, U.S. Environmental Protection Agency (U.S. EPA), Washington, DC.

U.S.EPA. (1998). "MTBE Fact Sheet #2: Remediation Of MTBE Contaminated Soil And Groundwater, EPA 510-F-97-015." U.S. Environmental Protection Agency (U.S. EPA).

U.S.EPA. (2003a). "Cost and Performance Report: Electrical Resistive Heating at the Avery Dennison Site Waukegan, Illinois." U.S. Environmental Protection Agency (U.S. EPA), Washington, DC.

U.S.EPA. (2003b). "Electrical Resistive Heating at the Poleline Road Disposal Area, Arrays 4, 5, and 6 Fort Richardson, Alaska: Cost and Performance Report." U.S. Environmental Protection Agency (U.S. EPA), Washington, DC.

U.S.EPA. (2003c). "In Situ Conductive Heating at the Confidential Chemical Manufacturing Facility Portland, IN: Cost and Performance Report." U.S. Environmental Protection Agency (U.S. EPA), Washington, DC.

U.S.EPA. (2003d). "Steam Enhanced Extraction at the A.G. Communications Systems Site Northlake, Illinois." U.S. Environmental Protection Agency (U.S. EPA).

U.S.EPA. (2004). "How to Evaluate Alternative Cleanup Technologies for Underground Storage Tank Sites: A Guide for Corrective Action Plan Reviewers. (EPA 510-B-94-003; EPA 510-B-95-007; and EPA 510-R-04-002)." U.S. Environmental Protection Agency (U.S. EPA), Washington, DC., Washington, D.C. U.S.A.

U.S.EPA. (2009). "Technology Focus: In Situ Thermal Treatment." U.S. Environmental Protection Agency (U.S. EPA), Washington, DC., *http://www.cluin.org/techfocus/default.focus/sec/Thermal%5FTreatment%3A%5FIn%5FSitu/cat/Overview/*.

Unger, A. J. A., Sudicky, E. A., and Forsyth, P. A. (1995). "Mechanisms controlling vacuum extraction coupled with air sparging for remediation of heterogeneous formations contaminated by dense nonaqueous phase liquids." *Water Resources Research*, 31(8), 1913-1925.

van Dijke, M. I. J., and van der Zee, S. E. A. T. M. (1998). "Modeling of air sparging in a layered soil: Numerical and analytical approximations." *Water Resources Research*, 34(3), 341-353.

van Dijke, M. I. J., van der Zee, S. E. A. T. M., and van Duijn, C. J. (1995). "Multiphase flow modeling of air sparging." *Advances in Water Resources*, 18(6), 319-333.

van Genuchten, M. T. (1980). "A closed-form equation for predicting the hydraulic conductivity of unsaturated soils." *Soil Science Society of America Journal*, 44(5), 892-898.

van Hylckama Vlieg, J. E. T., and Janssen, D. B. (2001). "Formation and detoxification of reactive intermediates in the metabolism of chlorinated ethenes." *Journal of Biotechnology*, 85(2), 81-102.

Verce, M. F., Gunsch, C. K., Danko, A. S., and Freedman, D. L. (2002). "Cometabolism of cis-1,2-dichloroethene by aerobic cultures grown on vinyl chloride as the primary substrate." *Environmental Science & Technology*, 36(10), 2171-2177.

Vermeulen, F., and McGee, B. (2000). "In situ electromagnetic heating for hydrocarbon recovery and environmental remediation." *Journal of Canadian Petroleum Technology*, 39(8), 24-28.

Vogel, T. M., and McCarty, P. L. (1985). "Biotransformation of Tetrachloroethylene to Trichloroethylene, Dichloroethylene, Vinyl-Chloride, and Carbon-Dioxide under Methanogenic Conditions." *Applied and Environmental Microbiology*, 49(5), 1080-1083.

Wiedemeier, T. H. (1998). "Technical protocol for evaluating natural attenuation of chlorinated solvents in ground water." National Risk Management Research Laboratory Office of Research and Development, Cincinnati, Ohio.

Wilson, D. J. (1992). "Groundwater Cleanup by Insitu Sparging .2. Modeling of Dissolved Volatile Organic-Compound Removal." *Separation Science and Technology*, 27(13), 1675-1690.

Wilson, D. J. (1995). *Modeling of in situ techniques for treatment of contaminated soils : soil vapor extraction, sparging, and bioventing*, Technomic, Lancaster, Pa.

Wilson, D. J., Gomezlahoz, C., and Rodriguezmaroto, J. M. (1994). "Groundwater Cleanup by in-Situ Sparging .8. Effect of Air Channeling on Dissolved Volatile Organic-Compounds Removal Efficiency." *Separation Science and Technology*, 29(18), 2387-2418.

Wood, B. D., Keller, C. K., and Johnstone, D. L. (1993). "Insitu Measurement of Microbial Activity and Controls on Microbial CO2 Production in the Unsaturated Zone." *Water Resources Research*, 29(3), 647-659.

WSRC. (2001). "Dynamic Underground Stripping-Hydrous Pyrolysis Oxidation at the Savannah River Site 321-M Solvent Storage Tank Area, Aiken, Georgia." Westinghouse Savannah River Company.

Yaws, C. L. (1999). *Chemical Properties Handbook: Physical, thermodynamic, environmental, transport, safety, and health related properties for organic and inorganic chemicals*, McGraw-Hill, New York.

CHAPTER 12

SOURCE CONTROL AND CHEMICAL REMEDIATION OF CONTAMINATED GROUNDWATER SITES

Natalie L. Cápiro and Kurt D. Pennell
Department of Civil and Environmental Engineering
Tufts University, Medford, MA

12.1 Introduction

The presence of synthetic organic contaminants (SOCs) was observed in groundwater as early as the 1970's. However, the widespread nature of the problem was not recognized until requirements for groundwater monitoring were expanded in the early 1980's (Schwille, 1988). Upon recognition of the extent of groundwater contamination in the United States (U.S.), the U.S. Congress amended the Safe Drinking Water Act (SDWA) in 1986, requiring the U.S. Environmental Protection Agency (USEPA) to establish national primary drinking water regulations (NPDWR) limiting the concentration of contaminants in drinking water supplied from public water systems (USEPA, 1987). Major amendments to the SDWA in 1986 set a goal of 83 contaminant standards to be established by 1989, and 25 more contaminants for every three years thereafter. The additional contaminants regulated by the 1986 Amendments included organic contaminants (e.g., pesticides and petroleum-based hydrocarbons) and chlorine disinfection byproducts. Additional amendments in 1996 further expanded the number of contaminant classes that were covered, establishing standards for microorganisms and radionuclides (Benner, 2004).

Currently, over 50 volatile organic compounds (VOCs) are regulated under the SWDA. These compounds are common groundwater contaminants due to their widespread use and improper disposal, either through accidental discharge or intentional release into the subsurface. A survey conducted by the USEPA found that VOCs were present in 17% of small-scale, and 28% of large-scale, public drinking water treatment facilities using groundwater as their water source (USEPA, 1990). The VOCs found in groundwater are characterized by their relatively low aqueous solubilities (i.e., less than 1000 mg/L), and therefore, often exist as a separate organic phase, or non-aqueous phase liquid (NAPL), in the subsurface. Dissolution of NAPLs or NAPL constituents into the aqueous phase can serve as a long-term source of groundwater contamination that may persist for years or even decades. Mass transfer limitations and dilution effects often prevent the observed concentrations of a contaminant in groundwater from reaching the aqueous solubility value. Nevertheless, the concentration of these compounds frequently exceeds the maximum concentration level (MCL) as specified by the USEPA by several orders of magnitude.

Over the past several decades, environmental scientists and engineers have developed a number of *in situ* (in the subsurface) remediation technologies that are designed to contain and/or treat groundwater and soil contamination. Development of these technologies has been largely driven by federal and state regulations, such as the Resource Conservation and Recovery Act (RCRA) and the Comprehensive Environmental Response, Compensation and Liability Act (CERLA), which establish risk based thresholds on contaminant concentrations within and/or emanating from waste sites. This chapter provides an overview of several of the most promising physical-chemical groundwater remediation technologies designed to treat contaminant source zones. Such *in situ* technologies may include source zone isolation or containment, mass removal and/or destruction, and treatment of the dissolved-phase groundwater plume. It is generally acknowledged that remedial goals may need to be addressed through a combination of management approaches and/or remediation technologies. This chapter focuses on *in situ* remediation of source zones impacted by two of the most widespread contaminant classes, petroleum hydrocarbons and chlorinated solvents. Information regarding the treatment of additional contaminants, such as pesticides, heavy metals, radioactive wastes, polychlorinated biphenyls (PCBs) can be found in other resources (NRC, 1994, 1997; USEPA, 1998b; NRC, 1999; USEPA, 2003, 2004b). In addition, the discussion presented herein will primarily focus on unconfined aquifer formations that have been impacted by surface and near surface contaminant releases.

12.2 Non-aqueous Phase Liquid (NAPL) Source Zones

When present in the subsurface as a distinct liquid phase or NAPL, organic compounds are further subcategorized on the basis of their density relative to the aqueous phase (i.e., water). If the organic liquid phase (e.g., gasoline and mineral oils) is less dense than the aqueous phase, it is referred to as light non-aqueous phase liquid (LNAPL); whereas, if the organic liquid phase has a greater density (e.g., chlorinated solvents) relative to the aqueous phase, it is referred to as a dense non-aqueous phase liquid (DNAPL). In either case, when NAPLs is released to (water-) unsaturated surface soils (i.e., above the water table) they will tend to migrate downward due to gravitational and capillary forces acting on the continuous body of NAPL. As the NAPL body migrates vertically through the soil profile, a portion of NAPL will become entrapped in pore spaces due to capillary forces and snap-off phenomena (Lake 1989). Entrapped NAPL ganglia are considered to be immobile in the subsurface under normal pressure and flow regimes, as is often referred to as the residual saturation. This residual NAPL has been reported to occupy between 12 to 60% of soil pore volume (Hoag and Marley, 1986; Hunt et al., 1988), but typically ranges from 10 and 25% in fine and medium-textured sands (e.g., Pennell et al. 1996).

If the volume of continuous phase NAPL is sufficient to reach the water table (water-saturated pore space), subsequent migration behavior will depend primarily on density of the NAPL with respect to water. More specifically, DNAPLs will continue to migrate vertically through the subsurface, displacing a portion of soil pore water, leaving behind entrapped residual saturation NAPL, higher-saturation "pools" above lenses of lower permeability media or a confining layer (Figure 12.1), or until the organic liquid source is exhausted. In contrast, LNAPLs will tend to spread laterally within the capillary fringe above the water table. However, LNAPL can also migrate into the previously water-saturated zone if the continuous body of LNAPL above the water table provides sufficient entry pressure to displace water. In addition, fluctuations in the water table can also result in redistribution or "smearing" of LNAPL within the upper regions of the water-saturated zone. Such intermittent raising and lowering of the water table will induce LNAPL to migrate upward and downward, with residual LNAPL ganglia being entrapped by capillary forces. The specific behavior of a NAPL following release into subsurface is influenced by several factors specific to the NAPL as well as the porous media, including dynamic viscosity, interfacial tension (IFT), heterogeneity, and wettability of the porous media (Mercer and Cohen, 1990). Interactions between these variables in the subsurface, as well as the effects of scale, contribute to the large range of residual saturation values reported in the literature and the complexity of NAPL saturation distributions (i.e., architecture) that is commonly observed at field sites.

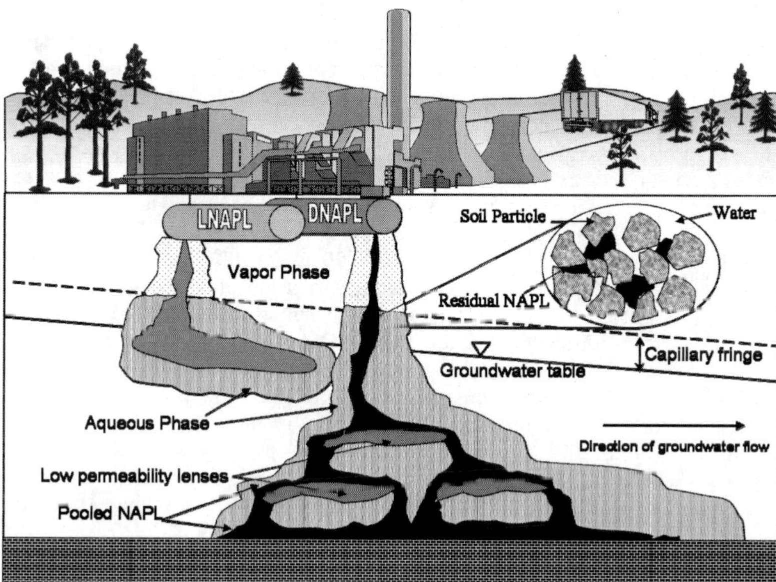

Figure 12.1: Schematic diagram of hypothetical DNAPL and LNAPL releases into the subsurface.

12.2.1 NAPL Constituent Properties

Among the most common LNAPLs that serve as groundwater pollutants are petroleum-based hydrocarbons, consisting of alkanes, alkenes, heterocyclic and aromatic constituents. These compounds frequently enter the subsurface as a free product released from oil production sites, underground storage tanks (USTs), and refineries. One of the most common LNAPLs released to the subsurface is gasoline, a complex hydrocarbon mixture consisting primarily of monoaromatic hydrocarbon compounds including benzene, toluene, ethylbenzene and xylenes (BTEX). In general, BTEX compounds are sparingly soluble in water and possess a relatively low tendency to interact with the solid phase via sorption processes. These characteristics are reflected in their aqueous solubilities, which range from 150 to 1,800 mg/L, and moderate organic carbon partition coefficients (K_{oc}), which range from 57 to 1,100 mL/g (Table 12.1). In comparison, compounds that are considered to be strongly hydrophobic typically possess aqueous solubilities in the part per billion (ppb) range, with K_{oc} values greater than 1×10^6 mL/g (Hornsby et al. 1996; MacKay et al. 2006). Although all of the BTEX constituents pose health risks, the most important from a toxicological perspective is benzene, which is classified as a potential human carcinogen (ATSDR 2007). Benzene possesses a relatively high aqueous solubility (1,800 mg/L), a correspondingly low organic carbon partition coefficient, and therefore is relatively mobile in groundwater systems. While the majority of BTEX constituents are susceptible to biotic degradation, benzene is most recalcitrant of these compounds. For these reasons, benzene is the most highly regulated of the BTEX compounds, and often serves as the quantifiable constituent for cleanup goals (i.e., MCL = 0.005 mg/L). In addition, gasoline source zones are often located in close proximity to residential properties and/or sources of drinking water due to the ubiquitous nature of gasoline stations, further increasing the priority of such sites for remedial action.

In the U.S., there are still an estimated 30,000 to 50,000 sites with groundwater contamination, excluding petroleum contamination from underground storage tanks (Kavanaugh et al., 2003). About 80% of these sites are contaminated with organic chemicals, and of these, 60% likely have DNAPLs present. Of the DNAPLs, chlorinated organic solvents, such as trichloroethene (TCE) and tetrachloroethene (PCE), are the most common contaminants due to their widespread use at military installations, industrial facilities, and dry cleaning operations. Similar to the BTEX contaminants discussed above, TCE and PCE are sparingly soluble in water and possess moderate K_{oc} values (Table 12.1). In general, chlorinated solvents are considered to be more toxic than gasoline constituents, with the exception of benzene, as reflected in the consistently low MCL values shown in Table 12.1.

Table 12.1: Relevant properties of organic compounds that commonly exist as NAPLs in the subsurface.

Compound Name	Molecular Weight (g/mol)	Aqueous Solubility (mg/L)	Liquid Density (g/cm^3)	K_{oc}[a] (mL/g)	MCL[b] (mg/L)
Benzene	78.1	1,780	0.88	83	0.005
Toluene	92.4	515	0.87	300	1.0
Ethylbenzene	106.2	150	0.87	1,100	0.7
m-Xylene	106.16	158	0.86	57	1.0
o-Xylene	106.2	175	0.88	57	1.0
Carbon tetrachloride (CT)	153.8	850	1.59	110	0.005
Trichloroethene (TCE)	131.4	1,100	1.46	126	0.005
Tetrachloroethene (PCE)	165.8	200	1.68	364	0.005
Chlorobenzene (CB)	112.6	450	1.11	330	0.100

[a]K_{oc} = organic carbon-water partition coefficient, which is equal to the solid-water distribution coefficient (K_D) divided by the organic carbon (OC) content of the solid phase; [b]MCL = maximum concentration limit in drinking water. Data obtained from LaGrega et al. (2000), (Riddick and Bunger, 1970), (Schwarzenbach et al., 2003), and (Wiedemeier et al., 1999).

12.2.2 NAPL Dissolution from Source Zones

The distribution of contaminants at a typical NAPL site can be divided into two main regions: a highly concentrated *source zone* (discussed above), and a lower concentration zone that includes a *dissolved solute plume* and *solid- or sorbed-phase contaminant*. As uncontaminated water moves through a NAPL source zone, dissolution of contaminants from the organic liquid phase into the aqueous phase occurs, resulting in down-gradient movement of the contaminant. This dissolved-phase contamination is generally referred to as the groundwater plume, in which contaminant concentrations are no greater than the aqueous solubility, and are generally much lower. These lower dissolved phase concentrations are often the result of pore-scale mass transfer limitations from the organic liquid into aqueous phase and dilution due to mixing with uncontaminated groundwater. In addition, the presence of heterogeneity in the source zone may create zones of bypassing and/or limited contact where there is little opportunity for the transfer of mass from the organic liquid phase to the aqueous phase (Powers et al., 1991). The specific surface area of the NAPL also limits contaminant dissolution, where regions of the source zone that contain NAPL pools exhibit lower overall mass transfer than regions that are contaminated by NAPL ganglia or droplets, which possess relatively high specific surface areas (Rathfelder et al., 2001).

At the pore-scale, mass transfer of contaminants from an organic liquid phase into groundwater can be conceptualized as consisting of organic liquid phase (as NAPL ganglia or pools), in contact with a stagnant film of water, which is in contact with the bulk mobile or flowing aqueous phase (Figure 12.2).

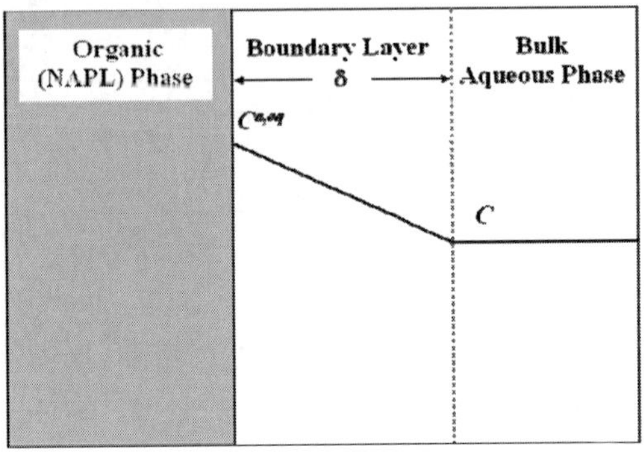

Figure 12.2: Conceptual description of mass transfer from organic phase (NAPL) to bulk aqueous phase.

Here, $C^{a,eq}$ (M/L^3) represents the concentration of the contaminant at the equilibrium aqueous solubility in immediate contact with the organic liquid; C (M/L^3) is the contaminant concentration in the bulk aqueous phase; and δ is the thickness of the stagnant water film. Based on this conceptual model, the dissolution of NAPLs into groundwater can be described using a linear driving force expression (Powers et al., 1991; Taylor et al., 2001c, b):

$$q\frac{dC}{dx} = ka_o(C^{a,eq} - C) \tag{12.1}$$

where q is the Darcy velocity (L/T); k is the NAPL-water mass transfer coefficient (L/T); and a_o is the specific surface (interfacial) area of the NAPL (L^2/L^3). Here, the "driving force" is the difference between the contaminant concentration in the bulk aqueous phase (C) and the aqueous solubility of the contaminant ($C^{a,eq}$). In most cases, the interfacial area of the NAPL is not known, and thus, an effective or lumped mass transfer coefficient (K_l, 1/T) is often employed, where $K_l = k \times a_0$.

Under natural groundwater flow gradients, mass transfer of NAPL constituents into the aqueous phase often approaches equilibrium conditions at the local or pore scale (i.e., bulk concentration equal to solubility) (Powers et al., 1991). Despite such observations, measured concentrations of contaminants within the groundwater plume are almost always well below the solubility limit due to flow bypassing and dilution effects. Although these processes serve to reduce observed groundwater concentrations in the vicinity of NAPL source zones to part per million (ppm or mg/L) levels, these lower concentrations remain well above the MCL for many contaminants, which are typically at the part per billion (ppb or μg/L) level (Table 12.1). As a result, the presence of NAPL in a contaminant source zone can render large quantities of groundwater unsuitable for use as a source of drinking water. For example, one 55 gallon drum of TCE-DNAPL has the capacity to contaminate 1.61×10^{10} gallons (6.08×10^{10} L) of clean water to the current drinking water MCL of 5 ppb. In addition, many source zone remediation technologies rely on NAPL dissolution to achieve contaminant removal or reactivity (e.g., pump-and-treat, chemical oxidation, bioremediation), and are therefore limited by the relatively low solubility and mass transfer limitations discussed above. Thus, the combination of relatively low aqueous solubility and orders-of-magnitude lower drinking water MCLs present substantial challenges with respect to treatment of NAPL source zones and restoration of contaminated groundwater to levels that are considered safe for human consumption.

12.2.3 In-Situ Remediation Technologies

Treatment of aquifer formations contaminated with NAPLs is an arduous task made difficult by issues associated with subsurface characterization, source zone architecture, sorption of contaminants by aquifer materials, and migration of dissolved-phase and NAPL contaminants into inaccessible regions of the subsurface. The relative difficulty associated with the cleanup of a contaminated aquifer depends upon both the contaminant properties, such as aqueous solubility, sorption, degradability and volatility, as well as subsurface properties, especially the level of heterogeneity with respect to low-permeability layers or lenses and fractured media. The factors can be expressed in rubric form to provide a relative assessment of the ease with which an aquifer formation can be restored (Table 12.2). For example, relatively homogeneous sites contaminated with dissolved-phase contaminants are the easiest to clean up (rankings of 1 and 2), whereas aquifer formations containing complex hydrogeology that are contaminated with strongly sorbed contaminants or NAPLs are the most difficult to clean up (rankings of 3 and 4). As one would expect, most of the sites that fall under the 1-2 ranking scale have been treated and received site closure, while the more difficult sites (3-4 rankings) have often been subject to prolonged and/or failed remediation efforts, and continue to pose a threat to human health and the environment. In some cases, these sites are considered to be "technically impractical (TI)" to remediate based on current practices and technologies, and are now subject to long-term monitoring and plume control. As a result, considerable effort has been expended to develop *in situ* technologies, or combinations of technologies, that can achieve substantial contaminant mass reduction within an acceptable time frame. To this end, a number of innovative technologies have been developed, including hydraulic containment, physical and reactive barriers, chemical oxidation, thermal treatment, air sparging, cosolvent flushing and surfactant flushing. In the sections that follow a description of several promising *in situ* physical-chemical treatment technologies is provided, along with examples of their applications in laboratory and field settings. In addition, a discussion of the advantages and potential limitations associated with the implementation of each technology is presented.

Table 12.2: Relative ease (1 = easiest and 4 = most difficult) of cleaning up contaminated aquifers as a function of contaminant chemistry and hydrogeology (adapted from NRC (1994)).

Hydrogeologic Conditions	Contaminant Characteristics					
	Mobile, dissolved (degrades or volatilizes)	Mobile, dissolved	Strongly sorbed, dissolved (degrades or volatilizes)	Strongly sorbed, dissolved	LNAPL present	DNAPL present
Homogeneous, single layer	1	1-2	2	2-3	2-3	3
Homogeneous, multiple layers	1	1-2	2	2-3	2-3	3
Heterogeneous, single layer	2	2	3	3	3	4
Heterogeneous, multiple layers	2	2	3	3	3	4
Fractured media	3	3	3	3	4	4

12.3 Hydraulic Controls and Pump-and-Treat Systems

One of the most common remedial technologies used to limit and/or control the migration of dissolved-phase contaminations from NAPL source zones is pump-and-treat. This approach involves the extraction of contaminated groundwater from the subsurface via a network of wells. The extracted water is then treated aboveground to remove contaminants, often by air stripping for VOCs and/or by contaminant adsorption onto granular activated carbon (GAC) (Figure 12.3). The treated water may then be either re-injected into the aquifer through a separate well or network of wells, released to a surface water body, or discharged to a waste-water treatment facility. As of 1998, pump and treat was employed as the sole cleanup technology at 89% of groundwater contamination sites in the United States (USEPA, 1998b).

12.3.1 Plume Containment

Pump-and-treat remediation is an effective means for plume containment and dissolved-phase contaminant recovery, and in the short-term is economically attractive due to its relatively low initial costs when compared to other *in situ* treatment technologies (Ramsburg and Pennell, 2001). Since conventional pump-and-treat processes focus on the aqueous phase, successful source zone remediation is dependent on contaminant mass transfer from either the solid (sorbed) and organic liquid (NAPL) phase into flowing groundwater. Hence, the effectiveness of dissolved-phase contaminant mass recovery is strongly dependent upon the intrinsic permeability and heterogeneity of the subsurface, which may render groundwater extraction and/or hydraulic control systems infeasible. However, if an aquifer formation has sufficient permeability for the engineered system to maintain extraction/injection pumping integrity, analytical and numerical models are available to guide well design and placement, and to provide estimates of the hydraulic capture zone.

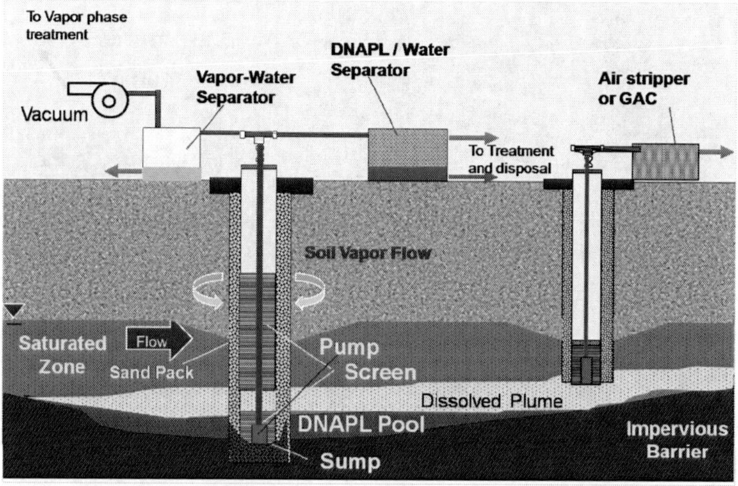

Figure 12.3: Schematic diagram of a multiphase extraction well (left) in combination with a down-gradient plume capture well (right) treatment system for a contaminated aquifer formation where DNAPL is pooled above a lower confining layer.

Ideally, the capture zone created with one or more wells should be larger than the contaminant plume to prevent further uncontrolled plume migration (Figure 12.4). The balance between well spacing and pumping rate is critical to prevent contaminant bypassing and to achieve complete plume capture. The USEPA provides a useful guide for design considerations when performing capture zone analysis associated with groundwater pump-and-treat systems (USEPA, 2008). Javandal and Tsang (1986) developed an analytical solution for the design of recovery well systems, based on the concept of capture zones. Following site characterization to determine parameters such as aquifer thickness and groundwater velocity of the flow field, the model aids in the design of an effective pump-and-treat system, including pumping rates, well number and spacing. Re-injection systems for treated groundwater are analogous to extraction systems in many respects, and thus, similar models can be employed to develop these design specifications.

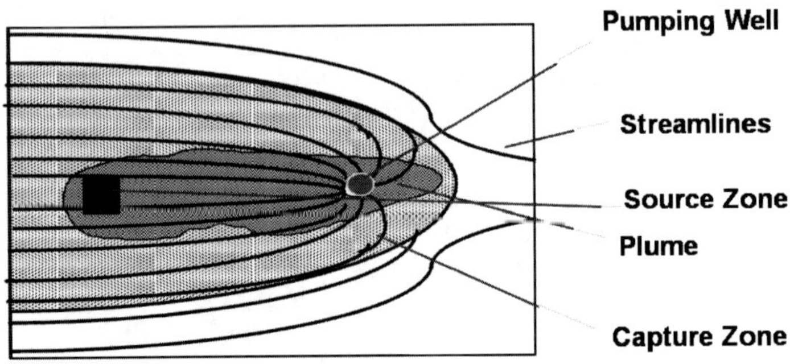

Figure 12.4: Plume capture zone (water pathways indicated by streamlines) of a hydraulic containment pumping well.

12.3.2 Multi-Phase Extraction (MPE)

When contaminants are present in the aqueous, NAPL, and gas phases, a more robust multi-phase extraction (MPE) system can be used to treat source zones. MPE systems were initially developed for the remediation of VOCs in low to moderate permeability aquifer formations. The MPE process is based upon a modification of conventional soil vapor extraction (SVE) technology, which is applied to the (water) unsaturated zone above the water table. In general, SVE systems involve extraction of VOCs from the subsurface by inducing advective air flow through the subsurface by applying a vacuum to the extraction wells. The extracted gas, which contains VOCs, is either vented to the atmosphere or, more commonly, treated to remove contaminants from the gas stream (e.g., GAC, catalytic oxidation). A more complete description of SVE systems, including fundamental principles and practical implementation, can be found in (USEPA, 2004b)). As noted above, MPE involves the modification of traditional SVE systems to address multiple phases; for example, simultaneous extraction of both groundwater and soil vapors. In many MPE applications, groundwater extraction acts to lower the water, resulting in dewatering of the saturated zone so that vapor extraction can be applied to the newly unsaturated soil (Figure 12.3). In general, MPE systems that utilize localized dewatering coupled with vapor extraction greatly improve the efficiency of contaminant mass removal, particularly in source zones containing substantial NAPL contamination.

To achieve water and gas extraction from a single well, two technologies are applied: two-phase extraction (TPE) and dual-phase extraction (DPE). The TPE approach employs a high vacuum (approximately 18 to 26 inches Hg) pump to extract both groundwater and soil vapor from the extraction well. A pipe is lowered into the extraction well to extract the soil vapor and groundwater from the subsurface. For some TPE methods, turbulence generated within suction pipes facilitates the transfer of aqueous phase contaminants to the vapor phase (up to 98% stripping). By comparison, the DPE technology employs a submersible or pneumatic pump to extract groundwater, along with a high vacuum (approximately 18 to 26 inches of Hg) or low vacuum (approximately 3 to 12 inches Hg) extraction blower to recovery soil vapors. For DPE wells using a submersible pump, a sump is installed at the bottom of the well to prevent cavitation of the submersible pump (Figure 12.3).

MPE is often combined with traditional pump-and-treat extraction wells to achieve a more comprehensive remediation strategy that addresses NAPL source zone treatment as well as plume containment and groundwater treatment. This combined approach has been applied at dozens of low- to moderate-permeability sites, and has consistently proven to be more effective at removing subsurface VOCs than either conventional pump-and-treat or soil vapor extraction systems alone (USEPA, 1997). The enhanced remedial performance is realized through a combination of increased removal rates of contaminated groundwater and soil gas, and the volatilization of NAPLs from previously water-saturated soils (e.g., residual NAPL ganglia). Such increased mass removal rates typically result in shorter total cleanup times and reduced costs. A summary of pilot- and full-scale implementation of MPE at field sites contaminated with chlorinated solvents (e.g., Travis Air Force Base) is provided in a Supplemental Bulletin on MPEs (1997). These examples demonstrate the success in relatively permeable media (sandy silts and gravels) in terms of mass of VOCs removed per day, yielding a cost estimate of approximately $235 per pound of VOC extracted. However, MPE has shown to be less effective for sites consisting of very high permeability media composed primarily of gravels or cobbles, where costs increase to $32,000 per pound VOC removed. The latter sites exhibited very high groundwater flow rates, which prevented effective lowering of the water table, limiting the benefits of SVE. Thus, it is important to determine the appropriateness of MPE at a particular site, which is strongly dependent on subsurface conditions (e.g., permeability) and contaminant characteristics (e.g., vapor pressure).

12.3.3 Limitations of Pump-and-Treat and MPE Technologies

Due to low relatively low aqueous solubility of many priority organic contaminants in water and their slow rates of mass transfer into aqueous phase, NAPL contamination can persist in aquifer formations for decades or even centuries. In addition, as the complexity of the subsurface geology increases, the likelihood for success with traditional pump-and-treat extraction systems decreases markedly (Table 12.2). Costs associated with maintaining and operating pump-and-treat operations for extended periods of time can become prohibitive, and the impacted groundwater is not available for public use until remedial efforts are complete. For NAPL source zones, particularly in areas of subsurface heterogeneity, pump-and-treat has been shown to be unsuccessful at removing significant fractions of contaminant mass in reasonable time frames (years to decades) while operational, maintenance, and monitoring costs escalate (Mackay and Cherry, 1989; MacDonald and Kavanaugh, 1994; NRC, 1994, 1997). For example, the median cost to operate a groundwater pump-and-treat system is on the order of $180,000 per year, with a range of $30,000 to $4,000,000 depending upon the size and complexity of the site (Kavanaugh et al., 2003). Considering only the 3000 chlorinated solvent sites managed by the Department of Defense, over $100 million annually, and $2 billion life-cycle, could be spent to manage and remediate these sites using traditional hydraulic controls, such as pump-and-treat (Stroo et al., 2003). Furthermore, it is now widely accepted that pump-and-treat is largely ineffective in restoring groundwater contaminant levels to drinking water standards for NAPL-contaminated sites, and thus, from a practical perspective this approach only provides containment of the source zone (Mackay and Cherry, 1989; MacDonald and Kavanaugh, 1994).

12.4 Physical and Reactive Barriers

12.4.1 Physical Groundwater Barriers

In an effort to overcome the limitations of conventional pump-and-treat systems, physical containment methods were developed to isolate the source zone and minimize plume migration. Barriers are typically classified as either vertical or horizontal, where vertical barriers are used to block lateral flow of water into or from a contaminated zone and horizontal barriers are used prevent vertical infiltration or flow. Physical barriers prevent the flow of contaminated groundwater, typically through the installation of slurry walls, grout curtains, sheet piling, compacted liners or geomembranes (NRC, 1997). Although physical barriers do not result in contaminant mass removal or destruction, barriers may be used to contain contaminated groundwater, NAPL, and leachate, and to prevent the flow of clean groundwater into contaminant zones. They may also be used in combination with pumping or collection systems (discussed in Section 12.3) to improve hydraulic control and groundwater recovery.

12.4.2 Permeable Reactive Barriers

Permeable reactive barriers (PRBs) consist of subsurface units constructed to intercept contaminated groundwater and react with dissolved-phase contaminants. As the contaminated groundwater passes through the permeable barrier media, the contaminants are either immobilized through sorption or precipitation reactions, and/or react to form less toxic compounds. Typical immobilization mechanisms include physical, chemical and/or biological processes such as sorption, oxidation-reduction and precipitation. Selection of the reactive media is dictated by a number of considerations, including site hydrogeology and geochemistry, contaminant properties, concentration and reactivity, and size of the groundwater plume.

Much of the PRB research and development has focused on chlorinated organic compounds. Gillham and O'Hannesin (1993; 1994) describe some of the first laboratory and field trials of a solid-phase barrier material, elemental iron filings, for *in situ* treatment of dissolved-phase chlorinated solvents. To date, the majority of installed PRBs employ zero-valent iron, Fe(0), as the reactive media for converting contaminants to nontoxic or immobile species. This relatively inexpensive form of iron serves as a source of electrons for reductive dehalogenation of hydrocarbons, such as PCE and TCE. Halogen ions (e.g., Cl^-) are replaced by hydrogen species in reductive dehalogenation, ultimately yielding ethane (Gillham and O'Hannesin, 1994; Orth and Gillham, 1996). The reaction pathways shown in Figure 12.5 have been reported based on batch reactor studies conducted with zero-valent iron and dissolved-phase PCE at room temperature (Arnold and Roberts, 2000). The reaction byproducts of TCE and PCE reductive dechlorination are primarily ethane and ethene; however, accumulation of intermediates is possible if contaminant reactivity and/or residence time within the barrier is not sufficient. Thus, it is important to have knowledge of reaction pathways under site specific conditions, especially when the reaction products may be more mobile and/or more toxic than the parent compounds (e.g., dichloroethene and vinyl chloride). In addition, field-scale monitoring efforts designed to assess PRB effectiveness should include known and potential reaction products, rather than only the parent compound (e.g., PCE or TCE).

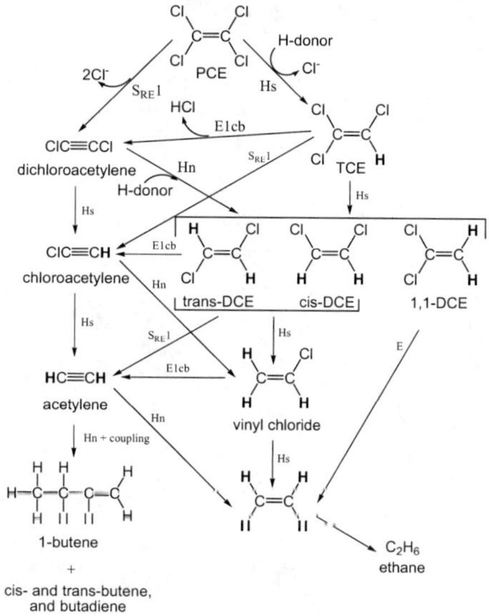

Figure 12.5: Summary of possible PCE reaction pathways in reducing environments (adapted from Arnold and Roberts (2000)).

The ability to achieve dehalogenation in a reactive barrier is significant because the sources of DNAPL contamination, such as trapped residual ganglia, are often not easily located and may generate a continuous plume of dissolved-phase contamination. Although such plumes can often be controlled by pump-and-treat, these systems require extensive maintenance and continual energy input unlike passive PRB systems (NRC, 1999). Typically designed to control downward and lateral migration of plumes, PRBs may extend the full width of the contaminant plume or be combined with sheet piling or low-permeability slurry walls to "funnel" contaminated groundwater through the reactive wall, known as a funnel-and-gate

(Figure 12.6). A funnel-and-gate system typically results in a rise in the up-gradient water table, which induces some downward flow at the gate interface. The depth of the PRB should therefore contact or penetrate an impermeable layer, or be sufficiently deep, so that the dissolved-phase plume does not flow beneath the PRB funnel and gate system. In most field applications, PRB technologies are used to treat dissolved-phase plumes, rather than DNAPL in source zone, in relatively shallow, well characterized aquifers. Additional details of the PRB technology and applications can be found in reports released by the USEPA (1998a) and the Interstate Technology and Regulatory Council (ITRC) as part of the Permeable Reactive Barriers Working Group (ITRC, 1999).

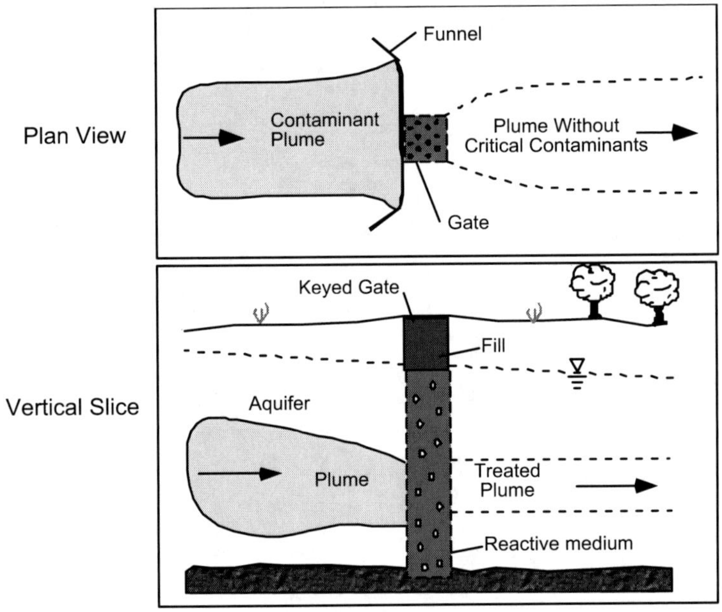

Figure 12.6: Funnel and gate reactive barrier (adapted from Pankow and Cherry (1996)).

12.4.3 PRB Field Application and Limitations

PRBs have been installed extensively in the field for the treatment of chlorinated solvents, hydrocarbons, and other contaminants such as metals, radionuclides, and nutrients. As with all technologies, PRBs have limitations and certain criteria should be met to achieve successful remediation. In general, the application of PRBs is restricted to sites with relatively shallow aquifers (<13 m), well-delineated plumes, and primarily dissolved-phase contamination. In addition, spatial and temporal variations in the permeability and the presence preferential flow pathways within the barrier medium may alter local contaminant residence times within a barrier, which can negatively impact barrier effectiveness. Additionally, barrier longevity, in terms of loss of reactivity, is a concern for long-term treatment, and may require periodic regeneration or replacement of barrier media.

In a recent study, the U.S. Department of Defense, U.S. Department of Energy, USEPA and ITRC coordinated efforts to examine issues of longevity and hydraulic performance of zero-valent iron PRBs (FRTR, 2002). As part of this study, two sites, the former Naval Air Station (NAS) Moffet Field and the former Lowry Air Force Base, were selected for the evaluation of PRB reactivity over time. These two sites were selected because PRB installation occurred at both locations a few months apart, and both PRBs had been in operation for approximately 4 years. Furthermore, the groundwater at these sites contained relatively high levels of total dissolved solids (TDS), a factor that may accelerate precipitation and reduce PRB effectiveness. A complementary hydraulic performance evaluation focused on four sites: Former NAS Moffett Field (funnel-and-gate), Former Lowry AFB (funnel-and-gate), Seneca Army Depot (continuous reactive barrier), and Dover AFB (funnel with two gates). These sites provided a range of PRB designs and hydrogeologic characteristics that could be studied in an effort to develop guidance for future PRB applications. The groundwater at all of the sites was contaminated with chlorinated solvents, and all of the PRBs showed evidence of dechlorination, typically to levels below MCLs.

Column tests conducted with groundwater from the field PRBs at former NAS Moffett Field and former Lowry AFB showed that after flushing 1,300 pore volumes, the half-life of TCE increased approximately by a factor of 2 and 4 in the Moffett Field and Lowry AFB columns, respectively. While some effects of aging may be intrinsic to the iron itself, or to the manufacturing process, the observed differences in reduced reactivity toward TCE may be due to the inorganic content of the water and the subsequent precipitation of dissolved solids. Former NAS Moffett Field has groundwater with a moderate level of dissolved solids (between 500 to 1,000 mg/L) and former Lowry AFB has groundwater with relatively high levels of dissolved solids (greater than 1,000 mg/L). This observation is consistent with the greater decline in TCE reactivity observed in the Lowry AFB column relative to the Moffett Field column. These studies suggest that the rate of decline in iron reactivity was negatively impacted by geochemical constituents of the groundwater and the native level of dissolved solids (e.g., alkalinity, sulfate, calcium, magnesium, and silica) in the groundwater. These findings suggest that under these conditions barriers are likely to exhibit reduced reactivity before the entire zero-valent iron mass is consumed, indicating that some form of PRB regeneration or replacement will be necessary.

In general, the groundwater studies conducted at PRB sites indicate that hydraulic gradients within PRBs are relatively small, which is expected given the relatively high permeability of the PRB media. A few transient flow reversals were reported, for example, at the Moffett Field site, but these occurrences appear to have been temporary and generally within measurement error (Battelle, 1998). In addition, seasonal groundwater level fluctuations were determined to play an important part in

PRB design, demonstrating the importance of ensuring that hydraulic capture zone width and residence time were sufficient for contaminant detoxification. For example, at Dover AFB, historical measurements indicated that groundwater flow direction changed by about 30° on a seasonal basis (Battelle, 2000). This seasonal change in flow direction had a considerable effect on the design of the PRB, which was orientated so that groundwater flow was perpendicular to the PRB at most times of the year.

Based on this multi-agency report, several recommendations were reached regarding the long-term performance of PRBs: (1) adequate site characterization, especially knowledge of the hydraulic flow regime at a prospective PRB location, is imperative; (2) low-flow or passive sampling approaches are required to collect representative data from PRBs; (3) over long periods of groundwater exposure, the reactivity of the granular zero-valent iron declines due to precipitation of native groundwater constituents; (4) in geologic settings where low flow (fine textured formations) conditions dominate, extra care should be taken to insure good hydraulic connection between the native aquifer material and the permeable reactive zone; (5) increased microbial activity and biomass in the immediate vicinity of a barrier wall may contribute to loss of reactivity and/or permeability over time; and (6) additional studies are needed to monitor PRB performance over time and improve predictions of PRB longevity based on site-specific hydrologic, geochemical, and microbiological conditions (FRTR, 2002). Comprehensive summaries of additional PRB field applications can be found at in guidance documents released by the USEPA (1998a; 1999; 2002), ITRC (1999), and the Federal Remediation Technologies Roundtable (FRTR) (USEPA, 1998a; ITRC, 1999; USEPA, 1999; FRTR, 2002; USEPA, 2002). In addition, a recent review of the long-term performance of zero-valent iron PRBs provides additional insights to the causes of PRB failures (Henderson and Demond 2007)

12.5 *In Situ* Chemical Oxidation

In situ chemical oxidation (ISCO) involves the introduction of chemical oxidants that react with organic contaminants in the subsurface, with the goal of transforming potentially toxic contaminants into benign reaction products. Several different oxidants have been utilized for ISCO filed applications including permanganate (MnO_4^-), Fenton's reagent (hydrogen peroxide (H_2O_2) and ferrous iron (Fe^{+2})) or catalyzed hydrogen peroxide (CHP), ozone (O_3), and persulfate ($S_2O_8^{2-}$). The most frequently used oxidants are permanganate (Schroth et al., 2001; MacKinnon and Thomson, 2002) and Fenton's regent. However, Fenton's reagent offers the highest potential for oxidation at a reasonable cost (Basel and Nelson, 2000). With Fenton's reagent, a concentrated hydrogen peroxide solution (10-50%) is used in conjunction with a ferrous iron catalyst (e.g., iron sulfide) to produce highly reactive hydroxyl radicals (Zoh and Stenstrom, 2002) according to

$$H_2O_2 + Fe^{2+} \rightarrow Fe^{3+} + OH^- + OH^\bullet \qquad (12.2)$$

Hydroxyl radicals, along with free radicals produced during side reactions such as $O_2^{\bullet}$, are powerful, non-specific oxidants that react with a variety of organic compounds. ISCO has been used extensively for *in situ* treatment of dissolved-phase organic contaminants, such as chlorinated solvents and petroleum hydrocarbons. A recent report release by the ITRC (2005) offers a comprehensive guide that outlines the technical and regulatory requirements that are relevant to ISCO.

12.5.1 ISCO Design Considerations

Chemical oxidation technologies are predominantly used to treat contaminants in the (water-) saturated zone and capillary fringe. Cost concerns often preclude the use of chemical oxidant technologies to address large, dilute contaminant plumes. More commonly, chemical oxidation is employed to treat relatively small areas, where the contaminant mass is more concentrated. However, when mobile NAPL mass exists in a source area, other remedial technologies, such as free product recovery, may be applied prior to chemical oxidation to reduce costs and improve safety (USEPA, 2004b). A general guidance matrix for the applicability of ISCO to different contaminant concentration ranges is provided in Table 12.3. Detailed knowledge of site hydrogeology is an important factor when considering chemical oxidation technologies because these conditions often control the extent of contact with the contaminant. For example, chemical oxidants may not penetrate into low permeability media, which often contain a large fraction of the contaminant mass.

Reactivity of chemical oxidants with native soil constituents, as opposed to reactions with the target contaminant, is another important consideration in the design and application of ISCO. For example, substantial losses of chemical oxidant can occur through reactions with natural organic matter (NOM), which may preclude the use of ISCO in near surface applications or groundwater with high levels of dissolved organic matter. However, chemical oxidants can be selected to be most appropriate for site-specific conditions. For example, Fenton's reagent is less effective in groundwater that contains high concentrations of carbonates (USEPA, 2004b). The carbonate ion preferentially scavenges the hydroxyl radicals created by Fenton's reactions before they are able to react with the target contaminants. In contrast, the presence of carbonate minerals in the geologic matrix generally has a positive effect on permanganate oxidation.

12.5.2 ISCO Application and Limitations

Numerous laboratory studies have undertaken to investigate Fenton's reagent as a means for remediation of chlorinated solvents. The mechanism of oxidizing TCE in groundwater and soil slurries with Fenton's reagent, and the feasibility of delivery in a sandy aquifer were examined in a series of column and batch experiments (Chen et al. (2001). The results of this study demonstrated that the effectiveness of Fenton's reagent or H_2O_2 (as a Fenton-type oxidant) is strongly dependent upon the soil oxidant demand, presence of sufficient quantities of ferrous iron in the application area, and the proximity of the injection area to the zone of high aqueous contaminant concentrations. In addition, these researchers found that the application of H_2O_2 could result in a gas-sparging of dissolved VOCs in groundwater, requiring careful attention to account for off-gas collection in remedial system design. Teel et al. (2001) completed a study comparing four Fenton's systems: a standard Fenton's system; a modified soluble iron system with a pulse input of hydrogen peroxide; and two modified mineral-catalyzed systems (pH 3 and 7) using dissolved TCE (0.1mM) as the contaminant. Results of this work demonstrated that the most efficient stoichiometry was 78% for the standard Fenton's reagent, while in a goethite system of pH 3, greater than 99% of the initial TCE reacted. However, the goethite system at pH 3 was approximately 3 times less efficient than the standard Fenton's system because of the need for a one-time addition of H_2O_2.

Table 12.3: Applicability of ISCO technology as a function of contamination scenario (adapted from ITRC (2005)).

Extent of Organic Contamination	ISCO Applicability	Design Considerations
Mobile NAPL or NAPL pools	Low, may not be effective	Requires high oxidant dose, multiple applications
Entrapped residual NAPL	Moderate	Likely to require high oxidant dose, multiple applications
High groundwater conc. (> 10 mg/L)	High	Standard application
Low groundwater conc. (< 1 mg/L)	High, may not be cost-effective	Application should account for matrix oxidant demand and plume size

In addition to extensive laboratory investigations, a number of ISCO field studies have been performed to date that have achieved mixed results (USEPA, 1998b). At a former solvent recovery and distribution operation, Fenton's reagent was used remediate a mixture of chlorinated organic compounds at concentrations indicative of the presence of DNAPLs in the saturated zone (DeHghi et al., 2002). Following a two-week field treatment, it was estimated that Fenton's reagent injection achieved greater than 85% contaminant mass destruction and concentration reduction. This estimate was based on a comparison of pre- and post-treatment direct push soil and groundwater samples. While groundwater and soil samples collected immediately after the treatment indicated significant reductions in contaminant concentrations, a rebound in dissolved concentrations was observed at later time, a common observation after ISCO treatment. An evaluation of permanganate injection was conducted at Launch Complex 34 (LC34) at Cape Canaveral Air Station, Florida (USEPA, 2003). During the test, contaminant mass recovery/destruction estimates ranged from 62-84%; however, DNAPL migration to regions outside the test cell was observed (Stroo et al., 2003).

Despite the challenges faced in the field, ISCO remains a promising remediation technology because it transforms the contaminant *in situ*, which can minimizes the need for post-treatment contaminant disposal that is typically required for technologies that remove mass from the subsurface (e.g., cosolvent flushing, soil vapor extraction). In addition, oxidation reaction rates are rapid, potentially shortening remediation times. The major drawback of ISCO is that the reactions are nonspecific, and can involve oxidation of native organic carbon, including microorganisms, soil and dissolved organic matter. Such nonspecific reactions can lead to oxidant depletion prior to reaching the treatment area and negative impacts on subsurface microbial communities and geochemical conditions (Basel and Nelson, 2000). A second limitation is that chemical oxidation is most effective for treatment of dissolved organic species, limiting the utility of ISCO for NAPL source zone remediation at the field-scale (Chen et al., 2001; USEPA, 2003, 2004b). Finally, ferric iron and permanganate species present during chemical oxidation will precipitate if pH increases above 4, resulting in solid precipitates that may reduce the permeability of the treated zone, leading to flow bypassing (Chen et al., 2001; Schroth et al., 2001).

12.6 Surfactant Enhanced Aquifer Remediation (SEAR)

Surfactant enhanced aquifer remediation (SEAR) is an *in situ* flushing technology based on the ability of surfactants to increase aqueous solubility of NAPLs due to micellar solubilization (miscible displacement) and/or displace NAPLs as a result of interfacial tension reductions (immiscible displacement). The initial development of surfactant flushing technologies was largely based on research conducted by the petroleum industry for enhanced oil recovery (EOR). In EOR applications, surfactant formulations often contain cosolvents, polymers, and/or salt (brine) to improve the mobilization of residual crude oil from petroleum reservoirs that have been subjected to water or brine injection (Nelson, 1989). In order to adapt EOR-based surfactant strategies for the remediation of NAPL-contaminated aquifers, a number of important issues had to be addressed, including NAPL density, subsurface heterogeneity, as well as surfactant cost, sorption losses, and toxicity issues. These factors were addressed through a combination of laboratory, pilot-scale and mathematical modeling studies that are discussed below.

12.6.1 Surfactant Classification

Surfactants, an abbreviation for surface active agents, are amphiphilic molecules typically composed of a hydrophilic (polar) head group and a hydrophobic (nonpolar) tail. The hydrophobic group often consists of a long chain hydrocarbon, while the hydrophilic group may contain ionic or polar groups, such as sulfate or ethylene glycol (Rosen, 1989). Surfactants are typically classified as anionic, cationic, or nonionic based on the composition of the hydrophilic group. The hydrophilic group of anionic surfactants possesses a negative charge, (e.g., sulfate or carboxyl). Anionic surfactants have been used extensively in EOR applications because of their low sorption tendency, stability at high temperature and pressure, and improved performance in the presence of salts (Lake, 1989). As a result, a number of studies investigated the potential use of anionic surfactants for SEAR-based remediation of chlorinated solvents (Pennell et al., 1994; Dwarakanath et al., 1999). The hydrophilic group of cationic surfactants possesses a positive charge; typical examples include amine or quaternary ammonium groups. Due to their strong tendency to adsorb to negatively-charged soil constituents, cationic surfactants have been investigated as a means to increase the *in situ* sorption capacity of clay liners (Xu and Boyd, 1995). However, this strong sorption tendency, as well as high cost and toxicity, of cationic surfactants render them unsuitable candidates for most source zone remediation applications. In contrast to anionic and cationic surfactants, the hydrophilic group of nonionic surfactants does not possess an ionic species, but rather consists of polar ethylene oxide (EO) groups. Several attractive features of nonionic surfactants are their relatively low cost, low critical micelle concentration, relatively large solubilization capacity for most hydrophobic organic compounds, insensitivity to background electrolytes, and potential to be biodegraded in the subsurface (Ramsburg et al., 2004; Amos et al., 2007). While a number of nonionic surfactants exhibit toxicity to aquatic species (e.g., alkylphenols), the majority of nonionic surfactants are relatively non-toxic and are widely used in food-grade applications, pharmaceuticals, and household products (Pennell and Abriola, 1997; Pennell et al., 1997). As a result, considerable research has been conducted to evaluate the potential use of nonionic surfactants for remediation of NAPL-contaminated aquifer formations (Ramsburg and Pennell, 2001; Taylor et al., 2001a; Ramsburg et al., 2004; Ramsburg et al., 2005; Suchomel and Pennell, 2006; Suchomel et al., 2007).

12.6.2 Surfactant Recovery Mechanisms

Due to their hydrophobic and hydrophilic structure, surfactants tend to accumulate at interfaces, and can substantially alter the properties of liquid-liquid, air-liquid and solid-liquid interfaces in subsurface systems. A second important characteristic of surfactants is their ability to self aggregate to form micelles above a specific concentration, referred to as the critical micelle concentration (CMC). The CMC of nonionic surfactants is on the order of 10-100 mg/L, while the CMC of anionic surfactants is usually greater than 1000-2000 mg/L, and can be strongly dependent upon system properties. At surfactant concentrations that are less than the CMC, surfactants exist as individual molecules or monomers. When the concentration exceeds the CMC, the monomers begin to aggregate to form micelles (Figure 12.7). In aqueous solutions, these surfactant micelles typically consist of a hydrophilic shell or mantle that surrounds a hydrophobic core, into which nonpolar organic compounds readily partition. As the surfactant concentration is increased above the CMC, the number of micelles continues to increase while the number of monomers remains constant, and thus, the overall or apparent solubility of a hydrophobic organic compound (HOC) tends to increase linearly with surfactant concentration as shown in Figure 12.7.

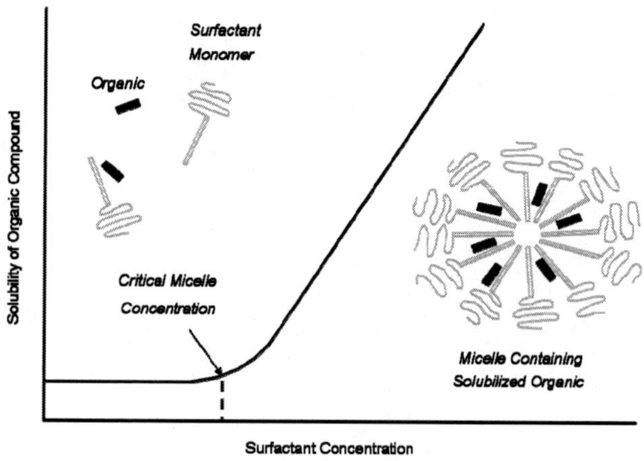

Figure 12.7: Influence of surfactant concentration on the aqueous solubility of hydrophobic organic compounds. At surfactant concentrations greater than the CMC, a linear increase in the aqueous solubility is typically observed (adapted from Pennell and Abriola (1997)).

As noted above, surfactants tend to accumulate at interfaces due to their amphiphilic nature, thereby reducing interfacial forces, such as the surface tension of water (air-water interface) and the interfacial tension of NAPLs (liquid-liquid interface). The latter phenomenon is particularly relevant to the remediation of aquifer formations containing NAPLs, since a reduction in interfacial tension can lead to the displacement or "mobilization" of entrapped NAPLs. Pennell et al. (1996) developed a method to estimate, *a priori*, the potential for NAPL mobilization based on the relative contributions of gravitational (buoyancy), viscous and capillary forces during surfactant flushing. This approach is based on combining the dimensionless capillary number, N_{Ca}, which accounts for the ratio of the viscous forces to the capillary forces:

$$N_{Ca} = \frac{q_w \mu_w}{\gamma_{ow} \cos\theta} \tag{12.3}$$

where q_w is the Darcy velocity of the aqueous phase (positive in the upward direction); μ_w is the aqueous phase viscosity; γ_{ow} is the interfacial tension between the organic and aqueous phases; and θ is the contact angle, with the dimensionless Bond number (N_B), which accounts for the ratio of the buoyancy forces to capillary forces:

$$N_B = \frac{\Delta p g k_i k_{rw}}{\gamma_{ow} \cos\theta} \tag{12.4}$$

where $\Delta\rho$ is the difference between the aqueous and organic phase densities (i.e., $\rho_w - \rho_o$); g is the gravitational constant; k_i is the intrinsic permeability of the porous media; and k_{rw} is the relative permeability of the aqueous phase. When expressed in this manner, the Bond and capillary numbers can be directly combined to yield a total trapping number (N_T):

$$N_T = \sqrt{N_{Ca}^2 + 2N_{Ca}N_B \sin\alpha + N_B^2} \tag{12.5}$$

where α is the angle the system flow makes with the horizontal axis.

For flushing strategies employing horizontal flow (α equal to 0°), Equation (12.5) reduces to:

$$N_T = \sqrt{N_{Ca}^2 + N_B^2} \tag{12.6}$$

For vertical flushing systems in the direction of the buoyancy force (α is equal to 90°), Equation (12.5) becomes:

$$N_T = \left|N_{Ca} + N_B\right| \tag{12.7}$$

A series of experiments was conducted by Pennell et al. (1996) to determine values of the total trapping number corresponding to the initiation of NAPL ganglia mobilization, as well as complete NAPL displacement. These experiments were conducted in soil columns using several different surfactant formulations, flow rates and size fractions of quartz sand to provide a range of initial DNAPL saturations and trapping number values. In all cases, PCE mobilization was observed once the total trapping number exceeded 2×10^{-5}, with nearly complete PCE displacement when the value of N_T exceeded 1×10^{-4} (Figure 12.8). Thus, if calculated value of N_T is less than 2×10^{-5} NAPL mobilization is not expected to occur during surfactant flushing, whereas, if the value of N_T is greater than 1×10^{-4} substantial mobilization of entrapped NAPL should be expected. All of the parameters needed to calculate the value of N_T can be measured independently or estimated from reference data, allowing for *a priori* estimation of the total trapping number for specific NAPL-soil-surfactant systems. In aquifers contaminated with LNAPLs (e.g., gasoline), mobilization may be desirable because the displaced NAPL will tend to migrate upward toward the water table. In sources zones contaminated by DNAPLs (e.g., chlorinated solvents), however, mobilization may not be desirable because this could result in uncontrolled downward migration of DNAPL.

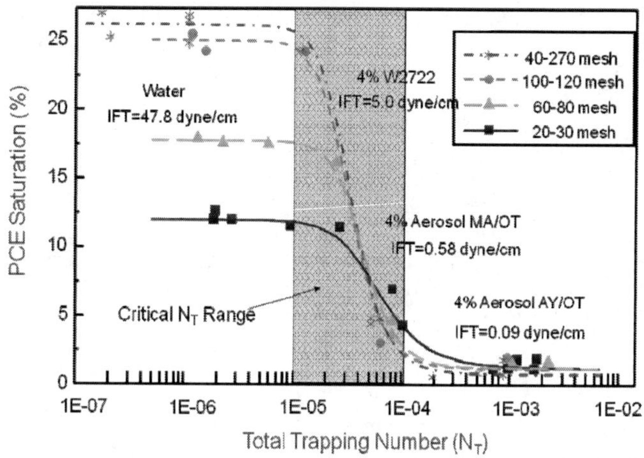

Figure 12.8: Displacement of PCE-DNAPL as a function of total trapping number. The shaded rectangle corresponds to the critical region, below which DNAPL displacement does not occur and above which nearly complete NAPL displacement is observed (adapted from Pennell et al. (1996)).

12.6.3 Surfactant-Based Remediation Technologies

In the early 1990's, a number of surfactant formulations were tested to enhance the removal of strongly-sorbed contaminants, such as PCBs and polycyclic aromatic hydrocarbons (PAHs) from contaminated soils via micellar solubilization (Abdul and Gibson, 1991; Edwards et al., 1991). In addition, success with the displacement of free phase hydrocarbons from soils was realized by the Texas Research Institute (API, 1985). In these experiments, approximately 80% of a gasoline free product was recovered a from soil column packed with Ottawa sand using a mixture of an alkyl benzene sulphonate (2% Richonate YLA) and an ethoxylated nonylphenol (2% Hyonic PE-90). (Pennell et al., 1993) reported that residual dodecane NAPL was readily solubilized by a nonionic surfactant (4% Witconol 2722), but the authors also observed that dodecane concentrations in the column effluent were far less than the measured equilibrium solubility. These results demonstrated the potential effects of rate-limited micellar solubilization on contaminant recovery, a process that was described using a linear driving force expression (Abriola et al. 1993). Subsequent studies conducted in aquifer cells confirmed that both rate-limited solubilization and subsurface heterogeneity contribute to reduced recovery of PCE-DNAPL during surfactant flushing when compared to batch and column studies (Taylor et al.(Taylor et al., 2001c).

As noted above, surfactant-based mobilization of NAPLs offers an efficient means to recover LNAPLs from the subsurface; however, downward migration of DNAPLs during surfactant flushing is rarely desirable. To overcome this potential problem, vertical flushing (Lunn and Kueper 1997) and the use of dense brine solutions (Miller et al. 2000) have been proposed, yet both approaches introduce complexities that are problematic in field applications (e.g., vertical confinement and residual fluids). To address these issues, a density modified displacement (DMD) method was developed by Ramsburg and Pennell (2002), where an alcohol (e.g., butanol) is delivered to the source zone and partitions into the DNAPL, converting it to an LNAPL. The density conversion step is then followed by a low-interfacial surfactant flood that displaces the newly converted LNAPL from the subsurface (e.g., Ramsburg et al. 2003). The key benefit of this approach, compared to other mobilization techniques, is that can be applied in an unconfined aquifer system using conventional horizontal flow field configuration, without the need for vertical barriers.

12.6.4 SEAR Field Studies

While surfactant-based source zone treatment technologies have been shown to be highly effective under controlled laboratory conditions, field-scale applications initially yielded mixed results. A number of early field demonstrations were conducted at sites contaminated by petroleum hydrocarbons, PCBs, and wood-treating creosote NAPLs. A few of these early field demonstrations showed success with LNAPL mobilization (AATDF, 1997), enhanced recovery of sorbed-phase hydrophobic compounds (Abdul et al., 1992; Abdul and Ang, 1994), and removal of residual DNAPL (Fountain et al., 1996a; AATDF, 1997) in subsurface systems that were relatively homogeneous and underlain by a confining layer. However, a significant number of early studies faced difficulties due to insufficient site characterization and limited laboratory evaluations using site contaminants and soils (Roote, 1998). Technical challenges included contaminants perched on clay lenses, low permeability zones, improperly screened surfactant solutions, reductions in aquifer permeability due to surfactant-induced dispersion of fines, and formation of highly viscous emulsions, all of which can contribute to low contaminant mass recoveries (Fountain et al., 1996a; AATDF, 1997; Roote, 1998).

Based on the information learned from these early studies, subsequent field trials showed substantial improvement. For example, a number of well-controlled, field-scale tests of surfactant flushing indicate that DNAPL recoveries in the range of 60-70% can be expected (Rao et al., 1997; Holzmer et al., 2000; Jawitz et al., 2000; Brooks et al., 2004; Soga et al., 2004), and that mass recoveries of greater than 90% are achievable (e.g., Hasegawa et al. 2000; Londergan et al. 2001; Ramsburg et al. 2005). For example, after flushing a DNAPL-contaminated source zone at Hill Air Force Base (AFB) OU2 with approximately 2.5 pore volumes of surfactant solution, researchers from the University of Texas, in collaboration with INTERA, Inc., and Radian, Inc., estimated that 99% of the DNAPL mixture was recovered from the test cell at an average cost of \$793/L (Londergan et al. 2001). Representative examples of SEAR field demonstrations, including the surfactant formulation and estimated mass recovery, are summarized Table 12.4. Here, it should be noted that assessments of remedial performance are typically accomplished through some measure of mass recovery, which often involve considerable uncertainty with respect to the initial mass of NAPL present in the treated source zone.

Table 12.4: Representative examples of SEAR field demonstrations.

Field Site	Surfactant Formulation	DNAPL	Mass Recovery	References
Dover Air Force Base	3.3% Aerosol MA + 3.3% isopropanol + 0.4% $CaCl_2$	PCE	68%	(Childs et al. 2006)
Bachman Road Oscoda, MI	6% Tween 80	PCE	> 90%	(Abriola et al., 2005; Ramsburg et al., 2005)
Alameda Point, CA	5% Dowfax 8390 + 2% Aerosol MA + 3% NaCl + 1% $CaCl_2$	PCE, TCE, DCA, DCE	97%	(Hasegawa et al., 2000)
Camp Lejeune, Marine Corps Base	4% Alfoterra 145-4PO sulfate + 16% propanol + 0.2 % $CaCl_2$	PCE	72%	(Delshad et al., 2000; Holzmer et al., 2000)
Hill Air Foce Base OU1	3% Brij 97 + 2.5% pentanol	Jet Fuel, Chlorinated Solvents	72%	(Jawitz et al., 2001)
Hill Air Force Base OU2	7.6% Aerosol MA + 4.5% isopropanol + 0.7% NaCl	TCE, TCE, PCE, CT	98%	(Londergan et al., 2001)
Hill Air Force Base OU1	4.3% Dowfax; 2.2% Aerosol OT + 2.1% Tween 80 + 0.4% $CaCl_2$	Jet Fuel, Chlorinated Solvents	40-50%; 85-95%	(Knox et al., 1997)
Coast Guard Station, Traverse City, MI	3.6% Dowfax 8390	PCE , Jet Fuel	3.3 g PCE + 45 kg TH*	(Knox et al. 1997)
Thouin Sand Quarry, Quebec, Canada	9.2% butanol + 9.2 % Hostapur SAS 60 + 13.2% toluene + 13.2% d-limonene	TCE, PCE, Waste Oil	86%	(Martel et al., 1998)
Canadian Forces Base Borden	2% 1;1 Rexophos 25/97 + Alkasurf NP10	PCE	69%	(Fountain et al., 1996a)

* % recovery not reported, mass of PCE and total hydrocarbons (TH) recovered.

12.6.5 SEAR Application and Limitations

Although surfactant-based remediation strategies have shown great promise as an effective strategy to rapidly treat NAPL source zones, the technology is not without limitations. Perhaps the most critical issue, which impacts all technologies that involve *in situ* flushing of active ingredients, is the ability to deliver the surfactant solution to the target zone, and more importantly, to extract the mobile contaminant (and surfactant). Due to their excellent emulsifying properties, surfactants tend to disperse fine particles, which can lead to particle mobilization and pore clogging (Liu and Roy 1995; Rao et al. 2006). In addition, surfactant losses may occur due to adsorption on the solid phase and/or partitioning into NAPL; the latter process may lead to the formation of viscous emulsions that can reduce soil conductivity (Jain and Demond 2002; Zimmerman et al. 1999). These issues are typically addressed through laboratory and/or pilot-scale treatability tests conducted prior to full-scale field implementation, and serve to highlight the need for careful selection of a surfactant formulation that is both compatible and effective for the field site. A second important consideration is the cost of the surfactant formulation. As with all aggressive source zone remediation technologies, SEAR entails significant upfront costs, which are largely driven by the cost of the surfactant and the effluent treatment system. To minimize surfactant costs, it is advisable to consider surfactants that are used in other commercial applications, such as food products, detergents, and pharmaceuticals. One such class of surfactants is the sorbitan ethoxylates (e.g., Tween 80®), which typically cost less than $2/lb (Ramsburg and Pennell 2001), and are widely used in whipped toppings and other food products. These surfactants have the added benefit of low toxicity and biodegradability, an important consideration for subsurface applications. In contrast, specialty surfactants can cost $20/lb to $40/lb, and formulations that include alcohols, salts, and other organics can substantially increase material costs. Thus, SEAR is usually best suited to highly-contaminated, permeable source zones that can be treated in a relatively short period of time (e.g., < 1 year).

12.7 Cosolvent Flushing

Cosolvent (alcohol) flushing is similar to SEAR in objective, mode of action, and field application. However, cosolvents do not form micelles, but rather increase contaminant dissolved-phase concentrations by making the aqueous phase less polar (NRC, 2004). During cosolvent flooding, a concentrated alcohol solution, typically comprised of up to 70-95% alcohol in water, is injected into the subsurface. At high concentrations, alcohol solutions act as cosolvents, greatly increasing the aqueous solubility of many organic-phase contaminants due to the miscibility of the alcohol-NAPL system. In addition, highly concentrated alcohol solutions may also lower the IFT between the aqueous and organic phases, resulting in low-IFT displacement (mobilization) of the NAPL. Taken together, these factors make cosolvent flooding a potentially very attractive remediation technology. Laboratory studies (Imhoff et al., 1995; Lunn and Kueper, 1997) have demonstrated the effectiveness of this technology in remediation of NAPL source zones. Lunn and Kueper (1997) reported efficient recovery of greater than 90% of the PCE initially present in a two-dimensional aquifer cell during flushing with a 90% ethanol cosolvent formulation. Effluent PCE concentrations of nearly 100,000 mg/L were observed, and no DNAPL displacement was observed. Conversely, while PCE recovery with a 90% 1-propanol cosolvent was also high (up to 96%), the primary contaminant recovery mechanism was PCE mobilization.

Several field-scale demonstrations of cosolvent flushing have been conducted, most notably at Hill Air Force Base, Dover Air Force Base, and the former Sages dry cleaner site in Jacksonville, Florida (Table 12.5). Two field demonstrations of cosolvent flushing were conducted at Hill Air Force Base Operable Unit 1 (OU1), which recovered approximately 80 and 85% of the NAPL mass from separate 3 m by 5 m test cells that were vertically confined by interlocking sheet pile walls (Rao et al. 1997; Falta et al. 1999). At the Sages dry cleaner site, injection of a 95% ethanol solution was able to achieve similar PCE mass recovery (62-65%) without test cell confinement, while down-gradient aqueous PCE concentrations were reduced by up to 92% following cosolvent flushing (Jawitz et al. 2000). A subsequent field demonstration of cosolvent flushing was conducted at the Dover Air Force Base, where a known amount of PCE was released into a test cell that was then flushed with a 70% ethanol solution, which recovered approximately 64% of initial PCE mass (Rao et al., 1997). Several important technical advances were realized in these studies, most notably the use of gradient injection (increasing the concentration of cosolvent delivered with time) to avoid density over-ride effects (Rao et al. 1997), placement of injection wells to promote upward migration of the miscible effluent in unconfined systems (Jawitz et al. 2000), and the potential for enhanced contaminant biodegradation in the down-gradient plume (Ramakrishnan et al. 2005).

Table 12.5: Representative field-scale demonstrations of cosolvent flushing.

Field Site	Cosolvent	DNAPL	Mass Recovery	References
Dover Air Force Base Test Cell	70% Ethanol	PCE	64%	(Rao et al., 1997)
Sages Dry Cleaner, Jacksonville, Fla.	95% Ethanol	PCE	63%	(Jawitz et al., 2000)
Hill Air Force Base OU1	80% t-Butanol + 15% Hexanol	Jet Fuel, Chlorinated Solvents	80%	(Falta et al. 1999)
Hill Air Force Base OU1	70% Ethanol + 12% Pentanol	Jet Fuel, Chlorinated Solvents	85%	(Rao et al. 1997)

Although cosolvent flushing has shown promise as an effective strategy for NAPL source zone remediation, the technology faces several challenges that have limited implementation on a broader scale. The first issue is material cost because the injected cosolvent solution is typically on the order of 80 to 95% active ingredient. Even when ethanol is purchased without federal and state alcohol consumption taxes, the cost of ethanol is still approximately $2/gallon, and can be as much as $5/gallon due to demand as an alternative fuel. Perhaps more importantly though, from an implementation perspective, are safety issues related to flammability of the injected ethanol solution and the extracted effluent stream, as well as the limited treatment options for the large volumes of a miscible waste stream that consists concentrated alcohol, the recovered organic contaminants, and water.

12.8 Thermal Treatment

The term "thermal treatment" encompasses a number of technologies designed to deliver thermal energy (heat) to the subsurface, including hot water injection, steam injection, conductive heating, resistive heating, and electromagnetic heating (Table 12.6). Compared to the other remediation technologies discussed in this chapter, thermal treatment provides two distinct advantages: (1) no chemical agents are introduced into the subsurface, and (2) the potential to efficiently treat low-permeability, heterogeneous porous media (Davis, 1997). The transfer or delivery of heat (thermal energy) to a contaminated soil matrix can be achieved through several techniques. In Table 12.6, five *in situ* thermal remediation technologies are compared in terms of operational factors, including heat transfer mode, upper temperature range, and recovery phase. Descriptions of each technology, as well as potential limitations and knowledge gaps, are discussed below. Additional information can be found through the USEPA that provides comprehensive reports discussing field studies and the current status of thermal technologies (USEPA, 1995c, 2004a).

Table 12.6: Comparison of *in situ* thermal treatment technologies.

Parameter	Hot Water Injection	Steam Injection	Conductive Heating	Electrical Resistive Heating	Electromagnetic Heating
Heat transfer	Convective	Convective	Conductive	Conductive	Radiative
Energy source	Hot Water	Steam	Steel well casing	Electrical	Electromagnetic
Temp. limit	100°C	100-120°C	500-800°C	100-200°C	100-300°C
Fluid injection	Liquid	Steam	None	None	None
Recovery process	Mobilize	Mobilize/ volatilize	Volatilize	Volatilize	Volatilize
Phase recovered	Liquid	Liquid + Gas	Gas	Gas	Gas
Technology example	Contained recovery of oily waste (CROW®)	Steam enhanced remediation (SER)	In-situ thermal desorption (ISTD)	Six-phase heating® (SPH)	Microwave; radio frequency
Advantages	No phase change, low cost	Large heat input, energy efficient	Low volatility contaminants	Low cost, low permeability layers	Uniform heating, low permeability
Limitations	Preferential flow, low efficiency	Preferential flow, DNAPL, safety issues	High element density, safety	Preferential heating, drying, safety	Low energy transfer, drying

12.8.1 Convective heating

Convective involves the transfer of heat via a mobile fluid, such as a liquid or a gas. Remedial technologies based on convective heating, most notably steam flushing and hot water injection, exhibit upper temperature limits in the range of 100-120°C (Udell, 1997). A hot water injection technique, referred to as Contained Recovery of Oily Wastes (CROW®), was developed in the late 1980's to displace and recover NAPLs from the subsurface (U.S. Patent No. 4,884,460; (Johnson Jr. and Suddeth, 1989). The CROW® technology was implemented at two Superfund sites: UGI Columbia Gas Plant Superfund site in Columbia, Pa. and the Brodhead Creek Superfund Site in Stroudsburg, Pa. Although substantial mass recovery was achieved at the Brodhead site (~1,000 gallons of coal tar recovered), the CROW® technology was unable to reduce the contaminant mass to residual (immobile) levels, and failed to achieve long-term remediation goals (USEPA, 2000). Injection of steam as the mobile convective heating fluid has been successfully implemented at both the pilot- and full-scale, and received considerable attention in the 1990's (Davis, 1997; Udell, 1997). Subsurface steam injection was initially investigated by petroleum engineers as a means to enhance oil recovery through viscosity reduction and distillation (Volek and Pryor, 1972). In the 1980's the technology was adapted for treatment of NAPL-contaminant source zones, and often incorporates aggressive extraction systems to capture contaminants in the gas, liquid water and NAPL phases (U.S. Patent No. 5,018,576;(Udell et al., 1991). The resulting process is often referred to as Steam Enhanced Remediation (SER)/Dynamic Underground Stripping (DUS). The SER/DUS technology has been tested at several field sites containing DNAPL, including the Savannah River Site, Aiken, S.C. and the Portsmouth Gaseous Diffusion Plant, Portsmouth, Ohio. However, delivery of mobile, heated fluids to a contaminant source zone is subject to preferential flow and density over-ride effects (Udell, 1997). Such phenomena may be especially problematic when contaminants exist in low-permeability zones and as DNAPLs, which tend to migrate downward through aquifer formations upon mobilization. To address uncontrolled DNAPL flow during steam flushing, several approaches have been proposed to minimize the formation of condensed DNAPL banks, including the creation of a lower steam boundary (Gerdes et al., 1998) and co-injection of air with steam (Kaslusky and Udell, 2002; Schmidt et al., 2002).

12.8.2 Conductive heating

Conductive heating occurs when two objects are in direct contact, where the molecular vibrations of one object cause the molecules of an adjacent object to vibrate. Due to the proximity of molecules in each phase, solids exhibit the highest heat conductivity, followed by liquids and gases. The technology was originally developed to enhance the removal of strongly sorbed contaminants from the subsurface in conjunction with gas phase extraction, coined in-situ thermal desorption (ISTD) (U.S. Patent No. 5,190,405; (Vinegar et al., 1993). In practice, heat is delivered to the contaminated soil matrix using either steel well casings or surface "thermal blankets," in conjunction with vacuum extraction systems that are used to recover volatilized/vaporized contaminants through a centrally-located extraction well or a network of combined heater-extraction wells (Iben et al., 1996; Conley and Lonie, 2000; Stegemeier and Vinegar, 2001). Relatively high *in situ* temperatures (e.g., 500-800°C) can be achieved with conductive heating, and thus the technology is frequently used to treat source zones containing relatively low volatility contaminants, such as PCBs and PAHs. At these elevated temperatures it has been suggested that PCBs and other chlorinated compounds undergo complete oxidative destruction (Stegemeier and Vinegar, 2001). However, PAH growth and formation of dibenzofurans from chlorophenols, which are possible human carcinogens, have been widely reported in combustion reactors at temperatures ranging from 500 to 600°C (Yang et al., 1998; Mulholland et al., 2000).

12.8.3 Electrical resistive heating

Electrical resistive heating (ERH), also known as "Joule" or "Ohmic" heating, involves the introduction of an electric current into the subsurface, from which heat is generated as electrons pass through the soil matrix, which functions as a resistor (U.S. Patent No. 4,957,393; (Buelt and Oma, 1990). The conduction of electrical current occurs primarily through soil pore water (Robain et al., 2003), and thus, soil resistivity varies strongly with water content, soil mineralogy and electrolyte content. Soil resistivity values can vary from greater than 10^9 Ω·cm for dry soils to approximately 10^2 Ω·cm for soils containing dissolved salts at soil water contents of 15-20% by weight (Kalinski and Kelly, 2003). In effect, dry soils serve as very efficient electrical insulators which can result in poor heat propagation and necessitate replenishment of water at the electrode to avoid melting (Heron et al., 1998). In field applications, electrical current has been delivered in either a three-phase or six-phase mode (U.S. Patent No. 5,330,291; (Heath et al., 1994). The latter process, commonly referred to as Six-Phase Heating,® has undergone pilot-scale (single array) testing at a number of DNAPL contaminated sites, including the Savannah River Site, Area M (Gauglitz et al., 1994) and the Cape Canaveral, Launch Complex 34 (USEPA, 2003). Although Six-Phase Heating® provides for efficient heating of the subsurface in relatively small, circular electrode arrays, three-phase heating is considered to be far less complex and safer to implement at larger scales.

12.8.4 Electromagnetic heating

Electromagnetic heating refers to the absorption of electromagnetic energy by an object, which results in the excitation and rapid vibration of molecules. For example, electromagnetic energy in the microwave region (10^8 to 10^{12} Hz) causes water molecules to vibrate vigorously. Although microwave heating has been studied at the laboratory scale for treatment of PCB-contaminated soils (Di et al., 2000) and other hazardous wastes (Wicks et al., 1999), it is unlikely that the technology can be implemented effectively at the field scale for source zone remediation. However, electromagnetic energy in the radio frequency range (e.g., 0.5 to 45 MHz) can be strongly absorbed by soils and other kinds of geological formations above the water table, which in turn generates heat (U.S. Patent No. 4,670,634;(Bridges et al., 1987). Specially designed electrode rods can be placed in vertical or directionally drilled holes for optimal "excitation" of the contaminated treatment zone, vaporizing hydrocarbons beyond their boiling points for capture in a vacuum extraction system. The vaporized contaminants are then transported to above-ground condensate and vapor treatment systems. This technology was evaluated under the USEPA Site program at several locations, most notably Kelly Air Force Base in San Antonio, Tex. (USEPA, 1995b, a). Initial results of this study indicated that systems reached desirable operational temperatures necessary to remove volatiles and semi-volatiles from clay; however, there has been little work on electromagnetic heating as a source zone remediation technology since the 1990's.

12.8.5 Thermal Treatment Applications and Limitations

Based on the above discussion, it is apparent that thermal remediation encompasses a range of *in situ* technologies, each with advantages and limitations (Table 12.6). A comprehensive review of thermal technologies applied to DNAPL source zones was developed by the USEPA (2004b), which provides information on the field applications, cost and performance analysis, and commercial vendors. Although cost data are often difficult to obtain, *in situ* thermal treatment technologies are expensive, primarily due to infrastructure (e.g., heater wells, extraction and above-ground treatment systems) and energy costs. For this reason, thermal treatment technologies are most often implemented at sites where other *in situ* remediation technologies have failed, or where low permeability media and/or subsurface heterogeneity clearly limits the effectiveness of competing technologies, particularly those that rely on delivery/flushing of active ingredients. Nevertheless, a modest, but steady, number of source zone sites are treated each year using *in situ* thermal remediation technologies. In addition, recent advances in the use of coupled thermal technologies (Heron et al. 2005; 2009) and the potential for enhanced contaminant reactivity (Truex et al. 2007) suggest that thermal remediation will continue to play an important role in treatment of highly-contaminated source zones.

12.9 Effectiveness of Source Zone Treatment Technologies

The *in situ* remediation technologies evaluated in this chapter have shown varied levels of success in treating source zones, especially when applied to sites with complex hydrogeology and contaminant distributions. Even under ideal conditions, it is now widely recognized that some portion of the initial contaminant mass will remain in the subsurface. As a result, there is a need to understand and quantify the long-term release of residual mass, existing as a separate organic liquid (NAPL) and/or associated with the solid (sorbed)-phase, from source zones that have been aggressively treated with physical- and/or chemical-based remediation technologies. Traditionally, regulatory agencies have focused on measurements of contaminant concentrations in the aqueous or solid phase over time at a particle location(s), often referred to as the point of compliance (e.g., a property boundary), to assess the effectiveness of remedial actions and the need for further action. More recently, researchers and several relevant organizations (e.g., ITRC) have considered the potential utility of complementary metrics, such as contaminant mass flux (mass per unit area per unit time) or mass discharge (mass per time), to provide additional information about source strength, remedial action effectiveness, and exposure assessment.

Mass flux and mass discharge measurements have the potential to provide site managers with a more complete understanding of the source zone and the corresponding down-gradient dissolved-phase plume. More specifically, this information can be used to assess the benefits of partial source zone mass removal and plume response following aggressive remediation. For example, spatially-resolved flux data can assist in: (1) the determination of contaminant distribution and strength within a heterogeneous source zone, (2) the extent of mass removal that will be necessary before transitioning to less intensive remediation technologies such as *in situ* bioremediation and PRBs, and (3) the capacity of natural processes to attenuate contaminant mass emanating from a source zone (i.e., monitored natural attenuation).

Even in cases where NAPL source zones can be accurately characterized and treated via physical-chemical methods to achieve mass removals exceeding 80%, most field-scale studies have not demonstrated the ability of aggressive source zone remedial actions to attain drinking water standards (e.g., MCLs) within, or immediately down-gradient from, the source zone. To date, critical assessments of partial mass removal from NAPL source zones have been based almost exclusively on mathematical predictions of dissolution, often leading to conflicting interpretations (McWhorter and Sale, 2003; Rao and Jawitz., 2003). Using an analytical solution for the dissolution of DNAPL ganglia and pools in a uniform aqueous flow field along with limited experimental data, Sale and McWhorter (2001) found that substantial removal of DNAPL mass from a source zone may not result in significant reductions in down-gradient dissolved-phase contaminant concentrations. The behavior described by Sale and McWhorter (2001) for a homogeneous aquifer with areas of discrete idealized source zone contamination is illustrated in Figure 12.9 as Line 1, where mass discharge remains relatively constant until more than 80% of the critical source zone mass recovery point is reached, at which time the reduction in contaminant mass discharge increases dramatically. One implication of sites exhibiting this type of behavior is that nearly complete mass removal would be required to approach regulatory drinking water standards, and that partial mass reduction will provide only minimal improvements in groundwater quality under this scenario.

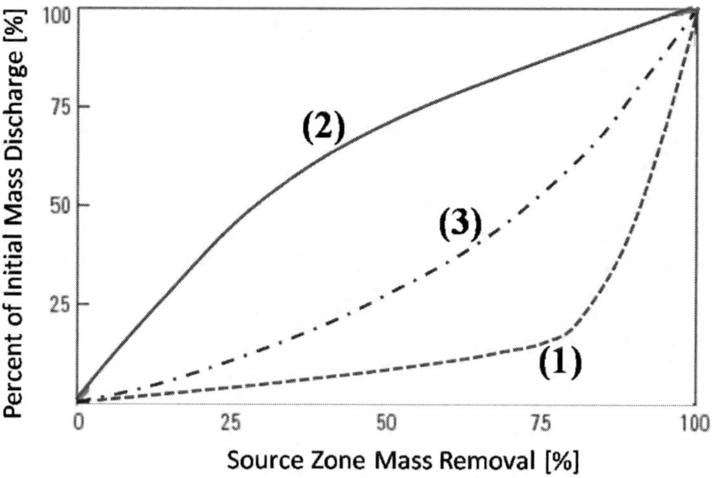

Figure 12.9: Conceptual descriptions of the relationship between the reduction in contaminant mass discharge (y-axis) and source zone mass removal (x-axis) (adapted from Stroo et al. (2003)).

Rao and Jawitz (2003) investigated source zone mass discharge as a function of contaminant recovery using an analytical model that consisted of "stream tubes" that contained a uniform NAPL distribution subject to a non-uniform flow. Results obtained from this model, indicated that up to 80% reductions in contaminant mass discharge may be possible after only 50% mass removal from the source zone (Figure 12.9, Line 2). The relationship obtained by Rao and Jawitz (2003) implies that modest (e.g., 50%) mass removal can lead to significant decreases in mass discharge, and thus, partial source zone treatment may positively impact groundwater quality, although not necessarily to regulatory endpoints. This work was later extended by Jawitz et al. (2005) to account for variability in both the NAPL distribution and the porous media permeability field, expressed as the stream tube travel time, into a single variable describing the overall spatial variability of the source zone. Using this approach, they determined that as the source zone spatial variability decreased, the amount of mass removal required to achieve a given reduction in contaminant mass flux is increased. For systems similar to that described by Sale and McWhorter (2001) (i.e., uniform NAPL distribution and flow fields that result in low spatial variability), contaminant discharge predicted by the model of Jawitz et al. (2005) was not significant until substantial (80-90%) source zone mass reduction had occurred (Figure 12.9, Line 1). The behavior illustrated in Line 2 (Figure 12.9) was also predicted by the model with more spatially variable source zones; however, overall source zone longevity increased due to flow bypassing and mass transfer limitations.

In practical settings, observed reductions in source zone mass discharge as a function of source removal may fall within the extremes described by the simplified analytical models described above. For example, Line 3 in Figure 12.9 was extrapolated from data measured during the Dover Air Force Base field demonstration of cosolvent flushing (Rao et al., 1997). The extrapolated line is consistent with source mass removal that is characterized by preferential dissolution of high interfacial DNAPL ganglia and small pools, with larger pools persisting in the source zone, serving as long-term contaminant sources. Under this scenario, moderate reductions in mass discharge are realized as DNAPL ganglia present in the source zone are depleted, followed by a gradual decrease in mass discharge as lower interfacial area pools are depleted at later time. Using a similar stream tube, Wood et al. (2005) predicted that 60% source zone mass removal from a test cell Hill Air Force Base OU1 flushed with cosolvent, would result in an 80% reduction in contaminant mass flux, which corresponded to a log reduction in flux-averaged concentration.

In addition to the analytical models described above, several numerical model simulations have been performed to evaluate mass flux reductions as a function of source zone removal. Parker and Park (2004) evaluated dissolution of non-uniform DNAPL distributions in a homogenous porous medium. In this scenario, dissolution rate was found to be strongly dependent on DNAPL distribution, with mass recovery significantly limited when the source zone was characterized by laterally extensive (i.e., pooled) DNAPL lenses. Lemke et al. (2004) performed a series of numerical simulations of dissolution of a non-uniform, PCE DNAPL distribution in a statistically homogeneous, non-uniform sandy aquifer. Results obtained from the model realizations indicated that mass removals ranging from 60 to 99% could result in a two order-of-magnitude decrease in dissolved-phase concentrations and mass flux. These findings demonstrate the potential variability in mass removal versus mass flux relationships, even in relatively homogeneous systems, prompting Lemke et al. (2004) to caution that predictions based on simplified conceptual models of DNAPL source zone architecture and flow may not accurately capture reductions in contaminant flux resulting from partial mass removal. The complexity of the DNAPL mass removal versus mass discharge relationships was confirmed by Suchomel and Pennell (2006) in a series of laboratory scale aquifer cell studies.

12.10 Combined *In Situ* Remediation Strategies

As noted above, it is now recognized even successful remediation technologies applications will not completely remove all of the contaminant mass from most source zones. In fact, aggressive source zone treatments are likely to increase the mobility and distribution of the residual mass, which may lead to increased aqueous phase concentrations in the short term. As a result, attention has shifted from developing the most effective stand-alone single technology, toward the development and testing of complementary *in situ* remediation technologies that can be combined, at the same time (in parallel) or sequentially (in series) to more efficiently treat contaminant source zones (e.g., Ramsburg et al. 2004; Christ et al. 2005; Costanza et al. 2005; Friis et al. 2006; Amos et al. 2007). In a sequential or "treatment train" approach, an aggressive *in situ* treatment technology, such as electrical resistive heating or surfactant flushing, is used to remove significant contaminant mass in a relatively short time-frame, while a second "polishing" technology is then applied to remove or detoxify the remaining contaminant mass. Such sequential remediation strategies have the potential to take advantage of efficient mass removal achieved by aggressive treatment technologies, while addressing limitations associated an individual technology (e.g., flow bypassing) that lead to incomplete mass removal.

The potential impacts of aggressive *in situ* treatment technologies on the effectiveness of combined remedies, and longer-term plume behavior have not been sufficiently explored to date (Stroo et al., 2003). For example, physical-chemical treatments may alter geochemical conditions and microbial ecology, which could be either detrimental (e.g., aquifer clogging, reduced microbial diversity) or beneficial (e.g., enhanced electron donor availability) to the overall secondary remediation process (Stroo et al., 2003; Christ et al., 2005). A frequently considered polishing technology is microbial reductive dechlorination, which has the potential to be coupled with thermal treatment, surfactant and/or cosolvent flushing. Recently, the potential for thermal remediation (ERH) to release electron donor has been demonstrated (Friis et al. 2005; Costanza et al. 2009), although in regions where the temperature exceeded 50°C for prolonged periods of time bioaugmentation may be required to achieve meaningful levels of microbial reductive dechlorination (Friis et al. 2006). Pilot-scale studies of both surfactant flushing and cosolvent flushing at former dry cleaning facilities provide evidence that these technologies can be successfully combined with reductive dechlorination to treat residual contaminants (Ramsburg et al. 2004; Ramakrishnan et al. 2005). In general, flushing with concentrated alcohol solutions has the potential to negatively impact microbial activity. However, long-term monitoring of the Sages site (3 years following cosolvent flushing) indicated that biological activity rebounded with decreasing alcohol concentration, although the behavior of populations critical to reductive dechlorination remains unclear (Mravik et al., 2003). Long-term monitoring of the Bachman Road site, which was flushed with a biodegradable, food-grade, nonionic surfactant (Tween 80), provided evidence of stimulated microbial reductive dechlorination. Within the treated PCE DNAPL source zone, fermentation of residual Tween 80, detected at concentrations of 50 to 2,750 mg/L 450 days after SEAR, provided suitable electron donors(s) that stimulated native microbial dechlorination activity in the oligotrophic aquifer (Ramsburg et al., 2004). These findings were complemented by laboratory-based studies that further defined the effects of Tween 80 on reductive dechlorination (Amos et al. 2007). However, selection of compatible surfactants and appropriate concentrations is critical due to potential surfactant toxicity or inhibition toward of microbial populations relevant to desired degradation pathway (Yeh et al., 1999; McGuire and Hughes, 2003; Ramsburg et al., 2004). The potential benefits of surfactant flushing coupled with subsequent bioremediation were evaluated by Christ et al. (2005). Results of this model prediction clearly demonstrate that bioremediation alone provides only minimal benefits when compared to natural gradient dissolution, while aggressive treatment of the source zone with SEAR followed by bioremediation dramatically increased the rate of mass removal and reduced source longevity by several orders-of-magnitude (Figure 12.10).

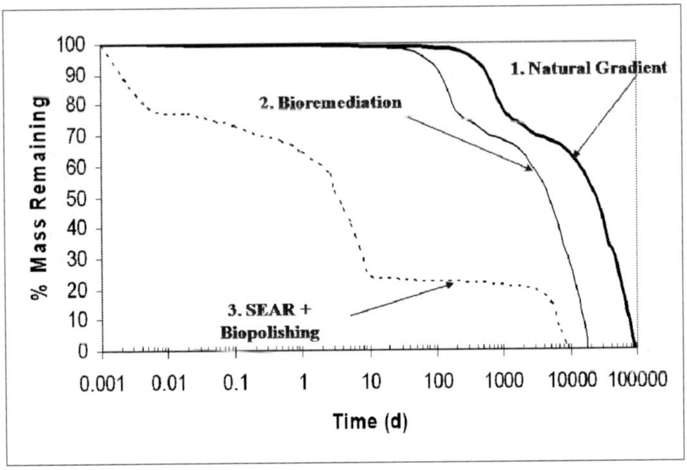

Figure 12.10: Percent DNAPL mass remaining as a function of time for three alternative remediation strategies: natural gradient dissolution alone; bioremediation alone; and SEAR (4% Tween 80) followed by bioremediation (Reproduced from Christ et al. (2005) with permission from *Environmental Health Perspectives*.)

12.11 Summary and Conclusions

Over the past decade, considerable resources have been directed toward the development and demonstration of source zone remediation technologies, including surfactant flushing, cosolvent flushing, thermal treatment and chemical oxidation. Although each of these innovative technologies provides distinct advantages and limitations over traditional pump-and-treat approaches (Table 12.7), considerable uncertainty remains regarding the effectiveness and long-term benefits of these aggressive source zone treatment methods. The successful application of any *in situ* remediation technology requires both the delivery of fluids or energy to the contaminated source zone, followed by contact and/or reactivity with the target contaminant. In addition, most *in situ* treatments technologies also require the subsequent recovery of contaminant mass from the gas, aqueous, and/or separate organic liquid phase. At the bench scale, these technologies have shown promise for reducing the overall remediation time while improving contaminant mass recovery/destruction compared to dissolution alone (i.e., pump-and-treat). However, the success of *in situ* technologies in the field is governed in large part by subsurface hydrogeology, in particular heterogeneity and the complexity of contaminant mass distribution. As a result, a number of early field trials of promising *in situ* remediation technologies have yielded mixed results and have failed to reach desired clean up goals (Fountain et al., 1996b; AATDF, 1997; Smith et al., 1997). As limitations were recognized and refinements were implemented, many of the *in situ* remediation technologies have achieved improved results (AATDF, 1997; Jawitz et al., 1998). Under ideal subsurface and operating conditions, aggressive *in situ* treatment strategies have been found to consistently recover greater than 80 to 90% of the initial contaminant mass. As the extent and complexity of a contaminated source zone increases, however, mass recovery rates on the order of 50-70% are more realistic. Nevertheless, such reductions in source zone contaminant mass are likely to substantially reduce source longevity, and may greatly reduce contaminant mass flux emanating from the treated source zone, thereby decreasing the risks of environmental and health effects (USEPA, 2003). In addition, recent efforts to combine aggressive, short-term remediation technologies (surfactant flushing, thermal treatment) with low impact, long-term strategies (e.g., bioremediation) hold great promise to both reduce treatment cost, reduce source longevity, and achieve remedial goals.

Table 12.7: Comparison of DNAPL source zone treatment technology outcomes.

Expected Outcome	Pump-and-Treat	Chemical Oxidation	Cosolvent Flushing	Surfactant Flushing	Thermal Treatment
Mass Removal	Low	Medium	High	High	High
Treatment Time	Long	Short	Short	Short	Short
Initial Cost	Low	Medium	High	High	High
Major Cost Components	Effluent treatment, O&M	Chemicals, injection system	Chemicals, effluent treatment	Chemicals, effluent treatment	Energy, injection system, effluent treatment
Limitations and Drawbacks	Ineffective, long-term monitoring	Nonspecific reactivity, Safety	Permeability, fluid density, safety	Permeability, complexity	Complexity, safety

12.12 References

Advanced Applied Technology Demonstration Facility (AATDF) (1997). *Technology practices manual for surfactants and cosolvents*, Houston, Tex.

Abdul, A. S., and Gibson, T. L. (1991). "Laboratory studies of surfactant-enhanced washing of polychlorinated biphenyl from sandy material." *Environmental Science and Technology*, 25(4), 665-671.

Abdul, A. S., Gibson, T. L., Ang, C. C., Smith, J. C., and Sobczynski, R. E. (1992). "In situ surfactant washing of polychlorinated biphenyls and oils from a contaminated site." *Ground Water*, 30(2), 219-231.

Abdul, A. S., and Ang, C. C. (1994). "In situ surfactant washing of polychlorinated biphenyls and oils from a contaminated field site: Phase II pilot study." *Ground Water*, 32(5), 727-734.

Abriola, L. M., Dekker, T. J., and Pennell, K. D. (1993). "Surfactant-enhanced solubilization of residual dodecane in soil columns: 2. Mathematical modeling." *Environmental Science & Technology*, 27, 2341-2351.

Abriola, L. M., Drummond, C. D., Hahn, E. J., Hayes, K. F., Kibbey, T. C. G., Lemke, L. D., Pennell, K. D., Petrovskis, E. A., Ramsburg, C. A., and Rathfelder, K. M. (2005). "Pilot-scale demonstration of surfactant-enhanced PCE solubilization at the Bachman road site: 1. Site characterization and test design." *Environmental Science & Technology*, 39(6), 1778-1790.

Agency for Toxic Substances and Disease Registry (ATSDR) (2007). *Toxicological profile for benzene (update)*, U.S. Department of Public Health and Human Services, Public Health Services, Atlanta, Ga.

Amos, B. K., Daprato, R. C., Hughes, J. B., Pennell, K. D., and Löffler, F. E. (2007). "Effects of the nonionic surfactant Tween 80 on microbial reductive dechlorination of chlorinated ethenes." *Environmental Science & Technology*, 41, 1710-1716.

American Petroleum Institute (API) (1985). *Test results of surfactant enhanced gasoline recovery in a large-scale model aquifer*, Publication 4390, American Petroleum Institute, Washington, D.C.

Arnold, W. A., and Roberts, A. L. (2000). "Pathways and kinetics of chlorinated ethylene and chlorinated acetylene reaction with Fe(0) particles." *Environmental Science & Technology*, 34(9), 1794-1805.

Basel, M. D., and Nelson, C. H. (2000). "Overview of in situ chemical oxidation: Status and lessons learned." *Treating dense nonaqueous-phase liquids (DNAPLs): Remediation of chlorinated and recalcitrant compounds*, G. B. Wickramanayake and A. R. Gavaskar, eds., Battelle Press, Columbus, Ohio, 117-124.

Battelle (1998). *Performance evaluation of a pilot-scale permeable reactive barrier at Former Naval Air Station Moffett Field, Mountain View, California*, Naval Facilities Engineering Service Center (NFESC), Port Hueneme, Calif.

Battelle (2000). *Design, construction, and monitoring of the permeable reactive barrier in Area 5 at Dover Air Force Base, Final Report*, Air Force Research Laboratory, Tyndall AFB, Fla.

Benner, T. C. (2004). "Brief survey of EPA standard setting and health assessment." *Environmental Science & Technology*, 38(13), 3457-3464.

Bridges, J. E., et al. (1987). "In situ decontamination of spills and landfills by radio frequency heating." U.S. Patent No. 4,670,634.

Brooks, M. C., Annable, M. D., Rao, P. S. C., Hatfield, K., Jawitz, J. W., Wise, W. R., Wood, A. L., and Enfield, C. G. (2004). "Controlled release, blind test of DNAPL remediation by ethanol flushing." *Journal of Contaminant Hydrology*, 69, 281-297.

Buelt, J. L., and Oma, K. H. (1990). "In situ heating to detoxify organic-contaminated soils." U.S. Patent No. 4,957,393.

Chen, G., Hoag, G. E., Chedda, P., Nadim, F., Woody, B. A., and Dobbs, G. M. (2001). "The mechanism and applicability of in situ oxidation of trichloroethylene with Fenton's reagent." *Journal of Hazardous Materials*, 87(1-3), 171-186.

Childs, J., Acosta, E., Annable, M. D., Brooks, M. C., Enfield, C. G., Harwell, J. H., Hasegawa, M., Knox, R. C., Rao, P. S. C., Sabatini, D. A., Shiau, B., Szekeres, E., and Wood, A. L. (2006). "Field demonstration of surfactant-enhanced solubilization of DNAPL at Dover Air Force Base, Delaware." *Journal of Contaminant Hydrology*, 82(1-2), 1-22.

Christ, J. A., Ramsburg, C. A., Abriola, L. M., Pennell, K. D., and Löffler, F. E. (2005). "Coupling aggressive mass removal with microbial reductive dechlorination for remediation of DNAPL source zones: A review and assessment." *Environmental Health Perspectives*, 113(4), 465-477.

Conley, D. M., and Lonie, C. M. (2000). "Field scale implementation of in situ thermal desorption thermal well technology." *Physical, chemical and thermal technologies: Remediation of chlorinated and recalcitrant compounds*, G. B. Wickramanayake and A. R. Gavaskar, eds., Battelle Press, Columbus, Ohio, 175-182.

Costanza, J., Davis, E. L., Mulholland, J. A., and Pennell, K. D. (2005). "Abiotic degradation of trichloroethylene under thermal remediation conditions." *Environmental Science & Technology*, 39(17), 6825-6868.

Costanza, J., Fletcher, K. E., Löffler, F. E., and Pennell, K. D. (2009). "Fate of TCE in heated Fort Lewis soil." *Environmental Science & Technology*, 43(3), 909-914.

Davis, E. L. (1997). *How heat can enhance in-situ soil and aquifer remediation: Important chemical properties and guidance on choosing the appropriate technology*, EPA/540/S-97/502, National Risk Management Research Laboratory, Ada, Okla.

DeHghi, B., Hodges, A., and Feng, T. H. (2002). "Post-treatment evaluation of Fenton's Reagent in-situ chemical oxidation." *Proc., Third International Conference on Remediation of Chlorinated and Recalcitrant Compounds, Monterey, California,* Battelle Press, Columbus, Ohio

Delshad, M., Pope, G. A., Yeh, L., and Holzmer, F. J. (2000). "Design of the surfactant flood at Camp Lejeune." *Physical, chemical and thermal technologies: Remediation of chlorinated and recalcitrant compounds*, G. B. Wickramanayake and A. R. Gavaskar, eds., Battelle Press, Columbus, Ohio.

Di, P., Chang, D. P. Y., and Dwyer, H. A (2000). "Heat and mass transfer during microwave steam treatment of contaminated soils." *Journal of Environmental Engineering*, 126, 1108-1115.

Dwarakanath, V., Kostarelos, K., Pope, G. A., Shotts, D., and Wade, W. H. (1999). "Anionic surfactant remediation of soil columns contaminated by nonaqueous phase liquids." *Journal of Contaminant Hydrology*, 38, 465-488.

Edwards, D. A., Luthy, R. G., and Liu, Z. (1991). "Solubilization of polycyclic aromatic hydrocarbons in micellar nonionic surfactant solutions." *Environmental Science & Technology*, 25(1), 127-133.

Falta R. W., Lee, C. M., Brame, S. E., Roeder, E., Coates, J. T., Wright, C., Wood, A. L., and Enfield, C. G. (1999). "Field test of high molecular weight alcohol flushing for subsurface nonaqueous phase liquid remediation." *Water Resources Research*, 35(7), 2095-2108.

Federal Remediation Technologies Roundtable (FRTR) (2002). *Evaluation of permeable reactive barrier performance, tri-agency permeable reactive barrier initiative*, Federal Remediation Technologies Roundtable, Washington, D.C.

Fountain, J. C., Starr, R. C., Middleton, T., Beikirch, M., Taylor, C., and Hodge, D. (1996). "A controlled field test of surfactant-enhanced aquifer remediation." *Ground Water*, 34(5), 910-916.

Friis, A. K., Albrechtsen, H.-J., Heron, G. and Bjerg, P. L. (2005). "Redox processes and release of organic matter after thermal treatment of a TCE-contaminated aquifer." *Environmental Science & Technology*, 39, 5787-5795.

Friis, A. K., Albrechtsen, H. J., Cox, E., and Bjerg, P. L. (2006). "The need for bioaugmentation after thermal treatment of a TCE-contaminated aquifer: Laboratory experiments." *Journal of Contaminant Hydrology*, 88, 235-248.

Gauglitz, R. A., Roberts, J. S., Bergsman, T. M., Schalla, R., Caley, S. M., Schlender, M. H., Heath, W. O., Jarosch, T. R., Miller, M. C., Dilek, C. A. E., Moss, R. W., and Looney, B. B. (1994). *Six-phase soil heating for enhanced removal of contaminants: Volatile organic compounds in non-arid soils, Integrated Demonstration, Savannah River Site*, PNL-101 84, Battelle Pacific Northwest Laboratory.

Gerdes, K. S., Kaslusky, S., and Udell, K. S. (1998). "Containment of downward migration of DNAPLs during steam enhanced extraction." *Nonaqueous-phase liquids: Remediation of chlorinated and recalcitrant Compounds*, G. B. Wickramanayake and R. E. Hinchee, eds., Battelle Press, Columbus, Ohio, 31-36.

Gillham, R., and O'Hannesin, S. (1994). "Enhanced degradation of halogenated aliphatics by zero-valent iron." *Ground Water,* 32(6), 958-967.

Hasegawa, M. H., Shiau, B. J., Sabatini, D. A, Knox, R. C., Harwell, J. H., Lago, R., and Yeh, L. (2000). "Surfactant enhanced subsurface remediation of DNAPLs at the Former Naval Air Station Alameda, California." *Physical, chemical and thermal technologies: Remediation of chlorinated and recalcitrant compounds*, G. D. Wickramanayake and A. R. Gavaskar, eds., Battelle Press, Columbus, Ohio, 219-226.

Heath, W. O., et al. (1994). "Heating solid earthen material, measuring moisture and resistivity." U.S. Patent No. 5,330,291.

Henderson, A. D., and Demond, A. H. (2007). "Long-term performance of zero-valent iron permeable reactive barriers: A critical review." *Environmental Engineering Science*, 24(4), 401-423.

Heron, G., Van Zutphen, M., Christensen, T. H., and Enfield, C. G. (1998). "Soil heating for enhanced remediation of chlorinated solvents: A laboratory study on resistive heating and vapor extraction in a silty, low-permeable soil contaminated with trichloroethylene." *Environmental Science & Technology*, 32, 1474-1481.

Heron, G., Carroll, S., and Nielsen, S. G. (2005). "Full-scale removal of DNAPL constituents using steam-enhanced extraction and electrical resistance heating." *Ground Water Monitoring and Remediation*, 25(4), 92-107.

Heron, G., Parker, K., Galligan, J., and Holmes, T. C. (2009). "Thermal treatment of eight CVOC source zones to near nondetect concentrations." *Ground Water Monitoring and Remediation*, 29(3), 56-65.

Hoag, G. E., and Marley, M. C. (1986). "Gasoline residual saturation in unsaturated uniform aquifer material." *Journal of Environmental Engineering*, 112(3), 586-604.

Holzmer, F. J., Hope, G. A., and Yeh, L. (2000). "Surfactant enhanced aquifer remediation of PCE DNAPL in low-permeability sand." *Physical, chemical and thermal technologies: Remediation of chlorinated and recalcitrant compounds*, G. D. Wickramanayake and A. R. Gavaskar, eds., Battelle Press, Columbus, Ohio, 211-218.

Hornsby, A. G., Wauchope, R. D., and Herner, A. (1996). *Pesticide properties in the environment*, Springer-Verlag.

Hunt, J. R., Sitar, N., and Udell, K. S. (1988). "Nonaqueous phase liquid transport and cleanup 1. Analysis of mechanisms." *Water Resources Research*, 24(8), 1247-1258.

Iben, I. E. T., Edelstein, W. A., Sheldon, R. B., Shapiro, A. P., Uzgiris, E. E., Scatena, C. R., Blaha, S. R., Silverstein, W. B., Brown, G. R., Stegemeier, G. L., and Vinegar, H. J. (1996). "Thermal blanket for *in-situ* remediation of surfacial contamination: A pilot test." *Environmental Science & Technology,* 30, 3144-3160.

Imhoff, P. T., Gleyzer, S. N., McBride, J. F., Vancho, L. A., Okuda, I., and Miller, C. T. (1995). "Cosolvent enhanced remediation of residual dense nonaqueous phase liquids: Experimental investigation." *Environmental Science & Technology*, 29(8), 1966-1976.

Interstate Technology Regulatory Council (ITRC) (1999). *Regulatory guidance for permeable reactive barriers designed to remediate chlorinated solvents*, 2nd ed., Washington, D.C.

ITRC (2005). *Technical and regulatory guidance for in situ chemical oxidation of contaminated soil and groundwater*, 2nd ed., Washington, D.C.

Jain V., and Demond, A. H. (2002). "Conductivity reduction due to emulsification during surfactant enhanced-aquifer remediation: 1. Emulsion transport." *Environmental Science & Technology,* 36(24), 5426-5433.

Javandal, I., and Tsang, C. F. (1986). "Capture zone type curves: A tool for aquifer cleanup." *Ground Water*, 24(5), 616-625.

Jawitz, J. W., Annable, M. D., Rao, P. S. C., and Rhue, R. D. (1998). "Field implementation of a Winsor type I surfactant/alcohol mixture for in situ solubilization of a complex LNAPL as a single phase microemulsion." *Environmental Science & Technology*, 32(4), 523-530.

Jawitz, J. W., Sillan, R. K., Annable, M. D., Rao, P. S. C. and Warner, K. (2000). "In situ alcohol flushing of a DNAPL source zone at a dry cleaner site." *Environmental Science & Technology*, 34, 3722-3729.

Jawitz, J. W., Annable, M. D., Rao, P. S. C., and Rhue, R. D. (2001). "Evaluation of remediation performance and cost for field-scale single-phase microemulsion (SPME) flushing." *Journal of Environmental Science Health*, A36, 1437-1450.

Jawitz, J. W., Fure, A. D., Demmy, G. G., Berglund, S., and Rao, P. S. C. (2005). "Groundwater contaminant flux reduction resulting from nonaqueous phase liquid mass reduction." *Water Resources Research*, 41, W10408.

Johnson, L. A., Jr., and Suddeth, B. C. (1989). "Contained recovery of oil waste, edited." U.S. Patent No. 4,884,460.

Kalinski, R. J., and Kelly, W. E. (2003). "Estimating water content of soils from electrical resistivity." *ASTM Geotechnical Testing Journal*, 16, 323-327.

Kaslusky, S. F., and Udell, K. S. (2002). "A theoretical model of air and steam co-injection to prevent downward migration of DNAPLs during steam-enhanced extraction." *Journal of Contaminant Hydrology*, 55, 213-232.

Kavanaugh, M. C., and Rao, P. S. C. (2003). *The DNAPL cleanup challenge: Is there a case for source depletion?*, EPA/600/R-03/143, National Risk Management Research Laboratory, U. S. Environmental Protection Agency, Cincinnati, Ohio.

Knox, R. C., Sabatini, D. A., Harwell, J. H., Brown, R. E., West, C. C., Blaha, F. and Griffin, C. (1997). "Surfactant remediation field demonstration using a vertical circulation well." *Ground Water*, 35(6), 948-953.

LaGrega, M., Buckingham, P., Evans, J., and Environmental Resources Management (2000). *Hazardous waste management*, 2nd ed., McGraw-Hill, New York.

Lake, L. W. (1989). *Enhanced oil recovery*, Prentice-Hall, Old Tappan, N.J.

Lemke, L. D., Abriola, L. M., and Goovaerts, P. (2004). "DNAPL source zone characterization: Influence of hydraulic property correlation on predictions of DNAPL infiltration and entrapment." *Water Resources Research*, 40, W01511.

Liu, M. W., and Roy, D. (1995). "Surfactant-induced interactions and hydraulic conductivity changes in soil." *Waste Management*, 15(7), 463-470.

Londergan, J. T., Meinardus, H. W., Manner, P. E., Jackson, R. E., Brown, C. L., Dwarakanath, V., Pope, G. A., Ginn, J. S., and Taffinder, S. (2001). "DNAPL removal from a heterogeneous alluvial aquifer by surfactant-enhanced aquifer remediation." *Ground Water Monitoring Remediation*, 21, 57-67.

Lunn, S. R. D., and Kueper, B. H. (1997). "Removal of pooled dense, nonaqueous phase liquid from saturated porous media using upward gradient alcohol floods." *Water Resources Research*, 33(10), 2207–2219.

MacDonald, J. A., and Kavanaugh, M. C. (1994). "Restoring contaminated groundwater: An achievable goal?" *Environmental Science and Technology*, 28(8), 362A-368A.

Mackay, D. M., and Cherry, J. A. (1989). "Groundwater contamination: Pump-and-treat remediation." *Environmental Science & Technology*, 23(6), 630-636.

Mackay, D., Shiu, W. Y., Ma, K.-C., and Lee, S. C. (2006). *Handbook of physical-chemical properties and environmental fate of organic chemicals*, 2nd ed., Vol. 4, Taylor and Francis, CRC Press, Boca Raton, Fla.

MacKinnon, L. K., and Thomson, N. R. (2002). "Laboratory-scale in situ chemical oxidation of a perchloroethylene pool using permanganate." *Journal of Contaminant Hydrology*, 56, 49-74.

Martel, R., Gélinas, P. J., and Saumure, L. (1998). "Aquifer washing by micellar solutions: 3. Field test at the Thouin sand pit (L'Assomption, Quebec, Canada)." *Journal of Contaminant Hydrology*, 30, 33-48.

McGuire, T., and Hughes, J. B. (2003). "Effects of surfactants on the dechlorination of chlorinated ethenes." *Environmental Toxicology and Chemistry*, 22(11), 2630-2638.

McWhorter, D. B., and Sale, T. C. (2003). Reply to Comment by P. S. C. Rao and J. W. Jawitz on "Steady state mass transfer from single-component dense nonaqueous phase liquids in uniform flow fields." *Water Resources Research*, 39, 1069.

Mercer, J. W., and Cohen, R. M. (1990). "A review of immiscible fluids in the subsurface: Properties, models, characterization, and remediation." *Journal of Contaminant Hydrology*, 6, 107-163.

Miller, C. T., Hill, E. H., III, and Moutier, M. (2000). "Remediation of DNAPL-contaminated subsurface systems using density-motivated mobilization." *Environmental Science & Technology*, 34(4), 719-724.

Mravik, S. C., Sillan, R. K., Wood, A. L., and Sewell, G. W. (2003). "Field evaluation of the solvent extraction residual biotreatment technology." *Environmental Science & Technology*, 37(21), 5040-5049.

Mulholland, J. A., Lu, M., and Kim, D. (2000). "Pyrolytic growth of polycyclic aromatic hydrocarbons by cyclopentadienyl moieties." *Proceedings of the Combustion Institute*, 28, 2593-2599.

Nelson, R. (1989). "Chemically enhanced oil recovery: The state of the art." *Chemical engineering progress,* 85(3), 50-57.

National Research Council (NRC) (1994). *Alternatives for groundwater cleanup*, National Academy Press, Washington, D.C.

NRC (1997). *Innovation in ground water and soil cleanup*, National Academy Press, Washington, D.C.

NRC (1999). *Groundwater and soil cleanup: Improving management of persistent contaminants*, National Academy Press, Washington, D.C.

NRC (2004). *Contaminants in the subsurface: Source zone assessment and remediation*, National Academy Press, Washington, D.C.

Orth, W. S., and Gillham, R. W. (1996). "Dechlorination of trichloroethene in aqueous solution using Fe0." *Environmental Science & Technology*, 30(1), 66-71.

Pankow, J. F., and Cherry, J. A. (1996). *Dense chlorinated solvents and other DNAPLs in groundwater*, Waterloo Press, Portland, Ore.

Parker, J. C., and Park, E. (2004). "Modeling field-scale dense nonaqueous phase liquid dissolution kinetics in heterogeneous aquifers." *Water Resources Research*, 40, W051091.

Pennell, K. D., Abriola, L. M., Weber, W. J., Jr. (1993). "Surfactant-enhanced solubilization of residual dodecane in soil columns: 1. Experimental investigation." *Environmental Science & Technology*, 27(12), 2332–2340.

Pennell, K. D., Jin, M., Abriola, L. M., and Pope, G. A. (1994). "Surfactant enhanced remediation of soil columns contaminated by residual tetrachloroethylene." *Journal of Contaminant Hydrology*, 16(1), 35-53.

Pennell, K. D., Abriola, L. M., and Pope, G. A. (1996). "Influence of viscous and buoyancy forces on the mobilization of residual tetrachloroethylene during surfactant flushing." *Environmental Science & Technology*, 30(4), 1328–1335.

Pennell, K. D., and Abriola, L. M. (1997). "Surfactant enhanced aquifer remediation: Fundamental processes and practical application." *Bioremediation: Principles and practice*, S. K. Sikdar and R. L.Irvine, eds., Technomic Publ., Lancaster, Pa., 693-750.

Powers, S. E., Loureiro, C. O., Abriola, L. M., Weber, W. J., Jr. (1991). "Theoretical study of the significance of nonequilibrium dissolution of nonaqueous phase liquids in subsurface systems." *Water Resources Research*, 27(4), 463-477.

Ramakrishnan, V., Ogram, A. V., and Lindner, A. S. (2005). "Impacts of co-solvent flushing on microbial populations capable of degrading trichloroethylene." *Environmental health perspectives*, 113(1), 55-61.

Ramsburg, C. A., and Pennell, K. D. (2001). "Experimental and economic assessment of two surfactant formulations for source zone remediation at a former dry cleaning facility." *Ground Water Monitoring and Remediation*, 21, 68-82.

Ramsburg, C. A., and Pennell, K. D. (2002). "Density-modified displacement for DNAPL source zone remediation: Density conversion and recovery in heterogeneous aquifer cells." *Environmental Science & Technology*, 36(14), 3176-3187.

Ramsburg C. A., Kennell, K. D., Kibbey, T. C. G., and Hayes, K. F. (2003). "Use of a surfactant-stabilized emulsion to deliver 1-butanol for density-modified displacement of trichloroethene." *Environmental Science & Technology*, 37(18), 4246-4253.

Ramsburg, C. A., Abriola, L. M., Kennell, K. D., Löffler, F. E., Gamache, M., Amos, B. K., and Petrovskis, E. A. (2004). "Stimulated microbial reductive dechlorination following surfactant treatment at the Bachman road site." *Environmental Science & Technology*, 38(22), 5902-5914.

Ramsburg, C. A., Kennell, K. D., Abriola, L. M., Daniels, G., Drummond, C. D., Gamache, M., Hsu, H.-L., Petrovskis, E. A., Rathfelder, K. M., Ryder, J. L., and Yavaraski, T. P. (2005). "Pilot-scale demonstration of surfactant-enhanced PCE solubilization at the Bachman Road site: 2. System operation and evaluation." *Environmental Science & Technology*, 39(6), 1791-1801.

Rao, P. H., He, M., Yang, X., Zhang, Y.-C., Sun, S.-Q., and Wang, J.-S. (2006). "Effect of an anionic surfactant on hydraulic conductivities of sodium- and calcium-saturated soils." *Pedosphere*, 16(5), 673-680.

Rao, P. S. C., Aimable, M. D., Sillan, R. K., Dai, D., Hatfield, K., Graham, W. D., Wood, A. L., and Enfield, C. G. (1997). "Field-scale evaluation of in situ cosolvent flushing for enhanced aquifer remediation." *Water Resources Research*, 33, 2673-2686.

Rao, P. S. C., and Jawitz, J. W. (2003). Comment on "Steady state mass transfer from single-component dense nonaqueous phase liquids in uniform flow fields." *Water Resources Research*, 39(3), 1068.

Rathfelder, K. M., Abriola, L. M., Taylor, T. P., and Pennell, K. D. (2001). "Surfactant enhanced recovery of tetrachloroethylene from a porous medium containing low permeability lenses: 2. Numerical simulations." *Journal of Contaminant Hydrology*, 48, 351-374.

Riddick, J. A., and Bunger, W. B. (1970). "Organic solvents: Physical properties and methods of purification." *Techniques of chemistry*, A. Weissberger, ed., Wiley-Interscience, New York.

Roote, D. S. (1998). *Technology status report in situ flushing*, TS-98-01, Ground-Water Remediation Technologies Analysis Center (GWRTAC), Pittsburgh, Pa.

Rosen, M. J. (1989). *Surfactants and interfacial phenomena*, 2nd ed., John Wiley & Sons, New York.

Sale, T. C., and McWhorter, D. B. (2001). "Steady state mass transfer from single-component dense nonaqueous phase liquids in uniform flow fields." *Water Resources Research*, 37(2), 393-404.

Schmidt, R., Gudbjerg, J., Sonnenborg, O., and Jensen, K. H. (2002). "Removal of NAPLs from the unsaturated zone using steam: Prevention of downward migration by injecting mixtures of steam and air." *Journal of Contaminant Hydrology*, 55, 233-260.

Schroth, M. H., Oostrom, M., Wietsma, T. W., and Istok, J. D. (2001). "In-situ oxidation of trichloroethene by permanganate: Effects on porous medium hydraulic properties." *Journal of Contaminant Hydrology*, 50, 79-98.

Schwarzenbach, R. P., Gschwend, P. M., and Imboden, D. M. (2003). *Environmental organic chemistry*, John Wiley & Sons, Inc., Hoboken, N.J.

Schwille, F. (1988). *Dense chlorinates solvents in porous and fractured media*, Lewis Publishers, Boca Raton, Fla.

Smith, J. A., Sahoo, D., McLellan, H. M., and Imbrigiotta, T. E. (1997). "Surfactant-enhanced remediation of a trichloroethene-contaminated aquifer: 1. Transport of triton X-100." *Environmental Science & Technology*, 31(12), 3565-3572.

Soga, K., Page, J. W. E., and Illangasekare, T. H. (2004). "A review of NAPL source zone remediation efficiency and the mass flux approach." *Journal of Hazardous Material*, 110, 13-27.

Stegemeier, G. L., and Vinegar, H. J. (2001). "Thermal conduction heating for in-situ thermal desorption of soils." *Hazardous and radioactive waste treatment technologies handbook*, C. H. Oh, ed., CRC Press, Boca Raton, Fla.

Stroo, H. F., Unger, M., Ward, C. H., Kavanaugh, M. C., Vogel, C., Leeson, A., Marqusee, J., and Smith B. P. (2003). "Remediation chlorinated solvent source zones." *Environmental Science & Technology*, 37, 224A-230A.

Suchomel, E. J., and Pennell, K. D. (2006). "Reductions in contaminant mass discharge following partial mass removal from DNAPL source zones." *Environmental Science & Technology*, 40, 6110-6116.

Suchomel, E. J., Ramsburg, C. A., and Pennell, K. D. (2007). "Evaluation of trichloroethene recovery processes in heterogeneous aquifer cells flushed with biodegradable surfactants." *Journal of Contaminant Hydrology*, 93(3-4), 195-214.

Taylor, T. P., Pennell, K. D., Abriola, L. M., and Dane, J. H. (2001). "Surfactant enhanced recovery of tetrachloroethylene from a porous medium containing low permeability lenses: 1. Experimental studies." *Journal of Contaminant Hydrology*, 48(3-4), 325-350.

Teel, A. L., Warberg, C. R., Atkinson, D. A., and Watts R. J. (2001). "Comparison of mineral and soluble iron Fenton's catalysts for the treatment of trichloroethylene." *Water Research*, 35(4), 977-984.

Truex et al., (2007). "In situ dechlorination of TCE during aquifer heating." *Ground Water Monitoring and Remediation*, 27(2), 96-105.

Udell, K. S., Sitar, N., Hunt, J. R., and Stewart, L. D. (1991). "Process for in-situ decontamination of subsurface soil and ground water." U.S. Patent No. 5,018,576.

Udell, K. S. (1997). "Thermally enhanced removal of liquid hydrocarbon contaminants from soils and groundwater." *Subsurface restoration*, C. H. Ward, M. R. Scalf, and J. Cherry, eds., Ann Arbor Press, Chelsea, Mich., 251-270.

U.S. Environmental Protection Agency (USEPA) (1987). Federal Register, Washington, D.C., 25,690-25,691.

USEPA (1990). *Handbook, ground water, volume 1: Ground water and contamination*, EPA 625/6-90/016a, Office of Research and Development, Washington, D.C.

USEPA (1995a). *IITRI radio frequency heating technology, Site technology capsule*, EPA/540/R-94/527a, Washington, D.C.

USEPA (1995b). *In situ remediation technology status report: Thermal enhancements*, EPA/542-K-94-009, Office of Solid Waste and Emergency Response, Technology Innovation Office, Washington, D.C.

USEPA (1995c). *KAI radio frequency heating technology, Site Technology Capsule*, EPA/540/R-94/528a, Washington, D.C.

USEPA (1997). *Supplemental bulletin multi-phase extraction (MPE) technology for VOCs in soil and groundwater,* OSWER Directive 9355.0-68FS, EPA/540/F-97/004, Washington, D.C.

USEPA (1998a). *Treatment technologies for site cleanup: Annual status report*, 9th ed., EPA-542-R98-018, Office of Solid Waste and Emergency Response, Washington, D.C.

USEPA (1998b). *Permeable reactive barrier technologies for contaminant remediation*, EPA/600/R-98/125, Washington, D.C.

USEPA (1999). *Field applications of in situ remediation technologies: Permeable reactive barriers*, EPA 542-R-99-002, Office of Solid Waste and Emergency Response, Technology Innovation Office, Washington, D.C.

USEPA (2000). *Western Research Institute Contained Recovery of Oily Wastes (CROW) Process*, EPA/540/R-00/500, Washington, D.C.

USEPA (2002). *Field applications of in situ remediation technologies: Permeable reactive barriers*, Office of Solid Waste and Emergency Response, Technology Innovation Office, Washington, D.C.

USEPA (2003). *The DNAPL remediation challenge: Is there a case for source depletion?* EPA/600/R-03/143, Washington, D.C.

USEPA (2004a). *How to evaluate alternative cleanup technologies for underground storage tank sites: A guide for corrective action plan reviewers*, EPA 510-B-94-003, EPA 510-B-95-007, EPA 510-R-04-002, Washington, D.C.

USEPA (2004b). *In situ thermal treatment of chlorinated solvents fundamentals and field applications*, EPA 542-R-04-010, Office of Solid Waste and Emergency Response Washington, D.C.

USEPA (2008). *A systematic approach for evaluation of capture zones at pump and treat systems*, EPA/600/R-08/003, Ground Water and Ecosystems Restoration Division, National Risk Management Research Laboratory, Ada, Okla.

Vinegar, H. J., Stegemeier, G. L., De Rouffignac, E. P., and Chou, C. C. (1993). "Vacuum method for removing soil contaminants utilizing thermal conductive heating." U.S. Patent No. 5,190,405.

Volek, C. W., and Pryor, J. A. (1972). "Steam distillation drive: Brea Field, California." *Journal of Petroleum Technology*, 24, 899-906.

Wicks, G. G., Clark, D. E., Schultz, R. L., and Folz, D. C. (1999). "Microwave technology for waste management applications including disposition of electronic circuitry." *Environmental issues and waste management technologies in the ceramic and nuclear industries IV*, J. C. Marra and G. T. Chandler, eds., 89-96.

Wiedemeier, T. H., Rifai, H. S., Newell, C. J., and Wilson, J. T. (1999). *Natural attenuation of fuels and chlorinated solvents in the subsurface*, John Wiley & Sons, Inc., New York.

Wood, A. L., Enfield, C. G., Espinoza, F. P., Annable, M., Brooks, M. C., Rao, P. S. C., Sabatini, D., and Knox, R. (2005). "Design of aquifer remediation systems: 2. Estimating site-specific performance and benefits of partial source removal." *Journal of Contaminant Hydrology*, 81, 148-166.

Xu, S., and Boyd, S. A. (1995). "Cationic surfactant sorption to a vermiculitic subsoil via hydrophobic bonding." *Environmental Science & Technology*, 29, 312-320.

Yeh, D. H., Pennell, K. D., and Pavlostathis, S. G. (1999). "Effect of Tween surfactants on methanogenesis and microbial reductive dechlorination of hexachlorobenzene." *Environmental Toxicology and Chemistry*, 18, 1408-1416.

Zimmerman, J. B., Kibbey, T. C. G., Cowell, M. A., and Hayes, K. F. (1999). "Partitioning of ethoxylated nonionic surfactants into nonaqueous-phase organic liquids: Influence on solubilization behavior." *Environmental Science & Technology*, 33(1), 169-176.

Zoh, K.-D., and Stenstrom, M. K. (2002). "Fenton oxidation of hexahydro-1,3,5-trinitro-1,3,5-triazine (RDX) and octahydro-1,3,5,7-tetranitro-1,3,5,7-tetrazocine (HMX)." *Water Research*, 36(5), 1331-1341.

CHAPTER 13

BIOREMEDIATION OF CONTAMINATED GROUNDWATER SYSTEMS

T. Prabhakar Clement
Department of Civil Engineering
Auburn University
Auburn, AL 36849-5337

13.1 Introduction

Bioremediation systems employ microbial processes to degrade environmental waste products. Technologies that utilize biological processes for treating waste material have been used by the mankind since the beginning of the recorded human history. In fact, well before the human evolution biological processes have been actively used by our mother nature to manage natural wastes in swamp and forest ecosystems. Perhaps, the very first environmental problem that our planet faced was the management of solid wastes generated from dead plants and trees. Interestingly, bioremediation, in the form of natural composting, was our nature's solution to this problem. In the distant past, it is quite possible that one of our ancestors observed this fascinating remediation process and learned from the nature a technology that can break down complex organic waste materials and convert them into a useful end product, later known as compost. This ancient knowledge would have been passed down to succeeding generations and over the years our ancestors must have developed methods to perfect the art of composting. There are Roman and biblical references to composting as well as numerous accounts of composting methods being practiced by people in different millennia (Rynk 1992). In the U.S., George Washington was perhaps the first "bioremediation engineer;" Washington owned a dung repository to make compost from animal manure (Arner 1995). Another well-known scientist who was interested in composing was Charles Darwin. At the age of 28, Charles Darwin presented a key lecture titled "On the formation of mould" to the Geological Society of London, which was perhaps the first key note address on bioremediation! In this lecture, Darwin illustrated the amazing churning effects of earthworms and their influence on soil. Darwin later documented his ideas in his book titled "The formation of vegetable mould through the action of worms with observations on their habits" (Darwin 1881); this could be considered as the first bioremediation manual!

The more "modern" use of the bioremediation technology began over 100 years ago when biological sewage treatment plants were used to treat wastewaters. Despite their widespread use, in the early 20th century, the design of wastewater treatment processes was more art than science. Although the development of microbiology as a scientific discipline has led to the discovery of organisms that are responsible for degrading waste products, little progress was made in the earlier part of the 20th

century towards the development of rational procedures for designing waste treatment systems. In the preface of a classic wastewater treatment textbook, the authors Benefield and Randall state that "(we) have observed during (our) years of experience that the majority of the biological processes actually put into operation are not designed using fundamental biological treatment principles, even though such principles have been available for many years (Benefield and Randall 1980)." It appears that the scientific understanding of sewage treatment process, one of the oldest bioremediation technologies, was at a development phase even in the early 80s, and perhaps could still be considered as an evolving technology.

Contamination of surface water systems by wastewater was one of the major environmental problems faced by 20th century environmental engineers. In the late 80s and 90s, contamination of groundwater systems by hazardous waste has emerged as one of the major environmental management issues. The term hazardous waste is used for indentifying wastes or a combination of wastes that have solid, liquid or semisolid forms, and because of its quantity, concentration, or due to its physical, chemical or infectious characteristics may: (1) cause or significantly contribute to an increase in mortality or an increase in serious irreversible or incapacitating reversible illness, or (2) pose a substantial present or potential hazard to human health or the environment when improperly treated, stored, transported or disposed of (Blackman 1993). While there are numerous hazardous wastes that are potentially harmful for human health and environment, fuel hydrocarbon compounds and chlorinated solvent wastes are the two major classes of hazardous chemicals that have contaminated thousands of groundwater aquifers throughout the U.S. These chemicals were inadvertently discharged to groundwater systems primarily from leaking underground storage tanks or from poorly designed hazardous waste disposal sites. The U.S. Environmental Protection Agency (USEPA) estimates that over one million underground storage tanks have been in service in the U.S. that are principally used for fuel storage at gasoline service stations. Of these, more than 439,000 leaking underground storage tank (LUST) sites have been identified as of September 2003 (USEPA 2004). So far, cleanups have been initiated at more than 403,000 of these sites. The cost of remediating these groundwater contamination problems at a LUST site can range from \$100,000 to over \$1 million depending on the extent of contamination (USEPA 2004). Chlorinated solvents are the second most frequently occurring type of groundwater contaminants. Based on a survey completed in 1997, USEPA estimated that cleanup of chlorinated solvent contaminated sites alone will cost over \$45 billion (USEPA 2000). Both of these contaminants can be treated using various types of bioremediations processes. To remediate petroleum and chlorinated solvent sites at a reasonable cost, the USEPA is currently promoting the use of multiple bioremediation alternatives including natural attenuation. Therefore, there is large body of published literature available that focus on bioremediation problems related to petroleum hydrocarbons and chlorinated solvents. In the past fifteen years, researchers have also identified various bioremediation processes that have the potential to remediate groundwater systems contaminated with metals and radionuclides (Palmisano and Hazen 2003; Wall and Krumholz 2006). However, in this chapter, we will only review bioremediation methods that are currently being used for restoring sites contaminated with petroleum products and/or chlorinated solvents.

13.2 Classification of Bioremediation Methods

Bioremediation processes used for remediating groundwater and the associated soils can be broadly classified into in-situ bioremediation methods and ex-situ bioremediation methods. The important difference between in-situ and ex-situ methods is whether the contaminated water and the associated soil are treated an in-place (in-situ) or transferred above ground and treated in an external (ex-situ) treatment system.

Facilities such as the traditional biological wastewater treatment plants, which can employ either aerobic or anaerobic processes to treat the contaminated groundwater, are ex-situ treatment systems. These systems are commonly run in conjunction with a pump-and-treat system, where groundwater is pumped from a set of wells and the extracted water is treated ex-situ in the wastewater treatment facility. Similarly, a variation of the traditional composting technology, known as biopiles, is commonly used to treat excavated contaminated soils ex-situ. Biopiles are also known as biocells, bioheaps or biomounds; in most site remediation applications, biopiles are used to reduce the concentrations of hydrocarbon contaminants in soils excavated from a petroleum-product contaminated site. Biopile treatment methods involve piling of contaminated soils into cells and stimulating aerobic microbial activity by supplying oxygen through aeration; in addition, required nutrients and moisture are also supplied to promote the microbial activity. Another variation of biopile technology is known as the landfarming approach. Landfarming methods are engineered systems that utilize oxygen from air to support the growth of aerobic microbial population, which can help degrade the petroleum constituents adsorbed to soils. The major difference between landform and biopile technologies is that the landfarms are aerated by tilling or plowing of soils, whereas biopiles are aerated by injecting or extracting air through slotted or perforated piping placed in soil piles. USEPA (2004) provides an excellent review of various ex-situ technologies, including biopile and landfarming methods used for treating petroleum contaminated soils.

Under the general category of in-situ bioremediation methods, there are numerous process variations available for treating contaminated groundwater aquifers. Broadly, in-situ bioremediation methods can be classified as active or passive bioremediation methods. In both methods, the focus is to employ microbial organisms to transform hazardous contaminants into less toxic or non-toxic forms. Active bioremediation methods require some type of engineered systems to actively control the progression of the biodegradation process, whereas the passive method (which is now identified as the monitored natural attenuation method) does not require active controls.

Active bioremediation methods can be further classified into in-situ biostimulation and in-situ bioaugmentation methods. In-situ biostimulation methods attempt to simulate favorable conditions in the subsurface to enhance the growth of native microbes that have the potential to degrade the contaminant of interest. The use of biostimulation methods assume: (1) microbes required for degrading various types of natural or anthropogenic contaminants are ubiquitously present in the subsurface, (2) many different types of microbes (or a consortia of microbes) can degrade a given contaminant, and (3) by supplying a growth-limiting nutrient or a substrate one can biostimulate a desired native microbe (or a consortia) to actively degrade the contaminant. The in-situ bioaugmentation methods, on the other hand, assume that a desired microbial population for degrading the contaminant of interest may not be present in the natural system. Therefore, bioaugmentation methods first augment the subsurface with a non-native microbial species and then provide favorable growth conditions to sustain the degradation activity of the non-native species. Presently, the scientific community appears to be still debating whether to use bioaugmentation of biostimulation for groundwater remediation. In a non-peer reviewed treatment technology opinion article, Nyer et al. (2003) suggested that the bioremediation researchers can be divided into two groups. The first group, the bioaugmentation group, "believes that they can find and name the bacteria responsible for degradation" and potentially grow this culture externally and use them for augmentation at other field sites. The second group, the biostimulation group, "believes that the key to (promoting) biodegradation in the field is to understand and create the correct environment." The authors appear to support the biostimulation idea and implied that subsurface environments naturally support a diverse group of microbes that can be stimulated to degrade any contaminant. This article sparked an active debate within the bioremediation community. It received highly critical comments from a team of twelve well-known academics and bioremediation practitioners (Major et al. 2003). In their note, Major et al. argued that while some simple petroleum hydrocarbons can be degraded by a variety of microorganism that are ubiquitously present, this may not be true for complex chemicals such as chlorinated solvents. Even though Major et al. (2003) rejected the supposition that that bioaugmentation is absolutely required to achieve complete degradation of chlorinated solvents, they did support the idea that a specific bacterial population known as "dehalococcoides" must be present at the site for complete degradation of chlorinated solvents through anaerobic pathways. Based on this conclusion they proposed an idea that chlorinated-solvent contaminated sites should be screened for the presence of a specific type of bacteria, known as dehalococcoides, and if they are absent the site should be augmented with this type of bacteria. This is a highly controversial conclusion since natural soils are known to support exceedingly high levels of bacterial diversity. For example, Torsvik et al. (1990) reported that there are likely more than 10,000 species of bacteria per gram of natural soil. Also, microbial ecologists tend to support the hypothesis that the native organisms preemptively colonize the niche and are also more fit for the niche (Tiedje 1993). Thus, externally introduced microbes, whether natural or genetically engineered, may stand little chance against these diverse indigenous populations. Also, designing engineered systems to uniformly deliver bacterial cells into deep heterogeneous aquifers without clogging pore spaces, and transporting them to mix with contaminants is a difficult task (Harvey et al. 1989; Taylor and Jaffe 1991; Clement et al. 1996a; 1996b; Harvey et al. 1997). Therefore, the larger question of whether bioaugmentation methods can be effectively used to treat large field sites still appears to be an unanswered question.

Passive bioremediation methods neither use the augmentation step nor the stimulation step. Instead they simply rely on the natural supply of substrates and nutrient as well as the natural degradation potential of the ambient microbial population to attenuate the contaminant plume. Passive biodegradation along with other attenuation mechanisms such as dilution and dispersion are collectively identified as natural attenuation processes. In recent years, researchers have recognized that managing natural attenuation processes require active monitoring to document the progression of various remediation mechanisms. Hence, the methodology is now indentified as monitored natural attenuation (MNA) system.

It should be noted that since all active in-situ bioremediation methods involve the use of some type of "engineering" measures to either add the nutrients, substrates and/or microbes they are often categorized as "engineered" in-situ bioremediation (Rittmann and McCarty 2001). This terminology often implies that the passive remediation systems are "non-engineered" systems. This could be a confusing terminology because even the passive remediation methods now requires "engineered" monitoring systems to track the effectiveness of remediation and, in some cases, active pumping to control the rate of plume migration. Therefore, the use of terms such as "engineered" or "non-engineered" to categorize bioremediation systems may not be a good idea.

13.3 Biochemical Principles of Bioremediation

The major component of any bioremediation method is bacteria, which are the living microscopic organisms that are found ubiquitously in subsurface soils. Natural microorganisms have the capability to break down human, animal, and plant wastes and recycle the nutrients stored in these wastes. Microorganisms can act as enzymes to catalyze complex contaminant transformation mechanisms. In several cases, microbes transform contaminants because they can utilize them as a material source for generating the required energy for supporting their reproduction and maintenance activities. The flow of material and energy in a typical biological system is depicted in Figure 13.1. As shown in the figure, all living organisms including microbes require an electron donor and an electron acceptor to generate sufficient energy to sustain life (known as the maintenance energy). In addition, they also need a carbon source, appropriate nutrients, and some energy to support the growth of new cellular material.

During degradation of an organic contaminant via the aerobic pathway, the contaminant will serve as the electron donor and the available dissolved oxygen will serve as the electron acceptor. In this reaction, energy is generated from a microbially-mediated respiration step that is commonly known as the oxidation-reduction (or redox) reaction. The energy generated from this step is stored in high energy compounds such as adenosine triphosphate (ATP), which are then transported and utilized for generating the energy required for cell growth and maintenance.

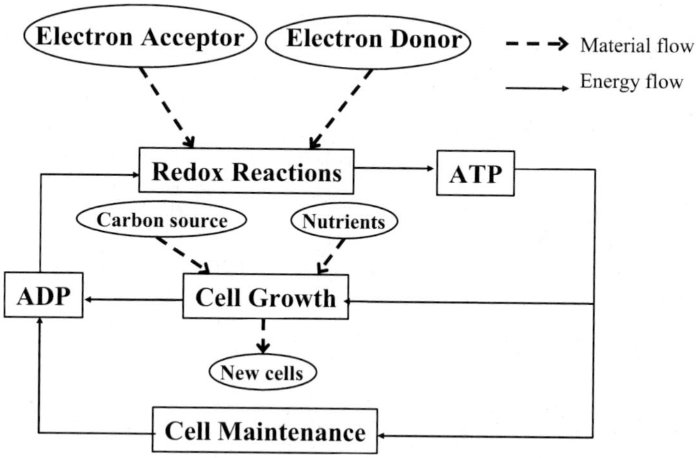

Figure 13.1: Conceptual model for material and energy flow processes during bioremediation.

Every redox reaction involves a series of steps where electrons are transferred from one atom to another using various coenzymes such as nicotinamide adenine dinucleotide (NAD) and NAD-phosphates (Rittmann and McCarty 2001). In the chemistry literature, loss of electrons from an atom is referred to as oxidation reaction since the atom that lost electrons is oxidized. Correspondingly, gain of electrons is referred as reduction and the atom that gained electrons is reduced. Further, an element that is capable of releasing electrons is often referred as the reducing agent or reductant, and an element that is capable of accepting electrons is referred as the oxidizing agent or oxidant. In the environmental microbiology literature, reductants are commonly referred as electron donors and oxidants are commonly known as electron acceptors.

The charge on an atom that participates in a redox reaction is known as the oxidation number or oxidation state. Some important rules and conventions used for assigning the oxidation state of various atoms are summarized in Table 13.1. The table also provides typical oxidization states of atoms involved in redox reactions of environmental significance. Using these data, let us analyze the electron exchange processes that occur during benzene (a common groundwater contaminant) degradation via various degradation pathways.

The stoichiometry for the benzene biodegradation reaction via aerobic pathway can be written as:

$C_6H_6 + 7.5\ O_2 \rightarrow 6\ CO_2 + 3H_20$

Since the oxidation state of H is +1, to maintain charge balance on the benzene molecule, the C atom should have a charge of -1 (which signifies that the oxidation state of carbon in benzene is -1). In the reaction product, since the oxidation state of O should be -2, to maintain charge balance on the CO_2 molecule, the C atom in CO_2 should have an oxidation state of +4. This implies that during benzene biodegradation, each carbon atom associated with benzene (which serves as the electron donor) goes from the oxidation state of -1 to +4 by donating five electrons (6 carbons in the benzene molecule donate a total of 30 electrons). On the other hand, each oxygen atom (which serves as the electron acceptor) goes from the oxidation state of zero to -2 by accepting 2 electrons (15 oxygen atoms in the reaction products would accept a total of 30 electrons).

Since electrons cannot exist alone in solution, during biochemical reactions the oxidation-reduction reactions involve transferring of hydrogen ions. When an electron is removed from a hydrogen atom it becomes a proton (H^+) and we need to account for the charge contribution from this proton. Associating the transferable electrons as a part of water molecules will help resolve this charge balance problem. Using water molecules as part of the electron transfer process, the balanced redox half reactions for benzene biodegradation can be written as:

$C_6H_6 + 12\ H_20 \rightarrow 6\ CO_2 + 30\ H^+ + 30\ e^-$
$7.5\ O_2 + 30\ H^+ + 30\ e^- \rightarrow 15\ H_20$

$C_6H_6 + 7.5\ O_2 \rightarrow 6\ CO_2 + 3H_20$

Unfortunately, the electron donating and electron accepting processes are not clear when the redox equations are written in the above form. However, the reactions can be further divided into electron transfer steps and product formation steps as shown below:

Electron transfer steps
$C_6H_6 \rightarrow 6\ C^{4+} + 6\ H^+ + 30\ e-$ (Electron donating half reaction)

$7.5\ O_2 + 30\ e^- \rightarrow 15\ O^{2-}$ (Electron accepting half reaction)

Product formation steps
$6\ H^+ + 3\ O^{2-} \rightarrow 3\ H_20$ (Formation of water)
$6\ C^{4+} + 12\ O^{2-} \rightarrow 6\ CO_2$ (Formation of carbon dioxide)

$C_6H_6 + 7.5\ O_2 \rightarrow 6\ CO_2 + 3\ H_2O$ (Net reaction)

The above reactions illustrate how electrons can be transferred from a donor to an acceptor in charge-consistent manner. In summary, the key to understanding biologically mediated oxidation-reduction processes is to keep track of the half reactions—there must always be one reaction involving electron donor and another reaction involving electron acceptor (Madigan et al. 2000).

Table 13.1: Conventions for assigning oxidation states of various atoms of environmental significance.

i. Oxidation state of an atom in the pure (uncombined) element is zero (for example, oxidation state of oxygen in O_2 gaseous form is zero; oxidation state of Na in a solid sodium sample is zero).
ii. In compounds, the oxidation state of H is +1 and oxygen is -2.
iii. The total of oxidation states of all the atoms in a molecule is zero; i.e., charge on a molecule must be equal to the sum of the charges on the constituent atoms. For example, the net charge of a H_2O molecule is zero.
iv. Mono-atomic ions have an oxidation state equal to its charge or valence.
v. Oxidation state of poly-atomic ions is determined by adding the oxidation states of atoms that make up the ion: e.g. in SO_4 , oxidation state of S = +6; O = -2, therefore, the oxidation state of sulfate ion is 2^- (hence, SO_4^{2-}).
vi. In biochemical reactions, C can have the oxidation state of -4 (CH_4), -1 (C_6H_6), 0 (natural organic matter) or +4 (CO_2); N can have -3 (NH_3 or NH_4^+), 0 (N_2), +3 (NO_2^-), or +5 (NO_3^-); Fe can have 0 (zero valent iron or pyrite), +2 (Fe^{2+}) or +3 (Fe^{3+}); and S can have -2 (H_2S), +4 (SO_2) or +6 (SO_4^{2-}).

13.4 Fundamentals of Petroleum and Chlorinated Solvent Biodegradation Processes

Natural microorganisms can degrade environmental contaminants using three distinct processes; aerobic respiration, anaerobic respiration, and fermentation. Aerobic respiration is a process whereby bacteria will extract energy from the contaminant by transferring electrons from the contaminant to oxygen. Anaerobic respiration is a process where anaerobic bacteria, such as denitrifiers, iron-reducers or sulfate reducers can extract energy from the contaminant by transferring electrons from the contaminant to other chemicals (other than oxygen) that act as external electron acceptors. Common external electron acceptors during anaerobic degradation include nitrate, sulfate and iron oxides [e.g., $Fe(OH)_3$]. Fermentation is a process where bacteria, such as methanogens, degrade contaminants without the use of an external electron acceptor. During fermentation, certain parts of the organic chemical (or its fermentation products such acids, alcohol, hydrogen or carbon-dioxide) essentially act as both electron donor and electron acceptor.

From a biochemical point of view, groundwater remediation problems can be classified into electron acceptor-limited systems and electron donor-limited systems. Sites contaminated with petroleum hydrocarbons are good examples of electron-acceptor limited systems, whereas sites contaminated with highly chlorinated solvents are good examples of electron-donor limited systems.

13.4.1 Petroleum Products Biodegradation Mechanisms

Several millions of underground storage tanks (UST) are used throughout the U.S. to store petroleum products. The USEPA estimates that a vast majority of these USTs have discharged petroleum products in the form of light non-aqueous phase liquid (LNAPL). When groundwater comes in contact with these NAPL products the water will be contaminated with the most soluble petroleum chemical species such as benzene, toluene, ethylbenzene and xylene, which are collectively identified as BTEX. Dissolved BTEX compounds are the most common groundwater contaminants associated with spilled or leaked fuels such as gasoline, diesel, jet fuel and heating oil.

The need to remediate BTEX contaminated sites played a major role in the development of bioremediation. In the early 1980s, researchers primarily focused on using aerobes for biodegrading petroleum hydrocarbons. It is now well understood that several other types of microorganisms can utilize BTEX compounds as electron donors. Also, it is well established that petroleum degraders are ubiquitously present in all natural systems and therefore if sufficient amount of appropriate electron acceptors are present, petroleum contaminants can be biodegraded by natural organisms. Oxygen, nitrate, iron, manganese, sulfate and carbon dioxide are the common electron acceptors widely available in groundwater systems. Among these electron acceptors, oxygen in the most favored electron acceptor since microorganisms gain more energy from aerobic respiration. A simple method of comparing biochemical reaction energetics of various electron accepting (or redox) reactions is to imagine the corresponding reaction as being a part of a vertical energy tower (Madigan et al. 2000). Table 13.2 summarizes some key redox couples and their corresponding energy. Note the redox pair with the most positive reduction potential (most likely acceptor) is placed at the bottom. These data clearly indicate that oxygen will be the most favored reaction process, followed by iron and nitrate reduction reactions. Sulfate reduction requires highly reduced conditions and methanogenic processes that employ the CO_2/CH_4 redox couple is the least favored reaction.

Table 13.2: The electron tower for selected redox couples at 25°C at pH 7 (Madigan et al. 2000).

Half Reaction	E_o (volts)
CO_2/CH_4	-0.25
SO_4^{2-}/H_2S	-0.10
NO_3^-/N_2	+0.74
Fe^{3+}/Fe^{2+}	+0.76
$\frac{1}{2}O_2/H_20$	+0.82

Almost all petroleum products can be biodegraded by aerobic bacteria. Alvarez and Vogel (1991) observed complete removal of mixtures of benzene, toluene, and p-xylene using a pure culture of bacteria. Chiang et al. (1989) observed rapid degradation of benzene, toluene, and xylene in soil microcosms amended with sufficient amount of oxygen. In natural groundwater aquifers the aerobic biodegradation process will be primarily limited by the solubility of oxygen in water. Consider the following theoretical stoichiometry for benzene degradation through aerobic pathway:

$$C_6H_6 + 7.5O_2 \rightarrow 6CO_2 + 3H_2O$$

During aerobic degradation of a benzene molecule, the carbon atoms in benzene will donate a total of 30 electrons which will be accepted by oxygen molecules. Based on this stoichiometry, for complete mineralization of 1 mg of benzene about 3 mg of oxygen is required. Under field conditions, groundwater can have dissolved oxygen concentration values ranging from 5 to 10 mg/l. Hence, aquifers supporting aerobic decay will become rapidly anaerobic when they are contaminated with high levels of petroleum hydrocarbon compounds.

Once all available dissolved oxygen is utilized the conditions will become anoxic; this environment will promote the growth of denitrifying bacteria that can degrade petroleum compounds using nitrate (NO_3^-) as the external terminal electron acceptor (note both nitrate and iron reactions have similar E_o values and hence these two processes may occur simultaneously). Several investigators have observed degradation of hydrocarbons under denitrifying conditions (Kuhn et al. 1988; Major et al. 1988). For benzene degradation, stoichiometry for the decay process can be written as:

$$C_6H_6 + 6NO_3^- \rightarrow 6CO_2 + 3\,N_2 + 6\,OH^-$$

The denitrification reaction uses nitrate ions as the external electron acceptor. In this reaction, the six carbon atoms in benzene donate a total of 30 electrons that are transferred to nitrogen atoms that change its oxidation state from +5 (when in the form of nitrate ion) to zero (when in the form of nitrogen gas). Based on the above stoichiometry, complete mineralization of 1 mg of benzene to carbon dioxide requires approximately 4.77 mg of nitrate. Since natural background levels of nitrate in groundwater aquifers are typically low, dissolved hydrocarbon plumes can quickly utilize and deplete nitrate.

Anoxic subsurface conditions can also support the growth of iron reducers that can utilize certain iron species as the external terminal electron acceptor. In groundwater systems, iron can exist in one of three forms: zero valent iron (e.g., pyrite minerals FeS_2), divalent iron (Fe^{2+}) in the form of soluble ferrous ions, or trivalent ferric iron (Fe^{3+}) in the form of iron hydroxides such as $Fe(OH)_3$. Ferrous iron is soluble under a wide range of normal pH conditions. On the other hand, if the pH is higher than 3.5 the ferric iron becomes insoluble and will precipitate ferric hydroxide as an orange/yellow precipitate. Note in the presence of oxygen, ferrous iron can be oxidized to form ferric iron. At petroleum-contaminated sites, the forms of natural iron material that can be easily utilized by microbes are poorly crystalline Fe(III) hydroxides, Fe(III) oxyhydroxides, and Fe(III) oxides (Lovley 1991). Natural systems may have large quantities of iron bound to the sediment phase which can potentially serve as the electron acceptor. Although the exact mechanism for microbially-mediated iron reduction is not fully understood, there is considerable field evidence to demonstrate the importance of iron reduction in mediating the biodegradation patterns of petroleum hydrocarbon plumes. Reduction of Fe(III), for example, would result in high concentrations of dissolved Fe(II), and several field studies have documented this increase (Wiedemeier et al. 1999). For benzene degradation, the overall stoichiometry for iron reduction can be written as:

$$C_6H_6 + 30Fe(OH)_3 \rightarrow 6CO_2 + 30Fe^{2+} + 60OH^- + 18H_2O$$

The above iron-reduction reaction uses Fe(III) ions as the external electron acceptor. Based on this reaction, the carbon atoms in benzene molecule will donate a total of 30 electrons which will be transferred to Fe(III) atoms bound in the solid hydroxide minerals and yield Fe(II) ions in dissolved form. From this stoichiometry, complete mineralization of 1 mg of benzene to carbon dioxide requires approximately 41.1 mg of $Fe(OH)_3$, and the reaction would also yield 21.5 mg of Fe^{2+}.

Researchers have shown that a variety of petroleum products can be degraded by sulfate reducers under strongly reducing conditions (Wiedemeier et al. 1999). Both laboratory and field studies have demonstrated hydrocarbon degradation under sulfate-reducing conditions (Thierrin et al. 1993; Beller et al. 1996). As illustrated in Table 13.2, from the thermodynamic point of view, sulfate reduction is less favorable than aerobic respiration, denitrification and iron reduction reactions. Therefore, to support sulfate reduction one would require highly reduced conditions and depletion of other electron acceptors such as dissolved oxygen, nitrate and iron. For benzene degradation, the overall reaction stoichiometry under sulfate reduction conditions can be written as:

$$C_6H_6 + 3.75\ SO_4^{2-} + \rightarrow 6CO_2 + 3.75H_2S + 7.5OH^- + 3H_2O$$

The sulfate reaction uses sulfate ions as the external electron acceptor; during the reaction, the six carbon atoms in benzene will donate a total of 30 electrons to sulfur atoms that change from oxidation state of +6 (when in the form of sulfate ion) to -2 (when in the form of hydrogen sulfide). Based on this stoichiometry, complete mineralization of 1 mg of benzene to carbon dioxide requires approximately 4.6 mg of sulfate.

Under methanogenic conditions the carbon present either in the substrate or in the degradation by-product, such as CO_2, would serve as the terminal electron acceptor. This is an important degradation pathway found in many anaerobic subsurface environments (Chapelle 1993). Methanogenesis is a fermentation process and it does not use any "external" electron acceptor. Instead the substrate (the contaminant source) or other simpler forms of carbon generated from the degradation reaction are used as the "internal" terminal electron acceptor. Fermentation is usually a multi-step process with each step mediated by different types of microbes or microbial population. For benzene degradation, the overall stoichiometry can be conceptually represented by the following two step process (Wiedemeier et al. 1999):

Step 1

$$C_6H_6 + 6H_2O \rightarrow 3CH_3COOH + 3H_2$$

This first step is known as the acid fermentation step and is mediated by acid forming bacteria. The fermentation products acetate and hydrogen are then utilized by different types of organisms to produce carbon dioxide and methane in the following step:

Step 2

$$3CH_3COOH \rightarrow 3CH_4 + 3CO_2$$
$$3H_2 + 0.75CO_2 \rightarrow 0.75CH_4 + 1.5H_2O$$

Note, the overall stoichiometry for this reaction can be written as:

$$C_6H_6 + 6H_2O \rightarrow 3.75CH_4 + 2.25CO_2 + 1.5H_2O$$

In this form, it is easy to see why the methanogenic reaction does not require an "external" electron acceptor. The electron transfer would occur internally exchanging electrons between various carbon atoms in the benzene molecule. Note when in the form of benzene, carbon has an oxidation state of -1; during the fermentation reaction, 2.25 carbon atoms are oxidized to form carbon dioxide and 3.75 carbon atoms are reduced to form methane. Since the oxidation state of carbon in CO_2 and CH_4 are +4 and -4, respectively, the carbon atoms associated with carbon dioxide loses a total of 11.25 electrons which are then accepted by the carbon atoms associated with methane. The terminal electron acceptors supporting this reaction come from an "internal" source. A total of 3.75 carbon atoms generated from the substrate (some of the carbon in the benzene molecule) accept all the electrons to produce the final degradation product methane.

13.4.2 Chlorinated Solvent Biodegradation Mechanisms

Chlorinated solvents are the second most commonly observed groundwater contaminants (Wiedemeier et al. 1999). These contaminants are normally disposed in the environment as dense non-aqueous phase liquids (DNAPLs). If sufficient amount of DNAPL is discharged at a hazardous waste site, it can rapidly migrate into deeper regions of the saturated groundwater zone contaminating large volumes of aquifer matrix. Within this contaminated region, DNAPL can be present at a residual saturation level or as isolated trapped blobs or as a concentrated pool trapped over a low permeable zone. Environmental transport processes such as advection, dispersion, and diffusion will cause the DNAPL contaminants to dissolve into water phase, which can then transport the dissolved contaminants over long distances (Clement et al. 2004a). This results in dissolved-phase groundwater plumes that can persist in the environment for a long period. Table 13.3 lists several widely used chlorinated solvents, which are also common environmental contaminants. The chemical properties of these solvents are also given in the table.

Table 13.3: Common chlorinated solvents [data from McCarty and Semprini (1994)].

Compound	Formula	Relative Density	Solubility (mg/L)	MCL (ug/L)
Carbon tetrachloride (CT)	CCl_4	1.595	800	5
Chloroform	$CHCl_3$	1.485	8,200	100
Trichloroethane (TCA)	CH_3CCl_3	1.325	950	200
1,2 Dichloroethane (DCA)	CH_2ClCH_2Cl	1.175	8,700	5
Tetrachloroethene (PCE)	CCl_2=CCl_2	1.625	150	5
Trichloroethene (TCE)	CHCl=CCl_2	1.462	1,000	5
Dichloroethene (cis-1,2 DCE)	CHCl=CHCl	1.214	400	70
Vinyl chloride (VC)	CH_2=CHCl	-	-	2

Biodegradation by bacteria is the most important removal process that occurs at many chlorinated solvent-contaminated sites (Wiedemeier et al. 1999; Clement et al. 2002). Bradley (2003) provides an excellent review of various chloroethene biodegradation processes. Chlorinated solvent biodegradation processes can be broadly classified into two types of mechanisms that include direct biodegradation processes and cometabolic processes. During direct biodegradation, the organic contaminant is directly metabolized by the microorganism as a carbon source and/or electron donor or as an electron acceptor. This degradation reaction would yield energy and/or growth benefit to the microorganism. On the other hand, cometabolic biodegradation processes involve fortuitous degradation of the contaminant of interest by an enzyme (or a co-factor) produced by the microbes while metabolizing another compound. Because microorganisms do not gain any direct benefit from utilizing the contaminant, the overall efficiency of cometabolic processes is relatively low. Rittmann and McCarty (2001) and Bradley (2003) provide a detailed review of various types of cometabolic processes that have been considered for use in bioremediation applications. Although several laboratory experiments have demonstrated the use of various cometabolic processes, it is difficult to distinguish and quantify the effectiveness of a cometabolic process under field conditions. In this section, we will only review the details of bioremediation mechanisms that can degrade chlorinated solvents using a direct process.

Chlorinated solvents can be directly biodegraded through oxidation or through anaerobic reduction (also known as reductive dechlorination). Highly chlorinated compounds tend to degrade through reductive chlorination and less chlorinated compounds degrade through oxidation. During oxidation, the chlorinated solvent compound is the electron donor and either oxygen (under aerobic condition) or another redox species such as iron (under anoxic conditions) is used as the terminal electron acceptor. Common chlorinated solvent compounds that can be directly oxidized include DCE, DCA, and VC (Bradley and Chapelle 1998, Bradley 2000; 2003). Bardley and Chappelle (1996) present data to show that VC can also be oxidized to carbon dioxide via Fe(III) reduction.

Anaerobic reductive dechlorination is by far the most efficient method for degrading highly chlorinated solvents such as PCE and TCE. During this process, anaerobic bacterial cells use hydrogen, a fermentation product of a complex carbon source, as the electron donor and chlorinated compounds as the electron acceptor (Maymo-Gatell et al. 1999; Bradley 2003). This reaction is known as halorespiration or dehalorespiration. This process will sequentially dechlorinate PCE in the following pattern: PCE $\rightarrow$ TCE $\rightarrow$ DCE $\rightarrow$ VC $\rightarrow$ ethene. Researchers have identified a specific species of bacteria known as *Dehalococcoides* that can use various solvent compounds as electron acceptors. For example, *Dehalococcoides ethogenes* strain 195 can obtain energy from all dechlorination steps except the final step (Maymo-Gatell et al. 1997; Maymo-Gatell et al. 2001). However, strain 195 can only degrade VC via cometabolic process and hence this strain would accumulate VC. Some recent studies have, however, shown that other strains of *Dehalococcoides* can directly obtain energy from VC dechlorination to ethene (Cupples et al. 2003; He et al. 2003). He et al. (2003) also report that their *Dehalococcoides* population can utilize both DCE and VC as direct electron acceptors. Further, the growth rate for using VC was about 50% faster than DCE, however, this culture failed to use PCE and TCE as a metabolic electron acceptor.

As pointed out in Nyer et al. (2003), it is unclear whether a certain species of bacteria, such as *Dehalococcoides*, need to be present at the site for supporting reductive dechlorination. Several microorganisms identified in the literature are capable of at least partially dechlorinating PCE or TCE to DCE while using acetate as an electron donor (McCarty 1997). However, Major et al. (2003) hypothesizes that a *Dehalococcoides* population must be present at a site to convert DCE to VC and eventually to ethene. Nyer et al. (2003) refutes this hypothesis and pointed out that review of non-*Dehalococcoides* literature indicates that many other microbial species are capable of dechlorinating several chlorinated compounds other than VC. Nyer et al. (2003) also pointed out several published studies that have used multi-species microbial consortium to rapidly dechlorinate PCE all the way to ethene (DeBruin et al. 1992; Tondai et al. 1994; Carbirol et al. 1998). More recently, Aulenta et al. (2003) report complete dechlorination of PCE to ethene by an anaerobic consortium isolated from brackish sediments collected at an industrial site.

Despite uncertainties, both researchers and practitioners do agree that adequate supply of electron donor is certainly the key for sustaining any type of reductive dechlorination process. McCarty (1997) states "frequently, PCE and TCE persist for years without change because electron donors required for halorespiration are absent." Therefore, it clear that degradation processes at chlorinated solvent sites will be primarily limited by the availability of electron donor. However, it is unclear what type of electron donor should be used to sustain the reductive dechlorination process. Since hydrogen is the ultimate electron donor that completes the dechlorination step, several practitioners focus on using "novel" hydrogen-release compounds that can slowly release hydrogen directly to the subsurface. Nyer et al. (2003) warns that "the use of slow release or limited-dose donors leads to incomplete dechlorination." Instead they support the use of complex organic carbon sources such as molasses, which will allow natural fermentation processes to supply hydrogen and other simple fermentation by-products such as acetate that can be used as electron donors. Studies with actual field sediments have shown that, in addition to molasses, other carbon sources such as methanol and lactate can also be used to promote reductive dechlorination in field soils (Gao et al. 1997; USEPA 2000). In summary, it appears that the key factor for promoting and sustaining the reductive dechlorination activity is to stimulate a carbon-rich anaerobic condition by delivering a readily biodegradable organic source which can serve as the electron donor.

13.5 Design of Bioremediation Systems

The objective of any active bioremediation design strategy is to engineer methods that can efficiently deliver and distribute the stimulatory material (either an electron donor or an electron acceptor) to the subsurface. All petroleum sites will be limited by the availability of an electron acceptor; hence, the design objective of most fuel-hydrocarbon bioremediation systems is to effectively deliver a suitable electron acceptor. Most chlorinated solvent sites are limited by the availability of an electron donor; hence, the design objective of chlorinated solvent bioremediation systems is to effectively deliver a suitable electron donor. In some instances, chlorinated solvent degradation might be limited by the availability of the electron acceptor. For example, direct oxidation of DCE or VC via aerobic pathway will be limited by the availability of dissolved oxygen. Hence, a combination of microbial reduction of PCE and TCE, using anaerobic reductive dechlorination processes, followed by microbial oxidation of DCE and VC, using an aerobic or anoxic process, is a possible alternative for complete degradation (Bradley 2000). In order to use this type of combined approach one has to deliver both electron donor and electron acceptor at different flow regimes of the plume. Also, certain cometabolic processes, such as TCE oxidation using methanotropic organisms, would require simultaneous delivery of both electron donor and electron acceptor (USEPA 2000).

Pumping and injection wells are the major engineered infrastructures used for distributing feed solutions containing appropriate stimulatory materials (such as electron donor or electron receptor, and other nutrients such as nitrate and phosphate) into the subsurface. To treat contaminated groundwater and the associated soils, the feed solution is delivered by circulating water through the contaminated saturated formation. In some cases, the electron donor or acceptor may be delivered in a gaseous form using a suitable well configuration that can effectively distribute the gaseous substrate throughout the subsurface. In sections below we provide the details of well designs that can be used for delivering stimulatory substances to the subsurface to treat petroleum or chlorinated solvent plumes.

13.5.1 Water Recirculation Systems

At BTEX-contaminated sites, water with high concentrations of oxygen or nitrate could be delivered to meet the required electron acceptor demand. In the case of oxygenated water, typically peroxide (H_2O_2) is used as a dissolved source of oxygen since it can provide dissolved oxygen levels substantially greater than air-saturated water (Rittmann and McCarty 2001). Figure 13.2 shows the typical configuration of a peroxide delivery system. As shown in the figure, in this system the stimulatory material (a mixture of peroxide and very low concentrations of nutrients) is delivered through an injection gallery located in the unsaturated zone closer to water table. One of the key design objectives of this water circulation system is to provide the necessary hydraulic controls to capture the injected water and allow continuous recycling through the contaminated zone. Therefore, recovery wells are used to provide hydraulic controls in these systems. The water extracted from the recovery wells is usually augmented with stimulatory substances and injected back into the system.

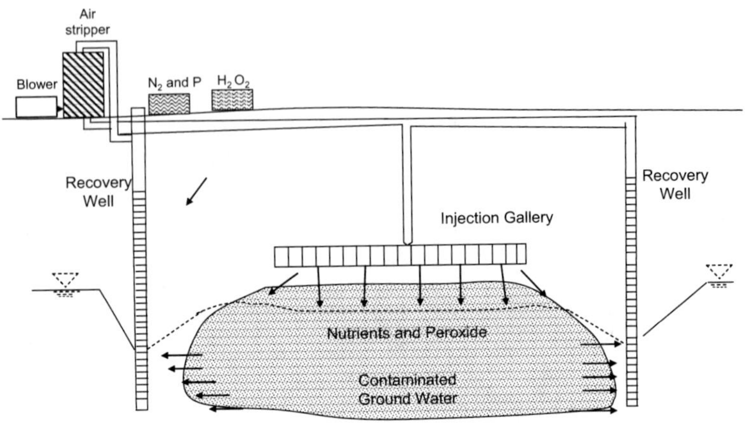

Figure 13.2: Nutrient infiltration and recovery system for bioremediation (modified from NRC (1994)).

Researchers at the Pacific Northwest National Laboratory demonstrated the feasibility of a two-well recirculation system to remediate a carbon tetrachloride plume at field site located in Hanford, Washington (Hooker et al. 1998). Figure 13.3 shows the conceptual cross section of the two-well, injection-extraction system used for treating a carbon tetrachloride plume at the Hanford site. As shown in the figure, the groundwater flow direction at this site is towards the right boundary and an extraction well is used to capture all the injected water. The extracted water was then amended with acetate or nitrate and was recirculated back through the well located in the right hand side. The objective of this bioremediation strategy was to stimulate a denitrifying population that can cometabolically degrade the carbon tetrachloride plume. Both the electron acceptor (nitrate) and the electron donor (acetate) were injected into the aquifer to biostimulate the bacterial activity. A process control system was used to alternatively pulse inhibitory levels of acetate and nitrate to prevent bioclogging of the pore spaces closer to the injection well. The progress of the bioremediation process was monitored using the well located in the center.

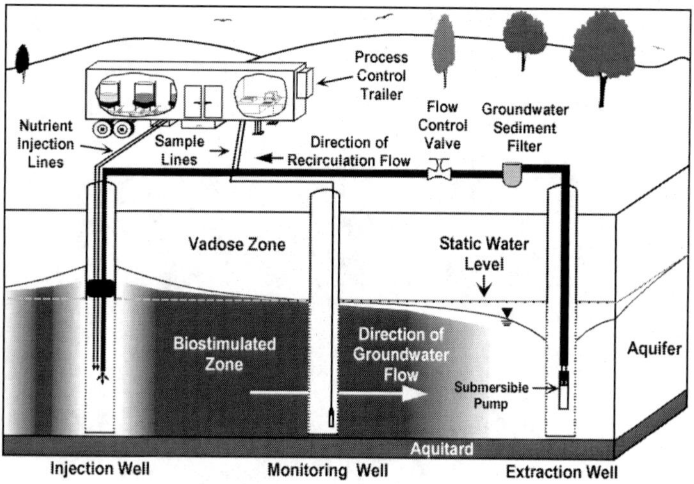

Figure 13.3: Two-well, injection-extraction system for bioremediation.

Stanford University researchers demonstrated the feasibility of using a direct/cometabolic process that utilized toluene degraders to treat a TCE plume at the Edwards Air Force Base (McCarty et al. 1998; USEPA 2000). The treatment system employed a novel recirculation system, similar to the design shown in Figure 13.4, which allowed circulation of groundwater between two screened intervals within a single well. As shown in the figure, a submersible pump placed in the well was used to draw contaminated water into the well through lower screen. Both electron donor (toluene) and the electron acceptor (oxygen) required for stimulating the cometabolic bioremediation process were introduced into the well via feed lines and mixed with the water using static mixers inside the well. The contaminated groundwater amended with stimulatory substances was discharged from the upper screen into the aquifer to promote development of a treatment zone. If required, the water recirculation process can be reversed by drawing water from the upper screen and discharging from the lower screen. MacDonald and Kitanidis (1993) developed a detailed analytical model for analyzing the hydraulic flow patterns of this single well recirculation system.

Stanford University researchers also designed a three-well, injection-extraction water recirculation system to test the feasibility of a direct/cometabolic aerobic oxidation process to treat a TCE plume at the Moffett Naval Air Station. This bioremediation field demonstration study evaluated the feasibility of oxygen and hydrogen peroxide as electron acceptors and methane, toluene and phenol as electron donors to biostimulate an indigenous methanotropic bacterial population (Hopkins et al. 1993; USEPA 2000). A schematic diagram of the recirculation well system used at the Moffett site is shown in Figure 13.5. As shown in the figure, contaminated groundwater was extracted from a well located in the middle of a 12 m long treatment zone. The extracted water was amended with required stimulatory compounds and injected via two wells that were located approximately 6 m away on either side of the extraction well. Several sampling wells were placed in between to monitor the progress of the bioremediation process. An automated data acquisition system was used to sample and analyze daily variations in concentration levels.

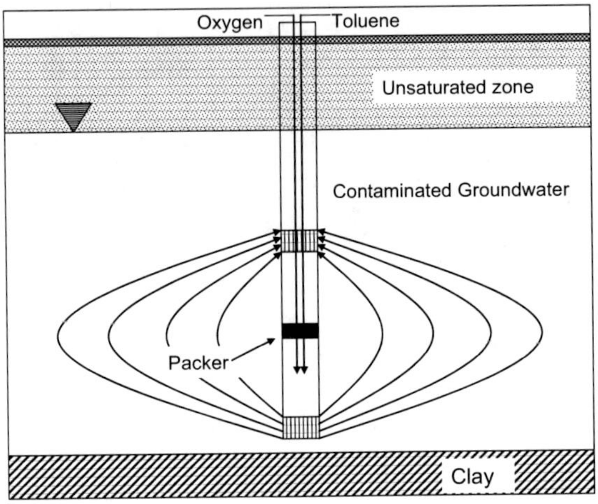

Figure 13.4: Single-well recirculation system for bioremediation.

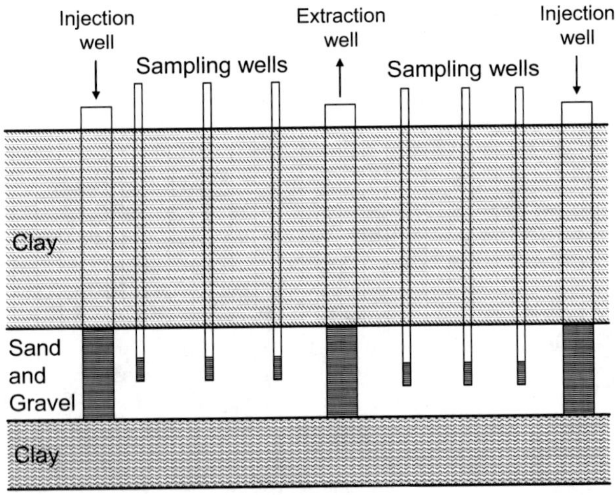

Figure 13.5: Three-well, injection-extraction system for bioremediation.

13.5.2 Gas Injection Systems

Injecting air to supply oxygen (the electron acceptor) to remediate petroleum contaminated groundwater aquifers was first introduced in Germany in early 1980s (Gudemann and Hiller 1988). Researchers quickly recognized that sparging air or oxygen directly into the saturated formation has a potential to deliver significant amounts of oxygen that can be utilized by microbes to support the biodegradation activity (Brown and Jasiulewicz 1992; Bianchi-Mosquera et al. 1994). In addition, air sparging systems also have the potential to strip volatile organic compounds from the contaminated zone (Panikow et al. 1993). Within the unsaturated zone, methods such as bioventing wells can be used to promote biodegradation. Figure 13.6 shows a typical design for implementing air sparging (or biosparging) systems. As shown in the figure, this design uses two wells to inject compressed air directly into the saturated formation. Further, two extraction wells are used to capture volatile contaminants and to provide air flow through the unsaturated zone. The required nutrients are distributed through a drain system located above the saturated zone. The overall success of an air sparging system would depend on its ability to transfer large amounts of oxygen from the air phase to the water phase where it can be utilized for biodegradation. It is important to recognize the practical limitations of air-water mass-transfer process when designing air sparging systems. Early researchers have ignored some of these mass-transfer issues, which has resulted in overly optimistic projection of success at many air sparging sites (Johnson et al. 1993).

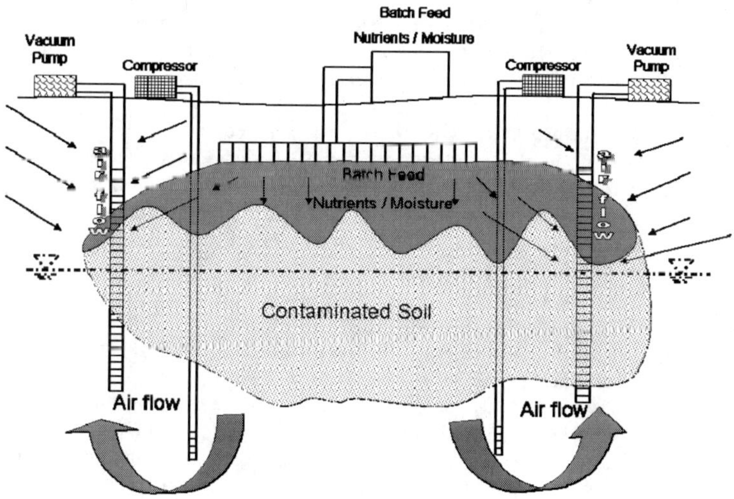

Figure 13.6: Typical air-sparging system design for bioremediation (modified from NRC (1994)).

13.6 Assessment of Bioclogging Effects During Bioremediation

When designing active in-situ bioremediation systems, careful attention must be given to the possibility of soil permeability reduction due to clogging of pore spaces by the biological material. During the operational phase of a bioremediation project, clogging will be a common occurrence in the vicinity of injection wells or near the infiltration points. Biological clogging is caused by the accumulation of cellular material in the pore spaces. In the published literature, several researchers have investigated the reduction in the permeability of soils due to biological growth. The work of Gupta and Swartzendruber (1962) was one of the first studies that documented the link between biological growth and soil permeability reduction. Mitchell and Nevo (1964) observed clogging of sandy soils after prolonged percolation of water containing organic matter. Macleod et al. (1988) stimulated the growth of starved bacterial cells to selectively plug high permeability formations to increase the efficiency of oil recovery wells. Ripley and Saleem (1973) studied clogging of aquifer soils during artificial recharge with treated wastewater. Soares et al. (1988) studied clogging due to accumulation of biogas during denitrification in a sand column. Taylor and Jaffe (1990a) performed laboratory experiments to study the effects of bioclogging in an aerobic soil column. Cuningham et al. (1991) performed bioclogging experiments that allowed optical in-situ measurements of biofilm thicknesses and biofilm-affected permeability. Taylor and Jaffe (1990b) developed a biofilm approach-based numerical model to analyze the laboratory results. Clement et al. (1996a) developed a macroscopic approach-based numerical model for predicting bioclogging effects. Baveye et al. (1998) provide a comprehensive summary of various microbial clogging processes observed in different types of bioclogging experiments. Thullner et al. (2004) and Thullner and Baveye (2008) provide a summary of various approaches available for modeling bioclogging processes.

Injection wells and infiltration galleries are most susceptible to bioclogging when both electron donor and acceptor are introduced simultaneously in a single injection well or infiltration gallery. A number of practical, common-sense operational strategies can be used to mitigate bioclogging effects. For example, when using infiltration galleries, clogging of unsaturated zone or the surface soils can be mitigated by allowing for a drying period.

Under saturated conditions, an approach that is often employed to prevent clogging is to inject either an electron donor or an electron acceptor at levels that may be toxic to the organism. For example, in aerobic treatment systems one could use hydrogen peroxide as the oxygen source and deliver high peroxide at concentrations (above 500 mg/L) to inhibit its growth near the injection zone. However, away from the injection point, concentrations will dilute to non-toxic levels due to hydrodynamic dispersion and reactions. Similar approaches can, in principle, be used in any type of bioremediation systems. For example, Franzen et al. (1997) evaluated the inhibitory effects of different types of concentrated substrate pulses used for bioremediating a carbon tetrachloride plume.

A second approach is to spatially separate the electron acceptor and donor using two separate injection wells. Biofouling will be suppressed since one of the constituents required for growth will not be available for the organisms growing near the injection point. This method attempts to separate the availability of electron donor and acceptor spatially near the injection points. However, due to hydrodynamic transport and mixing the biological growth will eventually occur deep within the formation, away from the injection points.

A third method is based on the sequential injection of an electron donor and acceptors into an aquifer. This method has been tested in field scale experiments (Semprini et al. 1992) and mechanistically modeled (Taylor and Jaffe 1991; Clement et al. 1996b; Franzen et al. 1997). This method attempts to separate the availability of electron donor and acceptor temporally at the injection point. This in turn inhibits growth near the injection point; however, due to hydrodynamic transport and mixing growth will be promoted away from the injection point.

It is important to note that the efficiency of the second and third methods would depend on the ability of the hydrodynamic dispersion processes to promote mixing of the constituents as they transport away from the injection points. However, our understanding of dispersion in heterogeneous porous media systems is limited, and therefore there is considerable uncertainty involved in quantifying this mixing process. During the design phase, if the mixing level was overestimated then the growth-enhancing constituents will not be available throughout the treatment zone of interest, resulting in significant loss of bioremediation efficiency. On the other hand if the mixing level was underestimated then significant growth will occur near the injection point leading to well clogging.

13.7 Monitored Natural Attenuation

In the past ten years, there has been increased interest in understanding various natural processes that control the migration patterns of groundwater plumes. This interest was motivated by numerous field studies indicating that natural processes can play a significant role in attenuating groundwater contaminants (Wiedemeier et al. 1998). Therefore, the USEPA has recognized the use of natural attenuation as one of the possible remedial alternatives (USEPA 1999). Other terminologies used in the literature to refer to natural attenuation include intrinsic remediation, intrinsic bioremediation, passive bioremediation, natural recovery, or natural assimilation. While some of these terms have similar meaning, the USEPA now recommends the use of the term "monitored natural attenuation" (MNA) in the 1999 OSWER directive (USEPA 1999). This directive emphasizes the need for careful monitoring at MNA sites, and also provides the scientific and regulatory basis for implementing natural attenuation at field sites. According to the USEPA, MNA refers to the reliance on natural attenuation processes (within the context of a carefully controlled and monitored site cleanup approach) to achieve site-specific remediation objectives within a time frame that is reasonable compared to that offered by other more active methods. The natural attenuation processes that are at work, in such a remediation approach, include a variety of physical, chemical, or biological processes that, under favorable conditions, act without human intervention to reduce the mass, toxicity, mobility, volume or concentration of contaminants in soil or groundwater. The in-situ processes include biodegradation, dispersion, dilution, sorption, volatilization, and radioactive decay. Further, it may also include chemical or biological stabilization, transformation, or destruction of contaminants. When relying on MNA for site remediation, the USEPA prefers to see documentation of processes that degrade or destroy contaminants. If implemented correctly, MNA has the following potential advantages (USEPA 1999):

- MNA should yield relatively low (in some cases zero) volume of remediation wastes.
- It has reduced potential for cross-media transfer of contaminants and hence pose low risk for human exposure to contaminants, and reduced disturbances to ecological receptors.
- Natural attenuation processes may result in complete in-situ destruction of contaminants.
- Depending on the site conditions and remedial objectives, MNA can be applied to all or parts of a given site.
- MNA can be used in conjunction, or as a follow-up to, other active remediation methods.
- Depending on the site conditions, when compared to other active treatment methods the MNA approach might offer a low cost remediation alternative.

The potential disadvantages of MNA include (USEPA 1999):

- MNA might require longer time frames to achieve remediation objectives.
- MNA would require more complex and costly site characterization efforts.
- During MNA, some contaminants might transform to more toxic or more mobile daughter products.
- MNA will require more extensive long-term monitoring effort.
- Institutional controls may be necessary to ensure long-term protection.
- Hydrological and geochemical conditions amenable to natural attenuation may change over time and could result in renewed mobility of a previously stabilized plume which might adversely impact remedial effectiveness.
- More extensive education and outreach efforts may be required to gain public acceptance.

The first step in demonstrating the efficiency of natural attenuation at a field site is collection of a detailed site characterization dataset. This dataset should include source data, hydrogeological data, chemical data, and microbiological data. These data should be integrated within a conceptual model that can help guide remedial actions.

The second step in the MNA assessment process is to use the site characterization data along with other auxiliary information to evaluate the feasibility of MNA as a remedial alternative. As part this step, the following three possible lines of evidences must be developed to demonstrate the efficiency of various natural attenuation processes active at the field site (USEPA 1999):

- Historical groundwater and/or soil chemistry data that demonstrate a clear and meaningful trend of decreasing contaminant mass and/or concentration over time at appropriate monitoring and sampling points. For groundwater plumes, this decrease should not be solely due to plume migration.
- Hydrogeologic and geochemical data that can be used to demonstrate indirectly the types of natural attenuation processes active at the site, and the rate at which such processes will reduce contaminant concentrations to required levels.
- Data from field or microcosm studies (conducted with actual field soils) that directly demonstrate the occurrence of a particular biological process at the site and its ability to degrade the contaminants of concern.

EPA requires demonstration of the first two lines of evidence. However, if the second evidence is inadequate or inconclusive then the third evidence may also be required.

Finally, the site characterization data along with the lines of evidence data should be integrated within a data analysis framework to predict whether site-specific "remediation objectives" can be achieved within a "reasonable time frame." This data analysis step would invariably involve the use of predictive computer models. Predictions resulting from the models must be used to address the following questions:

- Will potential receptors be impacted by the contaminant plume?
- What are the expected concentrations levels and the associated risk to the receptors?
- Is the plume stable? Will the plume stability be affected by hydrological changes over an extended time frame?
- How long will it take for the contaminant plume to completely degrade?
- What are the uncertainties regarding the estimated mass of the contaminants in the subsurface?
- What is the reliability of monitoring and institutional controls established for the site?

In order to answer these questions, one would need a mathematical model that is capable of predicting the simultaneous reactive transport of contaminants, substrates, and microbes within the subsurface porous media. In the following section, we review some of modeling tools available for supporting MNA studies.

Wiedemeier et al. (1998) provides the details of a USEPA natural attenuation protocol for completing various site investigation steps, discussed above, at a MNA field site. Clement et al. (2002) discusses a detailed case study for demonstrating the use of the USEPA protocol at a chlorinated solvent contaminated field site.

13.8 Numerical Modeling of Bioremediation Systems

Computer models can be valuable tools in the evaluation of natural remediation and for designing other active bioremediation systems. The models can also be used to predict the risks associated with various contaminant migration processes. The fundamentals of contaminant transport modeling in groundwater aquifers have been discussed in detail in Chapter 3. Analytical solutions to transport problems and the associated tools are reviewed in Chapter 4. In this section, we will review the reactive transport modeling basics required for developing numerical codes that can be used for modeling natural attenuation systems.

While there are several public-domain codes available for modeling field-scale transport in groundwater systems, only a few numerical codes are capable of simulating coupled, multi-species, reactive transport, especially when the reactions are mediated by complex microbial systems. Rifai et al. (1987) published the details of a two-dimensional code, BIOPLUME-II, which can be used for modeling the fate and transport of hydrocarbon plumes in groundwater aquifers. BIOPLUME-II was developed based on the instantaneous reaction concept that described the interactions between fuel-hydrocarbon and oxygen (Borden and Bedient 1986). The BIOPLUME-II code was later updated to BIOPLUME-III which allowed simulation of alternate electron acceptors such as nitrate, iron, and sulfate, for fuel hydrocarbon destruction (Rifai et al. 1998). The BIOPLUME-family of codes are primarily used for modeling fuel-hydrocarbon plumes.

de Blanc et al. (1996) modified the multi-phase code UTCHEM and developed a general purpose multi-species, bio-reactive transport code. Waddill and Widdowson (1998) presented a three-dimensional sequential electron acceptor model for modeling bioremediation of LNAPL-contaminated aquifers. Chilakapati et al. (2000) presented a detailed three-dimensional code that can be used for modeling multi-component transport and biogeochemistry in a rectangular saturated porous media systems. The more recent extension of the single-species USEPA transport code MT3D, now known as MT3DMS, has the ability to solve certain simple multi-species reactive transport problems (Zheng and Wang 1999). Prommer et al. (2003) coupled MT3DMS with a well-known geochemistry model PHREEQC-2 and developed a reactive transport code known as PHT3D. Clement (1997) developed a comprehensive three-dimensional reactive transport model known as the RT3D (reactive transport in 3-dimensions) model. RT3D solves the coupled partial differential equations that describe flow and reactive transport of multiple mobile and/or immobile species in three-dimensional saturated groundwater systems. RT3D employs the solution routine of MT3DMS to solve the basic transport problem. Similar to MT3DMS, RT3D uses the USGS groundwater model MODFLOW as its flow simulator.

The general reactive transport equations solved by the RT3D computer code can be written as (Clement 1997):

$$\frac{\partial C_k}{\partial t} = \frac{\partial}{\partial x_i}\left(D_{ij} \frac{\partial C_k}{\partial x_j} \right) - \frac{\partial}{\partial x_i}\left(v_i C_k \right) + \frac{q_s}{\phi} C_{s_k} + r_c - r_a + r_d, \text{ where, } k = 1,2,\ldots m \quad (13.1)$$

$$\frac{\partial \tilde{C}_{im}}{\partial t} = \tilde{r}_c + r_a - r_d, \text{ where, } im = 1,2,\ldots,(n-m) \quad (13.2)$$

where n is the total number of species; m is the total number of aqueous-phase (mobile) species (thus, n minus m is the total number of solid-phase or immobile species); C_k is the aqueous-phase concentration of the kth species [ML^{-3}]; $\tilde{C}_{im}$ is the solid-phase concentration of the im^{th} species [MM^{-1}]; D_{ij} is the hydrodynamic dispersion coefficient [L^2T^{-1}]; v is the pore velocity [LT^{-1}]; q_s is the volumetric flux of water per unit volume of aquifer representing sources and sinks [T^{-1}]; C_s is the concentration of source/sink [ML^{-3}]; r_c is the reaction rate that describes the mass of the species removed or produced per unit volume per unit time [ML^3T^{-1}]; $\tilde{r}_c$ is the reaction rate at the solid phase [$MM^{-1}T^{-1}$]; and r_a and r_d, respectively, are attachment (or adsorption) and detachment (or desorption) rates that describe the kinetic exchange of the transported species between aqueous and solid phases [$ML^{-3}T^{-1}$]. The RT3D code solves a set of multi-species reactive transport equations of the form (13.1) and (13.2). The code utilizes a reaction operator-split (OS) numerical strategy to solve the coupled transport problem (Clement et al. 1998). RT3D code is general code that can describe a variety of bioremediation reactions. The reaction terms in Equations (13.1) and (13.2) would depend on the kinetics of the contaminant degradation reactions considered in the study. Clement et al. (1998) solved several test problems using RT3D and validated the code performance by comparing the numerical results with several analytical solutions. Clement (2001) and Quezada et al. (2004) compare RT3D results against a set of analytical model results

Numerical codes, such as RT3D, have been used to solve several types of field- and laboratory-scale reactive transport problems including natural attenuation problems. Truex et al. (2002) provide the details of a modeling effort for predicting the fate and transport of a chlorinated solvent plume at a complex Superfund site location in Baton Rouge, Louisiana. Tartakovsky et al. (2002) discuss the use of RT3D to design a bioremediation system to treat a nitrate-contaminated site located within a Canadian airport. As a part of this field study, extensive groundwater data were collected at a nitrate-contaminated site within the airport. Ethanol was used as a carbon source to simulate the growth and activity of indigenous denitrifying bacterial population. The numerical model-based predictions were used to evaluate the nitrate distribution patterns during this bioremediation effort. Researchers at Michigan State University used a numerical model to design bioremediation system which employed denitrifiers to remediate a carbon tetrachloride plume (Phanikumar et al. 2002). Phanikumar and McGuire (2004) reported the development and application of reactive transport model that coupled RT3D with PHREEQC-2. The combined code was used to model test problems involving microbial transport in a laboratory column and redox zonation in a contaminated aquifer. Researchers at Cornell University developed a comprehensive reaction description for modeling an active biotreatment system for remediating chlorinated solvent plumes. The reactions were integrated within the RT3D code and the resulting tool was used to explore various optimal design strategies for managing solvent plumes (Willis and Shoemaker 2000). Huang et al. (2003) used a laboratory model to study electron acceptor and electron donor mixing patterns under instantaneous reaction conditions; they used a numerical model to recreate their experimental results. Clement et al. (2004b) present a numerical formulation to model a complex transport system that described multiple, coupled, rate-limited processes including DNAPL dissolution, rate-limited sorption and kinetic-biological reactions. Lee et al. (2006) used the RT3D code to model biotransformation and transport occurring within a remediation barrier used for treating nitrogenous waste products. Gomez et al. (2008) used RT3D to evaluate the effects of the common fuel additive ethanol on benzene fate and transport in fuel-contaminated groundwater systems. Rolle et al. (2008) used RT3D to model redox processes occurring under a landfill site in Piedmont, Italy.

13.8.1 RT3D Case Study - Model for Petroleum Hydrocarbon Degradation

Lu et al. (1999) describe the details of a numerical model that was used to predict natural attenuation scenarios of a fuel-hydrocarbon plume present at Hill Air Force Base (AFB). The model described sequential degradation pathways for BTEX utilizing multiple terminal electron accepting process including aerobic respiration, denitrification, Fe(III) reduction, sulfate reduction, and methanogenesis. All the sequential degradation reaction steps were modeled as rate-limited kinetic reactions. A non-linear Monod term was used to simulate the effects of electron acceptor (EA) limitations, and an inhibition model was used to simulate the sequential patterns of the electron acceptor utilization reactions. The fate and transport of BTEX and various EAs (or degradation products) observed at the Hill AFB were modeled using the following set of reactive transport equations (Clement et al. 1998, Lu et al. 1999):

$$R_{BTEX}\frac{\partial[BTEX]}{\partial t} = \frac{\partial}{\partial x_i}\left(D_{ij}\frac{\partial[BTEX]}{\partial x_j}\right) - \frac{\partial\left(v_i[BTEX]\right)}{\partial x_i} + \frac{q_s}{\varphi}[BTEX]_s + r_{BTEX} \quad (13.3)$$

$$R_{O_2}\frac{\partial[O_2]}{\partial t} = \frac{\partial}{\partial x_i}\left(D_{ij}\frac{\partial[O_2]}{\partial x_j}\right) - \frac{\partial\left(v_i[O_2]\right)}{\partial x_i} + \frac{q_s}{\varphi}[O_2]_s + r_{O_2} \quad (13.4)$$

$$R_{NO_3^-}\frac{\partial[NO_3^-]}{\partial t} = \frac{\partial}{\partial x_i}\left(D_{ij}\frac{\partial[NO_3^-]}{\partial x_j}\right) - \frac{\partial\left(v_i[NO_3^-]\right)}{\partial x_i} + \frac{q_s}{\varphi}[NO_3^-]_s + r_{NO_3^-} \quad (13.5)$$

$$R_{Fe^{2+}}\frac{\partial[Fe^{2+}]}{\partial t} = \frac{\partial}{\partial x_i}\left(D_{ij}\frac{\partial[Fe^{2+}]}{\partial x_j}\right) - \frac{\partial\left(v_i[Fe^{2+}]\right)}{\partial x_i} + \frac{q_s}{\varphi}[Fe^{2+}]_s + r_{Fe^{2+}} \quad (13.6)$$

$$R_{SO_4^{-2}}\frac{\partial[SO_4^{-2}]}{\partial t} = \frac{\partial}{\partial x_i}\left(D_{ij}\frac{\partial[SO_4^{-2}]}{\partial x_j}\right) - \frac{\partial\left(v_i[SO_4^{-2}]\right)}{\partial x_i} + \frac{q_s}{\varphi}[SO_4^{-2}]_s + r_{SO_4^{-2}} \quad (13.7)$$

$$R_{CH_4}\frac{\partial[CH_4]}{\partial t}=\frac{\partial}{\partial x_i}\left(D_{ij}\frac{\partial[CH_4]}{\partial x_j}\right)-\frac{\partial\left(v_i[CH_4]\right)}{\partial x_i}+\frac{q_s}{\varphi}[CH_4]_s+r_{CH_4} \quad (13.8)$$

where R is the retardation coefficient for various species; r represent the biodegradation rate term; D_{ij} is the dispersion tensor; v_i is the transport velocity; q_s is the fluid sink/source term; and φ is effective porosity. The concentrations of different species are represented by the square bracket placed around an appropriate notation to represent the species.

In the Lu et al. study, average BTEX concentration levels are assumed to represent the overall hydrocarbon contamination present at the site. Field observations at Hill AFB have shown that a complete sequence of microbially-mediated BTEX biodegradation processes are actively utilizing the electron acceptors O_2, NO_3^{-1}, Fe^{3+} or SO_4^{-2}, and producing Fe(II) and methane. Using benzene as an example reactant, the stoichiometry of different degradation processes can be described by the following set of reactions:

$$C_6H_6+7.5O_2\rightarrow 6CO_2+3H_2O \quad (13.9)$$

$$6NO_3^-+6H^++C_6H_6\rightarrow 6CO_2+6H_2O+3N_2 \quad (13.10)$$

$$30Fe(OH)_3+60H^++C_6H_6\rightarrow 6CO_2+78H_2O+30Fe^{2+} \quad (13.11)$$

$$3.75SO_4^{2-}+7.5H^++C_6H_6\rightarrow 6CO_2+3H_2O+3.75H_2S \quad (13.12)$$

$$C_6H_6+4.5H_2O\rightarrow 2.25CO_2+3.75CH_4 \quad (13.13)$$

A kinetic model was developed to describe the BTEX biodegradation reactions described above. The model describes the degradation pattern of BTEX under aerobic, denitrifying, iron-reducing, sulfate-reducing, and methanogenic conditions. The BTEX decay rates using five different EA are approximated by five first-order decay terms. A Monod term was used to account for the presence (or the absence) of various EAs, and an inhibition model is used to simulate inhibition due to the presence of any one of the earlier EA (i.e. an EA with higher free energy) in the reaction chain. The kinetic model can be written as (Clement et al. 1998, Lu et al. 1999):

$$r_{BTEX,O_2}=-k_{O_2}[BTEX]\frac{[O_2]}{K_{O_2}+[O_2]} \quad (13.14)$$

$$r_{BTEX,NO_3^-}=-k_{NO_3^-}[BTEX]\frac{[NO_3^-]}{K_{NO_3^-}+[NO_3^-]}\frac{K_{i,O_2}}{K_{i,O_2}+[O_2]} \quad (13.15)$$

$$r_{BTEX,Fe^{3+}}=-k_{Fe^{3+}}[BTEX]\frac{[Fe^{3+}]}{K_{Fe^{3+}}+[Fe^{3+}]}\frac{K_{i,O_2}}{K_{i,O_2}+[O_2]}\frac{K_{i,NO_3^-}}{K_{i,NO_3^-}+[NO_3^-]} \quad (13.16)$$

$$r_{BTEX,SO_4^{-2}}=-k_{SO_4^{-2}}[BTEX]\frac{[SO_4^{-2}]}{K_{SO_4^{-2}}+[SO_4^{-2}]}\frac{K_{i,O_2}}{K_{i,O_2}+[O_2]}\frac{K_{i,NO_3^-}}{K_{i,NO_3^-}+[NO_3^-]}\frac{K_{i,Fe^{3+}}}{K_{i,Fe^{3+}}+[Fe^{3+}]} \quad (13.17)$$

$$r_{BTEX,Me} = -k_{Me}[BTEX]\frac{[CO_2]}{K_{CH_4}+[CO_2]}\frac{K_{i,O_2}}{K_{i,O_2}+[O_2]}\frac{K_{i,NO_3^-}}{K_{i,NO_3^-}+[NO_3^-]}\frac{K_{i,Fe^{3+}}}{K_{i,Fe^{3+}}+[Fe^{3+}]}\frac{K_{i,SO_4^{-2}}}{K_{i,SO_4^{-2}}+[SO_4^{-2}]} \quad (13.18)$$

where r_{BTEX,O_2} is the BTEX destruction rate utilizing oxygen; r_{BTEX,NO_3^-} is the destruction rate utilizing nitrate; $r_{BTEX,Fe^{3+}}$ is the destruction rate utilizing Fe^{3+} (or producing Fe^{2+}); $r_{BTEX,SO_4^{-2}}$ is the destruction rate utilizing sulfate; $r_{BTEX,Me}$ is the destruction rate via methanogenesis; $[O_2]$ is oxygen concentration $[ML^{-3}]$; k_{O_2} is the first-order degradation rate constant for BTEX utilizing oxygen as the EA $[T^{-1}]$; K_{O_2} is the saturation constant for oxygen $[ML^{-3}]$; K_{i,O_2} is the oxygen inhibition constant $[ML^{-3}]$; and similar nomenclature is used for subsequent reactions. Note that by setting the half-saturation constants to small values, we can simulate zero-order dependency with respect to the electron donor and thus a first-order degradation model with respect to BTEX. The values of all the saturation constants were set at 0.01 mg/L. Similarly, the inhibition constants can be set to small values to simulate a pure sequential EA process. The inhibition function is used to represent the concept that the availability of any one of the EAs may inhibit the utilization of other EAs that provide less Gibbs free energy to the system. However, if a K_i is assigned a very large value (much larger than the maximum value of the EA species) then the inhibition function becomes one and the simultaneous use of EAs can be simulated.

Equations (13.3) to (13.8) were solved using RT3D to predict the fuel hydrocarbon natural attenuation patterns observed at the Hill AFB (Lu et al. 1999). The modeling exercise consisted of fitting the flow model to the observed groundwater head distribution and simulating the distribution of various chemical species as observed in July 1994 using the August 1993 data set as the initial condition. The hydraulic conductivity distribution and reaction rate constants were estimated based on this calibration process. The calibration results suggest that the location of LNAPL distribution coincides with the low hydraulic conductivity zone. This perhaps explains why the LNAPL has persisted for a long time in this area. The calibrated reaction rates for the BTEX biodegradations as calculated by the model are 0.051, 0.031, 0.0051, 0.003, and 0.002 day^{-1} for oxygen, nitrate, ferric iron, and sulfate reduction, and methanogenic degradation, respectively. Comparison of the calibrated model results to observed data indicated that the natural attenuation model was able to predict the observed transport patterns of BTEX and various electron acceptor plumes. In addition, the model also reasonably predicted the location of the plume front. Mass balance analyses showed that the total mass of BTEX in the aquifer at the end of one-year simulation period, as calculated by the model, was within 9.2% of the total mass estimated from field observations.

It is important to note the above model was developed based on several important assumptions. All the assumption and limitations of the model are described in Lu et al. (1999). One of the major limitations is that the amounts of Fe^{2+} and methane were restricted in every node at an empirically-defined field values, known as the expressed capacity values (Lu et al. 1999). Rolle et al. (2008) modified this Lu et al. model to develop a more comprehensive model that can simulate organic compound biodegradation directly coupled to solid phase iron (Fe^{3+}) utilization reaction and methane generation reaction.

13.8.2 RT3D Case Study - Model for Chlorinated Solvent Degradation

Clement et al. (2000) provide the details of a modeling effort that investigated the natural attenuation patterns of chlorinated solvent plumes present at the Dover AFB, Delaware. The conceptual biochemical model used for representing the biodegradation reactions observed at the site is illustrated in Figure 13.7.

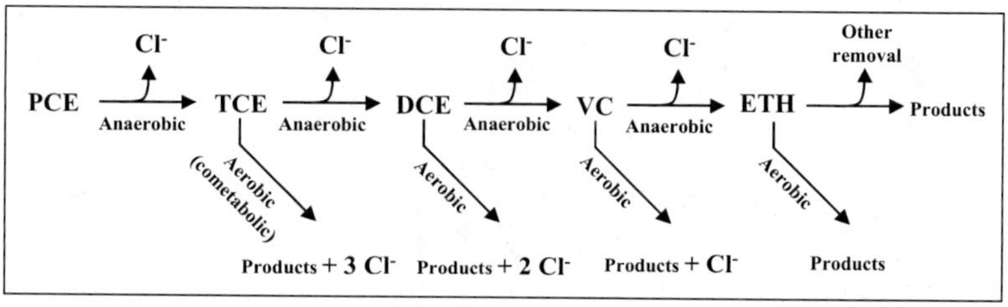

Figure 13.7: Biochemical conceptual model used for the Dover site (Clement et al. 2000).

The fate and transport of different chlorinated solvent plumes, mediated by the dechlorination reactions described in the figure, can be mathematically represented by a set of transport equations. Assuming first-order biodegradation kinetics, transport and transformation of PCE, TCE, DCE, VC, ETH, and Cl^- can be simulated by solving the following set of partial differential equations (Clement et al. 2000):

$$R_P \frac{\partial[PCE]}{\partial t} = \frac{\partial}{\partial x_i}\left(D_{ij}\frac{\partial[PCE]}{\partial x_j}\right) - \frac{\partial(v_i[PCE])}{\partial x_i} + \frac{q_s}{\phi}[PCE]_s - K_P[PCE] \quad (13.19)$$

$$R_T \frac{\partial[TCE]}{\partial t} = \frac{\partial}{\partial x_i}\left(D_{ij}\frac{\partial[TCE]}{\partial x_j}\right) - \frac{\partial(v_i[TCE])}{\partial x_i} + \frac{q_s}{\phi}[TCE]_s + Y_{T/P}K_P[PCE] - K_{T1}[TCE] - K_{T2}[TCE] \quad (13.20)$$

$$R_D \frac{\partial[DCE]}{\partial t} = \frac{\partial}{\partial x_i}\left(D_{ij}\frac{\partial[DCE]}{\partial x_j}\right) - \frac{\partial(v_i[DCE])}{\partial x_i} + \frac{q_s}{\phi}[DCE]_s + Y_{D/T}K_T[TCE] - K_{D1}[DCE] - K_{D2}[DCE] \quad (13.21)$$

$$R_V \frac{\partial[VC]}{\partial t} = \frac{\partial}{\partial x_i}\left(D_{ij}\frac{\partial[VC]}{\partial x_j}\right) - \frac{\partial(v_i[VC])}{\partial x_i} + \frac{q_s}{\phi}[VC]_s + Y_{V/D}K_{D1}[DCE] - K_{V1}[VC] - K_{V2}[VC] \quad (13.22)$$

$$R_E \frac{\partial[ETH]}{\partial t} = \frac{\partial}{\partial x_i}\left(D_{ij}\frac{\partial[ETH]}{\partial x_j}\right) - \frac{\partial(v_i[ETH])}{\partial x_i} + \frac{q_s}{\phi}[ETH]_s + Y_{E/V}K_{V1}[VC] - K_{E1}[ETH] - K_{E2}[ETH] \quad (13.23)$$

$$R_C \frac{\partial[Cl]}{\partial t} = \frac{\partial}{\partial x_i}\left(D_{ij}\frac{\partial[Cl]}{\partial x_j}\right) - \frac{\partial(v_i[Cl])}{\partial x_i} + \frac{q_s}{\phi}[Cl]_s + Y1_{C/P}K_{P1}[PCE] + Y1_{C/T}K_{T1}[TCE]$$
$$+ Y1_{C/D}K_{D1}[DCE] + Y1_{C/V}K_{V1}[VC] + Y2_{C/T}K_{T2}[TCE] + Y2_{C/D}K_{D2}[DCE] \quad (13.24)$$
$$+ Y2_{C/V}K_{V2}[VC]$$

where [*PCE*], [*TCE*], [*DCE*], [*VC*], [*ETH*], and [*Cl*] represent contaminant concentrations of various species [mg/L]; K_P, K_{T1}, K_{D1}, K_{V1} and K_{E1} are first-order anaerobic degradation rate constants [day^{-1}]; K_{T2}, K_{D2}, K_{V2}, and K_{E2} are first-order aerobic degradation rate constants [day^{-1}]; R_P, R_T, R_D, R_V, R_E, and R_C are retardation factors; $Y_{T/P}$, $Y_{D/T}$, $Y_{V/D}$, and $Y_{E/V}$ are stoichiometric yield constants for chlorinated compounds transformed under anaerobic conditions – their values are 0.79, 0.74, 0.64 and 0.45, respectively; $Y1_{C/P}$, $Y1_{C/T}$, $Y1_{C/D}$, and $Y1_{C/V}$ are stoichiometric yield values for chloride production under anaerobic conditions – their values are 0.21, 0.27, 0.37, and 0.57, respectively; and $Y2_{C/T}$, $Y2_{C/D}$, *and* $Y2_{C/V}$ are stoichiometric yield values for chloride production under aerobic conditions – their values are 0.81, 0.74, and 0.57, respectively. The yield values are estimated from the reaction stochiometery and molecular weights; for example, anaerobic degradation of one mole of PCE would yield one mole of TCE, therefore $Y_{T/P}$ = molecular weight of TCE/molecular weight of PCE (131.4/165.8 = 0.79). Note the reaction models presented above assume that the biological degradation reactions only occur in the aqueous phase.

Equations (13.19) to (13.24) were solved using RT3D to simulate the chlorinated solvent natural attenuation patterns observed at the Dover AFB. The flow and transport models were calibrated to reflect the field conditions observed at the site. The model recreated the flow conditions observed at the site. In addition, the concentration distribution of PCE, TCE, DCE, VC, and chloride plumes predicted by the model matched the observed data reasonably well. The field-scale degradation parameters estimated via the model calibration process were within the range of values measured by other laboratory and field methods. The calibrated model was used to generate a set of transport scenarios under natural attenuation conditions. Details of the calibration process along with a detailed sensitivity analysis are summarized in Clement et al. (2000).

One of the major limitations of the Clement et al. (2000) model is that the formulation assumed the spatial extent of the redox zones, which defined the level of degradation, are known and fixed. The user should delineate these zones *a priori* based on field observed geochemical data. In realistic field conditions, these redox zones evolve over time and their extent will vary depending on the amount of electron donors and terminal electron acceptors available in the systems. Rolle et al. (2008) proposed a more comprehensive model that can help track the fate and transport of the electron donor to simulate the redox zone evolution process. This type of advanced redox zone modeling approach can be coupled to the chlorinated solvent model to automatically vary the chlorinated solvent decay rates within different redox zones.

13.9 References

Alvarez, P. J., and Vogel, T. M. (1991). "Kinetics of aerobic benzene and toluene in sandy aquifer material." *Biodegradation*, 2, 43-51.

Arner, R. (1995). "By George! Composting at Mount Vernon." *Biocycle*, 36, 79-80.

Aulenta, F., Majone, M., Verbo, P., and Tandoi, V. (2003). "Complete dechlorination of tetrachloroethene to ethene in presence of methanogenesis and acetogenesis by an anaerobic sediment microcosm." *Biodegradation*, 13, 411-424.

Baveye, P., Vandevivere, P., Hoyle, B. L., Deleo, P. C., and Sanchez De Lozada, D. (1998). "Environmental impact and mechanisms of the biological clogging of saturated soils and aquifer materials." *Critical Reviews in Environmental Science and Technology*, 28,123-191.

Beller, H. R., Spormann, A. M., Sharma, P. K., Cole, J. R., and Reinhard, M. (1996). "Isolation and characterization of a novel toluene-degrading, sulfate-reducing bacterium." *Applied and Environmental Microbiology*, 62, 1188-1196.

Benefield, L. D., and Randall, C. W. (1980). *Biological process design for wastewater treatment*, Prentice-Hall Inc., Englewood Cliffs, N.J.

Bianchi-Mosquera, G. C., Allen-King, R. M., and Mackay, D. D. (1994). "Enhanced degradation of dissolved benzene and toluene using a solid oxygen-releasing compound." *Ground Water Monitoring & Remediation*, 9, 120-128.

Blackman, W. C. (1993). *Basic hazardous waste management*, Lewis Publishers, Boca Raton, Fla.

Borden, R. C., and Bedient, P. B. (1986). "Transport of dissolved hydrocarbons influenced by oxygen-limited biodegradation: 1. Theoretical development." *Water Resources Research*, 13, 1973-1982.

Bradley, P. M. (2000). "Microbial degradation of chloroethenes in groundwater systems." *Hydrogeology Journal*, 8, 104-111.

Bradley, P. M. (2003). "History and ecology of chloroethene biodegradation- A Review." *Bioremediation Journal*, 7, 81-109.

Bradley, P. M., and Chapelle, F. H. (1996). "Anaerobic mineralization of vinyl chloride in Fe(III)-reducing aquifer sediments." *Environmental Science and Technology*, 30, 2084-2086.

Bradley, P. M., and Chapelle, F. H. (1998). "Effect of contaminant concentration on aerobic microbial mineralization of DCE and VC in stream-bed sediments." *Environmental Science and Technology*, 32, 553-557.

Brown, R. A., and Jasiulewicz, F. (1992). "Air sparging: A new model for remediation." *Pollution Engineering*, July, 52-55.

Carbirol, N., Jacob, F., Perrier, J., and Foullet, B. (1998). "Complete degradation of high concentrations of tetrachloroethene by a methanogenic consortium in a fixed-bed reactor." *Journal of Biotechnology*, 62, 133-141.

Chapelle, F. H. (1993). *Groundwater microbiology and geochemistry*, John Wiley & Sons Inc., New York.

Chiang, C. Y., Salanitro, J. P., Chai, E. Y., Colthart, J. D., and Klein, C. L. (1989). "Aerobic biodegradation of benzene, toluene and xylene in a sandy aquifer." *Ground Water*, 27, 823-834.

Chilakapati, A., Yabusaki, S., Szecsody, J., and MacEvoy, W. (2000). "Groundwater flow, multicomponent transport and biogeochemistry: development and application of a coupled process model." *Journal of Contaminant Hydrology*, 43, 303-325.

Clement, T. P. (1997). *A modular computer model for simulating reactive multi-species transport in three-dimensional groundwater systems*, PNNL-SA-28967, Pacific Northwest National Laboratory, Richland, Wash.

Clement, T. P. (2001). "A generalized analytical method for solving multi-species transport equations coupled with a first-order reaction network." *Water Resources Research*, 37, 157-163.

Clement, T. P., Gautam, T. R., Lee, K. K., Truex, M. J., and Davis, G. B. (2004a). "Modeling coupled NAPL-dissolution and rate-limited sorption reactions in biologically active porous media." *Bioremediation Journal*, 8, 47-64.

Clement, T. P., Gautam, T. R., Lee, K. K., Truex, M. J., and Davis, G. B. (2004b). "Modeling of DNAPL-dissolution, rate-limited sorption and biodegradation reactions in groundwater systems." *Bioremediation Journal*, 8, 47-64.

Clement, T. P., Hooker, B. S., and Skeen, R. S. (1996a). "Macroscopic models for predicting changes in saturated porous media properties caused by microbial growth." *Ground Water*, 34, 934-942.

Clement, T. P., Hooker, B. S., and Skeen, R. S. (1996b). "Numerical modeling of biologically reactive transport near nutrient injection well." *Journal of Environmental Engineering*, 122, 833-839.

Clement, T. P., Johnson, C. D., Sun, Y. W., Klecka, G. M., and Bartlett, C. (2000). "Natural attenuation of chlorinated ethene compounds: Model development and field-scale application at the Dover site." *Journal of Contaminant Hydrology*, 42, 113-140.

Clement, T. P., Sun, Y., Hooker, B. S., and Petersen, J. N. (1998). "Modeling multispecies reactive transport in ground water." *Ground Water Monitoring & Remediation*, 18, 79-92.

Clement, T. P., Truex, M. J., and Hooker, B. S. (1997). "Two-well test method for determining hydraulic properties of aquifers." *Ground Water*, 35, 698-703.

Clement, T. P., Truex, M. J., and Lee, P. (2002). "A case study for demonstrating the application of US EPA's monitored natural attenuation screening protocol at a hazardous waste site." *Journal of Contaminant Hydrology*, 59, 133-162.

Cunningham, A. B., Characklis, W. G., Abedeen, F., and Crawford, D. (1991). "Influence of biofilm accumulation on porous media hydrodynamics." *Environmental Science and Technology*, 25, 1305-1311.

Cupples, A. M., Spormann, A. M., and McCarty, P. L. (2003). "Growth of a Dehalococcoides-like microorganism on vinyl chloride and cis-dichloroethene as electron acceptors as determined by competitive PCR." *Applied and Environmental Microbiology*, 69, 953-959.

Darwin, C. R. (1881). *The formation of vegetable mould through the action of worms with observations on their habits*, John Murray & Co., London.

de Blanc, P. C., McKinney, D. C., and Speitel, G. E. (1996). "A 3-D flow and biodegradation model." *Proc., Non-Aqueous Phase Liquid in the Subsurface Environment: Assessment and Remediation*, ASCE, Reston, Va. 478-489.

DeBruin, W. P., Kotterman, M. J. J., Posthumus, M. A., Schraa, G., and Zhender, A. J. B. (1992). "Complete biological reductive transformation of tetrachloroethene to ethane." *Applied and Environmental Microbiology*, 58, 1996-2000.

Franzen, M. F. L., Petersen, J. N., Clement, T. P., Hooker, B. S., and Skeen, R. S. (1997). "Pulsing as a strategy to achieve large biologically active zones during in situ carbon tetrachloride remediation." *Computational Geosciences*, 1, 217-288.

Gao, J., Skeen, R. S., Hooker, B. S., and Quezenberry, R. D. (1997). "Effects of several electron donors on tetrachloroethylene dechlorination in anaerobic soil microcosms." *Water Research*, 31, 2479-2486.

Gomez, D. E., de Blanc, P. C., Rixey, W. G., Bedient, P. B., and Alvarez, P. J. J. (2008). "Modeling benzene plume elongation mechanisms exerted by ethanol using RT3D with a general substrate interaction module." *Water Resources Research*, 44, W05405, doi:10.1029/2007WR006184.

Gudemann, H., and Hiller, D. (1988). "In situ remediaiton of VOC contaminated soil and groundwater by vapor extraction and groundwater aeration." *Proc., Petroleum Hydrocarbon and Organic Chemicals in Groundwater: Prevention, Detection, and Resoration*, NWWA/API, Houston, Tex, 147-164.

Gupta, R. P., and Swartzendruber, D. (1962). "Flow-associated reduction in the hydraulic conductivity of quartz sand." *Soil Science Society Proceedings*, 26, 6-10.

Harvey, R. W., George, L. H., Smith, R. L., and LeBlanc, D. R. (1989). "Transport of microspheres and indigenous bacteria through a sandy aquifer: Results of natural and forced-gradient tracer experiments." *Environmental Science and Technology*, 23, 51-56.

Harvey, R. W., Metge, D. W., Kinner, N., and Mayberry, N. (1997). "Physiological considerations in applying laboratory-determined buoyant densities to predictions of bacterial and protozoan transport in groundwater: Results of in-situ and laboratory tests." *Environmental Science and Technology*, 23, 289-295.

He, J., Ritalahti, K. M., Aiello, M. R., and Loffler, F. E. (2003). "Complete detoxification of vinyl chloride by an anaerobic enrichment culture and identification of the reductively dechlorinating population as a Dehalococcoides species." *Applied and Environmental Microbiology*, 69, 996-1003.

Hooker, B. S., Skeen, R. S., Truex, M. J., Johnson, C. D., Peyton, B. M., and Andersen, D. B. (1998). "In situ bioremediation of carbon tetrachloride: Field test results." *Bioremediation Journal*, 3, 181-193.

Hopkins, G. D., Munakata, J., Semprini, L., and McCarty, P. L. (1993). "Trichloroethylene concentration effects on pilot field-scale in-situ groundwater bioremediation by phenol-oxidizing microorganisms." *Environmental Science and Technology*, 27, 2542-2547.

Huang, W. E., Oswald, S. E., Lerner, D. N., Smith, C. C., and Zheng, C. (2003). "Dissolved oxygen imaging in a porous medium to investigate biodegradation in a plume with limited electron acceptor supply." *Environmental Science and Technology*, 37, 1905-1911.

Johnson, R. L., Johnson, P. C., McWhorter, D. B., Hinchee, R. L., and Goodman, I. (1993). "An overview of in situ air sparging." *Ground Water Monitoring & Remediation*, 13, 127-135.

Kuhn, E. P., Zeyer, J., Eicher, P., and Schwarzenbach, R. P. (1988). "Anaerobic degradation of alkylated benzenes in denitrifying laboratory aquifer columns." *Applied and Environmental Microbiology*, 54, 490-496.

Lee, M., Lee, K. K., Clement, T. P., and Hamilton, D. (2006). "Nitrogen transformation and transport modeling in groundwater aquifers." *Ecological Modeling*, 192, 143-159.

Lovley, D. R. (1991). "Dissimilatory Fe(III) and Mn(IV) reduction." *Microbiological Reviews*, 55, 259-287.

Lu, G. P., Clement, T. P., Zheng, C. M., and Wiedemeier, T. H.. (1999). "Natural attenuation of BTEX compounds: Model development and field-scale application." *Ground Water* 37, 707-717.

MacDonald, T. R., and Kitanidis, P. K. (1993). "Modeling the free surface of an unconfined aquifer near a recirculation well." *Ground Water*, 31, 774-788.

Macleod, F. A., Lappin-Scott, H. M., and Costerton J. W. (1988). "Plugging of a model rock system by using starved bacteria." *Applied and Environmental Microbiology*, 54, 1365-1372.

Madigan, T. M., Martinko, J. M., and Parker J. (2000). *Brock biology of microorganisms*, Prentice Hall, Upper Saddle River, N.J.

Major, D., Edwards, E., McCarty, P. L., Gossett, J., Hendrickson, E., Loeffler, F., Zinder, S., Ellis, D., Vidumsky, J., Harkness, M., Klecka, G., and Cox, E. (2003). "Discussion of Environment vs bacteria or let's play 'Name that bacteria'." *Ground Water Monitoring & Remediation*, 23, 32-48.

Major, D. W., Mayfield, C. I., and Barker, J. F. (1988). "Biotransformation of benzene by denitrification in aquifer sand." *Ground Water*, 26, 8-14.

Maymo-Gatell, X., Anguish, T., and Zinder, S. H. (1999). "Reductive dechlorination of chlroinated ethenes and 1,2-dichloroethane by 'Dehalococcoides ethenogenes' 195." *Applied and Environmental Microbiology*, 65, 3108-3113.

Maymo-Gatell, X., Chien, Y.-T., Gossett, J. M., and Zinder, S. H. (1997). "Isolation of a bacterium that reductively dechlorinates tetrachloroethene to ethene." *Science*, 276, 1568-1571.

Maymo-Gatell, X., Nijenhuis, I., and Zinder, S. H. (2001). "Reductive dechlorination of cis-dichloroethene and vinyl chlroide by 'Dehalococcoides ethenogenes'." *Environmental Science and Technology*, 35, 516-521.

McCarty, P. L. (1997). "Breathing with chlorinated solvents." *Science*, 276, 1521.

McCarty, P. L., and Semprini, L. "Ground-water treatment for chlorinated solvents." *Handbook of bioremediation*, R. D. Norris, R. E. Hinchee, R. Brown, P. L. McCarty, L. Semprini, J. T. Wilson, D. H. Kampbell, M. Reinhard, E. J. Bouwer, R. C. Borden, T. M. Vogel, J. M. Thomas, and C. H. Ward, eds., Lewis Publishers, Boca Raton, Fla, 87-116.

McCarty, P. L., Goltz, M. N., Hopkins, G. D., Dolan, M. E., Allan, J. P., Kawakami, B. T., and Carrothers, T. J. (1998). "Full-scale evaluation of in situ cometabolic degradation of trichloroethylene in groundwater through toluene injection." *Environmental Science and Technology*, 32, 88-100.

Mitchell, R., and Nevo, Z. (1964). "Effect of bacterial polysaccharide accumulation on infiltration of water through sand." *Applied Microbiology*, 12, 219-223.

National Research Council (NRC) (1994). *In situ bioremediation: When does it work?* National Academy Press, Washington D.C.

Nyer, E. K., Payne, F., and Suthersan, S. (2003). "Environment vs. bacteria or let's play 'Name that bacteria'." *Ground Water Monitoring & Remediation*, 23, 36-45.

Palmisano A., and Hazen, T. (2003). *Bioremediation of metals and radionuclides: What it is and how it works*, 2nd edition, LBNL 42595-2003, Lawrence Berkeley National Laboratory, UC-Berkeley, Calif.

Panikow, J. F., Johnson, R. L., and Cheery, J. A. (1993). "Air sparging in gate wells in cutoff walls and trenches for control of plumes of volatile organic compounds." *Ground Water*, 31, 654-663.

Phanikumar, M. S., Hyndman, D. W., and Criddle, C. S. (2002). "Biocurtain design using reactive transport models." *Ground Water Monitoring & Remediation*, 22, 113-123.

Phanikumar, M. S., and McGuire, J. T. (2004). "A 3-D partial-equilibrium model to simulate coupled hydrogeological, microbiological, and geochemical processes in subsurface systems." *Geophysical Research Letters*, 31, L11503.

Prommer, H., Barry, D. A., and Zheng, C. (2003). "MODFLOW/MT3DMS based reactive multicomponent transport modeling." *Ground Water*, 41, 247-257.

Quezada, C. R., Clement, T. P., and Lee, K. K. (2004). "Generalized solution to multi-dimensional, multi-species transport equations coupled with a first-order reaction network involving distinct retardation factors." *Advances in Water Resources Journal*, 27, 507-520.

Rifai, H. S., Bedient, P. B., Borden, R. C., and Haasbeek, J. F. (1987). *BIOPLUME II: Computer model of two-dimensional transport under the influence of oxygen limited biodegradation in groundwater*, Air Force Center for Environmental Excellence, San Antonio, Tex.

Rifai, H. S., Newell, C. J., Gonzales, J. R., Dendrou, S., Kennedy, L., and Wilson, J. T. (1998). *BIOPLUME III: Natural attenuation decision support system, user's manual, version 1.0*, EPA/600/R-98/010, National Risk Management Research Laboratory, U.S. Environmental Protection Agency, Cincinnati, Ohio.

Ripley, D. P., and Saleem, Z. A. (1973). "Clogging in simulated glacial aquifers due to artificial recharge." *Water Resources Research*, 9, 1047-1057.

Rittmann, B. E. (1993). "The significance of biofilms in porous media." *Water Resources Research*, 29, 2195-2202.

Rittmann, B. E., and McCarty, P. L. (2001). *Environmental biotechnology*, McGraw-Hill, Inc., New York.

Rolle, M., Clement, T. P., Sethi, R., and Molfetta, A. D. (2008). "A kinetic approach for simulating redox-controlled fringe and core biodegradation processes in groundwater: Model development and application to a landfill site in Piedmont, Italy." *Hydrological Processes Journal*, 22(25), 4905-4921.

Rynk, R. F. (1992). *On-farm composting handbook*, NRAES-54, Northeast Regional Agricultural Service, Ithaca, N.Y.

Semprini, L., Grbic-Galic, D., McCarty, P. L., and Roberts, P. V. (1992). *Methodologies for evaluating in situ bioremediation of chlorinated solvents*, EPA 600/R-92/042, Robert S. Kerr Environmental Research Laboratory.

Soares, M. I., Belkins, S., and Abelovich, A. (1988). "Biological groundwater denitrification: Laboratory studies." *Water Science & Technology*, 20, 189-195.

Tartakovsky, B., Millette, D., Isle, S., and Guiot, S. R. (2002). "Ethanol-stimulated bioremediation of nitrate-contaminated ground water." *Ground Water Monitoring and Remediation*, 22(1), 78-87.

Taylor, S. W., and Jaffe, P. R. (1990a). "Biofilm growth and the related changes in the physical properties of a porous medium: 1. Experimental investigation." *Water Resources Research*, 26, 2153-2159.

Taylor, S. W., and Jaffe, P. R. (1990b). "Substrate and biomass transport in a porous medium." *Water Resources Research*, 26, 2181-2194.

Taylor, S. W., and Jaffe, P. R. (1991). "Enhanced in situ biodegradation and aquifer permeability reduction." *Journal of Environmental Engineering*, 117, 25-46.

Thierrin, J., Davis, G. B., Barber, C., Patterson, B. M., Pribac, F., Power, T. R., and Lambert, M. (1993). "Natural degradation of BTEX compounds and napthalene in a sulfate reducing groundwater environment." *Hydrological Sciences Journal*, 38, 309-322.

Tiedje, J. M. (1993). "Bioremediation from an ecological perspective." *In situ bioremediation: When does it work?*, B. E. Rittmann, ed., National Academies Press, Washington, D.C.

Tondai, V., DiStefano, T. D., Bowser, P. A., Gossett, J. M., and Zinder, S. H. (1994). "Reductive dehalogenation of chlroinated ethenes and halogenated ethanes by a high-rate anaerobic enrichment culture." *Environmental Science and Technology*, 28, 973-979.

Torsvik, V., Goksoyr, J., and Daae, F. L. (1990). "High diversity DNA of soil bacteria." *Applied and Environmental Microbiology*, 56, 782-787.

Truex, M. J., Johnson, C. D., Spencer, J. R., and Clement, T. P. (2002). "Evaluating natural attenuation of chlorinated solvents at a complex site." *Proc., Third International Conference on Remediation of Chlorinated and Recalcitrant Compounds, Monterey, California*, Battelle Press, Columbus, Ohio.

Thullner M., and Baveye, P. (2008). "Computational pore network modeling of the influence of biofilm permeability on bioclogging in porous media." *Biotechnology and Bioengineering*, 99, 1337-1351.

Thullner M., Schroth, M. H., Zeyer, J., and Kinzelbach, W. (2004). "Modeling of a microbial growth experiment with bioclogging in a two-dimensional saturated porous media flow field." *Journal of Contaminant Hydrology*, 70, 37-62.

U.S. Environmental Protection Agency (USEPA) (1999). *Use of monitored natural attenuation at Superfund, RCRA corrective action, and underground storage tanks sites,* EPA/540/R-99/009 (OSWER Directive 9200.4-17P), Office of Solid Waste and Emergency Response, U.S. Environmental Protection Agency, Washington, D.C.

USEPA (2000). *Engineering approaches to in situ bioremediation of chlorinated solvents: Fundamentals and field applications*, EPA 542-R-00-008, Office of Solid Waste and Emergency Response, U.S. Environmental Protection Agency, Washington, D.C.

USEPA (2004). *How to evaluate alternative cleanup technologies for underground storage tank sites: A guide for corrective action plan reviewers*, EPA 510-R-04-

002, Office of Solid Waste and Emergency Response, U.S. Environmental Protection Agency, Washington, D.C.

Waddill, D. W., and Widdowson, M. A. (1998). *SEAM3D: A numerical model for three-dimensional solute transport and sequential electron acceptor-based bioremediation in groundwater*, Virginia Tech., Blacksburg, Va.

Wiedemeier, T. H., Rifai, H. S., Newell, C. J., and Wilson J. T. (1999). *Natural attenuation of fuels and chlorinated solvents in the subsurface*, John Wiley & Sons, Inc., New York.

Wiedemeier, T. H., Swanson, M. A., Moutoux, D. E., Gordon, E. K., Wilson, J. T., Wilson, B. H., Kampbell, D. H., Hass, P. E., Miller, R. N., Hansen, J. E., and Chapelle F. H. (1998). *Technical protocol for evaluating natural attenuation of chlorinated solvents in groundwater*, EPA/600/R-98/128, National Risk Management Research Laboratory, U.S. Environmental Protection Agency, Cincinnati, Ohio.

Wall, J. D., and Krumholz, L. R. (2006), "Uranium reduction." *Annual Reviews of Microbiology*, 60, 149–166.

Willis, M. B., and Shoemaker, C. A. (2000). "Engineered PCE dechlorination incorporating competitive biokinetics: Optimization and transport modeling." *Bioremediation and phytoremediation of chlorinated and recalcitrant compounds*, Battelle Press, Columbus, Ohio, 311-318.

Zheng, C., and Wang, P. P. (1999). *MT3DMS: A modular three-dimensional multispecies model for simulation of advection, dispersion and chemical reactions of contaminants in groundwater systems*, SERDP-99-1, Army Engineering Research and Development Center, Vicksburg, Miss.

CHAPTER 14

CLOSURE ON GROUNDWATER QUANTITY AND QUALITY MANAGEMENT

Mustafa M. Aral
Multimedia Environmental Simulations Laboratory
School of Civil and Environmental Engineering
Georgia Institute of Technology, Atlanta, GA 30332

14.1 Introduction

As this book goes into publication, it is important to review the current state of affairs in the field of groundwater quantity and quality management and to further elaborate on the purpose and goals of this document.

From an historical perspective, groundwater management paradigms have evolved over several decades as our understanding of groundwater quality and quantity issues and their effects on our society have become more clear. At this time we cannot say that we fully understand all water quantity and quality management issues and the societal impact of water management problems, but research findings in this area are increasing exponentially with very positive and encouraging results that may guide us through this maze. On the other end of this spectrum, the number of organic compounds that have been synthesized since the turn of the century now exceeds half a million, and some 10,000 new compounds are added to this list each year. As a result, many of these new compounds are now found in waste streams and also in groundwater. Given the ever increasing number of emerging contaminants and their illusive health effects, it maybe safe to say that the ultimate goal of completely identifying all societal issues associated with water quality and quantity management may not be fully achievable. This chapter is an attempt to provide a comprehensive review of the current state of affairs in this research and technology field.

14.2 Environmental Management Paradigms

Over the last several decades, environmental scientists, economists, physicists, social scientists, health scientists, and public health officials have been working on critical issues related to groundwater and environmental health management in order to find a feasible medium between limited resources, long-term demands, environmental impacts, health effects, and conflicting interest groups. In this process, in search of a solution to this multidimensional problem, our focus has shifted from one extreme to the other, which significantly affected groundwater management issues. For example, several decades ago in the Western world we first started with the ***Frontier Economics*** period. During this resource development period, little attention was paid to the environment or groundwater quality impacts associated with resource development activities. During this period it was assumed that the environment would yield abundant resources and supplies, and scientists and engineers concentrated on the development of these resources without regard to adverse groundwater quality or environmental effects. This approach had drastic environmental and also health effect consequences that are still influencing our society. During this period, the training, education, and development of individuals in relevant sciences concentrated on resource-based activities, and significant advances were made in the area of resource identification, resource utilization, and resource exploitation.

After the realization of the destruction caused by this approach, the pendulum swung to the other extreme and we entered the period that may be identified as ***Radical Environmentalism***. In this management philosophy the basic assumption was that environmental resources are limited and one should protect the environment without regard to economic and other considerations. During this period, scientific studies concentrated on the development of narrowly-based natural sciences, and significant scientific advances were made in fundamental topics of compartmentalized basic sciences. In training and education the emphasis was placed on environmental preservation and naturalism.

When the economic burden of the ***Radical Environmentalism*** period was realized, environmental and groundwater management policies shifted to the third phase, which may be identified as the ***Resource Management*** or ***Resource Allocation*** period. During this period, the environment was considered to be a subset of economics, in realization of the fact that, while we are developing our economic resources, we should also consider the environment and water quality issues. The development of concepts of environmental mitigation and assessment, and the "those who pollute will pay" mentality belong to this period of environmental management. Environmental legislation such as the Resource Conservation and Recovery Act (RCRA), Comprehensive Environmental Response, Compensation, and Liability Act (CERCLA), Superfund Amendments and Reauthorization Act (SARA), and others in the U.S. were the outcome of the policies of this period. In this period, multidisciplinary scientific specializations evolved that were a significant improvement over the traditional partitioning of sciences. Multidisciplinary programs and emphasis on multidisciplinary training and education can be considered to be an outcome of the policies of this period.

During the ***Resource Management*** period the environment still suffered because controls imposed on the system were materialistic and not naturalistic. Eventually, this philosophy did not fit well with the environmentalists, and we entered the period of ***Selective Environmentalism***. During this environmental management period, economic considerations were considered to be a subset of environmental issues. With this period we entered the era of environmental preservation and planning, and the development of environmentally-friendly technologies and products followed. Scientific developments concentrated on multidisciplinary specializations within traditional natural and basic sciences.

In the management philosophies described above, only two variables of concern were emphasized, i.e., the economy and the environment. In selecting a specific strategy one aspect was always given priority over the other. At this point, it became clear to scientists and also more importantly to the general public that neither of these management philosophies were considering the combination of these two variables in a harmonious way. A review and combination of the better parts of the earlier philosophies resulted in the concept of ***Sustainable Environmental Management*** as a resolution of the conflict between the environmental and economic considerations. The basic philosophy behind this approach is outlined in a report by the Brundtland World Commission on Environment and Development (Colby 1990). In this management philosophy the environment and economics are considered to be parts of mutually supporting ecosystem. Long-term issues and long-term solutions became key considerations of this management philosophy with an emphasis placed on sustainability.

14.3 Groundwater and Health Effects Management

Following this evolution, it is not very difficult to anticipate the next step if one asks the right questions. The proper question to ask may be: "Can there be a global or uniform groundwater and environmental health effects management policy?" Or "Based on their characteristics should different groundwater management issues, different regions, and different applications have their unique groundwater and environmental health effects management strategy?" Another question to ask may be: "Are we really worried about proper groundwater management for the sake of preservation of the environment or economics, or are there other reasons?" It seems that there is a more important reason behind this evolution that led us to the sustainability concept. We now realize that one of the main purposes of our search for a suitable environmental and groundwater management model is the protection of populations from adverse health effects. We now realize that adverse effects that impact us may be economic or environmental in nature, or they may be related to health effects. When one includes the concept of "health effects" into the overall picture and emphasize its importance as the primary concern, the policies, principles, and methods we have been working with will change considerably. In the earlier management models discussed above, the "health effects" issue was not out of the picture, but it was not emphasized as the primary policy issue or concern. In the earlier management philosophies, health effect issues appeared mostly as a concept, as something to worry about, measure, document, and possibly correct. Again, in the management philosophies summarized above, the emphasis on health effects appears to be more pronounced when the management model emphasizes environmental concerns. When we realize the importance and depth of the "health effects" concept, we may abandon the philosophy of ***Sustainable Environmental Management*** paradigm, mainly because it still reflects a two-dimensional picture of a three-dimensional problem.

Given our current understanding of groundwater management issues and their health effect outcomes, it is reasonable to adopt the following premise: ***Based on our current technological impacts, all human interventions to natural environments, our demand for built environments, and natural or forced disasters sooner or later will be associated with health issues.***

What is apparent in the premise above is that in the long-term environmental change is inevitable. Long-term environmental change has various effects on our society, the most important of which is the health effects. Our primary goal is then to limit the damage to human health caused by these changes. Based on this premise, the next stage of groundwater or environmental management philosophy we have to work with may be identified as ***Environmental Management for Sustainable Populations*** (Aral 1996; Aral 2005; Aral 2009). Here the term environment refers to built and natural environments. In this management philosophy, the goal will be the long-term harmonious management of economic resources and environmental preservation, for the health, safety, and prosperity of the populations. Policy decisions that will be made in this phase will now explicitly include a very complex element and also the very delicate "population" or the "human" element as a central point of departure. When populations are explicitly included in the overall framework as the primary variable of concern, then social policy, ethics, and health issues assume a very important role in the management strategy. It can be anticipated that, to identify and resolve the problems of this management style, scientists from diverse fields such as social sciences, public policy, health sciences, basic sciences, and engineering need to work more closely than they have in the past. To establish this working environment more barriers need to be broken, new rules need to be established, and more importantly, a common language has to be developed. Technological, scientific and holistic advances made in each field need to be translated into this common language and need to be put to use for the ultimate goal of maintaining sustainable populations. In this approach, economic incentives and environmental constraints have to be considered harmoniously as a subset of the human component. In this management philosophy the main emphasis is placed on protection and preservation of human health and sustainability of populations.

14.4 Integration of Scientific Fields and Educational Programs

Given the evolution of the management philosophies described above, our task is now to incorporate environmental health issues into the overall management philosophy as the primary concern, and in doing so we should now try to seek an answer to the question: Are the scientific fields we intend to include in this management philosophy ready for integration?

When answering this question we first realize that from engineering, basic sciences and health sciences perspectives: (i) our present information base is not adequate to fully incorporate environmental health concepts into environmental practices; (ii) emerging contaminants and their unknown behavior in the environment further complicate this picture; (iii) the relationship between hazardous groundwater quality exposures and their effect on various physiologies are not well understood; (iv) in most cases historical environmental data on hazardous materials and data on hazardous waste sites and their effect on groundwater quality are not well known or documented; and (v) health effect consequences of groundwater management decisions have not been fully characterized and evaluated. Thus, many pieces of this information do not yet fit and others remain elusive. When we find answers to parts of the questions posed above, we start struggling with other questions such as: What are the risks of human exposure along an environmental pathway, and what are the models to quantify that risk? What is the correlation between various segments of population to exposure and the health effects outcome? How do we evaluate, measure, and document this risk based on uncertainty and variability? What are the biomedical indicators of risk and toxicological profiles of chemicals to enumerate this risk? What are the management alternatives that need to be implemented to reduce this risk? And more importantly: How do we integrate progress made and knowledge accumulated in engineering sciences, basic sciences, and health sciences in finding answers to these questions? This is a very important task and understanding the interrelationships among these elements is of great importance to the environmental health community. Engineers, social, basic and health scientists have a significant role to play in formulating answers to these questions and providing the necessary integration platform.

For a successful integration effort we may have to start from our educational programs in universities, as well as educating ourselves. The research and education activities in engineering, basic sciences, and health programs at higher education institutions need to focus on the key questions and interrelationships partially identified above. Educational programs at these institutions are the starting point to establish this dialog. As a product of these integrated educational programs, the next generation of environmental engineers, groundwater scientists, and health scientists should come to the table with a basic understanding of each others strengths and weaknesses, and more importantly their language. This dialogue can then be extended to include knowledge dissemination to users and stakeholders on research and educational processes and products. The integrated programs between key parameters and components of health sciences field, health community, and engineering have a lot to offer in addressing the key research and education issues, critical concerns of a diverse community, and complex environmental problems.

To address these problems adequately, a variety of professional disciplines must be engaged in these activities. No single discipline can dominate these activities. We should expect to have engineers, basic and health scientists, epidemiologists, public policy specialists, educators, planners, and information technology specialists to be involved in these activities. This interdisciplinary approach is essential if the diverse issues associated with human health management are to be addressed fully and adequately. We should keep in mind that the decisions we make on health or groundwater management must consider how one is related to the other and what the consequences will be when policy is implemented.

14.5 Purpose and Goals

The management philosophies discussed above will continue to evolve in their own time frame at their own pace. There will be many contributions to this effort in years to come. We are already seeing important developments in the science and technology fields in reference to these new concepts and methodologies of analysis. One such new concept that is maturing is the concept of ***Resilience Analysis*** and its applications in groundwater management.

Resilience Analysis introduces a new perspective over existing methodologies that will be extremely useful in the integration of multiple concepts in environmental management. Current environmental models that we work with in groundwater management are mostly based on principles that are built to control change in systems that are assumed to be stable and also in equilibrium. The aim in that approach has been to manage the capacity of environmental systems to cope with, adapt to, and shape the change that will be imposed. Most of our groundwater resources management models and environmental applications follow this basic paradigm. However, there are risks with this "command-and-control" management style. Human induced changes have serious impacts on the environment. What is more, our earlier perception about the stability of ecosystems and that ***change is possible to control*** have proven to be false in many occasions in the past. Today we know that environmental systems do not respond to change in a smooth and predictable way; rather a stressed environment can suddenly shift from a seemingly steady state to a state that is difficult to reverse (Gunderson and Pritchard 2002). As a result, proper environmental management is becoming increasingly more complex.

Complex Systems Theory, introduced by Holland (1995), is the other concept that we should introduce in our new management paradigms. This analysis will be in contrast to the perspective of a world that is in near equilibrium and/or in steady state composed of one main control parameter (Gunderson et al. 1995), as they are defined in our current environmental models. A complex system, by definition, is a group of entities or organization that is made up of many interacting components. In complex systems the individual components and the interactions between them often lead to large-scale changes and behaviors that are not easily predicted from the knowledge only of the behavior of the individual components. We should acknowledge that a departure from the earlier "equilibrium" and "steady state" concept is necessary for the safe and satisfactory resolution of groundwater management problems. Resolution of this problem needs to be based on the adoption and the use of the new concepts developed in the literature such as ***resilience, vulnerability, adaptability and transformability of integrated complex environmental-social systems*** (Levin 1999; Scheffer et al. 2001; Gunderson and Pritchard 2002; Walker et al. 2004). Assessing and evaluating a design in the context of ***complex systems*** requires a shift in thinking and perspective as outlined in Ludwig et al. (2001). In this paradigm the earlier view of a system of interacting nature and society that is near equilibrium will be replaced by a dynamic view, which emphasizes the complex non-linear interactions between entities that are under continuous change and are facing discontinuities and uncertainty from numerous stresses and shocks. Managing for ***resilience*** rather than for ***control*** enhances the likelihood of sustaining development in changing environments where the future is unpredictable and uncertain (Levin et al. 1998; Holling 2001). Actually, that is where the control approach fails because it requires a predicted "future" input in a "design." The long-term unpredictable view of the future is the basis of the proposed resilience approach, and it is also the basis of the failure of the control approach. It seems these will be the frontiers of research and applications in the groundwater management field in the future. This effort can only be successful if the next generation of groundwater quantity and quality scientists, engineers, and other stakeholders of this community adopt these concepts and take them to the next level to address groundwater management and environmental health concerns for the benefit of the populations.

14.6 References

Aral, M. M., ed. (1996). "Advances in groundwater pollution control and remediation." *Proc., NATO Advanced Study Institute, Antalya, Turkey, May 20 - June 1, 1995*, Springer.

Aral, M. M. (2005). "Perspectives on environmental health management paradigms." *Proc., 1st International Conference on Environmental Exposure and Health*, M. M. Aral, C. Brebbia, M. Maslia, and T. Sinks, eds., WIT Press.

Aral, M. M. (2009). "Water quality, exposure and health: Purpose and goals." *Journal of Water Quality, Exposure and Health*, 1(1), 1-8.

Colby, M. E. (1990). "Environmental management in development: The evolution of paradigms." *World Bank Discussion Papers*, 80, Washington, D.C.

Gunderson, L. H., Holling, C. S., and Light, S., eds. (1995). *Barriers and bridges to the renewal of ecosystems and institutions*, Columbia University Press, New York.

Gunderson, L. H., and Pritchard, L. (2002) *Resilience and the behavior of large-scale systems*, Island Press, London.

Holland, J. H. (1995). *Hidden order: How adaptation builds complexity*, Addison-Wesley, Reading, Mass.

Holling, C. S. (2001). "Understanding the complexity of economic, ecological and social systems." *Ecosystems*, 4, 390-405.

Levin, S. A., Barrett, S., Aniyar, S., Baumol, W., Bliss, C., Bolin, B., Dasgupta, P., Ehrlich, P., Folke, C., Gren, I.-M., Holling, C. S., Jansson, A., Jansson, B.-O., Maler, K.-G., Martin, D., Perrings, C., and Sheshinski, E. (1998). "Resilience in natural and socioeconomic systems." *Environment and Development Economics*, 3(2), 222-262.

Levin, S. A. (1999). *Fragile dominion*, Perseus Books Group, Cambridge, Mass.

Ludwig, D., Mangel, M., and Haddad, B. (2001). "Ecology, conservation, and public policy." *Annual Review of Ecology and Systematics*, 32, 481-517.

Scheffer, M., Carpenter, S., Foley J. A., Folke, C., and Walker, B. (2001). "Catastrophic shifts in ecosystems." *Nature*, 413, 591–596.

Walker, B., Holling, C. S., Carpenter, S. R., and Kinzig, A. (2004). "Resilience, adaptability and transformability in social–ecological systems." *Ecology and Society*, 9(2), 5 <http://www.ecologyandsociety.org/vol9/iss2/art5/>.

Index